**In-Plant Practices
for Job Related
Health Hazards Control**

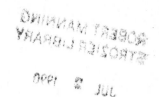

In-Plant Practices for Job Related Health Hazards Control

Volume 2   Engineering Aspects

Editors
Lester V. Cralley
Lewis J. Cralley

Associate Editors
Knowlton J. Caplan
John E. Mutchler
Paul F. Woolrich

WILEY

A WILEY-INTERSCIENCE PUBLICATION

JOHN WILEY & SONS

New York · Chichester · Brisbane · Toronto · Singapore

Library of Congress Cataloging-in-Publication Data:

In-plant practices for job related health hazards
   control.

   "A Wiley-Interscience publication."
   Includes bibliographies and index.
   Contents: v. 1. Production processes—v. 2.
Engineering aspects.
   1. Industrial safety—Handbooks, manuals, etc.
I. Cralley, Lester V. II. Cralley, Lewis J., 1911–
T55.I45   1988   670′.28′9   88-17200
ISBN 0-471-50121-2 (v. 2)

Printed in the United States of America

10 9 8 7 6 5 4 3 2 1

Contributors

Carl D. Bohl, ScD, CIH, PE
Monsanto Company
St. Louis, Missouri

Knowlton J. Caplan, CIH, PE
Pace Laboratories, Inc.
Minneapolis, Minnesota
(Formerly, Industrial Health
Engineering Associates)

Timothy J. Colliton, CIH, PE
Pace Laboratories, Inc.
Minneapolis, Minnesota

Lester V. Cralley, PhD, CIH
Fallbrook, California
(Formerly, Aluminum Company of
America)

Lewis J. Cralley, PhD, CIH
Cincinnati, Ohio
(Formerly, NIOSH)

John B. Feldman, CIH, PE
Raytheon Company
Newton, Massachusetts

Richard P. Garrison, PhD, CIH
University of Michigan
Ann Arbor, Michigan
(Formerly, Occusafe, Inc.)

Paul A. Kannapell, PE
Engineering Consultant

St. Louis, Missouri
(Formerly, Monsanto Company)

Lawrence W. Keller, CIH
PPG Industries
Pittsburgh, Pennsylvania

S. Lipton
Exxon Research and Engineering
Company
Florham Park, New Jersey

Jeremiah R. Lynch, CIH
Exxon Chemicals Americas
East Millstone, New Jersey

Donald R. McFee, ScD, CIH, PE
Occusafe, Inc.
Wheeling, Illinois

Michael A. Markowicz, CIH
Amoco Company
Chicago, Illinois
(Formerly, Occusafe, Inc.)

Dennis J. Paustenbach, PhD, CIH
McLaren Environmental Engineering
Rancho Cordova, California
(Formerly, Stauffer Chemical
Company)

Jack E. Peterson, PhD, CIH, PE
Peterson Associates
Brookfield, Wisconsin

v

Thomas J. Slavin, CIH
Navistar International Corporation
Chicago, Illinois

Robert D. Soule, CIH
Indiana University of Pennsylvania
Indiana, Pennsylvania
(Formerly, Clayton and Associates)

Charles H, Stevens, CIH, PE
Spotts, Stevens, and McCoy, Inc.
Wyomissing, Pennsylvania

Thomas J. Walker, CIH, PE
The FDE Group
Lafayette, California
(Formerly, Thomas J. Walker, Inc.)

Paul F. Woolrich, CIH, PE
Consultant
Mattawan, Minnesota
(Formerly, The Upjohn Company)

Preface

The core of an in-plant program for the cost-effective control of job related health hazards is engineering innovation that focuses on the characteristics of the production process and the building that houses it. This avoids trade-offs that may shortchange critical elements of the program with subsequent reoccurring maintenance and repair costs. Thus, an overall cost-effective installation mandates a team approach in which industrial hygiene engineering participates fully from the beginning.

Information in this volume will guide engineers and plant managers to this goal. All chapters are authored by engineers highly qualified by personal knowledge and experience. The scope includes both basic engineering concepts and unit operation controls that constitute the cornerstone of a cost-effective program.

LESTER V. CRALLEY
LEWIS J. CRALLEY

Fallbrook, California
Cincinnati,, Ohio
January 1989

Contents

Notation

The wide scope of this volume has made it necessary to impose restraints to keep it within manageable limits. Thus, coverage does not include basic industrial hygiene aspects, ventilation control design, air-cleaning methods, and the like, that are readily available. Instead, authors have drawn upon their personal knowledge and experience to present information as a guide to those charged with this responsibility.

Metric conversion of measures in this volume are approximate.

CHAPTER ONE

Introduction

Lester V. Cralley
Lewis J. Cralley

1 GENERAL COMMENTS

The impact of the Occupational Safety and Health Act of 1970 has been profound. Previously, engineering controls were installed as a necessity, instigated by the need for compliance with regulatory codes, internal company standards, workmen's compensation regulations, lawsuits, and the like. The attention given was somewhat related to the financial conseqeunces to be incurred by these actions and were handled on a case-by-case basis. Even new construction facilities frequently included minimum controls with the proviso that additional controls would be added as required by governmental regulations. The cost-effect end result of this approach, though considered rational at the time, is no longer tenable. Ever-changing governmental regulations and competitive inroads in industry now require major attention to this aspect. The main thrust has been to make health hazard controls, to the extent achievable, a basic part of process equipment design.

2 ENGINEERING CONTROL STRATEGY

Before a decision can be properly made on a cost-effective control system that offers adequate protection and is in compliance with governmental regulations, full information must be developed that defines the problem and the options that are open.

1

2.1 Existing Production Equipment

Improvements in health hazards controls in existing equipment, at best, is a compromise in which the objective is to come into compliance with governmental regulations. Since each instance is a situation unlike any other, few generalizations can be made. Basic, of course, is a list of options available that will satisfy the noncompliance problem. This list, and the assessments needed to reach a decision, need the fullest input of the industrial hygiene engineer. Engineering practicality and innovation are the basic ingredients of a successful answer.

2.2 New Equipment

New process equipment design offers the best opportunity for industrial hygiene engineering input. Inadequate attention to equipment selection, layout, and building design will undoubtedly lead to permissible exposure level (PEL) noncompliances that will become more serious in time, even to the extent of requiring major expensive modifications. New process equipment design must be based on a full knowledge of the health stress agents associated with each step of the operation with respect to both identification and emission profiles. Thus, the control program must consider:

- Chemical and physical properties of the substances being controlled.
- Selection of equipment less likely to involve subsequent industrial hygiene problems.
- Adequate emission capture and air cleaning.
- Noninterference with production.
- Adequate general room ventilation and makeup air.
- Equipment layout complementary with the maintenance requirements and the like.

Awareness of the overall spectrum of requirements will relate closely to the degree of control achieved, initial capital expenditure, and subsequent maintenance expense. Thus, it involves equipment selection, layout, and building design that are a proper meld of the requirements for both cost-effective production and worker protection. Health hazards are best controlled when control is an integral part of the process. This requires indentification of the health hazards at the the laboratory bench stage and the selection of equipment at the pilot stage. Information from these steps permits controls to be designed into the semiworks and their subsequent refinement into the commercial production unit. It must be emphasized that the inherent characteristics of process equipment have a direct bearing upon the emission pattern of the substances to be controlled. Thus, less expensive equipment may prove more expensive in the long run because of the increased environmental controls that may be needed. While the ideal is the closed system in which the integrity of the unit is not lost, except as planned, many systems cannot achieve this goal. The necessary opening of reactors, line clean-outs, and the like, can be planned with ancillary controls.

On the other hand, many processes are characterized as open systems in which the various operating steps are somewhat discrete and require individual consideration. Again, the input of engineering production and the industrial hygiene professional is needed. The lack of such input may result in add-on controls that are always more expensive and less satisfactory than those incorporated as an integral part of the process. In addition, significant production downtime may be required for installations.

2.3 Catastrophes

There is a new awareness today of the potential for catastrophic situations that may result from spills and equipment failure. Consequently, this potential is receiving major attention in process equipment design. Equally important are the ongoing maintenance, repair, and inspection programs to assure proper equipment performance.

3 SUMMARY

It is obvious that the total control program will involve some mix of engineering control, administrative control, and personal protective equipment. This may vary with the process, age of the plant, production rate, geographical location, and a host of other factors. This mix, though, must take into account the reliability of the control program in worker protection.

The proper starting point of the cost–achievement aspect of captial investment is a thorough understanding of the type and degree of control required for worker protection. However, governmental regulations do change, and adequate control of job-related health hazards is the responsibility of industry. Therefore, good engineering practices must be overriding.

Inasmuch as there are a host of high-priority factors that enter into equipment selection, layout, and building design, many trade-offs exist. Whether these trade-offs properly accommodate job-related health hazard protection can best be resolved by making the industrial hygiene professional a member of the management team that approves both the early planning and the final plans for the production unit.

GENERAL REFERENCES

Cralley, L. V., and L. J. Cralley (eds.), *Industrial Hygiene Aspects of Plant Operations*, Vol. 1, *Process Flows*, New York: Macmillan, 1982.

Cralley, L. J., and L. V. Cralley (eds.). *Industrial Hygiene Aspects of Plant Operations*, Vol. 2, *Unit Operations and Product Fabrication*, New York: Macmillan, 1984.

The Industrial Environment, Its Evaluation and Control. National Institute for Occupational Health, Superintendent of Documents, Washington, D.C.: U.S. Governmental Printing Office, 1973.

Perkins, J. L., and V. E. Rose (eds.), *Case Study in Industrial Hygiene*, New York: Wiley-Interscience, 1987.

CHAPTER TWO

Materials and Their Characteristics

Donald R. McFee
Michael A. Markowicz

1 INTRODUCTION

The process material type determines the degree of control and general basic features of equipment selection, layout, and facility design. Quantity, of course, affects size and extent of handling or movement. All factors must be integrated if optimum productivity and control of potential hazards are to be successful.

Material types are classified primarily by the physical state of the material, that is, gas, liquid, or solid. However, this is only the beginning. Many physical and chemical properties must be considered prior to determination of the specific system involved. Not only do these properties affect the system, but they also influence productivity and hazard potential, including the toxicity and dose to the individual (exposure).

The first part of this chapter addresses the physical and chemical properties and describes how materials are categorized and classified. The latter sections discuss exposure control considerations. These will be useful to the industrial building and facilities design engineer. These can be applied for appropriate equipment selection and workplace layout.

2 PHYSICAL STATES

2.1 General Comments

Whether a material is a gas, liquid, or solid is the first consideration in selecting any piece of equipment or completing any design or layout. Gases and liquids must be contained and can be either forced (pumped) or drawn (sucked). It is frequently more convenient to pump. But in the case of highly toxic and corrosive materials, it is safer to draw. These help to preclude a leak from spurting the material toward personnel or sensitive materials.

Solids generally retain their bulk configuration and contain themselves, except when finely divided. Very finely divided solids can be made to flow like liquids and/or diffuse (disperse) to fill the space of a container. These, of course, must be fully contained.

With liquids and solids the extent of division frequently determines the potential for mixing, dispersion, reactivity, and general capacity to leak or be released into the working environment.

Every imaginable type, form, or state of material is handled or processed, in a variety of quantities or forms. Each involves special building and facility design parameters. For the sake of simplicity, only the major general classes of materials typically encountered in industry will be discussed here. These include gases, vapors, liquids, mixtures or slurries, and solids.

The following classes and definitions will be useful in the subsequent discussions:

Gases

Normally formless fluids that occupy the space of enclosure and that can be changed
   to liquid or solid state by combined effect of increased pressure and decreased
   temperature. Gases diffuse.

Vapors

The gaseous form of substances that are normally in the solid or liquid state at
   standard temperature and pressure and that change to these states by either
   increasing the pressure or decreasing the temperature alone. Vapors diffuse.

Liquids

Substances that seek a uniform level due to gravity and take the shape of their
   container.

Mists

Suspended liquid droplets generated by condensation from the gaseous to the liquid state or by breaking up a liquid into a dispersed state, such as by splashing, agitation, foaming, and atomizing.

Fumes

Finely divided solid particles generated by condensation from the gaseous state, generally after volatilization from molten metals and often accompanied by a chemical reaction such as oxidation. Fumes may disperse by being easily carried in air currents. In this manner they behave almost like vapors. However, they also may agglomerate and form large particles that behave like dusts.

Dusts

Solid particles generated by handling, crushing, grinding, rapid impact, detonation, and decrepitation (roasting or calcining) of organic or inorganic materials such as rocks, ore, metal, coal, wood, grain, and the like. Dusts do not tend to agglomerate, except under electrostatic forces. They do disperse and diffuse in air, but settle under the influence of gravity.

Solid Pieces

Materials that have their own size and shape, and are not finely divided. Generally, solid pieces are materials with individual masses larger than 100 μm in all dimensions.

Others

Living agents, such as viruses, bacteria, molds, and other parasites constitute another group that should appear in a comprehensive classification of potential industrial health hazards.

2.2 Gases and Vapors

This class of materials may be organic (e.g., gases—methane, acetylene; vapors—acetone, trichloroethylene) or inorganic (e.g., gases—hydrogen, oxygen; and vapors—water). Important physical and chemical properties of gases and vapors include density, molecular weight, diffusional characteristics, and compressibility.

The density and molecular weight (MW) of a gas or vapor provide an indicator of the control requirements in the workplace. Taking normal air as a typical example with an average MW of about 29, any gas or vapor with a significantly higher or lower MW than air will tend to settle or rise, respectively. This is provided the concentration is sufficient to make mixtures that are more or less dense than the surrounding atmosphere. This may be a critical consideration for the placement of an exhaust ventilation system pickup point. For example, where Freons are being handled (MW ranging from 100 to 200), vapors evolved in air are apt to settle, and placement of general exhaust hoods near the floor may be appropriate. Conversely, where hydrogen (MW of 2) is handled or evolved, such as when charging lead–acid batteries, placement of general exhaust ventilation at the highest point near the roof, ceiling, or top of a wall may be appropriate. Other diffusional characteristics of a gas or vapor may be highly influenced by thermal, forced, and random air currents. All of these factors must be considered in the design and placement of control systems.

Gases and vapors generally diffuse uniformly to the limits of their containment, especially for those with an MW close to that of the medium (usually fluid) in which they are present. However, they are unique in their ability to be compressed. Because of these characteristics, leakage attributed to pressurized materials must be dealt with and controlled. Leakage may be inward or outward from a container, the direction being dependent on pressure differentials.

It is preferable to keep highly toxic materials under a negative pressure so they cannot leak out. However, in the case of materials that may be reactive with the surrounding material, it is preferable they be pressurized (positive pressure) so leakage is outward. The surrounding material then is not drawn in to contaminate and react with the primary process material.

Flow is always from higher to a lower pressure. Storage of gases and vapors under pressure requires specially constructed vessels capable of safely venting any over-pressurization and certified as to their ability to safely contain a given pressure requirement. Types of suitable containers include certified pressure vessels, tanks, and closed piping systems.

Rupture disks and pressure relief valves are often involved. Their burst or relief pressure rating should be appropriately matched with the assigned vessel pressure certification. The vessel vent discharge should be to a remote location well away from personnel. In case of noncondensing gases and vapors, this may be above and well away from equipment and personnel. If condensing materials, liquids, or solids are involved, it should be down and through traps or other collecting devices. "Knockout" chambers may be desirable to trap solids, liquids, and gas–vapors, respectively.

A characteristic of gases and vapors is their ability to condense. Condensation may be defined as a gaseous or vapor phase of a material that is transformed to its liquid phase. This may occur from dropping temperatures or by impingement onto a solid surface, especially when this surface is cooler. Condensability may be an important control factor such as when cooling coils are used in solvent degreasers to recover vaporized solvent.

All gases and vapors are a potential concern in industrial hygiene because they can be inhaled and may be the type that can be absorbed into body tissues. The extent of potential concern and required level of control is highly dependent upon the toxicity, mode of action, and physiological activity of each specific gas or vapor. These material characteristics are described in Section 3.

2.3 Liquids

This class of materials may be organic (e.g., industrial solvents—methyl chloroform), inorganic (e.g., mineral acids—sulfuric or hydrochloric; bases or caustics—sodium hydroxide) or neither (silicones, nonflammable oils). Important physical or chemical properties of liquids requiring consideration in building and facilities design include volatility, density, viscosity, and corrosivity or reactivity.

All liquids have a characteristic vapor pressure, which is the tendency of the liquid to vaporize at a given temperature and usually at atmospheric pressure. Vapor pressure is commonly expressed as millimeters of mercury (mm Hg) as compared to standard atmospheric pressure of 760 mm Hg. The vapor pressure of a given material will increase with increases in the temperature of the material. As an example of the significance of vapor pressure, methyl ethyl ketone (2-butanone) is a highly volatile organic liquid with a vapor pressure of 71.2 mm Hg at 20°C. This means that sufficient liquid methyl ketone in a sealed container is capable of developing an air concentration of 9.37% or 93,700 ppm (71.2 mm/760 mm $\times$ 100%), given enough time to reach full equilibrium at this temperature. Similarly, an organic liquid of low volatility, such as ethylene glycol with a vapor pressure of 0.05 mm Hg at 20°C, is only capable of reaching an air concentration of 0.0066% or 66 ppm (0.05 mm/760 mm $\times$ 100%). From a plant design standpoint, this means that a higher degree of control is required for highly volatile liquids. This further requires that plant design specifications must be established to operate and handle the most volatile feed, by-product, or production materials involved.

The density or viscosity of a liquid will be a major factor in how it can be transferred. Some liquids, such as No. 6 fuel oil, can only be pumped when heated to about 140°F (60°C) or more. Other liquids, such as acetone, cannot be pumped by conventional means because they are so highly flammable and potentially explosive that special pumps with nonsparking moving parts must be used for safe transfer. Where possible, vertical flow arrangements for liquid systems are generally preferable to take advantage of the effects of gravity. Most liquids can easily be pumped against gravity to heights when required.

The corrosivity or reactivity of a liquid is also an important consideration in building and facility design. Strong acids and bases attack most common metals. Special coatings, metal alloys, or plastic containers and plumbing fixtures will be required to combat these effects. Exclusion of water or water vapor may be required to eliminate corrosive activity. Addition of chemical agents such as neutralizers and oxygen scavengers can be used.

A unique aspect of liquids worth noting is the potential for azeotropic mixtures of two or more liquids in solution. An azeotrope is a unique phase mixture of liquids in which there is weak molecular bonding. In combination at the proper proportions this mixture boils at a different temperature than the boiling point of either constituent. At these azeotropic temperatures, the proportions remain constant. This phenomenon plays an important role in the selection of flammable and nonflammable mixtures. If an azeotrope is involved, the proportions will remain constant and flammability of the mixture does not change. On the other hand, if there is no azeotrope, then the more volatile of the components will vaporize first. A nonflammable mixture can become flammable if the nonflammable material is allowed to evaporate.

Another important factor is the slow degradation of a material in the presence of another. A third material in minute quantities may be used to block this reaction, and this is included in the mixture. An example of this is a small percentage of glycidol in methyl chloroform, or epichlorohydrin in trichloroethylene. These agents stabilize the respective halogenated hydrocarbon by suppressing or inhibiting the breakdown of the material in the presence of certain metals.

Industrial hygiene concerns related to liquids are primarily those of direct skin or eye contact and inhalation of airborne vapors. Toxicity and physiological activity of the liquid set the parameters for the degree of control required to minimize personal exposure potential.

2.4 Mixtures

The only mixtures to be discussed in this chapter will be those involving solid particles in a liquid matrix or a slurry. Of course, there are several other mixtures of materials, such as liquid–vapor or vapor–gas, but no discussion specific to these will be made. Mixtures or slurries can be either organic solvent or water based. Paints are an example to illustrate this difference. There are the traditional petroleum distillate solvent-based varieties and the more recently developed latex-base types. The water-based latexes offer the advantages of reduced flammability, easy cleanup with water, and faster drying time. It is of interest to point out that a new wave of technology is well underway that may make petroleum distillate solvents almost obsolete for use in paints, adhesives, and similar industrial coatings and synthetic lubricants.

Granular, pulverized, and powdered materials have a high surface area to weight ratio. When mixed with liquids, there may be a substantial release of heat from mixing. Explosive exothermic reactions may develop. Depending on the hazard potential, especially toxicity, and the amount of airborne particulate material generated, enclosure, venting, and other appropriate contaminant controls may be required.

If pressure relief valves, rupture disks, or other relief devices are involved, they should discharge to a safe location. Solids and liquids should be discharged downward at a safe location where there will be no personnel contact. Drop-out chambers or traps may be advisable.

Processes with vent stacks may require stack sampling or measurement of fugitive emissions. The responsible and aware designer will incorporate provisions to expedite the task. These include access ports, stairs, platforms, and the like. This foresight will likely pay for itself. The lack of such provisions too frequently makes the cost of subsequent stack measurement relatively high.

Built-in controls should be used to preclude direct skin or eye contact and inhalation of mists, dusts, and other contaminants associated with mixtures.

2.5 Solids

Solids possess a specific form or given size and shape and may be organic (e.g., synthetic—plastic resins, or natural—grain) or inorganic (e.g., mineral—quartz, granite; or refined metals and alloys). Sizes may vary from superfine powders (submicrometer diameters) to massive blocks or boulders (many meters in a single dimension). In contrast to small particles, larger solid pieces usually are of low potential industrial hygiene concern. For example, processing large pieces of many toxic metals such as lead, cadmium, or chromium does not pose a significant potential health hazard until there is work on the material that generates particles. The finer the particle the greater the hazard potential and the necessity to design or select equipment and systems that will contain the particles.

Containment may be difficult and must be given special attention when the smaller size range of solids, such as smoke, fumes, dusts, granules, and fibers are involved. These are of primary concern in industrial hygiene because of their smaller and readily respirable size. The respirable size range is commonly defined as less than 10 μm nominal diameter. However, particles in the 0.5–5 μm range are most commonly deposited and retained in the deep lung when inhaled. Particles less than about 0.5 μm are so small they tend to remain suspended in air and a vast majority are exhaled. Particles greater than 15–25 μm are filtered by the nasal passages, while those in the 5–20 μm range are generally impacted upon bronchiolar surfaces and rarely reach the deep lung. An illustration of common particle sizes and their characteristics is depicted in Fig. 2.1.

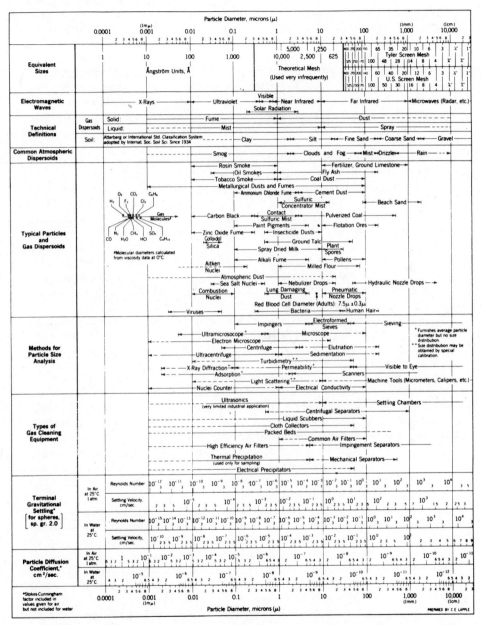

Figure 2.1. Common particle dispersions and methods of size measurement. (Courtesy of Stanford Research Institute, Menlo Park, CA, now known as SRI International.)

In still ambient air, dispersed and suspended particles settle at a rate inversely proportional to the average aerodynamic diameter. That is, smaller particles settle at a slower rate than larger ones. This is a function of size, shape, and density of the particle. It must be noted that thermal, mechanical (forced), and random air currents will also have a significant effect upon actual particle settling rates and movement.

Mechanical action or processing of solids is the primary factor determining the rate of dust or particle generation. Generally, the smaller and less dense the solid, the larger the volume and greater the length of time the generated dust will remain airborne. Operations such as chipping, grinding, cutting, and other machining operations create larger particles, while pulverizing, sifting, buffing, and polishing create finer particles.

Processes involving high temperatures create smoke, fumes, and gases in varying amounts depending on the type of process. Much of the particulate matter from such processes is in the critical respirable range. For this reason control is desirable and usually necessary if the toxicity or quantity of airborne material is high. Heat will tend to cause the materials to rise. Designers have frequently taken advantage of this property in selecting control methods. Examples are overhead exhaust fans and vents in foundries and canopy hoods over ovens or furnaces. To be effective, however, such ventilation control systems require air volumes that are extremely high and may position the worker into the flow of contaminants. For these reasons, a designed local exhaust system drawing contaminants away from the worker's breathing zone are preferable to provide a cost-effective primary means of control.

Some solids are highly hygroscopic and will absorb significant amounts of water or other vapors from the air. This can create substantial operating difficulty for processing powders, granules, and similar materials where it is important to keep materials dry and to retain their free-flowing characteristics (e.g., powdered detergents). However, many times this property of solid materials may be used as an advantage in minimizing dust generation. With some particulates, it is necessary to retain a minimum moisture content to prevent or minimize electrostatic attractions and permit efficient handling (e.g., gunpowder). Condensation of gases or vapors onto dusts or other particulates may occur also and may require attention.

There is now a wide variety of equipment available for measuring and adding moisture to materials and spaces. Frequently, with proper controls, the designer can adequately control airborne particulates by moisture content alone. If so additional ventilation controls may be avoided or limited.

Another characteristic of particulates is the potential for surface changes. Many materials oxidize rapidly in air, especially in the presence of moisture. This is a chemical change forming a new compound on the surface of individual particles. These surface effects can render a material highly flammable, as in the case of organic dusts, and even toxic as in the case of zinc fumes. For example, zinc fume forms a surface of zinc oxide that is toxic upon acute exposure. Metallic zinc particles with low surface area are not toxic. Other surface changes, such as mechanical and chemical bonding or coating may also substantially alter the toxicity of a material and the degree of control required.

Solid pieces have a relatively low surface area and if the toxicity of the material is no greater than that of lead, the potential for generation of toxic levels in the air is low. Equipment and procedures need to be selected that avoid excessive handling and piece-to-piece contact that may result in generation of toxic levels of airborne particles.

Not to be overlooked is noise created by handling the piece. Noise also is likely to be generated by the machinery and equipment required for handling, as opposed to the piece itself, unless the engineer selects quiet equipment.

Solid materials should be handled horizontally with a minimum of transfer points. Conveyors may require covering to minimize particle release. Transfer points usually represent some vertical motion. Vertical motion requires lifting, if upward. If downward, gravity can be utilized to allow free fall. Either situation usually generates dusts and control is often necessary. In the case of larger pieces where dust is not a problem, other health and safety concerns may arise. Even when the number of pieces to be moved is small, lifting devices such as cranes and hoists may be used. These require special procedures from a safety viewpoint.

Cranes and hoists require special design considerations for ventilation control duct work arrangement. Underground tunnels or plenums may be necessary for duct work to allow proper crane access and clearances. These should be significantly oversized in design to accommodate likely future changes and allow access for cleaning and maintenance.

Other handling devices may include mechanical vehicles, such as fork trucks, front-end loaders, earthmovers, and the like. Use of vehicles indoors should include provisions for general ventilation to prevent contaminant buildup and for special local exhaust systems to remove combustion products and air contaminants at their source of generation, where possible.

Principal industrial hygiene concerns related to solid pieces are the inhalation of respirable airborne particles, noise, and skin abrasion or dermatitis from direct contact. As before, types and level of controls required will be dependent on the hazardous properties of the material being processed. These characteristics are discussed in the next section.

3 HAZARDOUS PROPERTIES

3.1 General Comments

Characteristic physical and chemical properties of materials determine the hazard potential and consequently the criteria for equipment selection. These innate features of a material will in large part determine the rquired methodology for a building or facility design and for safe and efficient handling, processing, and use. The physical state of a substance—that is, whether it is encountered as a gas or vapor, liquid, particulate material, mixture, or a solid—is usually the starting point. This is followed by the type of potential hazards that may be encountered with the material.

With liquids and solids the extent of division determines the hazard potential. The more finely divided a material, the greater the potential for leakage, dispersion, reactivity, and dose intake. Surface area becomes an active factor. Other physical characteristics, such as density, aerodynamics, and flow capability will also play a significant role. Chemical characteristics of a material, such as flammability, pH, reactivity, stability, concentration, purity, and toxicity are also major factors in hazard potential determination.

Since transportation and storage usually are involved, one needs to give first consideration to hazard classification for these purposes. This then can be followed by classification for chronic health hazard potential over extended time periods. Finally, for hazardous waste control purposes consideration must be given to Environmental

Protection Agency (EPA) classifications—extraction procedure (EP), toxicity, flammability, reactivity, and corrosivity.

The National Fire Protection Association (NFPA) "Hazard Diamond" provides a system of rating that is easily understood and gives reference data on the hazardous properties of materials. This system was developed for use under adverse conditions encountered with fires or other emergency situations that could involve chemicals with varying degrees of toxicity, flammability, and reactivity (instability and water reactivity). For a full description of the NFPA classifications, see NFPA Standard No. 704M, *Identification of the Fire Hazards of Materials*. A fairly complete listing on given chemicals can be found in NFPA Standard No. 49, *Hazardous Chemicals Data* and in Standard No. 325M, *Fire Hazard Properties of Flammable Liquids, Gases and Volatile Solids*. Use of this system in the design and layout of plant facilities will serve as a comparative guide for the degree of controls necessary to maintain personnel health and safety.

A typically useful NFPA Hazard Diamond symbol is shown in Fig. 2.2. Since this book is primarily concerned with the industrial hygiene aspects of plant operations, health-related concerns are covered first. Starting at the left and going clockwise in Fig. 2.2, health hazard information is presented on a blue field. At the top, fire hazard information is given on a red field. At the right, reactivity information is presented on a yellow field. In each of the upper three quadrants of the diamond a numerical ranking from 4 (indicating the greatest hazard potential) down through 0 (indicating the least hazard potential) is shown. See Fig. 2.3 for a detailed explanation of these numerical rankings. In the remaining quadrant at the bottom, other specific hazard information is given on a clear field. Any special hazards, such as corrosivity, oxidizing potential, radiation, or incompatibility with water is depicted, as shown in Fig. 2.2. This system of categorizing the hazardous properties of materials is relatively simple, yet can provide specific information at a glance when needed. Further elaboration of these four general hazardous material property categories is described in following subsections.

3.2 Health Hazard Potential

Several systems to rank the toxicity of substances are in common use. These include a ranking relating the lethal dose or lethal concentration from acute exposure required to kill 50% of a test animal population (LD_{50} or LC_{50}). LD_{50}s are usually reported in milligrams of material per kilogram of body weight (mg/kg) of the test species. LC_{50}s are usually reported as airborne concentrations in parts per million (ppm).

Another toxicity rating system ranks a material on a scale from 1 to 6 based on the probable amount required to produce death when swallowed by an average 70-kg man (150 lb). A numeral 1 is considered practically nontoxic with more than a quart or 2 lb of material required to produce death. A numeral 6 would be considered supertoxic, with only a taste (less than 7 drops) required to produce death. See Table 2.1 for an illustration of this toxicity rating chart.

Other sources of information for ranking the hazardous properties of materials include the Occupational Safety and Health Administration (OSHA) Hazard Communication Standard, the EPA rankings under the Toxic Substances Control Act (TSCA), and the Resource Conservation and Recovery Act (RCRA), the Department of Transportation (DOT) shipping regulations, the National Institute for Occupational Safety and Health (NIOSH)—Registry of Toxic Effects of Chemical Substances

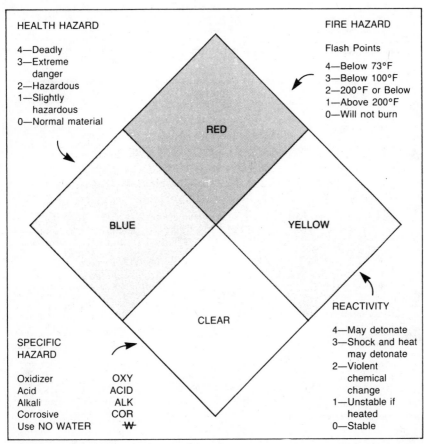

Figure 2.2. NFPA Hazard Diamond. (Reprinted with permission from NFPA 704-1980, Standard System for the Identification of the Fire Hazards of Materials, Copyright© 1980, National Fire Protection Association, Quincy, MA 02269. This reprinted material is not the complete and official position of the NFPA on the reference subject, which is represented only by the standard in its entirety.)

Table 2.1. Toxicity Rating Chart

	Probable Lethal Dose (human)	
Toxicity Rating or Class	mg/kg	For 70-kg (150-lb) man
6 Super toxic	Less than 5	A taste (less than 7 drops)
5 Extremely toxic	5–50	Between 7 drops and 1 tsp
4 Very toxic	50–500	Between 1 tsp and 1 oz
3 Moderately toxic	500–5 gm/kg	Between 1 oz and 1 pt (or 1 lb)
2 Slightly toxic	5–15 gm/kg	Between 1 pt and 1 qt
1 Practically nontoxic	Above 15 gm/kg	More than 1 qt

Courtesy Williams and Wilkins Co., Baltimore, MD.

BLUE	RED	YELLOW
IDENTIFICATION OF HEALTH HAZARD	**IDENTIFICATION OF FLAMMABILITY**	**IDENTIFICATION OF REACTIVITY**
Type of Possible Injury	**Susceptibility to Burning**	**Susceptibility to Release of Energy**
Signal	Signal	Signal
4 Materials which on very short exposure could cause death or major residual injury even though prompt medical treatment were given.	**4** Materials which will rapidly or completely vaporize at atmospheric pressure and normal ambient temperature, and which will burn.	**4** Materials which are readily capable of detonation or of exposive decomposition or reaction at normal temperatures and pressures.
3 Materials which on short exposure could cause serious temporary or residual injury even though prompt medical treatment were given.	**3** Liquids and solids that can be ignited under almost all ambient temperature conditions.	**3** Materials that are capable of detonation or exposive reaction but require a strong initiating source, or that must be heated under confinement before initiation, or react exposively with water.
2 Materials which on intense or continued exposure could cause temporary incapacitation or possible residual injury unless prompt medical treatment is given.	**2** Materials that must be moderately heated or exposed to relatively high ambient temperatures before ignition can occur.	**2** Materials that are normally unstable and readily undergo violent chemical changes but do not detonate; also materials that may react with water violently, or that may form potentially explosive mixtures with water.
1 Materials which on exposure would cause irritation but only minor residual injury even if no treatment is given.	**1** Materials that must be preheated before ignition can occur.	**1** Materials that are normally stable, but that can become unstable at elevated temperatures and pressures, or that may react with water with some release of energy, but *not* violently.
0 Materials which on exposure under fire conditions would offer no hazard beyond that of ordinary combustibles.	**0** Materials that will not burn.	**0** Materials that are normally stable even under fire explosive conditions, and that are not reactive with water.

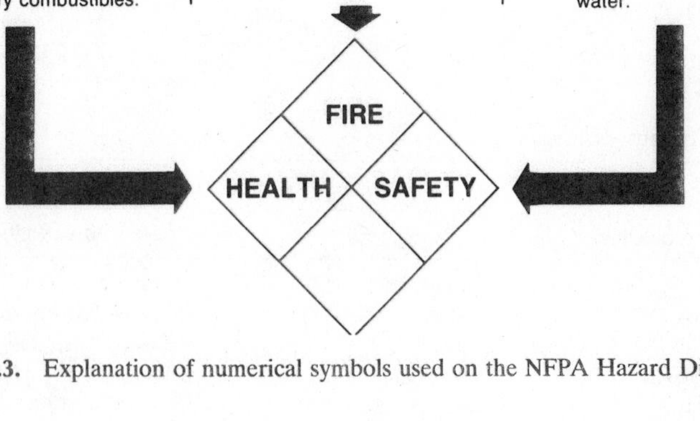

Figure 2.3. Explanation of numerical symbols used on the NFPA Hazard Diamond.

(RTECS), various data sheets published by the National Safety Council, the Chemical Manufacturers Association, the American Insurance Alliance, the American Industrial Hygiene Association—Hygienic Guide Series, Workplace Environmental Exposure Levels (WEELs), and individual chemical manufacturers' data sheets. Recently, there has been a movement to classify chemical hazards by their potential to adversely affect specific organs in the body. Such target organ effects may provide a sufficient additional warning to workers who may have a known predisposition or susceptibility for damage to a certain organ.

Ranking the toxicity of a material seldom provides adequate information or guidance for design and selection of equipment for plant operational purposes. One must consider all aspects and contact potential. A closed system can nearly eliminate a toxicity problem.

3.2.1 Routes of Exposure. The primary routes of exposure or absorption of toxic substances are *inhalation* (breathing airborne contaminants into the lungs), *ingestion* (swallowing into the gastrointestinal tract), and *skin contact* (primarily liquids, solids, or mixtures). Other routes of exposure include *eye contact* (primarily liquids) and *injection* (not usually associated with occupational settings). Inhalation of contaminated air is by far the most important means by which industrial toxicants gain entry to the body.

3.2.2 Modes of Action of Toxic Substances. There are various modes in which substances can manifest their toxicity. *Acute* effects are those that occur suddenly and in a short time (seconds to hours) following exposure, generally to higher levels or concentrations. Acute effects include nonspecific symptoms (such as headache, nausea, dizziness, rapid pulse, and irritation) and more serious symptoms (such as convulsions, coma, and death). *Chronic* effects are those that occur gradually over a long period of time (weeks to years) following repeated and prolonged exposure to relatively low levels or concentrations of a substance. There are a myriad of chronic health effects. Some of the more important ones are irritation, reproductive system effects, blood disorders, systemic diseases, and cancer. Another unique health effect is the phenomenon known as *sensitization*. This occurs when an initial exposure to a substance begins a sequence of events whereby subsequent exposures to the same substance (oftentimes at a much lower concentration) initiate a severe physiological reaction.

3.2.3 Physiological Activity. Toxic substances may exhibit adverse health effects in several major ways. *Asphyxiation* is the exclusion of oxygen within an organism. It may be simple, such as that caused by excessive gas concentrations that deplete required oxygen levels, for example, high methane or carbon dioxide concentrations (greater than 5%) in a confined space. Or it may be chemical such as when excessive carbon monoxide levels are inhaled. Carbon monoxide preferentially binds to blood hemoglobin preventing adequate oxygen uptake or consumption.

Another adverse health effect is *irritation* that may occur at the site of the eyes, skin, or respiratory and digestive tracts, dependent on the route of exposure. Irritation is most commonly encountered when the pH of the substance differs widely from that of the body part surface affected. *Systemic* effects are those that cause abnormal functions of one or more of the body's major physiological systems; for example, excessive chronic lead exposure has been shown to affect the blood cells and the nervous system. Frequently the major adverse health effect is one that is *organ* or *site specific.* An example would be excessive chronic exposure to benzene, which has been demonstrated to damage the blood-forming tissues of the bone marrow.

3.2.4 Personal Exposure. Several sets of standards and guidelines have been developed to establish generally recognized "safe" levels of exposure to toxic substances. Some of these carry the force of law, such as the Occupational Safety and Health Administration's (OSHA) permissible exposure limits and the EPA Ambient Air Quality Standards. Others represent consensus standards, such as the American National Standards Institute's Hazardous Chemicals Series—Committee Z39. Still others, such as the American Conference of Governmental Industrial Hygienists—threshold limit values and short-term exposure levels and the American Industrial Hygiene Association's Workplace Environmental Exposure Levels, represent guidelines to protect nearly all persons. Most of these have been established to protect worker health with exposure at or below the recommended levels over an 8-hr work day, 40-hr work week, and up to a 40-yr working life. Additional guidelines, such as Emergency Exposure Levels and Immediately Dangerous to Life and Health Levels have been established by some organizations as an aid in preventing serious permanent health effects through designing and selecting equipment to handle the "once-in-a-lifetime" types of emergency situations.

There are numerous traditional and classical occupational health effects. These include the common pneumoconioses from excessive dust exposures, the development of chronic lung disease from excessive and prolonged exposure to crystalline silica, and lead poisoning from the inhalation and ingestion of excessive lead quantities. These potential hazards are generally well understood but are yet to be adequately controlled throughout industry. All new facilities or modernization of existing ones should be designed to minimize materials exposure to the greatest extent economically feasible.

In recent years, there has been a new array of industrial hygiene concerns. Of primary concern are the many new chemicals being developed each week. Many of these chemicals have not been adequately tested to determine their long-term toxicity. At times these materials are placed into use and workers unknowingly may be excessively exposed before the need for adequate control measures can be determined and appropriate controls developed and implemented. When designing facilities and selecting equipment for such new materials, it is prudent to incorporate controls to minimize personnel exposures.

The new wave of technological advancement involving the development of bacterial and biological agents, genetic engineering, drugs and pharmaceuticals, and other unique materials has brought with it new potential health hazards. Some are not clearly understood. Many will require tight systems that preclude any personnel exposure.

There also is an acceleration of public awareness that has been generated by the mass media regarding health concerns. These public concerns include cancer, genetic engineering, pesticides, and hazardous waste dump sites. Because of this, it is prudent and generally cost effective to go beyond minimum requirements to demonstrate no significant contact or exposure.

3.3 Fire Hazard Potential

Fire, combustion, and burning require three things (the absence of any one precludes a fire):

- A fuel (any oxidizable material)
- Oxygen (usually air)
- An ignition temperature (heat)

The following terms are useful in understanding fire hazard potential:

CLASS A FIRES. These are fires in ordinary combustible materials where the quenching and cooling effects of quantities of water or solutions containing large percentages of water are of first importance. Ordinarily combustible materials tend to produce glowing embers after burning, and these must be quenched to prevent rekindling. Water, foam, multipurpose dry chemical, or loaded stream-type extinguishers should be used on these fires, and extinguishers should be marked with an A on a green background.

CLASS B FIRES. These are fires in flammable liquids where a blanketing or smothering effect is essential to put the fire out. This effect keeps oxygen away from the fuel and can be obtained with CO_2, dry chemical (essentially sodium bicarbonate), foam, a vaporizing liquid type of extinguishing agent, loaded stream, or multipurpose dry chemical. CO_2, dry chemical, and vaporizing liquid extinguishers for these types of fires should be marked by a B on a red background.

CLASS C FIRES. These are fires of energized electrical equipment where the use of a nonconducting extinguishing agent is essential. Extinguishers for Class C fires should be marked with a C against a blue background.

CLASS D FIRES. These are fires in combustible metals, such as aluminum, lithium, magnesium, titanium, zinc, sodium, and potassium. Extinguishment of such fires requires special handling, materials, and techniques approved for that use. Since mishandling such a fire can cause explosions and spread of the area involved, combustible metals installations must be provided with the proper extinguishment materials. These include special dry powders, sand, and others. Such an extinguisher should be marked with a five-pointed star containing the letter D on a yellow background.

FLASHPOINT. This is the lowest temperature at which a liquid will give off enough flammable vapor at or near its surface such that in an intimate mixture with air it will burn in the presence of an ignition source. There are varying accepted test methods to determine the flashpoint of a liquid. However, the flashpoint of common flammable materials that flash at ordinary temperatures is usually determined with the Tag Closed Cup Tester (ASTM D56-52).

BOILING POINT. This is the temperature at which a continuous flow of vapor bubbles occur in a liquid being heated in an open container. The boiling point may be taken as an indication of the volatility of a material and may be used as a direct measure of the hazard involved with its use.

AUTOIGNITION TEMPERATURE. This is the temperature at which a material (gas, liquid, or solid) will self-ignite and sustain combustion in the absence of an ignition source (ASTM Designation D286-36). This value is influenced by the size, shape, and material of the heated surface, the rate of heating (in the case of a solid), and other factors.

EXPLOSIVE RANGE OR FLAMMABLE LIMITS. These values are used synonymously and are expressed in percent by volume of fuel–vapor in air. These are the concentration ranges at which a material will burn in the presence of an ignition source. If an air–fuel mixture within its flammability or explosive range is ignited, flame propagation will occur. If the flame spreads rapidly enough, it will be an explosion, otherwise only a fire will ensue. This range is indicated as the lower flammable limit (LFL) or explosive limit (LEL) to the upper flammable limit (UFL) or explosive limit (UEL). Below the LEL, there is not enough fuel to explode or burn. Above the UEL, oxygen content is not sufficient to support flammability. Examples of explosive ranges are acetone (2.6–12.8%), methane (5.2–14.0%), gasoline (1.4–7.6%), and trichloroethylene (12.5–90%).

FLAMMABLE LIQUIDS. A liquid that has a flashpoint (FP) less than 100°F (38°C). Examples of flammable liquids are acetone (FP 0°F; −18°C), methyl ethyl ketone or 2-butanone (FP 22°F; −6°C), toluene (FP 40°F; 18°C), and xylene (FP 90°F; 32°C). See Table 2.2 for a detailed illustration.

COMBUSTIBLE LIQUIDS. A liquid that has an FP equal to or greater than 100°F (38°C). Combustible liquids present a lesser fire hazard potential since they are less apt to become ignited at usually encountered ambient temperatures. This is because the volume of vapor generated at room temperature is too low to support burning. If the temperature is raised to the flashpoint or if the material becomes finely atomized such that there is sufficient vapor, then combustible liquids also can be highly hazardous. Examples of combustible liquids are Stoddard solvent (FP 106°F; 41°C) and mineral spirits (FP 108°F; 42°C). See Table 2.2 for a detailed illustration.

NONCOMBUSTIBLE LIQUIDS. A liquid that does not exhibit a flashpoint by conventional test methods. Examples of noncombustible liquids are some of the halogenated hydrocarbons such as trichloroethylene, perchloroethylene, and trichlorotrifluoroethane. However, some so-defined noncombustible liquids may support combustion in the presence of elevated temperatures (greater than 500°F; 260°C). For example, trichloroethylene has no flashpoint, however, it has an autoignition temperature of 770°F (410°C).

Regardless of the physical state of the material, the fire hazard potential of materials must be considered in the design and construction of production and plant facilities.

Table 2.2. Classification of Flammable and Combustible Liquids

Rating	Class	Flash Point	Boiling Point	NFPA Symbol
Flammable	IA	<73°F (23°C)	<100°F	4
	IB	<73°F	≥100°F	3
	IC	≥73°F, but <100°F (38°C)		3
Combustible	II	≥100°F, but <140°F (60°C)		2
	IIA	≥140°F, but <200°F (93°C)		2
	IIB	≥200°F		1
Noncombustible		Will not burn		0

Adapted from NFPA, Boston, MA, Volume 1, Chapter 30.

Most of the preceding discussion has been directed toward gases, vapors, and liquids since they usually exhibit the greatest fire hazard potential. However, some of these concerns also apply to solids, such as finely divided organic matter (grain and plastic resin dusts) that also exhibit considerable fire potential with definitive upper and lower explosive limits. Similarly, there are pyrophoric materials such as uranium and pyrotechnic materials (e.g., gun powder), that possess additional unique reactivity hazards. This will be the topic of the next subsection.

3.4 Reactivity Hazard Potential

This category of hazards deals with the degree of susceptibility of materials to be unstable, to react, and to release energy. Some materials are capable of a rapid release of energy by themselves (self-reaction) if they come into contact with oxidizing materials. Even water, other extinguishing agents, or certain other chemicals can react with some materials in an explosive manner.

The violence of reaction or decomposition of materials may be increased by heat or pressure (shock), by mixture with certain other materials to form fuel–oxidizer combinations, or by contact with incompatible substances, sensitizing contaminants, or catalysts. A specific source of ignition may or may not be necessary for a material to react quickly.

Reactivity is ranked as shown in Fig. 2.3, ranging from 0 to 4, with 0 being stable materials, that is, those that normally have the capacity to resist changes in their chemical composition, despite exposure to air, water, and heat or shock as encountered under unusual or emergency fire conditions.

Intermediate are the unstable materials that in the pure state or as commercially produced will vigorously polymerize, decompose, or become self-reactive and undergo other violent chemical changes.

Reactive materials are given a 4 ranking and include those materials that can enter into a chemical reaction with other stable or unstable materials. Reactive materials also include the explosives. These materials can decompose rapidly under certain conditions of temperature, shock, or chemical action to evolve either large volumes of gas or so much heat that surrounding air is forced to expand very rapidly. In either case an explosion results. These materials are particularly dangerous when involved in heat, shock, or heat-producing disaster conditions.

3.5 Specific Hazard Potential

In addition to the previously discussed hazard potentials that are represented symbolically on the NFPA hazard diamond with numerical rankings, there is another special means of designating unique hazards of a given material. The final quadrant of the hazard diamond is utilized to indicate unique hazards such as corrosivity (COR), acidity (ACID), alkalinity (ALK), oxidizers (OXY), incompatibility with water (⩊), and radioactivity (☢—the international symbol for radiation hazards).

3.5.1 Oxidizing Agents. Normal air with its approximate 21% O_2 content is the primary source. Many other materials are so constituted chemically that they can supply oxygen to a reaction, even in the absence of air. Some of these require heat before they will yield oxygen; others evolve significant amounts of oxygen at room temperature. Common materials that act as oxidizing agents are organic and inorganic peroxides, oxides, permanganates, chlorates, perchlorates, persulfates, organic and inorganic nitrates, iodates, periodates, bromates, perbromates, chromates, dichromates, ozone, and perborates.

3.5.2 Water-Sensitive Materials. These are materials that react with water, steam, or water solutions to evolve heat or flammable or explosive gases. Example of such materials are lithium, sodium, potassium, calcium, cobalt, rubidium alloys, and amalgams of these. Hydrides, nitrides, sulfides, carbides, borides, calcides, tellurides, selenides, arsenides, phosphides, and concentrated acids or alkalis are also in this category. These materials react exothermically with moisture and evolve hydrogen or other volatile, flammable hydrides. Storage and handling of these materials should be in areas where no water is present.

3.5.3 Corrosive Materials. These include acids, acid anhydrides, and alkalis. Such materials often destroy their containers and leak or diffuse into the atmosphere. Some are volatile; others react violently with moisture. Acid mists corrode structural materials and equipment. Such materials should be kept cool, but well above freezing. There should be sufficient ventilation to prevent accumulation of corrosive materials, and there must be regular inspection of containers in which these materials are stored.

Corrosivity, oxidizing ability, and water incompatibility have been discussed in Section 3.4. In addition, a given material may be incompatible with other materials.

3.5.4 Other Specific Hazards. These include radioactive materials that emit alpha, beta, gamma, or other forms of ionizing radiation. Such materials also require special placards, shielding, and warning signs that are beyond the scope of this chapter.

Some materials are chemically and biologically perishable and may require special attention. Chemicals that cause oxygen depletion and biological oxygen demand also must be given special storage and handling considerations.

This category of materials may also include the binary systems, that is, those systems where individual components must be handled separately to the point of use. Provisions must be made to keep such materials separate in all situations to the point where they are ready to be brought together. An example would be resin systems with separate curing agents and catalysts.

In some situations it is important that certain materials be physically separated and not shipped, stored, or handled together. An example would be blasting materials and the detonators. Another example would be acids that must be kept segregated from cyanide salts to preclude potential contact and release of highly poisonous cyanide gas.

3.6 Selection and Design for Storage

For most plants some provisions need to be made for storage of various categories of materials with potentially hazardous properties. Too frequently special storage provisions are made only for flammable materials, and this becomes synonymous with hazardous materials. The results can be catastrophic.

As can be understood from the preceding discussions, many hazardous materials definitely should not be located side by side with flammable materials. The wise designer will make separate provisions for each category of potential hazards. The following categories can be used as a basic storage breakdown.

FLAMMABLE MATERIALS. Strictly flammable liquids, that is, these liquids without any other potential hazard, should be provided storage for the purpose and hazard only. OSHA and NFPA guidelines for flammable materials should be followed.

COMPRESSED GASES. Tanks of compressed gases should be stored in a separate area and utilized in an upright and chained or otherwise secured condition. This is necessary to minimize the chance of their falling over and breaking or straining the valve or another part of the tank. Such tanks should be kept cool, out of the direct rays of the sun, away from hot pipes, and in ventilated areas where care has been taken in construction so that pocketing of gases that may escape from their containers will be kept to a minimum. Even here segregation provisions are necessary to separate incompatibles such as chlorine and ammonia and the flammables hydrogen or acetylene from the oxidizing materials such as oxygen.

ACIDS. These should have their own storage area.

ALKALIS. These include the caustic hydroxides and should have their own storage area.

OXIDIZERS. These also need separate storage facilities.

SOLID ORGANICS. These should be in a separate area from other categories and definitely away from flammable liquids, which if combined with the solid could cause spontaneous combustion.

TOXICS. This category is for special high-toxic materials that should not be mixed with any of the preceding materials. Examples include tetraethyl lead, pesticides, and carcinogens. Such toxic materials should also be segregated from each other.

Prudent plant layout, design, and equipment selection will provide separate areas for each of these types of materials and possibly others. It is preferable—and in some cases it may be required—that these areas be physically separated in different parts of the plant.

This approach often is not used and sometimes appears costly. Yet is could result in considerable savings. The aggregate of all these materials may be rather large. A single storage room to house them also will need to be large and will require special provisions to meet numerous criteria. The results can be a hazardous chemical storage building that is very costly.

Because the quantity of each different category of material frequently is limited, an appropriate cabinet may provide more than adequate space for that given category. Using this approach, individual cabinets may be sufficient for the storage of each category of material and may be small enough to fit conveniently into existing available spaces. The cost of an expensive storage room thus can be avoided in the majority of cases.

4 QUANTITIES

4.1 General Quantity Categories

The quantity of process materials may vary from as little as a few milliliters or milligrams to many thousands of liters or kilograms. Economic considerations within a given industry dictate the quantities processed. In mining, for example, kilogram quantities of material (e.g., diamonds) or thousands of metric tons (e.g., coal) may be mined for a given general class of material, in this case varying forms and states of carbon. These are also significant processed material quantity variations from one industry to another (e.g., mining vs. manufacturing).

The various process quantity categories may be classified in terms of the stage or development of a technique or industrial approach. In the early research stage, typical quantities of grams or milliliters are common for small-scale development work.

When proven successful and potentially economically feasible, pilot plant operations involving somewhat larger quantities are likely to be involved. These may be 125-ft³ (3540-L) cylinders for gases, 55-gal (208-L) drums for liquids, or 50-lb (23-kg) bags for solids. Once product market potential is sufficient and operating techniques have been developed, a company may wish to pursue full-scale production of a product or material. Enormous quantities of materials are usually encountered in full-scale production facilities because of economic incentives. Quantities in the range of numerous truck trailers or tankers, railroad boxcars or tankers, shiploads, or even continuous supply pipelines may become practical.

Research quantities of highly hazardous materials usually can be handled economically in a well-designed laboratory hood without concern for excessive exposure to the worker. The laboratory hoods should be selected to handle these highly hazardous materials, even if hazard testing of unknowns is not done until the materials are to be upscaled to pilot plant and production quantities. All materials should be handled inside the hood and assumed to be highly hazardous, unless known otherwise.

With larger quantities, hazard testing may be necessary and controls (which may become costly) will need to be selected accordingly. What may be satisfactory at the research quantity level may be too hazardous to put into production. Less hazardous substitutes may be required. Hazard potential may determine production quantity levels. Such is the case for fissionable materials. The wise designer/engineer will evaluate hazard potential versus quantity prior to completion of plans and upscaling to pilot plant or production quantities.

4.2 Packaging and Materials Handling

The more toxic or hazardous a material, the greater will be the need for examining unit quantities and the packages to be used. Ideally one selects a container type and unit size that not only minimizes waste and handling but also precludes the necessity for actions that generate exposure situations. Actions that involve manual transfer from one location or container to another are to be avoided. Situations that involve interim storage are also to be avoided, when practical.

Again, the physical, chemical, and toxicological properties of materials to be processed will limit what is practical. From this, economically feasible types of packaging and material handling methods that can be effectively utilized and will afford appropriate protection for personnel must be selected.

Typical types of packaging and shipping containers include cylinders, pressure vessels, and cans for gases; bottles, drums, vats, tubs, tanks, and piping systems for liquids or mixtures; and boxes, pallets, bags, bins, silos, and tanks for solids. Container sizes may vary from small "pills" or single-use packets to bulk quantities, up to supertankers.

Each part or unit creates a potential industrial hygiene exposure situation or condition. Generally, the smaller the package or unit size for a given process quantity, the greater the hazard potential. For example, ten 5-lb (2.3-kg) packages of powered material may present greater hazard potential than one 50-lb (23-kg) package due to the extra handling involved and the greater potential for increased broken packages, housekeeping, and individual packages being misplaced or stolen. In large part, this is why there has been emphasis in recent years on unitized loading systems. The market for a given product and distribution methods will be other factors determining package or unit size.

The needs and requirements of the processor or manufacturer and customers also will vary. A suitable packaging system for one may be unacceptable for the other. For example, it is common practice for some manufacturers to package powdered materials in 50-lb disposable bags. This may create serious excessive airborne dust exposures to personnel breaking, dumping, and compacting the bags for disposal. A cost-effective solution for one user was to require packaging of the material in collapsible, reusable, reinforced, and plastic-lined tote containers with a 500-lb (230-kg) capacity. This solution also afforded greater moisture resistance to the material than was previously achievable, resulting in less waste material at a substantial savings.

Many specific examples and detailed descriptions of processing and material handling methods are described in later chapters of this volume and in both earlier volumes of this series.

Material flow concepts and packaging techniques will play a major role delineating operation layout in a facility and may even dictate the type of building in which optimal efficiency may be obtained. For example, operation of a parts storage warehouse and distribution center can be optimized in a single-story flat-roof structure with multiple-layer racks equipped with a computerized storage/retrieval system. Such efficiency could not be obtained in a multistory or peaked roof structure with equivalent usable floor space and volume. See Chapter 6 on *Building Types* for further discussion.

In previous subsections, mention has been made of vertical versus horizontal flow arrangements. Gravity flow should be considered and taken advantage of whenever possible. The handling efficiency and container cost may determine whether disposable or reusable containers are the most practical. The degree of labor intensiveness of a process will be a function of labor costs (direct—salaries or hourly rates, and indirect—fringe benefits, accident and worker compensation costs) and need for consistency of product quality. People are better at interacting with complex and changing variables, whereas machines are far superior at performing repetitive tasks at a fast and consistent rate. Industrial hygiene or environmental air pollution exposure potential for a given facility location versus the degree of control required will often determine the level of appropriate automation.

Overhead placement of corrosive of hazardous materials should be avoided because leakage and downward spills can jeopardize the safety of personnel below. Where practical, liquids should be drawn by suction. Pressurization should be avoided to prevent spurts or leaks outward toward personnel or sensitive equipment.

Pressure relief line or vents should discharge high if the materials are vapors or gases. Discharge should be downward to ground level if the material is liquid or solid. Gravity flow can be used to avoid pumping in some liquid flow processes. This can help eliminate valve leaks and other situations that can result in excessive levels for personnel.

4.3 Designing Out Sources of Potential Exposures

Sources of potential exposures to toxic process materials are many and varied. However, major sources can be defined as the following:

- Points of measurement
- Material handling transfer points
- Packaging
- Specific process segments or operations
- Maintenance activities
- Process upset conditions

Points of measurement are common to all industrial processes in one form or another. These include anywhere materials are sized, weighed, gauged, or sampled. They occur in production, quality control, laboratory analyses, and other operations. Excessive exposures are usually the result of manual manipulation of materials in the absence of adequate control measures at such points. Cost-effective measurement means or control systems often can be incorporated into the design of a facility in the early planning stages.

Material handling and packaging activities present another major source of exposure. These activities include loading/unloading, transfer systems, use, and storage of materials are previously discussed. Again, the degree of potential exposure is associated with the extent to which materials are handled manually in the absence of adequate control systems and procedures.

Maintenance activities offer a somewhat unique exposure potential in many industrial facilities. This is because maintenance activities are more likely to involve infrequent encounters with toxic materials for short or extended durations. They also involve areas where personnel are not normally present and for which adequate occupational health control measures are frequently inadequate. In the design and layout of machinery, equipment, and facilities, great care must be taken to anticipate the control needs of maintenance personnel.

For example, when there is a requirement for a large tank greater than 4 ft (1.2 m) deep, sufficient drains, manway access openings, and even a ladder or stairs with provisions for fall protection should be incorporated if at all practical. Such a confined space may also require periodic entry by personnel wearing self-contained breathing devices and other appropriate protective clothing and equipment. Access openings need to be adequately sized for this purpose. Where possible, doors and stairs should be incorporated into the design in lieu of manholes and ladders. Provision should be made for either natural or mechanical ventilation.

Specific process segments or operations may involve discrete sources of exposure. For purposes of illustration, a brief presentation of those processes common to many industrial settings with the form of material of concern include:

- Welding (fumes, gases)
- Grinding (dusts)
- Mixing (dusts, vapors, gases)
- Machining (chips, smoke, oil mists)
- Painting or other spray coating (vapors, particulates)
- Plating (vapors, mists, fumes)
- Moving machinery and equipment (noise)
- Compressed air usage (noise)

Obviously, this list is far from comprehensive. Radiative stresses, such as ionizing and nonionizing radiation (including microwaves) and heat, also may present significant health concerns for selected operations.

Emergency or process upset conditions should be considered, for each system. These frequently create industrial hygiene concerns. These may be a result of burst piping, ruptured and leaking tanks or reactor vessels, or process stream overflows. They may be due to rare torrential downpours or other acts of God, overpressurization, runaway polymerization, or contaminated raw materials. Plant facilities should be designed to include adequate means of egress and personnel protection, and should allow for confinement of spills or leaks, and appropriate cleanup and disposal facilities.

5 HAZARD POTENTIAL CONTROL

5.1 General Controls

There are two primary and basic approaches to control: engineering controls and administrative or work practice controls. Engineering controls are those integrated and built into a system. In other words, by design the potential hazard is prevented or made significantly less probable to occur. Administrative or work practive controls are those that prevent or make potential hazards less probable or severe through management procedures and actions.

Engineering controls are generally recognized as the most effective since they take the human element out of the action. They are more positive, and permanent controls once properly designed, installed, and debugged, are usually far less dependent on human actions to be effective. Regardless, engineering controls or systems must be regularly maintained to ensure their continued effectiveness. More details regarding engineering controls are presented in a subsequent subsection.

Administrative and work practive controls can be as effective as engineering controls. However, this type of control possesses an inherent weakness in that reliance is place upon constant human action for the ultimate control potential to be realized. Administrative controls include employee training in proper safe and healthful work practices and their enforcement, scheduling of work to minimize personal exposures or the number of personnel exposed, employee selection and placement, and thorough recordkeeping. Work practice controls include: (1) proper supervision, care, and use of personal protective equipment; (2) written job descriptions with specific work procedures, such as always working upwind of contaminant generation sources; and (3) specific types of protective equipment to be utilized for potentially hazardous activites.

OSHA has followed the lead of many health and safety professionals by specifying the exclusive use of engineering controls where feasible. The courts have gradually defined the meaning of feasible to be technologically possible *and* economically practical. Engineering controls are available for almost all potential industrial hygiene exposure situations. However, there have been instances where one aspect or another of feasibility has been shown, but not both concurrently.

It is easy to understand some of the pitfalls inherent in administrative controls. However, they can be effective in the short term once a potential hazard has been recognized and until engineering controls can be proven effective. Administrative controls can serve as an added margin of safety in addition to feasible engineering controls or as the primary means of control when engineering controls are not feasible.

5.2 Engineering Control Concepts

First emphasis should be on the designing out or minimizing of hazard potential through the following approaches in the order given.

- Substitution/elimination (process or material)
- Enclosure
- Isolation/separation/shielding (especially radiative stresses)
- Redesign of person-machine interface (especially ergonomic stresses)
- Ventilation (airborne stresses/contaminants)

5.2.1 Substitution and/or Elimination. Where possible, hazardous processes, operations, materials, or pieces of equipment should be eliminated and substitute means employed. An example of an effective substitution would be replacing trichloroethylene in favor of inhibited 1,1,1-trichloroethane (methyl chloroform). Both of these nonflammable chlorinated solvents posses similarly effective properties in degreasing applications. However, 1,1,1-trichloroethane is considerably less toxic, thus lowering the hazard potential. Elimination of a hazardous process step, such as purchasing preassembled components instead of welding or soldering them, could eliminate or reduce the in-plant need for controlling fumes from the process.

5.2.2 Enclosure. Another common and effective engineering control is the enclosure of a hazardous process or operation. Locating compressors or other noisy pieces of equipment in a separate room can substantially reduce personal noise exposures to workers in the area. Processing of toxic chemicals in closed or sealed piping systems and reactor vessels is another example of this concept.

5.2.3 Isolation/Separation/Shielding. These methods are similar and compose a subset of the enclosure concept on a more limited scale. Isolation, separation, and shielding of hazardous operations can be achieved by distance, placing a barrier between the potential hazard and employees, or a partial barrier blocking the direct path of a hazard. An example of isolation would be locating a control room work station in a separate room to minimize employee exposure to process noise, dusts, and fumes. An example of separation would be remote location of an operator from a hazardous operation to minimize exposure to contaminants while maintaining visual contact and control via a television screen/computer terminal. Placement of a fixed barrier between a wood planer and the loading/unloading stations to reduce levels would be an example of effective shielding.

5.2.4 Redesign. The interaction of the machine and person can be studied with the objective of designing work stations and job demands in a manner that is compatible with human capabilities. Frequently, simply moving the person away from a given source will reduce or eliminate an exposure situation.

Ergonomic stress reducers are good examples of this and include special chair and tool configuration design to accommodate natural shapes and movements. Lifting aids, such as cranes and hoists, may also be classified as ergonomic stress reducers that prevent muscle strains and sprains. Ergonomic design of work stations and control display layouts consistent with optimal utilization including tool size, shape, color, contrast, and placement is also known as human factors engineering.

5.2.5 Ventilation. This is a most widely used engineering control technique for air-borne stress/contaminants. Ventilation may be defined as the controlled or predictable movement of air. There are two major types, natural and mechanical. Natural ventilation consists of wind, thermal, and random air currents. Out doors these are usually direct, while indoors these air forces are a function of the building design and movement of objects within the building. More details are provided in a subsequent chapter. Mechanical ventilation may be general (dilution) or local; these are further classified by their flow direction as supply or exhaust. Local exhaust ventilation has the greatest significance as an industrial hygiene engineering control method. Considerably more detail regarding ventilation and building design is provided in subsequent chapters.

Other engineering control concepts include general plant layout to optimize smooth flow of materials and operations efficiency, building types, design and construction, material storage, and maintenance requirements.

Hazard potential can be controlled through proper equipment selection and facility design. Appropriate controls are available for purchase as commercially available items for the control of most potential hazardous situations. When commercial items are not available it is incumbent upon the engineer and/or management to design and develop such controls in-house.

5.3   Control Factors

As in most engineering applications, there are cost trade-offs or compromises that will have to be made in the selection of a specific control method. Many factors enter into this matrix including regulatory standards, recognized guidelines, good engineering practices, and preservation or promotion of company image. For many applications simply reducing airborne contaminant levels to well below the maximum acceptable level, that is, the permissible exposure limit (PEL), or threshold limit value (TLV), is sufficient. But if practical and economically feasible to do so, it will be desirable to overdesign the control systems, so that control is obvious. This will aid in the future defense or prevention of exposure allegations and resultant worker's compensation or state industrial commission claims costs.

Equipment selection, layout, plant design, and controls to be incorporated will be largely dependent on physical and chemical properties, the hazardous property material classification, and the quantities, sizes, types, and distribution of the materials involved. Most of these topics have been discussed previously. However, the distribution of generation points has not. There frequently are numerous points of material or contaminant generation and distribution. These result in relatively low "background" levels, various source generation and spread rates, mainstream or direct exposure, incidental releases, and massive emergency emissions.

Many industrial operations generate nearly constant low concentration background levels of contaminants. Odoriferous or dusty releases are most obvious (e.g., hydrogen sulfide in a petroleum refinery or particulates in a quarry). Control of these contaminants cannot be ignored, and the designer should be selecting equipment and laying out operations to keep levels to their lowest feasible level. Adequate control usually can be achieved utilizing tight equipment and searching out points of material release in order to eliminate them or to control then insofar as possible. Just how tight equipment must be and how much elimination is required will depend on the key hazardous characteristics of the material and process. For highly toxic materials such as phosgene, each potential penetration/emission point is significant and must be controlled in some way. With such materials even incidental releases are unacceptable. With less toxic or otherwise less hazardous materials it is necessary to control only the sources of greatest magnitude and those that contribute most to the personnel doses. Emission at points where personnel may be located should nearly always be controlled.

Potential massive emergency releases present unique design situations. For these, design should be to an established or accepted frequency of occurrence and controls should be incorporated to limit the material levels to less than the emergency levels established for that material. At emergency levels temporary effects may occur and some risks are involved. Emergency levels have been developed for some military applications. Some situations may warrant the development of an emergency level for that particular situation.

Regardless, it is incumbent upon the engineer/designer to see that proper criteria are developed for the type of material involved and then to meet those criteria.

GENERAL REFERENCES

Clayton, G. D., and F. E. Clayton (eds.), *Patty's Industrial Hygiene and Toxicology, Vol. I, IIA, IIB and IIC: General Principles and Toxicology,* New York: Wiley, 1978, 1979, 1981.

Gosselin, R. E., H. C. Hodge, R. D. Smith, and M. N. Gleason, *Clinical Toxicology of Commercial Products,* 4th ed., Baltimore: Williams and Wilkens, 1976.

Cralley, L. J., and L. V. Cralley (eds.), *Patty's Industrial Hygiene and Toxicology, Vol. III: Theory and Rationale of Industrial Hygiene Practice,* New York: Wiley, 1979.

National Fire Protection Association Standard No. 704M, *Identification of the Fire Hazards of Materials;* Standard No. 49, *Hazardous Chemicals Data;* Standard No. 325M, *Fire Hazard Properties of Flammable Liquids, Gases and Volatile Solids;* Standard No. 321, *Basic Classification of Flammable and Combustible Liquids;* Standard No. 491, *Manual of Hazardous Chemical Reactions.* NFPA, Quincy, MA, current issues.

National Safety Council, *Accident Prevention Manual for Industrial Operations,* 7th ed., NSC, Chicago, 1974.

National Safety Council, *Fundamentals of Industrial Hygiene,* NSC, Chicago, 1979.

Sax, N. I., *Dangerous Properties of Industrial Materials,* 4th ed., New York: Van Nostrand Reinhold, 1975.

Steere, N. V., *CRC Handbook of Laboratory Safety.* Cleveland: Chemical Rubber CO., 1967.

Weast, R. E., *CRC Handbook of Chemistry and Physics,* 64th ed., Boca Raton, FL: CRC Press, 1983.

Windholz, M., *The Merck Index,* 9th ed., Rahway, NJ: Merck, 1976.

CHAPTER THREE

Control of Open Process Systems

Donald R. McFee
Richard P. Garrison

1 WHAT IS AN OPEN SYSTEM?

An open process is best defined as one that is not closed. A closed process is one where, except perhaps for changing raw material and/or packaging finished goods, all normal operations are conducted inside of airtight equipment and equipment is not opened routinely for production operations other than sampling.

Open systems can, and sometimes should, be "closed." This means implementing engineering controls to reduce or eliminate potential hazards. The need for control must be evaluated thoroughly and justified by the benefits to be derived.

Potential industrial hygiene hazards from open process systems include the full range of chemical and physical agents of concern to industrial hygiene. Open systems encompass a full spectrum of situations. One (most open) extreme would include unprotected work on open piles of toxic, dust-generating materials, such as asbestos-containing insulation from demolition of an old boiler facility. The other (least open) extreme would include briefly opening a process system to collect a liquid sample or make a quick measurement. Most process operations fall somewhere in between.

2 EXAMPLES OF OPEN SYSTEMS

2.1 Chemical Systems

There are a wide variety of industrial operations that involve the handling and processing of chemicals. Many of these are "open" in that they present opportunities for contact with material, both directly and when airborne.

2.1.1 Stockpiles. Uncovered stockpiles of solid material are a simple and sometimes extreme example of an open chemical system. Material characteristics such as personal and environmental toxicity, dustiness (windblown), water (rain) solubility, and cost (value) may make it necessary to protect the material. Protection may be needed underneath (e.g., an impervious surface), around (fencing, windrows of trees, embankments) or over (tarpaulins, buildings) the material. Enclosure of a stockpile by a building may create industrial hygiene concerns unless adequate general ventilation is provided. Provisions to minimize dispersive wind effects at stockpile building doorways also may be needed.

2.1.2 Tanks. Tanks containing liquids may be either covered or uncovered, inside or outside of a building. Covered tanks generally require venting, which may best be discharged outside of a building. For volatile liquids, the rate of evaporation and acceptable airborne levels should be used to confirm that adequate general ventilation is provided. For toxic and/or corrosive materials, it will generally be prudent to avoid locating tanks at elevated positions to minimize the possibility of leaks and spills onto people and equipment below. This may preclude the use of gravity flow for fluid transfer. The use of mechanical pumps may present additional industrial hygiene concerns during maintenance, unless thorough flushing and careful work practices are employed.

If tanks are to be located at elevated positions, they and their associated piping should be designed, operated, and maintained as though they were pressure vessels. If the tank is not actually a pressure vessel, adequate overflow provisions should be made.

2.1.3 Kettles. Kettles and similar processing containers involve the heating, and sometimes mixing and agitation, of materials. They are often open systems having the potential for significant emission to the surrounding workplace. Control frequently will require local exhaust ventilation because general ventilation cannot prevent the migration of fugitive airborne materials into work areas. Location of kettles near outside walls will facilitate installation and operation of local exhaust ventilation.

2.1.4 Chemical Reactors. Chemical reactors are most often closed or at least partially enclosed as a necessary aspect of their operation. However, there are occasions when they are open, such as when manways are opened for material additions, reactor washdowns, or sample collection. It is desirable to design the system to avoid opening a reactor unless there is a paramount reason to do so, or the materials inside are not toxic. Accepting that the need to open a reactor may exist, it is possible and advisable to design the reactor facility to take advantage of natural ventilation and otherwise minimize the potential for confinement of gases and vapors that might be released when the reactor is opened. From a process and vessel integrity viewpoint, one would like a vessel with no more openings than absolutely necessary. However, too frequently vessels do require cleaning and repair. Having adequate access and clearance from more than one opening, preferably at opposite ends, facilitates ventilation, cleaning, and repair. If material toxicity potential is high, provisions need to be made for mechanical ventilation, ensuring proper discharge of contaminated air and adequate, controlled makeup air.

2.2 Material Handling Systems

There are many ways by which materials used in industry are transported from one place to another. Significant industrial hygiene concerns are frequently raised in regard to the handling of potentially toxic materials, especially inside buildings. Environmental concerns may also be important for material handling outside of buildings.

2.2.1 Mechanical Conveyors. Mechanical conveyors frequently utilize belts for horizontal transfer and buckets for vertical transfer. Transfer points between belts or from belts to buckets (and vice versa) may cause significant dusting. Out of doors, this may not be a significant industrial hygiene concern, but inside a building it can be serious. It is possible to provide local exhaust ventilation around transfer points, and to even entirely enclosed belts and bucket conveyors.

Noise is another potential industrial hygiene concern, with belts usually quieter than buckets. Noise may be a good reason to locate conveyors outside a building if at all possible or to enclose and ventilate conveyors located inside a building.

2.2.2 Pneumatic Conveyors. Pneumatic conveyors have the potential for causing very heavy dusting at points of leakage. This is particularly true for finely divided (powdered) materials and is less significant for granular materials. All materials are, to some extent, abrasive and the potential for transport duct work erosion should be considered carefully. Transport velocities should be no higher than needed, long radius elbows should be used, and special replaceable surfaces may need to be designed into a system to transport highly abrasive materials.

Noise from air-moving equipment may present industrial hygiene concerns; location outside a building may be an effective solution depending also on environmental factors.

2.2.3 Augers. Augers, particularly of the screw type, can wear out very quickly when handling abrasive materials. This can result in leaks that pose significant industrial hygiene concern. For this reason, augers should be avoided for handling abrasive materials. Hopper-type feeds for augers also represent potential problems because of dust release. There are ways, however, to provide effective local ventilation control for hopper feeding done inside and to minimize adverse wind effects for hopper feeding done outside. It is also possible to eliminate manual hopper feeding by utilizing automated equipment to open and dump bags or otherwise feed hoppers.

2.2.4 Transfer Points. Material transfer points have been discussed in connection with mechanical conveyors and augers. In general, transfer points to and from containers and different types of transporting equipment represent the most significant potential for direct and airborne contact with materials. The free-fall distance should be minimized at a transfer point whenever possible. Failure to do so can create problems even for enclosed and ventilated transfer points because air currents induced by the falling material can interfere with airflow control at the transfer point. Bucket conveyors or inclined slides may be preferable approaches.

2.2.5 Containers. Material containers come in all shapes and sizes: big and little drums, boxes, paper bags, large tote bags, pallets, pressurized cylinders, and many more. All containers should be constructed of sufficiently sturdy materials, and specific work practices for handling each type should be developed and enforced. All containers can be spilled, and this should be taken into account in the layout of a material handling process. Provisions for containing spillage and performing prompt cleanup should be made. This should include the drainage system.

The potential for material incompatability (i.e., reaction) should be considered; it may be necessary to drain into a holding tank rather than the regular sewer. Emergency exhaust ventilation may also be needed for leaks or spills involving highly toxic materials.

2.2.6 Vehicles. Vehicles used for material handling are many and varied. Industrial hygiene concerns derive primarily from operations and maintenance of the vehicles and characteristics of the environment in which they are used. For example, carbon monoxide will be released from vehicles driven by fuel combustion. For areas with a lot of vehicle traffic, adequate general ventilation is very important, especially in cold weather when the buildings are closed. The importance of effective maintenance cannot be overemphasized because vehicles must be in proper tune-up condition to minimize emission of carbon monoxide. Vehicles usually are cleaned periodically, and cleaning agents can present industrial hygiene concerns. Careful work practices should be emphasized. Possible ways to minimize contact with chemicals, such as automated washing facilities, should be pursued if large numbers of vehicles are cleaned on a frequent basis.

2.3 Manufacturing Systems

Many manufacturing operations involve open, partially open, or sometimes closed systems. The range of variation in these systems is far too broad to be discussed completely in this chapter. Other chapters discuss a variety of examples. The following discussion provides general guidelines for several basic open system manufacturing operations: spray painting, welding, grinding, drying, plating, and cleaning.

In general, most industrial hygiene concerns resulting from manufacturing operations can be addressed by recognized work practices and engineering controls. Control equipment is available commercially for many situations involving open manufacturing systems. These can be very effective if properly installed and maintained. There are also many situations for which new and possibly creative design will be required. For these, final design of engineering controls should be reviewed and accepted by a qualified and experienced industrial hygiene engineer prior to installation. This helps ensure against costly retrofit due to faulty design.

2.3.1 Spray Painting. Spray painting operations should be designed and operated to facilitate two things: (1) local exhaust ventilation and (2) orientation of the painter with respect to the ventilation air flow and the object being painted. Local exhaust ventilation should provide a minimum of 100 ft (30 m) per minute of air flow velocity in the area being painted. The cross-sectional area that is ventilated should be kept as small as possible to minimize overall air flow requirements. This will mean the workpiece should be painted in a fixed orientation that permits the smallest cross-sectional area for air flow. This is especially important for larger objects such as vehicles and large equipment. Smaller objects can often be placed on turntables so that all surfaces can be painted without requiring the painter to move from the optimum painting position, upstream of the object. Air cleaning to remove particulate material, such as filters or flowing water, generally is necessary.

2.3.2 Welding. Welding operations that are performed in fixed locations can and should be ventilated by local exhaust ventilation. An effective system should provide air flow control at the point of welding, without disturbing shielding gases that may be in use. Turntables and movable exhaust inlets can be particularly helpful. Work practices should emphasize maintaining proper position during welding. For welding operations performed in large work areas, without local exhaust ventilation, it is important to provide adequate general ventilation. Generally, it is not necessary to clean exhaust air from welding operations before discharge to the atmosphere.

2.3.3 Grinding. Grinding operations can present significant industrial hygiene concerns. These may result from the metals or other object materials being ground, and/or from lubricating oils and the grinding abrasive. It is usually necessary to provide partial enclosure of the grinding wheel and to concentrate local exhaust ventilation air flow to take advantage of the release trajectory of air contaminants. The ventilation manual published by the American Conference of Governmental Industrial Hygienists (ACGIH) provides excellent design guidelines for a variety of grinding situations.

Hand-held grinders used in open work areas can be particularly difficult to control. High-velocity/low-volume exhaust ventilation can be applied successfully for many of these situations. Inlet nozzle design is critical, as are providing sufficient air flow to the exhaust nozzle. maintaining the balance of the tool, and otherwise avoiding unnecessary encumbrance of the operation.

It is usually less expensive to use conventional grinding tables and booths wherever possible. Air cleaning generally is necessary, sometimes to protect the local exhaust air mover. Primary collection by cyclone or settling chamber can be quite effective for the usually relatively large-size metal and abrasive grinding fragments.

2.3.4 Drying. Drying (baking and curing) ovens can present significant industrial hygiene concerns unless emissions are controlled effectively. For ovens that remove only water, there are no significant air pollution problems, and system design should emphasize minimizing corrosion. For ovens that remove solvents, there may be significant air pollution concerns and solvent value, but corrosion is not usually a major problem. Flammability needs to be taken into account for solvent systems.

Air flow should be sufficient to ensure that concentrations are far below the lower limit (LEL) of evaporated solvents. Exhaust ventilation systems should have sensing devices, warning alarms, and interlocks with oven heating controls to prevent operation when there is failure of the air-handling system. Air movers should be of the non-sparking type. Fans are best located at the end of the oven ventilation system to minimize pressurizing duct work inside the building.

2.3.5 Plating. Plating operations involve the potential emission of a wide variety of chemical agents. Proper selection of materials, enclosures, and ventilation systems can provide adequate control. There are two types of basic ventilation duct work systems: above and under the floor. Under-the-floor systems permit greater flexibility for materials handling in and around the tanks, but are more difficult to maintain. They should be oversized to provide access and permit work to be performed within them. Layout of the plating tank system is extremely important, and can have a profound effect upon the effectiveness of the ventilation control system.

Compatibility of materials is another important consideration. Some acids such as perchloric and nitric are not compatible with organic materials such as plastic duct work, tapes, and/or other sealants (forming potentially explosive perchlorates and nitrates). Cyanide solutions cannot be allowed to mix with acids (forming hydrogen cyanide). This is more likely to occur in common drainage systems. It is less likely in duct work, unless there is heavy misting and agitation, failure of the exhaust ventilation system, or spillage into underground systems.

2.3.6 Cleaning. Cleaning operations involve both aqueous and organic cleaners and solvents. Caustic cleaning agents for aqueous systems generally are not highly volatile.

Ventilation controls usually are not necessary, but careful work practices to avoid contact with hot and caustic materials are important.

Organic cleaning systems can involve volatile and toxic vapors. Degreasing tanks are sometimes provided with refrigerated vapor traps to minimize vapor loss, but unacceptable vapor emissions are still possible. Local exhaust ventilation is frequently required across the top of solvent cleaning tanks and sometimes in dripping/drying areas near the tanks. It is sometimes possible to utilize enclosed washers to minimize vapor release from open tanks.

Cleaning vessels will periodically require cleaning themselves. When this occurs, the potential exists for creating a hazardous confined-space entry situation. It is possible to avoid or at least minimize industrial hygiene concerns by cleaning vessels more frequently without entry, and perhaps utilizing special tools or vessel designs to facilitate routine cleaning.

3 WHEN IS A SYSTEM OPEN?

Chemical, material handling, manufacturing, and other systems are open, from the standpoint of industrial hygiene, wherever there is uncontrolled opportunity for contact with potentially hazardous conditions. The closer the point of release is to personnel, that is, the interaction of the person with that point in the system, the more significant will be openness of the system. Depending on the specific operation involved, systems may be open at all times, during specific routine operations, during routine maintenance, and/or during nonroutine activities both planned and unplanned. Identification of when the systems are open and how potential exposures should be controlled when the systems are open is often the key aspect of industrial hygiene management for plant operations.

3.1 Routine Operations

Routine operations have the potential for significant exposure situations in many ways. This is due to the longer periods of time and greater numbers of personnel who may come in contact with potentially hazardous conditions while the system is open. However, controls can be provided for routine operations, and these controls often can be justified more easily than for nonroutine activities because they contribute directly to productivity, product quality, and other benefits.

Buildings should be designed and equipment selected so as to minimize contact with toxic materials and to reduce physical and mental stress. This can be accomplished effectively for new construction by anticipating potential hazards and designing appropriate controls. This will not preclude the need for evaluation during operation, with the possibility of additional or modified engineering controls.

Effective planning against potential industrial hygiene concerns during routine operations also should include written standard operating procedures. This act of planning frequently highlights needs and control requirements not apparent otherwise. Where administrative controls may be applied to supplement engineering controls, written procedures are a necessity. Procedures, however, should not be regarded as substitutes for engineering controls. They are dependent on the actions of people with their inherent weaknesses and are less desirable than feasible engineering controls.

3.2 Maintenance

Maintenance activities can be either routine, including preventive maintenance, or nonroutine reactive work, including emergencies. As stated previously for routine operations, the more routine the activity, generally, the easier it will be to design and justify the cost of engineering controls.

Maintenance activities have very significant industrial hygiene considerations. They frequently involve the greatest extent to which a system can be "opened." Potential hazards may be greatest during these activities for several reasons, including higher airborne contaminant levels, the need to work quickly, difficulties in ventilating, and the potential for skin contact without adequate protection.

Effective planning and design are necessary to minimize potential hazards cost effectively. This includes providing facilities to accommodate safe maintenance work, such as bypass piping, drains, pump/filter flushing, and a wide variety of other means. Other important facility design considerations include ventilation, access to equipment, and lighting (illumination). There will often be a need for the extra margin of safety that can be provided by personal protective equipment.

Preventive maintenance, which is usually productive, scheduled, and nondisruptive, can be better planned to incorporate appropriate engineered protective provisions for minimizing potential industrial hygiene problems that can arise when "reactive" maintenance is needed. Process systems are usually less open, and less potential exposure time will be involved during planned preventive maintenance as opposed to unplanned reactive maintenance.

3.3 Accidents

Planning and design of building, equipment, and layout also take into consideration the potential for accidents. The occurrence of an accident, by definition, is not planned. An accident is a breakdown in planning and procedures. Accident possibilities can and should be anticipated and evaluated thoroughly. If this is done well enough, there will be no accident because the potential situation becomes one covered by appropriate procedures and equipment.

Systems may be open to the maximum possible extent during an accident. As a result, exposure situations may be very great. Systems that can be hazardous when opened by accident can also be controlled to some extent. Effective controls may include, but not be limited to, automatic shut-off devices for equipment, additional drainage with material holding capability, emergency general ventilation, escape routes, warning alarms, communication means, emergency lighting, emergency utilities, personal protective equipment, and a wide range of other means to combat or diminish an emergency.

Accidents may have serious, even catastrophic, implications that go beyond the plant "fence line." Response planning and actions must take this into account and should comply with federal, state, and local regulations that may apply.

3.4 Modifications

The need to modify manufacturing systems can bring with it serious industrial hygiene considerations because systems may be more open than usual. There frequently will be the need for temporary engineering controls to minimize potential hazards. Here again the emphasis should be on planning and appropriate built-in control measures in lieu of personal protection.

As stated previously for all other times when the system is open, planning is necessary and can be very effective in minimizing potential hazards. Planning for system modifications usually will emphasize designing flexibility into the system. This may mean additional utilities, additional piping to permit bypassing a process stream, rather than "hot tapping" into it, or providing means for effective temporary ventilation.

The question should always be asked, "Does the change or modification increase the potential for exposure to hazardous materials and/or conditions?" If the answer is "yes," then appropriate action should be taken to evaluate the hazard fully and control it as needed. Some OSHA standards, such as for inorganic lead, require that this be done in connection with changes in process systems. Potential hazards should be reevaluated after modifications are completed.

3.5 Start-Up/Shutdown

Start-up and shut-down operations usually are routine, although they may be prompted by nonroutine circumstances such as unscheduled maintenance and accidents. Sometimes these situations involve working under conditions that are more open to potential hazard situations than during routine operations.

Manufacturing process systems should be designed to consider the special situations that can arise during start-up and shut-down. This may require mechanical material transfers and additions, additional ventilation, and personal protective equipment. Standard written operating procedures also should be prepared. Some complex systems, such as chemical manufacturing and petroleum refining, may require additional specific and unique procedures for each start-up/shut-down. If so, these should be documented in writing so that they can be implemented with other necessary procedures.

The key to minimizing potential hazards when systems are open is planning. This should lead to more appropriate design of facilities and equipment and development of safe and complete work procedures.

4 EVALUATING OPEN SYSTEM CHARACTERISTICS

The characteristics of industrial hygiene concern associated with open systems are a function of the type, quantity and hazardous nature of materials involved, and characteristics of the process. For proper equipment selection, layout, and design, these must be relative to the characteristics of the system whether open or closed.

System characteristics need to be evaluated to determine the extent of potential hazards that may exist, and to establish the level of control needed and how it may be achieved. Results of this evaluation may have significant impact upon the design of manufacturing and processing facilities and the selection of process equipment.

The evaluation involves a variety of factors including the potential hazards, whether or not they are airborne, operating conditions, feasibility of change, and the benefits to be derived. The process of evaluating an open process system also is the first step in establishing effective hazard control.

4.1 Potential Hazards

Potential hazards from open process systems must be put into proper perspective before selection of equipment and facilities begins. Otherwise the result may be costly under- or overdesign. This requires having specific criteria against which characteristics of a given system can be compared and measured.

A variety of criteria available to assist in the evaluation of potential hazards and the design of controls. Most of these address the "safe" or acceptable level concept, meaning that the risk associated with a potential hazard is acceptable or unacceptable at levels above or below the guideline level.

No specific criteria is absolute. There is no significant difference between being slightly below versus slightly above a limit value (except for fine points of the regulations regarding compliance with a statutory requirement such as an OSHA standard). However, without a planned objective it is very difficult to design controls and document that proper action has been taken to minimize industrial hygiene hazards.

Sources of criteria include well-recognized agencies such as the Occupational Safety and Health Administration (OSHA), the National Institute for Occupational Safety and Health (NIOSH), the Environmental Protection Agency (EPA), the American National Standards Institute (ANSI), and the American Industrial Hygiene Association (AIHA). Industry and trade associations frequently provide guidelines for their members. Insurance companies sometimes set requirements that directly affect insurance premiums. A company may—and many do—adopt its own internal industrial hygiene criteria or guidelines. This is frequently the case when other criteria do not exist or are subject to question, such as for materials of relatively unknown toxicity.

Data for comparison against these criteria may include manufacturer/supplier information such as that found on material safety data sheets, the NFPA Hazard Diamond (see Chapter 2), and product labels. For industrial hygiene applications, a primary source of information for evaluating a hazard is the data obtained from measurements in the workplace environment. Measurements made on personnel are used for evaluating personal hazard potential and compliance with health-related criteria. Monitoring results for work areas should go beyond personal levels and also provide useful diagnostic data to help identify sources needing control and the level of control that is needed.

If monitoring data does not exist or cannot be obtained, as may be the case for unbuilt facilities, it is sometimes possible to obtain reasonable estimates of levels by calculation from estimated emission rates and ventilation rates. If similar facilities exist elsewhere, it may also be useful to monitor them to obtain estimated levels.

4.2 Airborne Materials

The airborne nature and the physical state of a material in an open system usually will determine its potential hazard and how effectively it can be controlled. Physical state affects the potential route of entry for a toxic material, including its potential for dispersion and ability to remain airborne.

For example, relatively large pieces of a solid material are not likely to be of significant industrial hygiene concern. However, the same material finely divided into a powder or vaporized and oxidized to a fume, could have serious industrial hygiene implications. The same material, fragmented by a high-speed abrasive grinding wheel would have still other aspects of industrial hygiene significance.

Perhaps the most significant difference is the extent to which the material is divided and becomes airborne. *Airborne* is defined by Webster's as supported wholly by aerodynamic and aerostatic forces, transported by air. As a large solid piece, it does not become airborne. As a fine powder, it may be airborne almost immediately and may remain so for long periods of time. It also is apt to be highly respirable. As chips or granules, it may pass through the air, due to its own kinetic energy, but may never really be airborne.

The control method to capture and remove the material will be different depending on these characteristics. If the material is airborne, then the airstream that contains it must be controlled. If the material is ejected and "just passing through" the air, then the material but not the air must be caught and transported elsewhere.

Liquids may be potential hazards due to skin absorption, aspiration as a spray (not airborne) or as a mist (airborne). The nature of potential contact will indicate the appropriate control approach. Toxic or corrosive chemicals should not be moved overhead in open systems where they may leak, drop onto, or spray someone or something below. Mist generation should be minimized as much as possible, but ventilation control still may be needed in addition. There usually are practical and effective ways to "tighten" a system to reduce leaks and to reduce misting.

Vapors and gases mix intimately in air and are therefore airborne. Some materials may stratify in air if they are substantially more or less dense ("heavier" or "lighter") than air. However, this concept is frequently misapplied because diffusion and mixing commonly take place. "Still" air occurs very rarely, and it is unlikely to exist if there is significant personnel or vehicular traffic, or natural or mechanical ventilation. Confined spaces are a notable exception for which poor ventilation may be a defining characteristic favoring contaminant stratification.

4.3 Operating Conditions

Careful examination of the operating conditions that characterize an open system is an essential aspect of evaluating the potential hazards and control opportunities of the system. This aspect of the evaluation relates closely to the release of airborne materials because such release can be altered by operating conditions.

Examination of process requirement tolerances, existing conditions, and potential industrial hygiene concerns will indicate where flexibility exists in controlling or "closing" an open system. Where the flexibility is found, it is advisable to make process adjustments accordingly and thereby minimize potential hazards.

For example, reducing temperature can reduce volatility and resulting vapor or gas emissions into a workplace. Increasing pressure can accomplish the same thing. Placing a material in solution may also be used to reduce its potential for hazardous emissions.

The release of a material from an open process will often be a function of whether it is being used in a batch mode or continuous mode process. Continuous systems frequently are tighter and less open; however, if poorly designed they may lead to excessive exposure and cumulative effects having serious health consequences.

Agitation is another controllable condition; generally, it should be reduced to minimum acceptable process requirements. When agitation is necessary, there may be other ways to satisfy process needs and still minimize the release of materials to the workplace.

To reemphasize, the physical state and condition of materials in open process systems determine the potential health hazards and possible means of control. In evaluating an open process, it is always cost effective to identify where flexibility may exist and how it may be applied to reduce potential hazards.

4.4 Physical and Radiative Stresses

Industrial hygiene associated stresses include physical and radiative-type stresses, heat and cold, noise vibrations, and ergonomic stresses.

Radiation involves the generation and propagation of energy and/or particles in the form of a wave front. The wave front passes through air; it is not airborne and, therefore, is not affected by air movement or ventilation controls. Radiation normally is further subdivided into ionizing and nonionizing (including light) electromagnetic energy.

An open system, as applied to radiation, is characterized by the strength of the radiating source, the physical space into which it radiates, the extent of personnel interaction with the radiation field, and the nature of the potential health hazard. Radiation can be reflected in many directions within a work space, and can also propagate in a very diffuse or directional manner; and it can be contained and absorbed. The industrial hygiene evaluation should address all of these aspects.

Engineering controls focus upon the source and the physical environment in which it operates. Primary means for controlling radiation are by blocking its movement and absorbing its energy. Both of these can be accomplished directly at the source and within the surrounding environment.

Specifications for equipment selection should limit the strength of the radiating source to no more than necessary for process requirements, and should ensure that adequate shielding is provided to contain the radiation. For example, maximum acceptable radiative levels should be specified for potentially radiative equipment prior to purchase, and final payment should be contigent upon satisfying the specification.

Equipment layout is another important factor. The alignment of and spacing between radiation sources with respect to personnel work and traffic areas should be evaluated carefully. Lasers are an example of equipment for which layout is extremely important, not only with direct respect to the radiation sources, but also to reflective surfaces of any type in the work area.

Building design and structures can have profound effects on radiation control. Wall, ceiling, and floor surfaces can range from highly reflective to highly absorptive. They can be used to create a mazelike path through which radiation levels may be greatly reduced. Characteristics of the building design and materials should be evaluated carefully for existing radiation sources and applied carefully in designing new facilities.

The preceding basic control principles also apply to most other physical stresses. This is because most also have radiative characteristics.

Heat and cold have radiative aspects, but unlike other forms of radiation, they can be affected substantially by air movement. Moisture content of the air can affect the perception or feeling of temperature. Control of heat and cold, therefore, involves ventilation as well as shielding, absorption (insulation), and humidity adjustment.

Noise from open systems frequently results from mechanical looseness, most economical design, and the movement of solids, which can involve numerous sources of noise emission. The noise is frequently a function of mechanical elements and machine design, and magnitude of point-to-point contact between the different elements. Ideally, control yields best for redesign or modification of the generating sources. When and where this cost is excessive, either initially or in a retrofit situation, enclosure becomes necessary.

Physical stresses include those related to ergonomics and human factors. Interaction of the person with the system determines whether there is or is not a stress or hazard situation. With ergonomic and physical safety situations, the interrelationship of the person and the system is the element to be examined and optimized to avoid a hazard and to achieve optimum productivity.

4.5   Control Feasibility

Feasibility of controls, especially closure, is another important aspect of open system evaluation. This applies to controlling potential hazards that have been identified. This aspect of the evaluation is important because it will establish practical limits upon what can and cannot, should and should not, be done to "close" an open system.

The preceding aspects of evaluating the characteristics of an open process system have focused upon the potential hazards involved and factors that could be applied for hazard control. These factors determine what is technically feasible. There will always be constraints related to the process and the effectiveness of the controls. If the process constraints cannot be satisfied, it is not feasible to implement the control. If there does not exist a way to control a hazard sufficiently, then it is not technically feasible to do so. However, this is not a subject for snap decision because it is rarely impossible to achieve an acceptable level of control.

Cost is the second primary aspect of feasibility. If a technical solution is prohibitively costly, this may be sufficient reason not to implement the control. There are no specific guidelines on what is economically feasible. There is no ready answer to the question, "What is it worth not to expose an individual to a potential hazard?" However, all process systems, whether open or closed, generally exist for one purpose—to make money. If they do not accomplish this, there is no reason to have the system. Careful assessment of all of the benefits from a process system, including a need to close or control it, is very important in assuring economic feasibility.

Compliance with regulatory standards is another aspect of evaluating feasibility. In some cases, such as when acting under an OSHA citation or consent agreement, control feasibility has been decided by the agency, and it is no longer up to the company. The company is not in control. Being in control can profoundly affect feasibility, and this is one reason to take appropriate action in a planned and budgeted manner, without being "under the gun." Feasibility regarding compliance is sometimes resolved by court action. This has the obvious disadvantage of applying requirements that may pertain to one solution (company, industry) but not generally to others. What is feasible for compliance can sometimes be negotiated, and variances can be obtained to avoid trying to do what is not feasible.

The final concept to be considered is very nonspecific, almost vague. It could be described by various words, perhaps "prudence" is the best. This refers to a company's philosophy regarding industrial hygiene and the control of potential hazards. The primary reason to take action may be because "it is the right thing to do." Responsibility, morality, ethics, emotion, public image, and other factors may enter into this important but rarely identified aspect of evaluating and deciding on the feasibility of a given action.

4.6 Benefits

Although covered last in this discussion, determination of the benefits to be enjoyed from evaluating and implementing appropriate controls for open process systems should be one of the first things to be considered. Without determining benefits, and without there being sufficient benefit, there may very well not appear to be good reason to proceed further. Benefits provide justification for expending the time and resources to evaluate potential hazards and then to control them.

The most important benefit to be derived from acting upon an industrial hygiene concern may be control of the health hazard, but other benefits may be profoundly important also. These may be particularly helpful in "selling" the need to act upon industrial hygiene matters to other parts of company management.

A partial listing of additional benefits would include employee attitude and morale, reduced absenteeism, less training and retraining, higher production, less waste material, fewer environmental concerns, improved product quality, reduced insurance costs, and less risk and liability. All of these can help to increase profitability, which is the primary reason for being in business. These benefits also impact upon the image of a company—the image it has and the image it wishes to convey. Company image is an important part of the "prudence" aspect of evaluating feasibility.

Every effort should be made to place some dollar value on each benefit. For some benefits, this can be a fairly specific value, for others only a rough estimate is possible, and for others (such as image, prudence) it may be only a sense or feeling of relative value. Without some common basis, and dollars are the most quantitative and meaningful, it is difficult fully to assess the benefits to be derived and how they impact on plans to take further action.

5 CONTROLLING/CLOSING OPEN SYSTEMS

Process equipment should be selected and laid out, and the building that houses it designed, to best accommodate the control strategy that has been developed. This generally means miminizing the extent to which process systems are "open." If an open system has been evaluated thoroughly, then potential hazards, control opportunities, feasibility, and the benefits that justify further action will have been identified.

5.1 Engineering and Administrative Controls

The two basic control approaches for resolving industrial hygiene concerns are engineering and administrative. Engineering controls are the primary focus of this discussion. Administrative controls will be addressed primarily to the extent that equipment and facility considerations may impact upon them.

Engineering controls have the direct objective of eliminating or at least reducing potential hazards. From a control standpoint, they are preferable to administrative controls. This has been well recognized by regulatory agencies such as OSHA. However, engineering controls are not always feasible, nor are they always entirely effective.

Administrative controls do not directly reduce a hazard. They do provide protection against it, satisfying the need for an extra margin of safety. The inherent weakness of administrative policies and procedures is that they can be expected to fail because people make mistakes.

5.2 Ventilation

Ventilation, the controlled movement of air, is one of the most effective ways to control potentially hazardous airborne materials. This primary control technique can be, and too often is, misapplied with the end result of ineffective hazard control and costly redesign and modification. This is a problem with open systems that characteristically may lack closures for control and that are highly dependent on ventilation to maintain an acceptable safe atmosphere.

5.2.1 Ventilation Philosophy. Ventilation controls are most effective when they are applied in a manner consistent with an overall "ventilation philosophy" for a given building. The basic objective of the ventilation philosophy is to establish the most cost-effective utilization of controlled and conditioned air flow. This, in essence, means supplying fresh air to "clean" areas (offices, control rooms, lunch/locker areas) and exhausting air from the most contaminated process areas. In between, the general flow of air should not create any secondary problems or be wasteful of energy.

5.2.2 Local Exhaust. Local exhaust ventilation usually provides the best means to control airborne contaminants from specific source locations within an open process system. Air flow control at the contaminant source prevents migration of a potential hazard to other areas and does so for a minimum amount of air flow.

Some equipment is provided with a complete, ready-to-go ventilation system—for example, an abrasive blasting booth or box. Some equipment is provided with a local exhaust hood built into it, ready for connection to an air-moving system to be supplied by the user. Examples of this would include laboratory hoods, some liquid drum-filling machines, and some hand-held grinders. These systems usually will be effective when installed according to manufacturer guidelines. However, this is not always the case and, as with all controls, it will be necessary to evaluate performance after installation.

Many process systems require enclosure and ventilation by the user of the equipment. The design of an effective add-on local exhaust ventilation system will vary with every different situation. Detailed discussions are beyond the scope of this chapter, but many specific applications are provided in other chapters. The basic approach always is to create an air flow capable of capturing contaminants at the source. The most critical component in the ventilation system, relative to this purpose, is the local exhaust hood or inlet.

After air flow and static pressure requirements for each hood in the local exhaust system have been established, the layout of the system can determine the final design and performance of the system. There are many considerations to be accounted for; one primary objective is to achieve a "balanced" system that requires little or no adjustment. Locating hoods having greater air flows and greater static pressure losses closer to the fan is advisable. Location near exterior walls or on upper levels can minimize duct work length. Positioning the fan near the discharge points will minimize the potential for leakage from the duct work into the workplace. Open windows should be avoided to prevent cross drafts that can interfere with air flow control at a hood; the same is true for makeup air diffusers.

5.2.3 General Ventilation. General ventilation can be used to dilute air contaminants and to control their movement through a work area, between work areas, and through a building. It is often achieved through the use of wall and roof exhaust fans, which move relatively large volumes of air without duct work. The location of the fans and, accordingly, the location of open process equipment are primary factors affecting how well air contaminants are controlled. Natural air flow currents, such as convection over a hot process, should be recognized and utilized. Heated air rises and will carry airborne materials upward. However, while particulates may remain at a higher level, gases and vapors diffuse and will fill the space if not removed. Cold walls result in cooling and contaminants carried aloft at one location may drop to personnel levels at other locations.

Location of personnel between a contaminant source and a general exhaust fan should be avoided. Passageways should be sized and oriented to avoid problems such as strong pressure differentials across doors and "wind tunnel" effects through narrow corridors.

5.2.4 Makeup Air. Makeup air flow replaces what is removed from a building or work area by operation of local and general exhaust systems. Without makeup air, these systems, and natural ventilation such as for flue stacks, cannot perform as they are designed. Makeup air should be supplied in a manner consistent with the ventilation philosophy for the building. If it is conditioned (heated, cooled, cleaned), then it will be expensive and should be utilized in the most efficient manner. Large makeup air handling/conditioning units frequently are less efficient than several smaller ones having the same total capacity. Makeup air should be provided so as to avoid creating drafts. The layout of supply air systems should be carefully coordinated with the exhaust systems and desired general air flow patterns.

Frequently, there are significant differences between "summer" and "winter" ventilation modes. Makeup air may be drawn directly from the outside through various openings (windows, doors, transfer grills) during warm weather. During colder weather, there is the temptation to utilize heating units at 100% recirculation to conserve heat. However, as such they do not provide direct makeup air. Air that can "leak" into the building will still need to be heated, and it is more efficient to do so directly by fresh air intake at the makeup air unit. It is advisable to have some minimum amount of fresh air intake at all times, such as 20% of the unit capacity. This is particularly important for some newer buildings that are built very "tight" to make them highly energy efficient. Without deliberate fresh air intake, there is the possibility of a gradual buildup of contaminants, resulting in discomfort and complaints sometimes referred to as the "tight building syndrome."

5.3 Barriers

Another basic engineering control approach is to interrupt the path between a person working and the condition and/or materials that present a potential hazard. This is accomplished by placing a barrier across the path. Barriers for open process systems masy be attained in several ways: as an integral part of purchased equipment (inherent to the process), as a purchased accessory item (specified by the user), or as an add-on item designed and installed by the user. There are a variety of basic types of barriers, these include guards, covers, surface treatments, and partitions.

5.3.1 Guards. Guarding is the technique of placing a physical surface around or in front of the source of potential hazard. An example would be placing a spray or splash shield in position to protect against a possible sudden leak of material. Enclosures are more extensive guards, and might be used to contain an oil mist from a high-speed grinder or flux and solder fumes from a wave solder machine. Construction of an enclosure also may require provision of exhaust ventilation to prevent a buildup of airborne levels within the enclosure.

5.3.2 Covers. Covers simply are barriers placed over open tanks or process vessels such as mixers and blenders. They can effectively reduce the emission of volatile materials into the work area. There also may be the potential for significant contact with airborne materials when a cover is removed. Local exhaust ventilation may be needed, at least at the time of cover removal.

5.3.3 Surface Treatment Some open process systems, such as tanks and kettles, can be "covered" or surface treated to minimize airborne contaminant release, to insulate against heat loss, or to achieve other process objectives. Examples of this approach include "ping pong balls" floating on the surface of a chromic acid tank or a coke blanket located directly on top of molten lead in an annealing tank.

5.3.4 Partitions. Building surfaces (walls and ceilings), and extensions of them can be effective barriers. They can isolate operations from one another and be used to prevent the propagation of potential hazards such as noise, heat, and electromagnetic radiation. Partitions also can be useful for directing air flow currents through and between work areas.

5.3.5 Distance. Although not a definitive physical barrier, distance can accomplish the same basic objective by dispersing and diluting potentially hazardous conditions and materials. This approach can have a profound effect upon equipment layout if greater separation between machines is required. This principle can also affect building size, floor area, and ceiling height. Larger-volume rooms will support natural dilution of airborne contaminants.

5.4 Material Handling

Controls for material handling situations must address the specific nature of how the materials present a potential hazard. This can vary for a wide range of possibilities, some of which have been discussed previously (see Section 3.2). Basic categories include points where people manually handle materials or interact with the system, system transfer points that result in fugitive emissions, and other system characteristics that create conditions of concern throughout the work area.

5.4.1 Personnel Interaction. Points at which people interact with an open material handling system can be enclosed and ventilated to minimize potential hazards. Examples of this include laboratory hoods for handling toxic chemicals and the ventilated cab of an overhead crane operator in a foundry.

5.4.2 Transfer Points. Conveyor transfer points (such as belt drops, elevators, plows) and transfer points in general for open material handling systems must be evaluated closely. Control actions to minimize industrial hygiene concerns include reducing drop distances, eliminating free-falls, and slowing belts down by using more and/or wider belts. It is also possible to enclose and ventilate transfer points.

5.4.3 Work Areas. Industrual hygiene concerns about general work area conditions may arise during material handling. These include air contaminants from vehicle exhaust. These contaminants can be controlled by maintaining the material handling vehicles in good operating condition and ensuring that there is adequate general ventilation.

5.5 Physical State/Conditions

Basic ways by which the physical state of materials and the process conditions that establish it can be evaluated and controlled have been discussed previously. The opportunity to exercise control depends to a large extent upon the flexibility of process constraints. It is vitally important, therefore, to understand thoroughly how a process works in order to determine if and how potential hazards can be controlled by affecting the physical state of materials and the conditions of the process.

5.5.1 Basic Examples. One example of controlling an open process system by physical state change is a vapor degreaser with a chilled "trap" to condense vapors and return them to reprocessing. It is also possible to place materials in solution to reduce their volatility and resulting emissions.

5.5.2 Encapsulation. Encapsulation is a similar technique. This involves coating or binding materials so that they are less likely to cause industrial hygiene concerns. One example of this is painting and/or plastering asbestos-containing surfaces to prevent the release of fibers. Another example would be a vapor-tight floor covering over areas where substantial amounts of mercury have been spilled.

5.6 Policies and Procedures

Open systems are characterized by fewer controls. With a given hazard potential, the lesser the degree of control, the greater will be the need for policy and procedures to achieve the same degree of health and safety. Some aspects of facility design and equipment selection can help to assure the effectiveness of administrative controls. These should be anticipated insofar as possible prior to construction of new facilities for open process systems.

5.6.1 Personal Protective Equipment. Personal protective equipment is needed to provide an extra margin of safety for some operations and for emergency situations. In either case, it is usually necessary to have the equipment located close to the process facility. This means that appropriate storage locations, legible identification, and sometimes special utilities such as electricity and breathable air may also be needed. Access to emergency equipment must be free of obstructions and well marked.

5.6.2 Confined Spaces. Work in confined spaces requires very strict work practices in order to be accomplished safely. These can be supported by proper facility design. First and foremost, every effort should be made to eliminate the need for confined-space entry by designing the system to provide for easy access, egress, and ventilation, and thereby eliminate the confined-space element. If entries will be necessary, then manways should be large enough to accommodate a worker and safety gear, including respiratory protection. Special communication, lighting, and rescue capabilities may be needed. Both the primary and backup routes of emergency escape should be considered carefully. Special tools may be useful for accomplishing tasks during confined-space entries in the most efficient matter. Better yet, is to provide tools that eliminate the need for entry. There will need to be provisions for mechanical ventilation, both supply and exhaust throughout the full time of entry. Such considerations will have significant impact upon facility design and equipment selection.

5.6.3 Lockout. All sources of energy (mechanical, electrical, steam, air) and process flows (water, fuel, chemicals) are potentially hazardous when maintenance is being done on equipment associated with them. Lockout is one of the principal means by which accidents can be prevented by the inadvertent "turning on" of a switch, opening a valve, or actuating a lever. All devices can be locked to prevent their operation. This can be facilitated by proper selection and/or design of process equipment. The need to do so should be anticipated.

5.6.4 Work Practices. A wide range of other safe work practices can be facilitated by the proper design of process facilities. These can be identified through careful review of operating procedures, such as "job safety analysis" and "system safety analysis" techniques. Such analyses should be a must for every system, whether open or closed.

6 DESIGNING AND IMPLEMENTING CONTROLS FOR OPEN SYSTEMS

The process of hazard evaluation has been completed and basic control concepts have been considered. The tasks remaining to be accomplished are completion of design and implementation of specific engineering controls. It is at this point that the final decisions regarding equipment selections, layout, and building design will be made.

Open systems too frequently are designed and built without adequate controls and retrofitting is necessary. Commercially available controls are usually the choice when they fit the situation. Otherwise it is up to the engineer working closely with the industrial hygienist to accomplish the design. It takes the knowledge, skills, and experience of both specialists to achieve the desired results.

6.1 Preliminary Design

The early stages of control design should focus upon developing several alternative concepts. Each alternative will have somewhat different objectives and anticipated results. Each will satisfy different priorities and process constraints.

Preliminary design alternatives may range from taking essentially no action to implementing a "Cadillac" plan, that is, doing everything of possible benefit. The most cost-effective approach generally will fall somewhere in between.

A key aspect in developing each preliminary design concept is making rough, "budget" cost estimates. These early estimates should be conservative (on the high side) or include a substantial contingency factor. The basic nature of estimation (cost, time, etc.) is such that error is most likely to result from omissions. This is particularly true during the preliminary design stage, when engineering details have not yet been worked out fully.

The preliminary design stage is a time for keeping an open mind. Creativity should be emphasized so as to reduce the hazard or stress potential to acceptable levels without the necessity of controls, especially enclosing types. This is not the time for drawing conclusions, but rather to follow through on ideas even though not all of them are likely to be implemented.

6.2 Selection

The selection of a preliminary control design for development into the final design should be based on clear, objective criteria. These include the performance (level of control) requirements, process constraints, and an evaluation of the costs and anticipated benefits (on a dollar basis).

Modeling can be an effective approach to help select between alternatives. This can involve developing an actual model or using a computer program to approximate the performance of the system. Parameters characterizing performance, expressed as functions of other parameters that characterize the design concepts, must be determined. Computer modeling has the advantage of permitting many variations to be "tested" and evaluated without purchasing equipment or beginning installation. However, modeling has the disadvantage of being approximate at best, and of becoming a relatively involved project of its own. Modeling is most appropriate for complex and very costly systems.

Frequently, priorities establish a logical sequence of action, by a step-by-step or phased approach. This is often the case where cost is high and results are somewhat unpredictable, as is true of some noise control applications. The order of priority generally follows from control of the source of potential hazard itself, to enclosure of the source, to treatment of nearby locations, and to treatment of the general work area and perhaps the entire building.

It is also advisable not to dispose entirely of preliminary design concepts that have not been selected, at least until the final design is confirmed to be effective and indeed "final." There may be an unanticipated need to "go back to the drawing board."

6.3 Final Design

Successful installation and operation of an engineering control depends to a large extent on how well (correctly and completely) it has been thought out. Important aspects of the final design process include drawings, specifications, cost estimates, and contracts/bids.

6.3.1 Drawings. Design drawings should provide sufficient detail to ensure proper construction and installation of the process control. Notations should be used liberally to convey information, even information that might reasonably be assumed to be "common knowledge." Drawings need to be checked carefully by someone who is technically qualified but also who is not part of the final design personnel, so that one's mind will be as open as possible.

Drawings should be marked to indicate "as installed" variations from the original design. If these variations are great, design plans should be redrawn to show all modifications. Many problems in subsequent work on the facilities can be avoided by having accurate drawings.

Some experienced contractors can do a good job of constructing an equipment or facility control, such as a local exhaust hood, from a simple sketch and a few instructions. This can simplify the design process and sometimes is acceptable. However, without detailed drawings and specifications, there is almost no way to hold a contractor accountable for errors made.

6.3.2 Specifications. Specifications are written documents that provide detailed requirements regarding the equipment, materials, installation methods, performance requirements, and other aspects pertaining to the final design.

Control aspects should be well detailed and delineate how controls are to perform. Where possible, performance levels should be noted.

6.3.3 Cost Estimates. Final design cost estimates should be detailed and as accurate as possible. There are a variety of resources available for estimating cost. The *Means* cost data manuals, for example, are updated every year. These texts usually provide detailed unit costing for a wide variety of facility installations, breaking down costs into basic materials and labor categories. Manufacturers can provide very useful information. Consultants with experience in the specific types of controls being installed can also provide useful assistance.

It is advisable to pursue more than one approach in estimating costs. This will help to establish the credibility of the estimate. Two basic approaches are (1) actual estimated cost for materials and labor and (2) unit cost estimates or "rules-of-thumb." The first will be more accurate and should always be performed. Unit cost estimates are approximate but can be useful as a check on the detailed estimates.

6.3.4 Contracts and Bids. With appropriate engineering plans and specifications, one is free to seek fabrication and installation on a competitive basis and thus keep costs down. If the work is to be done by an outside contractor, then it is important to have a proper contract. This should include provisions relative to work rules and health and safety, in addition to requirements for insurance, meeting deadlines, and cost limits. A basic contract usually is prepared for all projects, with possible additions and deletions as needed for a specific project.

Bids should be obtained and evaluated in an objective, unbiased manner. All prospective bidders should receive the same information. The more detailed this information is, the more closely the bids can be prepared to satisfy specific needs, and the better they can be evaluated accordingly. Detailed drawings and specifications provided to contract bidders and detailed cost estimates used in bid evaluation can go a long way toward assuring successful installation of engineering controls for open systems.

6.4 Implementation

Important steps in the installation of engineering controls for open systems include planning, scheduling, ordering materials, and installation.

6.4.1 Planning. The essential first step for implementing controls, as indeed for any major activity, is planning. Proper planning can prevent many pitfalls. It is important to set specific, realistic objectives in the planning stage. These should be written down, and specific responsibility for accomplishing them should be assigned to project personnel. As someone once said, "Do not start vast projects with half-vast ideas."

6.4.2 Scheduling. All key aspects of the project should be included in the overall project schedule. Every effort should be made to establish reasonable and accurate time estimates to accomplish the work. Everyone involved should be advised of how much they have been allocated to do their tasks. It is also appropriate to allow for some "windows" or breaks in the schedule to permit catching-up as needed. Important aspects that are sometimes left out include deadlines for progress reporting and conducting follow-up evaluations to assess performance after the installation is completed.

6.4.3 Ordering Materials. The final design, as discussed previously, should include relatively detailed specifications for equipment and materials. In addition, there should be clearly stated requirements for when equipment and materials are to be ordered, and what price and delivery time shalll be. Many times, equipment must perform at a specific level in order to be acceptable. An example would be maximum allowable noise levels from a machine. It will sometimes be appropriate to make final purchase of the materials contingent upon satisfying performance requirements.

6.4.4 Installation. The installation of a control facility, or modification of equipment or a building to affect control, will require close coordination. This coordination will involve personnel within the company, and sometimes outside contractors, utility companies, and governmental agencies. The communication and follow-up required for this coordination are usually administered best by assigning one individual to have primary responsibility.

6.5 Checkout

This important aspect of installing controls is sometimes neglected. There may be many reasons for this, but none can really justify going to all the effort and expense of designing and implementing, and then not following through to see if controls work properly.

Often, there is no follow-up evaluation because it was not planned, or no one was given the responsibility to do it. Someone should have this responsibility and be accountable for acting accordingly.

It will usually be necessary to conduct some type of testing and measurements to evaluate the performance of the control. This can include ventilation measurements, air sampling, noise level measurements, and a wide range of other specific tests.

If the final installation satisfies design objectives, then the control is finished. One important remaining step is to let everyone know about the accomplishment. This will benefit all concerned, especially those who are responsible for the success and who deserve the credit.

If the final installation does not meet the desired or required performance objectives, then it will be necessary to make adjustments and changes, move to the next phase, or start over again. At this point, the value of saving some of the preliminary design concepts that were not selected for the final design will become apparent.

Follow-up also involves future maintenance and periodic performance confirmation. There may also be a need to schedule ongoing routine testing and documentation of satisfactory performance. An example of this would be periodic checking of laboratory hood performance.

6.6 Performance Standards

With hazard controls installed and checked out, it is appropriate to establish performance standards to help assure acceptable ongoing operation. Performance standards should make it clear when a process is "in control" and "out of control." Performance standards should prescribe tests and criteria to determine that an open process remains under control. The standards must provide a clear image and understanding of what must be done to confirm control. They must be measurable and attainable. If possible, standards should include some visual criteria that can be checked quickly, such as the absence of gross emissions, spills/leaks, or dusting of surfaces.

Performance standards for control of open process systems should be incorporated into a working document. As for any policies and procedures, it is important there be clear assignments of individual responsibility and that individual job performance evaluation and reward (salary) be tied to the implementation of performance standards for hazard controls.

Many provisions of performance standards will be company specific and process specific. There may also be important considerations and specific criteria related to compliance with government regulations, for example meeting OSHA permissible exposure limits (PELs), implementing a hearing conservation program, and providing chemical information under federal, state, and local right-to-know legislation.

6.6.1 Routine Operations. Performance standards for hazard controls during routine operations should be included in the written standard operating procedures for a given process. Some performance standards may be drawn from the final design specification for the controls. Others may be obtained from job safety analysis, and still others may be the direct result of necessary compliance with governmental regulations.

6.6.2 Inspection/Maintenance/Repair. Process controls will almost always be incomplete and subject to failure with potential for a hazardous exposure without effective preventive maintenance. Process systems may be the most open and potentially hazardous during inspection and maintenance. Process controls should be designed to minimize risks during routine preventive maintenance and during unplanned maintenance and repair. This may include providing access to equipment and facilities for illumination, communication, ventilation, and personal protection. Performance standards should assure facilities needed for safe inspection, maintenance, and repair work are adequately designed and operational.

6.6.3 Housekeeping. It is beyond the scope of this discussion to conclude if, in fact, "cleanliness is next to godliness." However, it is a well-established fact good housekeeping is conducive to hazard control and general safety, and vice versa. Performance standards for hazard controls should emphasize the work area be as neat and clean as practical, with frequent visual inspections.

6.6.4 Accidents and Emergencies. Performance standards must address criteria that avoids accidents. Some criteria may be obtained from thoughtful what-if considerations. These include job safety analysis and formalized system safety approaches such as failure-mode and fault-tree analyses. Extremely valuable information may result from close attention to accident near misses. Growing public awareness and concern over industrial accidents has lead to various regulations that call for prompt notification of leaks and spills, provision of information on toxic and physical hazards of chemicals, and preparedness for action and cooperation with local agencies during emergencies. Performance standards should address responsible emergency action to conceivable extremes, including catastrophe accidents such as severe weather (flooding, tornados/hurricanes, lightening), major fires, and major hazardous chemical spills.

6.7 Retrofit

It may be necessary to modify and retrofit hazard controls for open process systems. Reasons for this include changes in the manufacturing process, changes in criteria for acceptable performance (e.g., individual exposure limits), and failure of the installed controls to meet design specifications. It is important to view the need for retrofit not as a problem, but as an opportunity to make improvements and enjoy additional benefits. Accordingly, retrofit controls should be given the same complete process of alternatives, cost estimates, selection, final design, installation, checkout, and performance standards as followed for all controls. Under ideal circumstances, retrofit should not be needed. Under real circumstances, it is often needed and should be undertaken seriously, carefully, and positively to avoid having to do it again.

7 THE IDEAL OPEN SYSTEM

The ideal open system, from the standpoint of industrial hygiene, would be one that would involve nontoxic and nonvolatile materials and conditions. In this extreme case, there would be no hazard despite the extent to which the system might be open or exposed. When a potential hazard does exist, the ideal open system would be one that was controlled to the extent that no actual hazard existed. Other benefits from controlling an open system, as discussed previously, also would be maximized in ideal open process systems.

Control is inherent to closed systems, except when they are opened for special purposes. This form of control is obvious, even to the untrained eye, because there are no circumstances involving release or interaction with materials or conditions, whether hazardous or not. A properly designed closed system is and appears to be basically "clean."

Control is less obvious for open systems. Engineering controls for these systems may perform effectively, but this may not be obvious even to an experienced industrial hygienist. The effectiveness of open system controls must be tested and documented. Persons working with the system must be trained to understand how the controls are operated and maintained.

The appearance or perception of effective control can be very important for company image. In this way, the "ideal" open system will also be that for which the control of potential hazards is apparent, understood, and not subject to doubt by the operators and observers of the system.

GENERAL REFERENCES

Burgess, W. A., *Recognition of Health Hazards in Industry—a Review of Materials and Processes,* New York: Wiley, 1981.

Clayton, G. D., and Clayton, F. E. (eds.), *Patty's Industrial Hygiene and Toxicology,* Vol. I, *General Principles,* Vol. II, *Toxicology,* Vol. III, *Theory and Rationale of Industrial Hygiene Practice,* 3rd rev. ed., New York: Wiley, 1978.

The Industrial Environment, Its Evolution and Control, National Institute for Occupational Safety and Health, Washington, D.C.: U.S. Government Printing Office, 1973.

Industrial Ventilation—A Manual of Recommended Practice, 19th ed., Cincinnati: American Conference of Governmental Industrial Hygienists, Committee on Industrial Ventilation, 1986.

Means Building Construction Cost Data, 41st annual ed., Kingston, ME: Robert Snow Means Company, 1983.

Means Mechanical & Electrical Cost Data, 6th annual ed., Kingston, ME: Robert Snow Means Company, 1983.

Olishifski, J. B., (ed.), *Fundamentals of Industrial Hygiene,* Chicago: National Safety Council, 1971.

CHAPTER FOUR

Process Characteristics— Closed Systems

Jeremiah R. Lynch
S. Lipton

1 INTRODUCTION

Closed, continuous processing systems are typical of the petrochemical, high-volume chemical, and petroleum refining industries. In a characteristic closed processing unit, raw materials in liquid or gaseous form are fed continuously into the process through piping systems at various temperatures and pressures. Chemical reactions that occur within the process unit are either exothermic or endothermic. These reactions are often conducted at high pressure and temperature and usually in the presence of catalysts. The type of catalysts and the reactions systems vary widely in the process industry, but in high-volume chemical and refining processes the catalysts are generally solid. The reaction systems are typified by fixed-bed, fluid-bed, or slurry processing operations.

The individual steps in a continuous process encompass the classical unit operations of chemical engineering. In these processes, intermediates, by-products, and products are moved from one vessel to another with pumps, compressors, and pressure level differences. End product or material from any stream may be stored in a pressurized container, refrigerated container, or an atmospheric tank. Although tankage is often termed atmospheric, tankage emissions are controlled through a variety of methods and can be considered closed systems. Ovens and dryers normally do not achieve the same degree of tightness or containment as higher pressure processing systems. Nevertheless, these units can be designed for a tight seal that minimizes leakage. Based on the engineering definitions in this chapter, tightly sealed ovens and dryers are considered closed systems.

While enclosure is often used as a means of controlling specific occupational health hazards, closed systems aim at total process containment, and health hazards normally arise when there is a failure in maintaining containment. In contrast, for less controlled or open systems, emissions are usually greater than in a closed system and engineering controls of health hazards are principally directed at the prevention of worker exposure through contaminant interception with local exhaust ventilation systems before the contaminant reaches the worker. The use of control rooms with associated remote or automatic control systems is a major tool in reducing operator exposure. Such control rooms should be pressurized with clean air. For additional information see Chapter 6, Section 4.3. Although the primary method of preventing hazardous exposure in a closed system is through maintenance of containment, ventilation controls may be used in a specific application within a closed system. This chapter will review potential loss of containment or breaching, general engineering design considerations and operating characteristics intended to prevent loss of containment, methods of monitoring hazard control effectiveness, and certain cost considerations.

2 MAINTENANCE OF CONTAINMENT

The primary safeguard against excessive employee exposure to hazardous contaminants present in a closed system is containment of the materials within the piping, vessels, and equipment. However, containment is never absolute and minor or trace releases will occur even under ideal conditions from valves, pumps, flanges, and so on. These trace releases can be controlled to levels that preclude any significant hazard or risk in the workplace as described in other sections of this chapter. Although concern for exposure is a primary consideration, small releases may be acceptable in exchange for cost reductions if the risk is carefully assesed.

In the past, safety was the primary objective of leak prevention in chemical plants and refineries. Emission controls reduced the risk of fire and explosion, albeit average contaminant concentrations in the workplace were considerably higher than currently acceptable levels. As the value of processed chemicals increased, considerable emphasis and effort were directed at loss reduction where improvements in loss control were reflected in immediate economic benefits.

As a result of these safety and loss control efforts, worker exposure was also reduced. However, even with the most effective loss control procedures (where very small losses are masked in the inaccuracy of precise material balance methodology), it is still possible to create a hazardous workplace environment in the vicinity of a leak or a number of leaks. Exposures occur when containment is lost and a contaminant is released into the workplace where it is inhaled by a worker or comes into contact with the skin. These release events fall into several categories that are shown in Table 4.1. The prevention of an unacceptable risk to workers often requires further improvements in emission controls for containment. For potent carcinogens, such as bis(chloromethyl) ether, a degree of containment is achieved that essentially eliminates releases. This effective containment system coupled with work practices and personal protective precautions can prevent exposure should a release occur.

2.1 Fugitive Emissions

The largest quantitative loss from a closed system is through fugitive emission or process leaks. While it is possible to essentially eliminate leaks, complete containment is very expensive and most processing units have a number of small and potential leak sources. For example, unless all pipe connections are welded, a typical plant will have hundreds of flanged piping connections that can be small leak sources. Of greater importance, in terms of mass leakage emissions, are valve stems, drains, and pump seals. Compressor, agitator, and scraper shafts may also leak along with sources in off-site areas. The large number of emission point sources result in a background concentration level of contaminant throughout a process unit. Worker exposure is directly related to this general area concentration and leakage from specific sources may result in high short-term exposure levels for operating personnel and maintenance workers. For specific toxic compounds the Environmental Protection Agency (EPA) has issued regulations requiring control of fugitive emissions.

Table 4.1. Loss of Containment

Conditions	Continuous	Intermittent
Routine operations	Vents (rarely)	Vents Filter or strainer changes Transfer losses Sampling Maintenance
Failures, accidents, and emergencies	Leaks from pumps, valves, flanges, mixer shafts, etc. Drains	Equipment failure Safety valve releases Spills

2.2 Fugitive Emission Regulations

Although exposures to several highly toxic compounds are controlled through specific regulation, the EPA also issued general regulations covering control of fugitive emissions in chemical plants and petroleum refineries. The purpose of the general regulations is to reduce the photochemical reactions of volatile organic compounds (VOC) that result in high ozone ground-level concentrations in populated areas adjacent to the plants. While designed to minimize general population exposures, these regulations also reduce exposures in the workplace.

The EPA regulations requiring emission control of specific toxic substances are termed National Standards for Hazardous Air Pollutants, or NESHAPs. Only benzene and vinyl chloride are currently regulated, but regulations for other toxic compounds are under review. The fugitive emission controls required in the NESHAP regulations are similar to the general fugitive emission control regulations issued by the EPA for VOC compounds. These latter regulations termed Standards of Performance for New Stationary Sources, or NSPS, essentially cover most compounds in petroleum refineries and a large number of itemized compounds in the chemical industry. The only compounds not covered by these regulations in the refining industry are methane, ethane, and several chlorofluorocarbons.

VOC and NESHAP regulations minimize workplace exposures through engineering controls and frequent emission source monitoring requirements. A number of controls along with monitoring are described in this chapter. Regulatory requirements are detailed in the federal regulations and in recent publications. The latter also describe the types of controls desired by the EPA to meet the regulations. All of the controls, along with monitoring, significantly reduce fugitive emissions, and some emission rates are discussed in Section 2.3.

The current regulations cover fugitive emissions from valves, pump and compressor seals, pressure relief valves, flanges, manual sampling, product accumulator vessels, and tanks. Other EPA regulations will be issued shortly covering fugitive emissions from drains, sewers, and wastewater treating equipment. In addition, fugitive emission regulations also cover various tank truck loading installations with barge and ship loading regulations expected within the next few years.

In considering the NSPS regulations covering VOC or fugitive emissions, the basic application requirements must be understood. Although termed New Source Performance Standards, the regulations are also applicable to existing process units under reconstruction or modification clauses. The latter has a very low criteria threshold that frequently requires application of the regulations to older process units. The status of regulatory compliance must be ascertained to assist in defining control requirements.

2.3 Fugitive Emission Mass Rates

Although there are a number of fugitive emission sources, valves, drains, flanges, and pump seals are the major emission sources. In this grouping, valves are the greatest source of emissions as described in various EPA and National Institute of Occupational Safety and Health (NIOSH) reports. Methods of estimating emission rates, particularly from valves, have been devised by the EPA, and an emission rate for a specific valve can presumably be determined by a concentration measurement at a valve stem. However, a comparison of emission rates from various studies indicates that large

Table 4.2. Mean Equipment Emission Rates from Various Studies (g/hr)

Equipment Type	Study of Mean Emission Rates				
	Radian 1979	Monsanto 1979	Los Angeles 1958	Bierl 1977	Fawley[a] 1981
Pump seal-all	56.8(14.7)	22(64.4)	79(89)	—	10.9(3.8)
Single mech.	—	—	—	6	7.1(4.2)
double mech.	—	—	—	0.02	0.03(0.02)
Valve seal	9.5(3.2)	4.5(29)	3.8(168)	2.3	
Compressor seal	293(113)	30(33)	161(510)	13	—
Flanges	0.24(1.72)	1.4(20.8)	NL	0.016	—
Relief valves	62.7(20.9)	5.1(14)	55	2.8	—

Note: Figures in parentheses are the logarithmic mean of leaking equipment; they do not include the nonleakers. Figures not in parentheses are the arithmetic mean of total equipment in that category, including nonleakers.
[a]Fawley Chemicals Plant, Esso Chemical Ltd.

differences exist in emission rates for individual items of equipment, as shown in Table 4.2. These variations are due to differences in packing types, packing age, maintenance, number of rings, valve stuffing box configurations, and stream conditions. Similar explanations cover differences in other equipment emission rates. The determination of fugitive emission rates for a specific control improvement requires field testing. Although precise emission data are not available, the values shown in Table 4.2 provide a basis for estimating concentrations and potential reductions with new controls.

2.4 Periodic Releases

Periodic releases from vents, safety valves, samplers, and the discharge of spent samples from continuous on-line process analyzers add to the background level of contamination in a process unit. While emissions from these sources may not be avoidable, they can be contained and, where permitted, released at locations a distance away from the work areas. EPA regulations cover fugitive emissions from vents, safety valves, and manual samplers. Periodic discharges from safety valves should be investigated to ensure ground-level concentrations do not exceed allowable exposure levels and workers on nearby platforms are not overexposed.

2.5 Operating Procedures

Plants are designed to avoid routine operations that require the opening of subsystems to the stmosphere, but this objective may be difficult to achieve in all cases. Wherever solids are added or removed from the system, there is a possibility of a contaminant release. Periodic filter and strainer changing and filter cake removal may be a manual operation that can result in contaminant releases.

Centrifuged solids movement and processing are often contaminant emission sources. In certain existing sampling systems, the process stream may be discharged on the ground or pad to purge the sample line. As a result, spilled material drains into an effluent line, partially evaporating contaminant into the workplace atmosphere. Improved system designs in these various examples release very small contaminant quantities into the environment.

2.6 Maintenance

Whenever equipment is opened for maintenance, the release of contaminants may result in a potential worker exposure situation. Block valves separating adjacent pressured portions of the unit may leak or traces of the process streams may remain in pumps, exchangers, or control valves. Turnarounds or unit shutdowns, where the entire process unit is opened for inspection and maintenance, require elaborate precautions to avoid worker exposure.

2.7 Transfer Operations

In practice, feeds or products are received or shipped via tank truck, tank car, or barge and as a result rundown or product storage tanks are necessary to provide storage for varying amounts of liquids. All of these shipping and storage facilities are sources of fugitive emissions and are also sources of potentially large releases.

When volatile liquids are pumped into an empty tank, the tank vapor containing a mixture of air and contaminant is displaced and vented. Environmental regulations generally require control of tankage emissions, although the controls vary with locality, age, and the contaminant or material. These controls, while effective in reducing emissions, may still be a potential source of worker exposure unless an enclosed vapor recovery system is used. Flexible connections to transport equipment and the ancillary equipment associated with tankage, pumping, and metering are also release sources requiring controls.

2.8 Emergencies

Major loss of containment may result in an event such as the disaster at Flixborough, which is the subject of a large body of literature on fire and explosion hazard risk technology. However, lesser emergencies such as spills or complete failure of a pump seal can result in lethal clouds of toxic materials.

3   ENGINEERING CONTROL OF WORKER EXPOSURE

The engineer has the responsibility of selecting design options permitted in the regulations or even more stringent ones to reduce worker exposure to acceptable levels. This requires a review of the alternatives available for control of emissions from a number of small emission sources. These sources, although characterized by low emission rates, may result in relatively high background concentrations that can contribute to local overexposure due to the large number of emission points. In contrast, large emission sources are readily identified and controls are generally selected early in the design stage to meet environmental regulations or worker exposure criteria.

For each equipment item in an operation that can result in worker exposure, there is a hierarchy of increasingly effective options accompanied by increasing costs. In addition, effective controls may add maintenance complexity and operating difficulty. The effect of improved emission controls on various operations can be readily observed in Table 4.3. Available options for achieving a desired level of control in various applications associated with closed processes are described in the following paragraphs. Other information on design strategies is presented in Section 5.

Table 4.3. Engineering Control Effectiveness

Activity	Maximum Peak Concentrations[a]	
	Before	After
Process area		
Process sampling	1800	6
Tank car sampling	1300	5
Tank car disconnect	750	2
Maintenance		
Clearing pump	3800	4
Opening compressor	9	9
Repairing safety valve	9600	4
Lab analysis		
Cylinder transfer	1900	4
GC work	40	5

[a]Ethylene or propylene oxide

3.1 Shaft Seals

A significant source of emissions in closed processes is shaft seals in rotating or reciprocating machines. Centrifugal-type pumps or compressors and agitators are examples of the former, and reciprocating-type compressors or pumps represent the latter. Typical emission rates obtained during various test programs are shown in Table 4.2. The significant disparity in emission rates is a function of various factors including, but not limited to, seal types, seal installation, maintenance, stream physical properties, and pump operating conditions. Various options for containment of contaminants in pumps and compressors are also discussed.

3.1.1 Centrifugal Pumps

PACKING. The traditional, and simplest, sealing system for a centrifugal pump is composed of packing rings around the shaft with leakage controlled through adjustment of packing pressure. The pressure is maintained with an outside gland and adjustments are frequently required to ensure a small flow of stream fluid along the shaft, which lubricates the packing. This system cannot contain contaminants, and a lantern ring sealing oil system is frequently installed to reduce emissions. However, the lantern ring system does not normally reduce emissions to the levels desired in the workplace.

MECHANICAL SEALS. Mechanical seals significantly reduce pump emissions and are the major sealing systems in continuous, closed processing operations. The seals are installed as single or dual seals in a variety of configurations in accordance with general industry recommendations and mechanical practices. Typical petroleum industry seal installations and auxiliary piping arrangements for various services are given in API Standard 610. A general description of options follows.

Single mechanical seals improve containment compared with seal packing through a reduction in emissions. The single seals shown in API Standard 610 also have a seal end plate on the outboard side of the seal as shown in Fig. 4.1. The seal end plate installation provides a cavity where trace liquid leakage is collected and drained. These

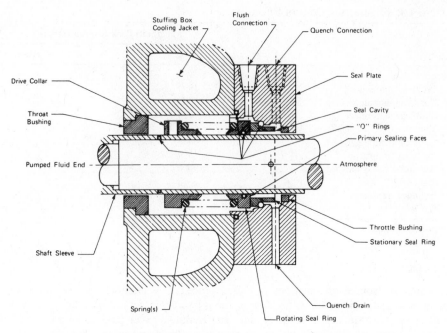

Figure 4.1. Centrifugal pump, single mechanical seal with throttling bushing.

plates provide an additional degree of containment and exposure control for single seal installations. A throttle bushing with relatively very close clearance in the seal plate provides worker protection should a seal fail completely.

Since some liquid may leak along the shaft to the atmosphere during normal operation, the throttle bushing is often replaced with a few rings of packing, termed an auxiliary seal, to aid in the containment of leakage or quench fluid. A quench fluid may be sprayed against the seal face for mechanical reasons, and the liquid will collect in the seal plate cavity. The auxiliary packing may be a graphitic type for high temperatures, while nongraphite packings are installed at temperatures < 400°F (204°C). Alternatively an elastomeric lip seal may be substituted for a throttle bushing.

Dual mechanical seals refer to either two opposed mechanical seals on a shaft (double) or two mechanical seals in series that are aligned in one direction (tandem). In some instances a face-to-face seal configuration is preferred. As shown in Table 4.2, there is a significant improvement in containment with double mechanical seals compared to single mechanical seals. In contrast, emission data on tandem seals is extremely limited and containment with this sealing arrangement may not be as effective as emission control demonstrated for the double-seal installation. However, tandem seal emissions are considerably less than those of single mechanical seals, and provide definitely improved containment.

The seal cavity between the seals in either installation is filled with a barrier fluid. The seal fluid pressure in a double-seal installation is normally higher than the stuffing box pressure at the pumped or impeller end. A typical double-seal installation is shown in Fig. 4.2. Seal leakage or failure of the inboard seal results in a flow of barrier fluid into the process stream. Since the barrier fluid dilutes the process stream, a

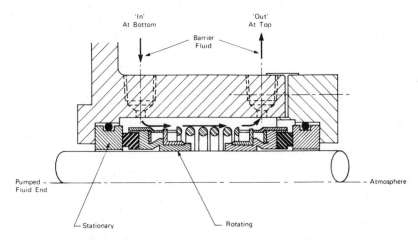

Figure 4.2.   Centrifugal pump, double mechanical seal.

barrier fluid compatible with the process stream is required. The tandem seal cavity pressure is lower than the pressure at the impeller end, and process fluid leaks into the barrier fluid under low leakage rates or seal failure. Barrier fluid compatibility with the process stream is not required in a tandem seal installation.

Although barrier fluid installations perform satisfactorily, additional considerations are necessary to ensure containment during normal operations and seal failures. The double mechanical seal barrier fluid circulates at a pressure level determined by the desired stuffing box pressure. The general barrier fluid system consists of a storage drum, a pump that circulates barrier fluid, and associating piping. In certain installations one pump may supply barrier or seal fluid through several process pumps. Separation of the barrier fluid stream into the various pump seals can be accomplished with standard control procedures. However, block valves must be installed to ensure isolation of a failed sealing system. In addition, instrumentation is necessary to indicate potential seal failures through flow rate changes in individual pumps or by other measuring techniques. A level indicator–controller on the barrier fluid drum is also necessary to maintain a satisfactory level of barrier fluid under seal failure conditions through automatic control of makeup barrier fluid to the storage drum. The storage drum may be vented to the atmosphere, and vent location is determined based on potential worker exposure during normal operating conditions or seal failure.

Tandem seal barrier fluid systems, in contrast to the pumped fluid system previously described, do not have a barrier fluid pump to circulate fluid through the tandem seals. The seal fluid system is a syphon-type system that circulates fluid from a small storage drum under gravity, heat, and impulse from the rotating pump tandem seal. Seal leakage or failure of the seal at the impeller results in a flow of stream fluid into the syphon system. The effect on the barrier fluid storage drum depends on the physical characteristics of the stream fluid under drum conditions. If the stream fluid is above its bubble point or vaporizes, the vapor will discharge from the vent. Alternatively, stream material may remain in the liquid state, rapidly filling the storage drum. Under either condition, vapor or liquid discharge must be controlled to minimize potential worker exposures. As an example of control, vapor from the storage drum may be returned directly to the process at lower pressure levels or into a flare line. Instrumentation is necessary to determine when leakage becomes excessive or a seal failure occurs. The storage drum is normally provided with a level indicator, and a flow sensor may be located in a syphon line or in the vent line. A pressure sensor in the drum is another method of indicating seal leakage when high vapor flow occurs. A restriction in the vapor outlet is necessary to rapidly increase the drum pressure for this condition. Flushing and disposal provisions are necessary to purge and clean the syphon system when process fluid dilutes and contaminates the system.

Double or tandem seals installed in accordance with API Standard 610 also have seal end plates with throttle bushings. The seal end plate perform the same functions previously described.

Canned Pumps. Completely enclosed pumps are now available commercially where the pump and motor are encased within a single housing that permits complete containment of stream contaminants. Applications were restricted in large processing units due to flow limitations, but pump size has recently been increased. This pump provides complete containment for extremely hazardous operations.

Magnetic Drive Pumps. Magnets driven by an external motor, rotating around a closed can pump housing also provide a completely enclosed pump system. Complete containment exists in this unit and reported capacities, including pump heads, are apparently comparable to those provided in standard centrifugal pumps, with mechanical or packed seals.

Other Pumps. Several pump manufacturers commercially market centrifugal pumps with proprietary seals designed for hazardous service. The seals do not resemble the typical mechanical seals reviewed in prior paragraphs. In one design, the seals move along the shaft and block the opening to the stuffing box from the impeller when the pump begins to rotate. Normally, nitrogen or an inert gas provides the driving force that maintains seal face position during pump operation. The seal remains in position until shaft rotation ceases, and the seal is then retracted as the gas pressure that maintained seal position is released. Gas consumption during pump operation is essentially nil. Reports from various chemical operations indicate that these pumps successfully contain contaminants.

3.1.2 Reciprocating-Diaphragm Pumps. A number of small reciprocating-diaphragm pumps can be found in various processing plant services. Many of the streams contain hazardous materials and containment is necessary. Single-diaphragm pumps are typically installed, but failure of the single diaphragm results in oil contamination and leakage of the hazardous material. Double-diaphragm pumps ensure containment of the contaminant within the pump.

3.1.3 Ventilation. Although pump seal installations are staisfactory for many toxic contaminants and workplace concentrations can be maintained at the desired level, much more restrictive requirements are required in certain situations. Ventilation systems have been installed with exhaust hoods located near the pump seal to capture highly toxic emissions. This type of control system functions well, particularly in an indoor process installation. Outdoor installations or ventilating control systems present other difficulties, and little information or literature on the effectiveness of this type of outdoor control is available.

3.1.4 Centrifugal Compressor Seals. Many centrifugal seals are of the labyrinth type that restrict vapor leakage. The rather large emission rates from centrifugal compressors reported by NIOSH and the EPA are probably the result of testing older labyrinth seals. These seals have been improved substantially, and lower emission rates can be expected from these modified seals. Circulating oil seals now used in compressor seals significantly reduce emissions and provide improved containment for toxic contaminants.

The circulating seal oil often absorbs the contaminants, which then vaporize from one of the oil surge drums in the auxiliary oil system. In some installations, the circulating oil also provides bearing lubrication, and the auxiliary systems become quite elaborate. This requires control of all vents from the drums in the circulating oil systems. A solution for vent emission discharge control is to connect the vents to the compressor suction, which eliminates vent emissions and reduces losses. An alternative solution is to vent the drums in the auxiliary systems to a lower pressure level location in the process.

Dry seals are also available that can be purged with a gas to the process or a disposal point. As an alternate, the seal can be dead-ended with an inert or nontoxic gas that may leak at a slow rate to the atmosphere and into the process.

3.1.5 Reciprocating Compressor Seal. The piston rods in reciprocating compressors have a stuffing box for control of emissions and pressure maintenance. Emissions control can be improved by extending the stuffing box and, in certain applications, the installation of a lantern ring with a circulating oil system has reduced emissions. New packing types have also reduced emissions. However, in particularly critical toxic contaminant service, strict standards may require a reduction in emissions that cannot be achieved with packings. A control system that successfully reduces emissions closes the openings in the spacer piece, which is the section between the compression end and the crankshaft side. Emission leakage into this area from the packing is contained and the spacer area is continuously purged with an inert gas to an elevated atmospheric discharge point through tubing runs or to a disposal system (flare, incinerator, etc.). Venting to the atmosphere must be carefully analyzed to determine potential workplace concentrations.

3.2 Valves

Valves are the major source of fugitive emissions and require control or containment to reduce plant background concentrations and worker exposure. In continuous process operations, control, gate, plug, ball, globe, and butterfly valves are the typical emission sources. Control and gate valve emissions are significantly greater than emissions from the other valves, and emission reductions reviewed here only refer to control and gate valve installations.

3.2.1 Packing. For a number of decades the majority of valves were packed with an asbestos-based material. The asbestos packings were improved during this period and the latest packings are impregnated with lubricants, graphite, and other materials. In addition, metal wire within the packings provides corrosion resistance and the strength necessary for adequate service life and performance. Valves containing asbestos-based materials were the major source of emission test measurements conducted by the EPA in U.S. refineries.

The health concerns associated with asbestos resulted in packing manufacturers investigating alternative packing materials. Nonasbestos packings became commercially available in the late 1970s and several manufacturers stopped production of asbestos-containing packings. As a result of valve packing research, the number of different valve packing materials now available has increased along with some changes in mechanical configurations of the packing. Data from the packing manufacturers indicate valve emissions for the new packings are significantly lower than emissions from asbestos-packed valves. Although packing performance has apparently improved, actual field data are scarce on valve emissions with the new packings.

Based on a limited amount of information, the nonasbestos packings apparently are a significant improvement over the older asbestos-based packings in controlling emissions and extending service life. Although a number of different materials have been installed, most of the information available on performance pertains to pure graphite-type packings. These packings are provided commercially by a number of packing manufacturers.

Reductions in emissions and a reduction in workplace background concentrations can be achieved with graphite packing in gate and control valves. Moreover, the new packings further reduce potential exposures through a reduction in the number of leaking valves and a decrease in the number of valves requiring maintenance. Further improvements can potentially be obtained in control valves by increasing the amount of packing within the stuffing box. Normally, control valves are supplied with three large lantern rings. The substitution of packing for two lantern rings increases the number of packing rings in the stuffing box essentially providing two packing sets in the control valve. A further improvement in control valves can be obtained by placing a lantern ring adjacent to the gland and purging the lantern ring with an inert vapor or a barrier fluid.

The replacement of graphite packing rings in gate valves with a wedge-type interlock combination set of graphite packings now available commercially may further reduce emissions. If a greater degree of emission control is required, then lined plug, bellows, or diaphragm-type valves can be installed. However, certain valves are limited in size and in large plants a combination of valves and packings will usually provide the degree of containment necessary for control.

3.3 Flanges

Flanges in piping systems reportedly contribute about 6–8% of total fugitive emissions in petroleum refineries. Chemical plant flange losses in large, high-volume closed processing systems have not been defined in detail, and actual emission rates may be lower than the flange emissions observed in refineries. Lower valve emissions were observed in chemical plants compared with refinery valve emissions, and chemical plant flange emissions rates may also be lower compared to refinery rates. However, the portion of total fugitive emissions in chemical plants attributable to flange emissions is probably greater than the ratio reported for refineries.

The major flange gasket material in process plants through the years consisted of asbestos in various forms. The principal gasket form installed in large closed processing units consisted of asbestos impregnated with a rubber composition, which bound the asbestos fibers. The impregnated asbestos gasket, termed *compressed asbestos*, is normally specified in accordance with ASTM specification D2000. Flanges containing compressed asbestos gaskets were the major flange emitting sources tested by the

EPA in their investigation of fugitive emissions. Other gasket types commonly used include metal and plastic encased asbestos. In addition, elevated temperature and pressures may require ring-type flange joints, which have a cylindrical metal collar as a gasket rather than the typically flat gaskets normally installed in flat-faced flange assemblies. Data on emission factors for various types of gaskets and flange assemblies are not readily available.

As a result of asbestos health hazards, a variety of new gaskets and gasket materials are now available commercially. One gasket contains a series of concentric metal rings where each ring is separated by a distance of approximately one-eighth inch or less. The annular areas between the rings are filled with either asbestos or nonasbestos materials. Field information indicates this gasket performs well, but emission data have not been published in detail although graphite fill is considered superior in various tests. Other new gaskets consist of nonasbestos materials including plastics, fiberglass, and graphite. With the profusion of gasket materials, comparative emission rates or sealing efficiencies are difficult to obtain. However, several manufacturers are conducting emission tests on new gasket materials along with an industry group. Improved containment requires a thorough assessment of available manufacturer gasket test data prior to the selection of a gasket for a specific service.

Another source of flange leakage is improper bolting and flange fit. In critical services, flange fit can be significantly improved through improved bolt tightening techniques developed in the last decade. The revised bolt tightening procedures resulted from new instrument techniques applied to bolt stress and elongation analysis. For services where leakage must be contained, welded connections should be maximized. Valves are now supplied with weld fittings and can be directly welded to piping. In addition, various external flange sealing techniques are now available for control of flange leakage in critical locations and services. A combination of welding, improved bolting procedures, and improved gaskets will significantly improve containment in most instances.

3.4 Vents

In the past venting was either continuous or noncontinuous, in accordance with process requirements. Environmental regulations have essentially eliminated continuous venting to the atmosphere although noncontinuous vents currently present potential exposure problems. However, noncontinuous venting covers a wide number of functions, from minor instrument purges to venting large vessels, accompanied by a range of extremely small to large emission discharges. Noncontinuous vents also include vents that may operate infrequently or only during turnarounds. Therefore, a generalized conclusion regarding control of all vents is difficult. A single control option for vent control is difficult to achieve and in the majority of process installations is probably impractical. As an initial step, vents should be classified into function, flow rates, composition, and stream conditions. Small vapor vents may be grouped into a manifold and vented or recycled to the process where possible. If recycling or recovery is impractical, the vapor can be vented through a header to a common stack, a flare system, or an incinerator. Large discontinuous vapor vents may also discharge to the atmosphere, a flare system, or an incinerator. However, discharge to the atmosphere through a vent stack requires additional analysis as discussed later.

Vent stack discharge flow rates should be determined for maximum and minimum conditions, and the stack discharge elevation is defined by ground-level concentrations calculated for both flow rates and various contaminants to minimize potential worker exposure. In addition, potential concentrations of contaminants on nearby platforms should be determined to ensure a worker on any of the platforms is not overexposed. This analysis procedure assumes all vented material is a vapor. If partial condensation of the vapor occurs, rainlike droplets may occur, resulting in potential overexposure to aerosol mists and possible skin contact. Assessment of the condensation potential is required prior to calculating dispersion effects.

In certain processes, on-site tankage is provided that may require venting. However, environmental regulations may require a specific type of venting control or set a maximum loss rate to the atmosphere. Revised tankage emission factors were issued by the API in 1984, and the emission factors will aid in more accurately predicting emission vapor rates. If specific controls are not required, then various options may be investigated. Alternatives include a vent stack, an adsorber on the vent, a closed vapor recovery system, vent condensing, purge to an incinerator, an internal floating roof, an external floating roof, a pressurized tank, or a refrigerated tank. Costs vary according to the option and degree of containment necessary.

3.5 Safety Valves

Safety valves installed in all types of process operations discharge liquids, vapors, or combined liquid–vapor streams to relieve abnormal operating conditions. Control of these discharge streams is important in maintaining containment or minimizing potential worker exposure. A discussion of valve considerations and control options follows.

3.5.1 Safety Valve Construction and Installation. A typical spring-loaded safety valve is shown in Fig. 4.3. Since these valves are in direct contact with the process, removal of a safety valve during operation for testing or inspection following a safety valve discharge requires installation of valving necessary to isolate the safety valve. Moreover, in those services where the fluid is toxic or corrosive, purge connections between a block valve and a safety valve are required to protect a worker against overexposure during safety valve removal.

In certain installations, bellows-type safety valves are installed where the fluid is corrosive. A typical bellows-type safety valve is shown in Fig. 4.4. The bellows protects or shields the spring from the process fluid. However, the bellows valve bonnet must be vented for proper balancing of the valve disk. The bonnet is often vented to the atmosphere, but in situations where a toxic material can be released in a bellows failure, the vent should be connected to a low-pressure system or closed-drain system. In addition, the bellows safety valves should have block valves to permit safety valve removal in toxic or corrosive services as described in the previous paragraph.

As an alternative to bellows safety valves, rupture disks are often installed between the process fluid and a conventional safety valve in toxic or corrosive services. If pressure builds between the rupture disk and the safety valve due to leakage through the disk, the disk will not rupture. A vent or drain is installed in the space between the disk and safety valve permitting discharge of any liquid or vapor accumulation. Although the leak rate is small, the disposition of toxic materials from this space must be carefully investigated. In certain installations rupture disks may be installed as the safety valve or pressure relieving device. However, bursting of a rupture disk destroys the disk and pressure or operating conditions cannot be reinstituted unless a block valve is installed between the process and the disk. The ruptured disk is then removed and replaced for restoration of safety protection.

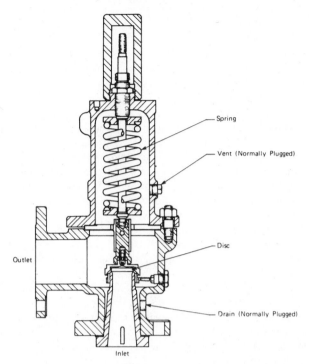

Figure 4.3. Typical conventional safety relief valve.

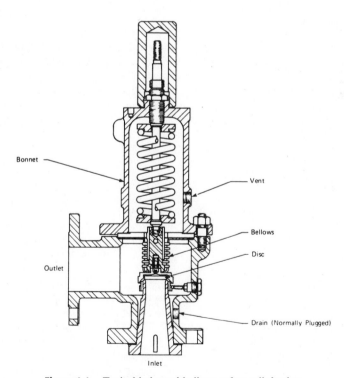

Figure 4.4. Typical balanced bellows safety relief valve.

3.5.2 Safety Valve Discharge. Since safety valves may discharge toxic liquid, vapor, or a liquid–vapor stream, disposal or containment of the safety valve discharge stream must be considered. Safety valves in liquid service should have a closed discharge with the liquid recycled to the process. Although relatively simple in concept, location of a receiver for the discharge stream requires a detailed examination of the process. If a disposal receiver is not available, then an alternate disposition site must be established that may be a blowdown drum, a closed discharge header, a process sump, an on-site or off-site tank, a toxic sewer, or an incinerator with a liquid surge system.

Vapor and liquid safety valve discharges must be analyzed to determine whether a vapor–liquid stream is formed or the vapor is at its dew point and condensation will occur under various alternate discharge situations. Assuming a mixed-phase discharge stream, the stream must be contained in a closed system. The main exhaust header that contains and directs large safety valve releases into a blowdown drum-flare header system can generally accept mixed-phase releases. However, the blowdown system at the inlet must be capable of separating and recovering the liquid portion of the safety valve release.

The discharge of vapor releases must be carefully analyzed to ensure the stream remains a vapor under disposal conditions. In addition, vapor flow rates must be accurately determined since the emergency flow rates in some processing units can be quite large. Certain vapor discharges can be sent into a closed flare header system and to the blowdown flaring system. However, the blowdown system may not be capable of processing a very large release added to an existing system. For large releases of toxic contaminants, several alternatives are potentially feasible.

Atmospheric Release.

Release of vapors from a very high elevation or location is feasible, providing the ground-level concentration (glc) criteria for worker exposure is not exceeded. In addition, the concentration of toxic contaminants on nearby platforms during a release should not result in potential worker exposure. If these criteria are exceeded, then a tall stack dedicated to large safety valve releases may be required that meets worker exposure criteria at ground level and on elevated working platforms. Elevated releases must also meet general environmental criteria.

Large Flare System.

The proposed flare system for normal operation can be expanded to accommodate a much larger release load. However, environmental criteria and equipment size must be thoroughly reviewed.

Ground Flaring.

In certain locations a ground flare may be acceptable for an emergency release. A ground flare dedicated for destruction of a specific stream can be effective in destroying the toxic contaminants since these flares are capable of processing large vapor loads. An incinerator for this service must have a rapid response to ensure complete combustion of all materials.

Equipment Design.

Certain process vessel and associated equipment design ratings can be increased that may have a significant effect on the vapor release rate through changes in the basis for estimating safety valve discharge conditions. Revised release rates will then require a detailed review of the entire safety system.

3.6   Manual Sampling of Streams

High potential inhalation or dermal exposures to contaminants exist when process streams are manually sampled although samples may only be drawn three or four times per day. These potential exposures occur in process operations where manual samples are obtained from simple drain-type connections or lines. The need for improved containment systems to control exposure has been recognized in the process industries where sampling systems are frequently designed to meet containment criteria.

An important design consideration is the fugitive emission (VOC) NSPS regulation that requires emission controls on manual sampling operations. The regulation stipulates that a closed container must be used and the container cannot be purged to the atmosphere. The sampling options in this section are based on the EPA regulations and a few modifications permitted in accordance with the stream liquid physical characteristics. Other types of manual sampling systems can be installed that control exposures to similar or lower exposure levels.

Typical vapor or liquid process stream sampling systems are presented in the following paragraphs. Sampling combined vapor–liquid streams is more complex and requires an extended commentary.

3.6.1 General Process Considerations. Prior to initiating a design, the location or requirements for manual samples should be firmly established with the process basis engineers and operating personnel. This requires identification of the specific stream that will be sampled, sampling frequency, and the type of analytical information necessary for process operations. The latter is required to define the container and associated sampling system needed to provide the correct sample to the laboratory. In addition, process stream temperatures are often higher than allowable manual sampling temperatures defined by the safety or operations groups.

Permissible manual sampling temperatures may be 125–150°F (52–66°C), which usually requires cooling of the withdrawn sample. As an example of the effect on sampling considerations of sample container temperature limitations, a vapor stream at elevated temperatures may partially or completely condense, making it necessary to choose a final sample container that is considerably different from that selected for collection of a vapor.

3.6.2 General Sampling Considerations and Intallations. A major consideration in containing contaminants and reducing potential exposures is minimizing or eliminating the need to flush the sampling line prior to withdrawing the sample. Flushing a line and possibly the container to the atmosphere with process material to obtain a typical sample is a major source of workplace exposure and VOC material into the air. The problem is essentially eliminated by the installation of a loop that circulates process material; a container may be inserted or a very short drain connection can be attached for sample withdrawal. A successful circulating loop installation requires that the loop be installed across equipment or a length of piping providing needed pressure drop. Typical circulation loops are shown in Figs. 4.5 and 4.6.

The design of a sampling system is a function of the sampling container recommended by the plant analytical laboratory and applicable regulations. These containers frequently reflect plant laboratory practice and capabilities. A few sampling containers and applications are:

CLOSED SAMPLE CONTAINER (I). These are closed cylindrical containers constructed of metal (generally stainless steel) with a valve at each end. The containers vary in size from a few hundred milliliters to 25 L. The simple sampling bomb described here is the manual container for vapor samples withdrawn from an operating process installation.

The bombs are normally cleaned and prepared for sampling in accordance with local practice. The bomb may contain an inert gas or can be evacuated, but either condition requires flushing the bomb to ensure obtaining a typical sample. A typical vapor bomb installation is shown in Fig. 4.6, where the bomb flush is returned to the process.

CLOSED SAMPLE CONTAINER (II). The metal sample bomb is modified to obtain a typical sample in those situations where light components may vaporize from the sample if an open container is used. Vaporization of the light component(s) affects analytical accuracy resulting in questionable process variable control. A properly constructed metal bomb for a liquid sample with light components includes a short dip tube that provides a liquid–vapor interface. The vapor zone is a safety measure to ensure the bomb is not completely filled with liquid.

The bomb is normally cleaned and evacuated or inerted prior to sampling. A typical bomb installation for liquid sampling is shown in Fig. 4.7. The bomb container is flushed with liquid until the overflow sight glass indicates the bomb is filled. The excess liquid can be returned to the process or connected to a closed drain system or sewer.

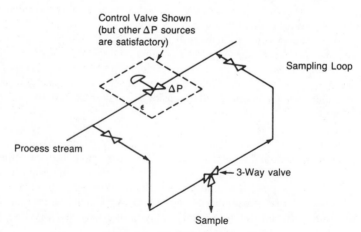

Figure 4.5. Typical circulation loop.

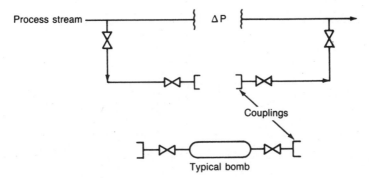

Figure 4.6. Typical vapor bomb installation.

If the excess liquid is returned to the process, a sight glass suitable for process service must be installed. Alternatively, the sight glass may be replaced with a local instrument indicator.

OPEN CONTAINER (I). Open containers are standard liquid sample receivers in many process plants. Samples are drained into the container from a sampling loop to provide typical stream material, and the container is closed before withdrawal from the sample box. Worker exposure can be a concern where the sampled material contains toxic contaminants and presents potential inhalation problems. However, the concentration of the toxic contaminants should be very small in this high boiling material. A proposed installation for contaminant control is shown in Fig. 4.8, which is similar to the installation reviewed by Emmel ("Control Technology Assessment of Petroleum Retinery Operations" presented at the 1983 NPRA Annual Meeting, March 20–22, 1983, San Francisco).

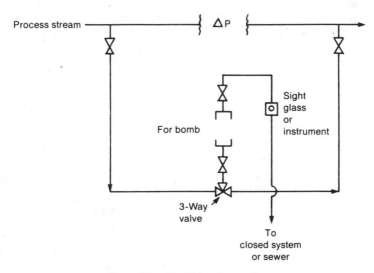

Figure 4.7. Liquid bomb sampling.

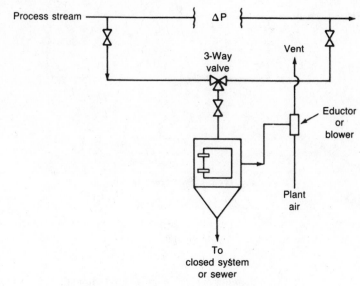

Figure 4.8.   Open container I—for liquids.

The sampling container is placed within a box or sample enclosure that contains the sampling liquid outlet. The box can be constructed of stainless steel or plastic with a hinged door containing safety glass. The container rests on an open grill that permits excess liquid to flow into the funnel at the bottom of the box. The funnel drains into a closed system or sewer to minimize exposure.

A flow of air through the box is maintained by a blower or eductor, with the exhaust discharging to a safe location as determined through dispersion calculations. Emmel indicates the air velocity through the open door was 60 fpm (18 m/min) in a refinery installation. Since the vapor material is discharged to the air, any toxic contaminant must be at a very low concentration.

OPEN CONTAINER (II).   When a liquid presents a potential dermal problem, but does not have a toxic vapor or inhalation potential, the sample container may be placed on an open funnel designed for splash protection. A typical installation is shown in Fig. 4.9 where the container is placed on the funnel grid. The sides of the funnel are extended vertically to provide splash protection and the funnel drains into a closed system or sewer.

OTHER SAMPLING TYPES.   A number of sampling systems have been described in papers presented at symposiums and in various journals. The systems frequently include specific containers and methods of obtaining typical samples. Many of the systems were designed for specific installations and hazardous areas. The sampling systems described in this chapter are quite general. Where problems requiring other solutions exist, a review of the literature is recommended.

Although manual sampling is described in this section, manual sampling should be replaced where possible by on-line analytical instruments. However, process analytical instrumentation requires careful review to ensure potential worker exposure is min-

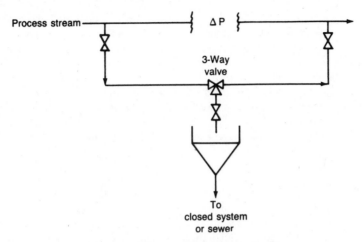

Figure 4.9. Open container II—for liquids.

imized. Instrument discharge vents, calibration, sampling systems, maintenance, and instrument room ventilation must be thoroughly investigated and controlled.

3.7 Filters

Filters and strainers reviewed in this section only refer to applications where precipitates, coke, or polymer present at relatively low concentrations in a process stream must be removed to clarify the stream. The filter media in this service are generally bags or metal baskets and are infrequently removed manually from the filter case for cleaning in contrast to high-volume filter systems that require cake removal at least once per day. The stream filters or strainers, although infrequently opened, may contain contaminants that present potential inhalation or dermal exposure hazards. This requires a closed outlet drain to a general drain system or sewer, provision for flushing, purging, steaming, inerting, and venting to a proper location. Excluding the permanent drain connection, shutdown and cleaning functions can be performed through hose connections. A more sophisticated system can be installed with permanent utility and vent connections controlled automatically.

The major objective in purging and flushing the filter is to minimize potential overexposures when the filter case is opened and the basket is removed for cleaning. Filter cases are not always cleaned and purged in operating plants for various reasons. Liquids remaining in the case or on the solids in the basket are a potential hazard exposure source. Moreover, liquid on the basket solids may present additional health problems when the solids are removed from the basket. Where possible, the formation of solids materials should be reduced.

3.8 Flush and Purge Connections

Pumps, control valves, heat exchangers, and other equipment that can be isolated for servicing while the unit is in operation need fitting and piping connections to permit drainage, flushing and purging, and venting to contain all contaminants during these steps and prevent overexposure when the equipment is opened. These connections can be simple valved fittings at suitable points and connectors to accept flexible hoses. Flexible hoses are often used for draining equipment with the discharge end connected to a closed drain header, closed system, or sewer. Closed, permanent piping connections for utilities, purging, and draining may be necessary for highly toxic materials where potential exposures are critical. This situation would exist where lighter contaminants that are well-defined health hazards may vaporize. In cold climates, steam tracing on permanent connections may demand a very complicated piping system and alternatives should be investigated before a purge and drainage system is selected. Hoses in these systems should have closed ends that do not permit leakage when disconnected.

3.9 Process Waste Drains

Chemical sewers that accept only hydrocarbon and process stream wastes should be closed drain systems to prevent the release of toxic vapors at the entry point. If the chemical sewer is a completely closed drain system, then the closed drain may discharge into a closed sump or drum and the liquor can be recycled to the process. Other disposal options may be available for a closed drain system but require a detailed knowledge of the overall processing unit scheme and disposal systems.

Mixed water and process materials are generally discharged into a sewer system where the stream is treated as wastewater. The sewer system is normally designed with sealed catch basins for rainwater and area flush water around the process units. The small drains are generally connected to the catch basins, and the contaminated water is sent to the main sewer that empties into the wastewater treating facilities. These sewer systems are vented to the atmosphere and can be a source of worker exposure. If the wastewater contains a significant fraction of materials immiscible with water, most of the process materials can be recovered and recycled to the process. Hydrocarbon–water separation in an API or plate separator are typical separation process steps. Miscible water–chemicals streams may require more sophisticated separation techniques such as distillation or reverse osmosis. Separators in wastewater facilities that have normally not been enclosed in the past should be enclosed to minimize worker exposure. Various enclosure systems will probably be required by the EPA in new VOC emission regulations under consideration. Safety is a major consideration in any enclosure design involving mixtures of vapor and air in the wastewater treating area, and control designs should be reviewed with safety personnel.

3.10 Transfer Operations

Transfer or loading of liquids with relatively volatile, toxic components into various containers is a potential source of worker exposure. Specifically, loading barges, tank cars, tank trucks, drums, or cans may present exposure problems during transfer. Containment systems have been developed for each loading or transfer operation but are generally designed based upon certain general principles:

- The container should be loaded from the bottom. If the container is a drum or barge, a loading spout should be extended to the bottom of the container to avoid splashing. Tank trucks and tank cars can readily be filled from the bottom.
- Vapor is collected from vents and hatches with steam-driven eductors, water-ring

compressors, blowers, or under a slight pressure and the vapor is sent to incinerators, furnaces, activated-carbon units, or condensers.
- Filling hoses are the self-closing dripless type.
- Barge vents are manifolded, and the vapors are sent to one of the disposal systems already described or discharged to the atmosphere from an elevated location where permitted.
- Exposure is also reduced by purging lines with an inert gas that discharges into the disposal unit before opening any lines to the atmosphere.

In addition to these items, small tanks in a loading area often have large activated-carbon canisters in the tank vents. Although various alternatives have been described, regulatory requirements may specify particular types of vapor control installations such as those required by the EPA in gasoline bulk terminals. Tanks may also accumulate a water level, which should be lowered with an automatic tank water separator valve. The valve discharge point is enclosed with this device, and the effluent from the valve is discharged into a closed drain system.

3.11 Additives

In many process plants, drums and bags containing additives or special chemicals are transferred to the processing systems. The additive quantities are small, but many are toxic materials and a potential source of overexposure. Initially, liquid additives are generally transferred to holding tanks and solids are transferred to solids storage facilities.

Liquids in drums are normally transferred to holding tanks with barrel pumps. However, the pumps are potential sources of inhalation and dermal exposure when the wetted pumps are removed from the drums. Alternative techniques for removal and transfer of liquid from drums to holding tanks can be accomplished through flushing or removal with an eductor. If a material is diluted with water, the liquid in the drum can be flushed with water into the tank. Alternatively, water or a process liquid can be pumped through an eductor connected to the drum. The eductor will then draw the drum liquor into the main stream, which is sent to the process or tankage. If these techniques are not feasible, drums can be supplied with connections that permit a tie-in with a small diaphragm pump. Other successful transfer systems that inexpensively minimize exposure have been described in the literature.

Dumping of dry solids from bags has been a consistent problem for a number of years. Recent equipment advances motivated by potential solids exposure problems have significantly changed dumping and bag handling techniques. In the new control systems, a bag is placed within a horizontal drumlike device and the lid is closed to prevent the escape of dust into the atmosphere. The bag is slit by remote operation within the enclosure and the bag contents are automatically dumped through a grate into a smaller solids hopper. Air drawn through the bag dumping device by a blower to control solids emissions is discharged to the atmosphere through a filter. The empty bags are discharged automatically and a separate bag bailing device can be obtained to compress and bind the empty bags, completing the operation.

Solids can be removed from the bagging hopper through a conveying system to a solids storage silo. If the solids are to be discharged into a tank partially filled with a volatile liquid, an air–solids separation unit above the tank separates the solids from the gas, and the solids enter the tank through a rotary valve or similar system. When the liquid is highly volatile and/or toxic, the rotary valve system can be improved to prevent vapor flow into the dry solids system.

4   DESIGN PRACTICES

In the chemical process industry, the design of major facilities includes a number of definable stages, commencing with planning and continuing through the process design basis, process design specifications, detailed design, construction, and start-up. Experience has shown that interaction of the design team and the industrial hygiene controls group at each stage of the project is an efficient and economic method of ensuring that health hazard controls are included in the final plant design. Retrofitting engineering controls following start-up is expensive, inconvenient, and increases potential workplace exposures. In providing advice and assistance to the design team, the industrial hygienist is guided by the principles shown in Table 4.4.

4.1   Exposure Prediction

Engineering controls of differing degrees of effectiveness when combined with a technique that defines the relationship between the control and workplace exposure can provide the design engineer with a general basis for selecting economical design options. Installation of controls without a method of estimating workplace concentration can result in overexposure or expensive control installations that may add insufficient worker protection for the added control complexity. An accurate exposure estimate is not necessarily required but is a method that provides a reliable prediction technique.

Exposure prediction begins with an understanding of vapor dispersion from a point source such as a pump seal or valve stem. A dispersion model calculation technique is the basis for defining the workplace concentration gradient in a specific area surrounding the emission source. The basic point source dispersion model can then be combined with a number of point sources to describe a general workplace concentration. The latter step is a more complicated modeling technique, but simplified methods of developing an area exposure model are available in the literature. However, model development for an overall process unit requires a more sophisticated approach to describe complex process process piping and equipment systems. An overall analytical model will probably be available in the future to define the average concentration in the workplace.

Table 4.4. Essential Characteristics of Successful Health Hazard Controls

- The levels of protection afforded workers must be adequate, consistent, and reliable.
- The efficacy of the protection for each individual must be determinable throughout the lifespan of the unit.
- The solution must minimize dependence on human intervention for its efficacy so as to increase its reliability.
- The solution must consider all routes of entry into workers' bodies and should not exacerbate existing health or safety problems or create additional problems of its own.

4.1.1 Dispersion Calculation Technique. Based on published information, certain pre-diction calculations can be used as a basis for estimating the efficacy of controls in a simplified situation until advanced models are available. The dispersion concentration from a point source can be calculated from the following equation:

$$C = \frac{Q}{kux^n}$$

where C = expected concentration, mg/m³
   Q = emission rate, mgms/sec
   X = distance between the worker and source, m
   k = constant (0.136)
   n = constant (1.84)
   u = wind speed, m/sec (0.5 m/sec minimum)

This equation requires a knowledge of the emissions rate from a point source to determine the concentration of various locations. Emission rates are available from various sources, and the EPA has published emission factors for valves, pumps, etc., for both petroleum refineries and chemical plants. However, the EPA data do not show emission reductions available from various valve packings, compressor seals, flange gaskets, or alternative pump seals. Consequently, a general approach can be developed that uses the EPA emission factors to define the expected ratio of emission rates to concentrations. The equipment item is modified, and a concentration is mea-sured at the emission face with the recommended EPA monitoring technique. This concentration can be converted to an emission rate with the EPA conversion method. The revised emission rate then results in an adjusted exposure concentration and a method of defining the efficacy of the improvement.

If the emission from a single mechanical seal in hazardous service results in a concentration greater than the threshold limit value (TLV) at a distance of 3 m from the seal, then a dual seal should be installed in the pump. For valves, a higher concentration than the TLV requires the replacement of standard packing with one of the new, low-emissions packings available from various packing manufacturers. Emission rates for other equipment can be calculated based on this general approach, with equipment changes to reduce potential exposure levels.

4.1.2 Emission Data. This simplified approach significantly reduces exposures by in-itially providing dispersion emission estimates from a known leak rate source and reducing the exposure estimate with an improved control system. The major problem with this approach is the lack of emissions data for preliminary exposure reduction estimates with various equipment control improvements. Emission factor information on control improvements would provide the industrial hygienist and engineer with a basis for selecting the optimum cost–benefit improvement. As an example, emissions data for a single mechanical seal are available from the EPA-sponsored studies. How-ever, emission rates for double mechanical or tandem mechanical seal installations available from the EPA studies vary considerably. As a result, realistic data on dual-seal performance are required to properly evaluate the emission reductions available with dual seals. The Bierl data on dual seals, which show significantly lower emission levels than the EPA data, are probably realistic, but additional verification from other sources is needed. Dual-seal emission data will probably become available from published sources during the next few years as various companies and organizations investigate these control improvements.

Similarly, valve emission rates have been investigated by the EPA and can provide an initial dispersion estimating basis. Data on emissions from the new nonasbestos packings are not available, but increasing acceptance of the new packings will provide emission information in the public domain within a relatively short time. Data from various manufacturers are an alternate basis for estimating emission reduction until published information is available. The situation is similar in the flange gasket area, but a task force is investigating the new gasket materials, and emission data or comparative analyses probably will be available in 1988 from one of the major engineering organizations.

4.1.3 Worker Location Exposures. Various publications present analyses estimating worker exposures based on known area concentrations. These analyses use an existing work pattern and can precisely define worker exposures in large petroleum or petrochemical units. Based on this analytical approach average area concentrations can be increased since exposure time is short. However, this analysis of work patterns is generally directed toward operating personnel rather than maintenance workers. Operators in large well-automated processing units may only spend a few minutes in potentially hazardous areas during a 8-hr shift. This time factor consideration results in an extremely low TWA (time-weighted average) exposure level. However, the work pattern approach does not cover all exposure situations, and the following considerations must be included in this type of analysis to ensure a satisfactory contaminant concentraton level is achieved.

- Maintenance worker exposures must be considered. Typical work patterns for an identical plant can potentially be defined through careful study of maintenance records, reviews with maintenance personnel, maintenance equipment repair practice, and reliability studies. This information provides a basis for determining exposure time periods and locations of maintenance personnel within a process area. However, releases associated with equipment maintenance must be estimated and concentrations for workplace exposures delineated.

- Emergency situations may require the presence of personnel at locations within a unit for time periods that may overexpose certain workers. Emergency requirements must be reviewed, work patterns established for the emergency condition, and probabilities of occurrence defined.

- Episodic releases occur, and an estimate of concentratons associated with the release and occurrence frequency are necessary.

- If a relatively high average concentration is acceptable within a process unit, the effect of this material on adjacent areas must be considered. Combining contaminants from one unit with another may result in unacceptable workplace exposures in specific locations.

Evaluation of work exposure patterns is difficult and depends on a constant method of working. This can be a potential problem with changes in manpower, process unit modifications, and major differences between anticipated probabilities and actual frequencies of equipment failure. Moreover, potential changes in exposure or health standards may require expensive retrofits.

Table 4.5. Manual Sampling Exposure Controls

A. Sample collected in a bomb connected to a closed loop
B. Sample outlet from a closed loop over a drain cell in a box with local ventilation exhausted to a safe location
C. Sample outlet from a closed loop over a drain
D. Sample outlet in a box for dust/splash containment draining to chemical sewer or closed waste drum
E. Any type

4.2 Equipment Selection Matrix

The outcome of the design-engineer/industrial-hygienist interaction and of the exposure estimation calculations is a matrix of design recommendations for worker exposure control. A typical example of a part of a matrix for manual sampling exposure controls is shown in Table 4.5. The letters (A, B, C, etc.) in the body of the table designate the level of control option for each category of equipment.

4.3 Costs

The costs of the closed system health hazard control engineering measures described in this chapter, while not insignificant, are usually a small fraction of the total capital cost of large processing units. For medium hazard manufacturing units handling volatile liquids and gases with TLVs that are generally in the range of 10–100 ppm, the total cost of health hazard controls is about 1% of the total capital cost of a large processing unit. For units that process highly toxic materials, such as (bis)chloromethyl ether, the engineering control costs will be much greater and may reach levels of 10% or more in certain processes. In addition, batch processing units and/or units with lower processing rates may have control engineering costs that are significantly larger than 1% of total investment. Moreover, cost calculations are approximate since a number of other plant requirements such as safety features have health spinoffs. The health hazard controls may also have economic benefits such as reduced product loss and improved operability, which often justify various expenditures.

5 EXPOSURE CONTROL PROCEDURES

Operating, maintenance, and emergency procedures must complement engineering controls to ensure containment and minimize releases to the workplace. Comments and recommendations on these areas are presented in the following subsections.

5.1 Work Practices

Engineering controls are installed to reduce the workplace exposure level, but the controls must be constantly used and properly operated to minimize worker exposure. Procedures must be developed prior to unit start-up to ensure satisfactory operation, and these procedures must be communicated to operating and maintenance personnel. The procedures are normally included in operating manuals and are reviewed in detail with assigned personnel during training sessions. The training sessions are conducted prior to unit start-up, but a periodic training update is often instituted to ensure that recommended engineering control operations continue during normal plant operations and turnarounds.

Several typical engineering control systems are presented in the following list as examples of control systems requiring a continuing commitment to approved operating procedures.

- When equipment is removed from service for maintenance, the equipment is blocked and isolated from the operating stream. Assuming an exchanger is the equipment of concern, the initial preparation step prior to bundle removal generally requires draining or purging the contents of the exchanger side containing the chemical contaminants. We have assumed the other side of the exchanger is water or steam. Following draining or purging, the chemical side of the exchanger is then flushed with a selected flush stream or solvent. The equipment is drained of flush material, followed by steaming and a water flush. The purge and flush liquids including chemical materials from the exchanger are drained through hoses or piping to a closed system. Following the water purge, the water is drained from the vessel under inert gas or air pressure. The vapor zone is tested to ensure the vapor mixture is well below the lower explosive limits. The contaminant concentration is then analyzed and if the concentration is at or below the recommended workplace exposure level, the exchanger is opened. If the contaminant concentration is excessive, the exchanger remains closed and the purge/cleaning procedure is repeated until a satisfactory concentration is achieved. An alternative control procedure is described in another paragraph for a situation where the proper concentration is not attainable.
- Filters in process streams are infrequently opened to remove the baskets for cleaning. The steps in draining, flushing, and purging the filter are generally similar to those described for the exchanger. However, in a filter, the basket contains solid or polymeric materials, and operations should ensure that basket contents are not contaminated with a material that may result in worker exposure. The basket and its contents are drained during removal from the filter and the water-wetted basket is placed in a container for disposal of the basket contents.
- Although manual sampling control systems were described in a previous section, operating personnel must be trained and encouraged to use the systems as designed to minimize potential worker exposure. Additional controls may be instituted to ensure that a sampling box is ventilated before the sample box door is opened. However, all the sampling systems are not automatic and a continuing commitment to recommended operating procedures is necessary to control exposure.

Although standard purging and flushing procedures should minimize worker exposure when opening equipment for maintenance or repair, certain contaminants may present exposure problems due to their toxic characteristics and the very low exposure levels recommended to minimize potential health effects. Contaminant levels are measured following purge procedures. If the proper level cannot be achieved with the recommended procedures, portable ventilation systems should be positioned near the equipment to control worker exposure levels, but an improved flush/purge system is preferred. This may require a separate flushing chemical system, higher flush temperatures, or additives in the flush chemical to remove the contaminants in the equipment. An effective flushing system design may require a separate investigation before the optimum system can be recommended to the design engineer. Portable ventilation systems are required for entry into vessels where the concentration target cannot be achieved and in certain situations a combination of forced ventilation and personal protective equipment may be required.

Although maintenance and repair are continuous programs, spills are a potential source of worker exposure, albeit on an infrequent basis. Spills may ocur with some frequency in certain processes, and these spill sources should be identified. Improved engineering controls should be installed along with containment walls that prevent a spill from spreading. Work practices and training procedures to cope with spills must be instituted to ensure exposures are controlled. This requires a combination of specific work practices and personal protective equipment for the emergency spill situation, but the response must be tailored for the particular chemical spill. A general outline of spill response procedure is beyond the scope of this chapter, but spill containment, drainage to a closed system, adsorbent removal and disposal, and personal protective equipment are elements in successful worker exposure control systems.

Liquids and vapors were considered in reviewing maintenance and spill control requirements. However, solids in the form of dusts are potential worker exposure problems in various processes and in those processes where solids are added to the system in small quantities. Small bagged solid addition systems were discussed previously in this chapter and controls were described. However, dust accumulates both outside and within the bag handling equipment. Work practices should be instituted that require vacuum removal of the dust on a scheduled basis. Dust is also a problem in solids powder storage or solids transfer movement operations, where equipment must be replaced or maintained. Exposure is particularly critical during maintenance, and a combination of personal protective equipment and vacuum systems is necessary to control exposures. Specific work practices must be devised and instituted for these dust or powder handling systems.

These examples of work practice controls indicate that engineering controls for the prevention of overexposure are not the complete control solution. While engineering controls provide the necessary contaminant exposure control equipment to achieve worker protection, the cooperation and participation of workers, gained through training and supervision, is also necessary.

5.2 Monitoring

The following are essential characteristics of a successful engineering control solution to a potentially hazardous workplace exposure:

The level of protection afforded workers is adequate, consistent, and reliable.

The efficacy of protection for each individual worker is determinable during the operating life cycle of the system.

For closed continuous process systems, the continued efficacy of protection is established through the following four levels of monitoring.

5.2.1 Inspection of Control Equipment Integrity. The specific process equipment features that prevent releases or capture releases if they occur should be inspected periodically to ensure they are intact. As an example, if local exhaust hoods have been removed or disconnected, system integrity has been destroyed. Obviously, pump seals with large leaks or "blown" seals are also readily observed. Conducting operations to ensure that engineering controls continually operate properly requires a change in work practices that should be accompanied by general industrial hygiene training. Thus, nonmeasurement inspection activity should be part of the continuous general process equipment inspections required to operate a safe and efficient unit.

5.2.2 Leak Detection. Small leaks, particularly vapor leaks, are difficult to observe. Leak detection surveys conducted with instruments indicate the location of small leaks and provide a method of focusing maintenance attention on critical equipment. The EPA now requires monitoring of valves, pumps, compressors and the like, on a continuous basis under the Monitoring and Maintenance (M&M) Regulations to control fugitive emissions. This requirement aids in controlling worker exposure by indicating the location of large leaks. However, this ongoing M&M program must be coordinated with industrial hygiene needs to determine whether critical leaks exist, although the leakage may be considerably lower than the regulation action level. Moreover, very low leak concentrations for specific contaminants may be required in certain situations.

Sensitive, convenient, portable detection instruments are used to "sniff" pump seals, valve stems, flanges, safety valves, and so on. Several have been recommended by the EPA for the M&M programs now required. The EPA regulations describe a recommended method of locating and measuring leaks. This procedure provides a consistent measurement technique and a reliable method of comparing concentrations and categorizing leaks as major or minor.

5.2.3 Continuous Automatic Area Monitors. A continuous automatic area monitor is a sophisticated leak detection system used in certain high-risk units to continuously measure concentrations in the workplace at various locations. The output can be shown in a control room accompanied by a visual and/or audible alarm in the control room or on the unit. These continuous systems permit quick response where leaks must be quickly repaired. They also provide input data to computer systems that perform statistical analyses, prepare reports, and maintain archives.

Although systems of this type have operated well, installation of an automatic outdoor area monitoring system for a large continuous processing unit can be an expensive investment compared to that required for an automatic monitoring system located within a process building. The outdoor system must be an all-weather unit, carefully engineered to detect releases or increased leakage rates. The following are some considerations involved in designing or planning an installation.

- What are the basic objectives of this system? Will the system be designed to monitor areas for general concentration level changes, thereby indicating a leak somewhere in a broad area or monitor specific equipment items including seals to indicate a leaking seal? The monitoring system may also be designed for a combination of general area monitoring and specific task.
- The areas and/or seals requiring sampling must be defined. If an area monitoring system is recommended, then the specific areas where monitoring is needed should be defined. In the same way, individual seals or equipment items requiring release alarms should be indicated.
- Sampling locations to achieve the general and specific objectives must be determined. The prevailing wind direction will affect sampling point location, and

sufficient points should be provided to ensure an indication of release if the wind shifts. In addition, a sample point location should not cover a broad area or the sampling point may not indicate a release. If seals or equipment require sampling points, then a monitoring sampling installation is required that does not interfere with maintenance.

- The frequency of sampling from a particular point must be defined. With a large number of sampling points, a considerable time period may elapse between repeat analyses at one location. The industrial hygienist must determine the elapsed time required between samples at all of the locations assigned by the designers. If the maximum time period between sample analyses at one location is 60 sec, but sample readouts occur every 15 min at the critical location on a multisample analyzer, then a second analytical device may be required. In analyzing the time period, the distance from the sampling point to the analyzer must be established since the sample may be drawn through a lengthy line affecting the analytical time frame. Moreover, back purging long lines prior to obtaining a fresh sample also increases sample time analysis.

- Service factor may be critical. If the contaminant in the unit is toxic, a service factor greater than the unit service factor may be necessary for a monitor. This requires consideration of a spare analyzer to ensure a malfunction or repair will not interrupt sample analysis. Sparing must be thoroughly investigated and a philosophy established for sparing each equipment item in the system. In addition, a loss of power may require an emergency power backup system. Service factor and analytical needs establish emergency power requirements.

A thorough evaluation of an automatic continuous analyzer system must be undertaken to determine whether the system has application in a particular process. A system covering the entire unit may not be required, which would reduce the investment. However, the continuous system can be an effective method of detecting a release episode.

5.2.4 Exposure Measurement. The final test of the effectiveness of exposure controls is the measurement of worker exposure on the unit. The statistics and strategies involved in measuring worker exposure are covered elsewhere, but it is generally necessary to obtain multiple, long-period (full-shift) samples to accurately characterize this highly variable quantity in a manner that is biologically relevant.

5.3 Maintenance

A detailed review of maintenance is presented in Chapter 10. However, several aspects of maintenance specific to closed systems are presented in the following paragraphs.

5.3.1 Maintenance Philosophy. While all competently operated plants conduct some preventive maintenance, other plants generally repair equipment only following equipment failure or when a failure is imminent. For continuing operation of effective engineering controls in closed processes, a maintenance program must be established that repairs or eliminates leaks while the leak or emission rate is still relatively small. There is probably a trade-off of lower leakage versuse increased maintenance, but generalizations are difficult since a number of factors need evaluation such as the toxic materials involved, average workplace exposure levels, area concentrations, frequency of repair, and so on. However, better estimates of this trade-off will become available in the future with improved emission rate data and more sophisticated engineering controls. Moreover, continuous monitoring procedures are improving, and this automatic analysis technique will be an important factor in providing cost-effective maintenance programs.

5.3.2 Repairs on the Run. Some major repairs are undertaken while the unit is operating on spare equipment and bypass systems previously installed. Particular operating precautions are necessary where piping and vessels near the deactivated equipment are hot and under pressure. Detailed comments on equipment flushing and purging are presented in Section 4.

5.3.3 Turnarounds. Many unit repairs are not undertaken during the run but await a periodic "turnaround" when the entire unit is shutdown and opened. While units are extensively flushed and purged prior to opening vessels, drums, and the like, small pockets of toxic chemicals may remain in the system, and coke deposits frequently retain significant amounts of toxic materials. Proper shutdown and start-up operating procedures are beyond the scope of this chapter, but detailed advanced planning of the safety and health aspects of any turnaround is extremely important.

GENERAL REFERENCES

Air Pollution Engineering Manual, 2nd ed., EPA No. PB-225-132, May 1973.

Asbestos Substitute Gasket and Packing Materials Seminar, Valve Manufacturers Assn. Houston, TX, August 6/7, 1986.

Assessment of Atmospheric Emissions from Petroleum Refining, EPA PB80-225253, 225261, 225279, 103830 and 225287.

Bierl, A., A. Stoeckel, H. Kremer, and R. Sinn, Leakage Rates of Sealing Elements, *Chemie Ingenieur Technik* **49**(2), 89 (1977).

Bussenius, S., Escape of Combustible Gases and Vapors from Equipment and Their Spreading over the Operating Space, *Chem-Techn.* **38**(6), 264 (June 1986).

Control of Emission from Seals and Fittings in Chemical Process Industries, NIOSH Pub. No. 81–118, 1981.

Control Technology in the Plastics and Resin Industry, NIOSH Symposium Proceedings, 1979.

Control Techniques for Volatile Organic Emissions from Stationary Sources, EPA 450/1-78-022, May 1978.

Control of Workplace Hazards in the Chemical Manufacturing Industry, Joint CMA/NIOSH Symposium, Philadelphia, 1981.

Electric Power Research Institute Study Final Report, EPRI NP 02560 1982.

Emissions to the Atmosphere from Petroleum Refineries in Los Angeles Country, Final Report, Southern California Air District, 1958.

Emmel, T. E., Control Technology Assessment of Petroleum Refinery Operations, presented at NPRA, San Francisco, 1983.

International Conference on Fluid Sealings (11th), British Hydraulic Research Assn, Cannes, France, April, 1987.

Jones, A. L., S. Lipton, and J. Lynch, Critical Review of Fugitive Emissions Data, presented at the AIChE meeting in Orlando, February, 1982.

Lipton, S., and J. Lynch, *Health Hazard Control in the Chemical Process Industry,* New York: Wiley, 1987.

Lovelace, B. G., Design of Material Sampling Systems, *Chem. Eng. Progr.* **75**(51), 1979.

National Emission Standards for Hazardous Air Pollutants; Benzene Equipment Leaks (Fugitive Emission Sources), (EPA) *Fed. Register* **49**(110), p. 23,498, June 6, 1984.

W. O'Keefe, Packing and Seals for Valves and Pumps, *Power,* August, 1984.

Standards of Performance for New Stationary Sources: VOC Fugitive Emission Sources; Petroleum Refineries, (EPA) 40 CFR Part 60. *Fed. Register* **48**(2), Jan. 4, 1983.

Standards of Performance for New Stationary Sources; Synthetic Organic Chemical Manufacturing Industry; Equipment Leaks of VOC, Reference Methods 18 and 22, (EPA) 40 CFR Part 60, *Fed. Register* **48**(202), October 18, 1983.

Standards of Performance for New Stationary Sources; Equipment Leaks of VOC, Petroleum Refineries and Synthetic Organic Chemical Manufacturing Industry, (EPA) 40 CFR Part 60, *Fed. Register* **49**(105), p. 22,607, May 30, 1984.

U.S. EPA, Compilation of Air Pollutant Emission Factors, AP-42, Part B, 3rd ed., PB 84–199744, January, 1984.

VOC Fugitive Emissions in the Petroleum Refinery Industry—Background Information for Proposed Standards. EPA 450/3-81-015a, November 1982.

VOC Fugitive Emissions in Synthetic Organic Chemicals Manufacturing Industry—Background Information for Promulgated Standards, EPA 450/3-80-0036, June 1982.

CHAPTER FIVE

Selection and Arrangement of Process Equipment

Jack E. Peterson

1 INTRODUCTION

1.1 Definition of Process Equipment

Exclusion is obviously necessary to reduce the scope of this chapter to manageable proportions. Specifically excluded is equipment used for such important operations as digging, mining, metal- and woodworking, transporting (outside of plant confines), and so forth. And, even within the plant confines, equipment used for purposes other than processing is generally excluded. With a few exceptions the excluded items present little or no health hazard potential. For purposes of this chapter, then, process equipment is that mechanical powered equipment used in the basic metallurgical and chemical processing of materials in plants and factories.

1.2   Scope of Chapter

The goal of this chapter is to aid in the selection and arrangement of process equipment to minimize health hazards or make them easier to control. To that end, a rather general scheme is followed. First, the process itself is briefly described, concentrating on function.

Following that, the general kinds of equipment used are described; these generalizations are then used as the main outline headings for discussions that emphasize the advantages and limitations with regard to hazard control and describe control methods.

Finally, some attention is given to equipment location and arrangement with respect to other equipment, especially when those factors have an impact on potential health hazards.

The following checklist is intended as an aid to the engineer when considering the selection and arrangement of process equipment. The items have not been ranked according to importance.

- Have the toxicity and hazards of by-products as well as those of raw materials and products been considered?
- Anywhere in the process have people been relied on to accomplish repetitive and/or routine tasks that might better be done by a machine?
- Do any governmental regulations apply to this process and/or equipment? Consider especially gaseous, liquid, and solid wastes and the possible presence of highly toxic and/or carcinogenic substances therein.
- Does all of the equipment selected for this process comply with governmental regulations? Consider especially liquid waste treatment equipment and gas (air) cleaning equipment.
- Keep ergonomics in mind. Where equipment must be operated by people, even small changes can have large effects—either good or bad. Consider the type and placement of meters, gauges, valve and switch handles, as well as how much of a load a person must lift and carry.
- Remember maintenance. As nice as the equipment looks in the catalog or on the brochure, it will break down some day. Be sure to locate the equipment so it is easily accessible for both routine and nonroutine maintenance.
- Is there sufficient headroom not only over aisles and walkways but also under any equipment where people must work?
- Are moving equipment parts guarded well? If guards must be removed for maintenance, can they be replaced easily?
- Has enclosure been used as much as is practical wherever ventilation appears to be necessary? Enclose, then apply local exhaust at sufficient velocities to assure that air moves in the proper direction. Finally, be sure to supply at least enough air to make up for that exhausted.
- Has an assumption been made that knowledge about fluid flow matches that regarding air flow? When in doubt, consult an expert.
- Have the effects of convection air currents around hot pieces of equipment and/or process material been considered when specifying (or designing) ventilation equipment?

- Have the effects of air crosscurrents on local exhaust ventilation hoods been considered? Sources of such air currents include open doors and windows (if they can be opened, they will be in hot—or even warm—weather). What about pedestal and other person-cooler fans? Unit heaters? Lift truck traffic? Pedestrian traffic?
- What kind of emergency is most likely to occur? Can its likelihood and potential severity be reduced by a better choice of equipment or location of equipment?
- What kind of emergency is likely to be most disastrous? Can its likelihood and potential severity be reduced by a better choice of equipment or location of equipment?
- In the choice of process equipment, has adequate attention been given to the likely effects of corrosion? Erosion? Wear? Sedimentation? Agglomeration? Dusting? Foaming? Sticking?
- Have interlocks been used where they can prevent misoperation?
- Has the use of interlocks been avoided where their presence can result in misoperation?
- Has recirculation of gases from stacks back into the plant been considered? If the plant (process) will have one or more stacks discharging contaminated air or gases to the environment, inadvertent recirculation becomes a possibility. If the effects of such occurrences can be serious (even psychologically), consider prevention by cleaning the air (gas) stream and/or by proper location and elevation of the stack(s).
- Has the use of noise (sound power) specifications been considered as a means of minimizing problems of noisy equipment? As an offset for the perhaps higher price of quiet equipment, consider not only the cost of complying with the Occupational Safety and Health Administration (OSHA) regulations but also the cost of workers' compensation for hearing loss.
- Have the effects of vibrating equipment on noise production, on nearby apparatus, and on the building structure as well as on people been considered?
- Has radiant heat shielding been designed into the process equipment whenever that equipment will operate at temperatures appreciably above ambient?
- Has the use of air-conditioned control rooms been considered where heat stress may be a real problem when operating process equipment?
- Where solids are to be handled have the effects of high humidity and/or roof leaks been considered?

2 CRUSHERS AND GRINDERS

2.1 The Process

Particle size reduction is the main function of industrial crushers and grinders but not the only one. At times, these kinds of equipment are used also to mix and even to initiate and/or perpetuate chemical reactions.

Choice of the method to use for size reduction is dictated by properties of the material, initial size of the particles, final desired size and size range (however specified), and even the kind of equipment available. There are very few generalizations that apply mainly because useful grinding theory is very limited in scope and application.

All crushers and grinders are noisy because some of the relatively large amounts of energy applied to the particles appears as noise. In addition to noise from the act of breaking the particles and lumps, there is the noise of the machinery itself and of larger particles striking shatter plates or walls of a cavity. The whole piece of equipment may vibrate as may the structure in which it is located, thereby creating more noise as well as possible vibration hazards.

Noise control for crushers and grinders is not a well-developed science. For the most part, vibration isolation is used where vibration is a problem. Some size reduction equipment is available with built-in noise insulation and/or features that reduce the amount of sound generated. However, isolation within an enclosure is still by far the most widely used means of control. Usually the equipment is isolated, but when that is impractical, the operator can be isolated.

2.2 Crushers and Pulverizers

One generalization is that crushers are always large and their main purpose is size reduction of raw material such as rock and ore.

Only those kinds of machines in relatively common use will be discussed in any detail here. The hazards and their control are quite similar throughout.

The word *pulverize* is not associated with any particular kind of force or equipment; it simply refers to the general purpose of making a dust or powder from larger (but relatively small) particles. Nevertheless, "pulverizers" are sold and used throughout industry (see Chapter 24).

2.2.1 Gyratory Crushers. The main workhorse in rock and ore size reduction is the "mortar and pestle" of the industry, the gyratory crusher. This device consists of a usually very large stationary conical bowl lined with a very hard material and, central to the bowl, a pestle that is caused to gyrate. Rock is dumped in at the top where the opening is largest and emerges at the bottom. The amount of size reduction performed is determined by the slope of the bowl sides and, of course, the size of the bottom opening.

Gyratory crushers may create much dust, most of which is not well controlled. These machines tend to be located out of doors or in a shed that is mainly roof and use the general ventilation of wind movement for dust "control." Crushed material exits at the bottom, usually to a belt conveyor, and that transfer point may be controlled with local exhaust ventilation especially if the dust is hazardous or creates much of a nuisance. Only where the dust is quite hazardous—as with asbestos ores and some hard rock (containing crystalline silica)—are extensive dust control efforts made, in which case total enclosure may be necessary.

One useful crusher/grinder hood is illustrated in Fig. 5.1. The hood has a permanent base pan so that dust collection duct work need not be disconnected for maintenance. The upper portion of the hood is arranged in two or more completely removable sections with split lines as necessary to accommodate the motor shaft, feed and discharge spouts, and so forth. When maintenance is necessary, the removable sections are unclipped and the sections are removed, completely exposing the machine. This kind of hood arrangement requires sufficient headroom for moving the hood sections and, of course, a crane. Crusher enclosures, when used, should be exhausted at a rate sufficient to cause indrafts of at least 60 m/min (200 fpm) through all openings.

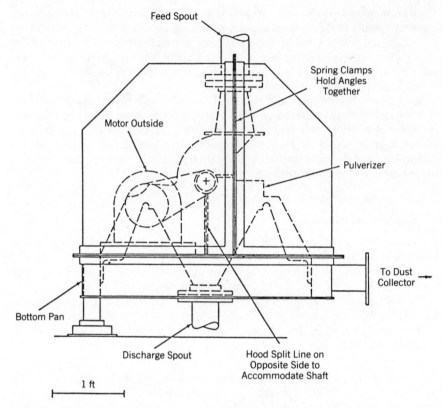

Figure 5.1. Pulverizer hood designed to be removed for servicing the pulverizer. (From *Uranium Production Technology*, Van Nostrand Reinhold, New York, 1959. By permission.)

The operator of a crusher may be stationed in an enclosure kept dust free by being pressurized with cleaned air. When practical, wet crushing may be used for partial dust control.

2.2.2 Jaw Crushers. Jaw crushers are used either as primary or secondary crushers, but most are found as primaries. Rock fed to them can range in size from about 0.3 to 1.5 m (12 to 60 in.), to be reduced in size by a factor of 3 to 5. Gyratory crushers are replacing jaw crushers in many applications because they have a larger throughput than jaw crushers for a similar size machine. Jaw crushers do not clog as easily as gyratory crushers, however, and in some cases are preferred.

Because they have a lower throughput than gyratory crushers, jaw crushers are not quite as dusty. Further, gyratory crushers clog easily if the material fed has a tendency to be cohesive when moist while jaw crushers do not; hence jaw crushers can be used more readily with moist or wet feed, thereby suppressing dust. Otherwise, problems and solutions are similar to those for gyratory crushers.

2.2.3 Heavy-Duty Impact Mills. Mill implies a finer particle size than *crusher* for both feed and product. Heavy-duty impact mills can be subclassified as rotor breakers and hammer crushers. A rotor breaker is a cylinder to which is fixed a tough metal (usually steel) bar lengthwise. Rotor breakers are very useful when the material handled is low in abrasion and shatters easily. Typically these machines are used for limestone and dolomite.

Hammer mills are similar to rotor breakers except the longitudinal bar is replaced with a few individual tough hammers, pivoted on a horizontal shaft.

Many other varieties of impact mills are available, each with its own areas of strength and weakness. There are reversible impactors, cage mills, ring-type granulators, and even autogenous crushers where the material being crushed is used to crush itself. Dust and noise hazards of all these machines are similar.

Rotor mills, hammer mills, and several others can be operated wet quite easily (suppressing noise as well as dust); they can clog if the water flow is reduced sufficiently. Many, however, are operated dry in which case much dust can be created that exits through the feed hopper, the discharge, and even through casing leaks at times (see Fig. 5.1). If the feed hopper is kept reasonably full, little dust escapes there, leaving the discharge as the most important control point. Here, as with other size reduction devices, local exhaust ventilation coupled with containment (an airtight enclosure to contain the draft created by action of the mill itself) can be used. The ventilation system must exhaust a volume equal to the "windage" created by the grinder, plus that caused by the falling material, plus that needed for the required indraft at openings—usually at least 60 m/min (200 fpm).

2.2.4 Roll Crushers. A roll crusher consists of two rolls positioned on a horizontal plane. Material is fed between them and is crushed as the rolls rotate toward one another. Each roll may be smooth, corrugated (ribbed), or toothed with the latter form being used most. One of the rolls rotates in fixed bearings; the other in movable bearings against strong springs.

Roll crushers wear far more rapidly than do gyratory and jaw crushers and for that reason are now rarely used as primary crushers for hard rock. However, this method of size reduction remains in use because it does a better job than other types on material that may be sticky. Its main limitation is that the compressive strength of the rock crushed should be similar to or less than that of limestone or dolomite.

Because they are less likely to clog with wet or damp material, roll crushers are more often used wet than other types of primary crushers. Water use tends to suppress dust and noise but may complicate further processing; when it does not, forming a slurry and pumping it may be the optimum means of transport between various processing devices (see Section 3.1).

2.2.5 Pan Crushers (Mullers). These crushers contain one or two (rarely more) heavy wheels that revolve in a circular container (the pan), or the pan may be rotated beneath the wheels. Resting on the bottom of the pan, the wheels are rotated by friction against that surface. The combination of rotating and revolving motion grinds and mixes material held in the pan.

This type of crusher is quite easy to enclose if necessary with a covering hood to contain the dust formed, as shown in Fig. 5.2. This and many other illustrations in Chapter 5 have been taken from the American Conference of Governmental Industrial Hygienists' *Industrial Ventilation: A Manual of Good Practice.*[1] There is very little noise associated with pan crushers because of the relative softness of the material handled and because size reduction takes place beneath a blanket of the material being crushed.

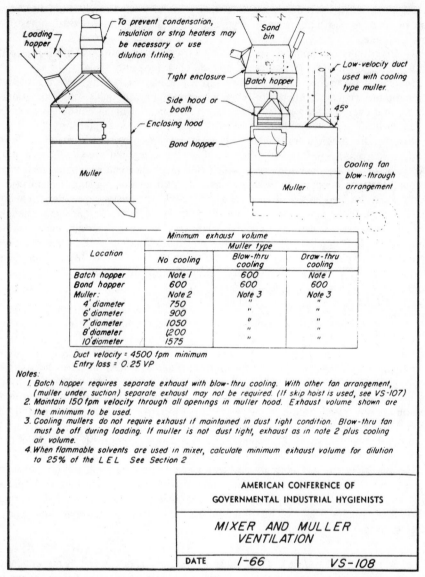

Figure 5.2. Pan grinder ventilation. In many mullers water is added to hot sand (and other material), creating a possible humidity problem in duct work. (From *American Conference of Governmental Industrial Hygienists,* Committee on Industrial Ventilation, P.O. Box 16153, Lansing, MI 48901. By permission.)

2.2.6 *Tumbling (or Media) Mills.* Inside a rotating cylinder (imagine a cement mixer) are the material to be crushed and balls, rollers, or pebbles, the "media" that do the crushing. Autogenous media mills are also found where the media is actually the material being crushed. This is the case where lead bars (or ingots) pulverize themselves in a rotating mill, oxidizing in the process to form lead oxide powder to be used in lead–acid battery manufacture. Otherwise the media may be made of almost any substance harder than the material fed to the mill.

Any of these mills can be operated wet or dry; of course, wet operation produces essentially no dust, but dust is not usually a severe problem even with dry operation. Because the mill is open only at the ends (usually only one end for a batch mill, both ends for a continuous-feed mill), local exhaust ventilation either at the opening(s) or applied to an enclosure can effect essentially complete dust control. See Fig. 5.3.

Although damping and/or sound-attenuating material can be applied to the outside surface of a rotating mill to reduce reverberation and noise levels, many of these mills are very noisy, indeed. This is particularly true when the milling is autogenous of metal parts, for instance. When the mill is used to produce a powder from granules, the noise level is usually not great because the powder formed acts as a cushion.

2.2.7 Nonrotary Ball or Bead Mills. Small balls, beads, or even sand can be used as the grinding media in a stirred or vibratory mill. Stirred (at up to 1500 rpm) mills are generally used wet on relatively soft materials, often with sand as the medium. Vibratory mills are often used dry on very hard materials with larger media (small balls rather than beads).

These mills are almost always secondary or even tertiary size reduction or polishing devices. Although the motor and any gears can be noisy, these mills rarely are hazardous from that standpoint, nor are they particularly dusty. If dust is a problem, it can usually be controlled with a perimeter hood at the mill opening.

2.3 Grinders

Grinders are smaller than crushers and produce smaller particles. They are usually secondary or tertiary in a chain of size-reducing equipment unless the material to be ground began as small particles. (See Chapter 24 for a discussion of some of this equipment.)

2.3.1 Dry Grinders. Almost any kind of grinder can be used for some materials. Criteria include lack of heat sensitivity and agglomeration tendency. Some materials cannot be ground wet; choosing the proper mill for them is sometimes troublesome. Many cereal and other vegetable products are in this category. Wheat, rye, and similar materials are usually ground into flour on roller mills. For some of these tasks hammer mills can be used; roll mills, disk attrition mills, and others have also been used.

When grinding vegetable products of any kind, one of the main hazards is dust explosion. This is especially true for starch and sugar, but fine dusts of any vegetable or organic origin should be regarded as explosive at some concentration range in air. Fortunately, the exposive range for many of these materials begins at a few grams per cubic meter of air, quite a high concentration. Even modest local exhaust ventilation can usually keep airborne concentrations much lower. If, in addition, sources of ignition (sparks, high temperatures, flames) are avoided, the possibility of an explosion is remote. (For additional information, see Volume 1, Chapter 28.)

General principles of exhaust ventilation apply here as elsewhere. First, dust sources should be enclosed by a solid structure as much as is feasible while still allowing access for operation and maintenance (see Figs. 5.1–5.3). Then, air should be exhausted either from that structure or from an external hood or enclosure at a rate sufficient to cause an indraft of 30–60 m/min (100–200 fpm) of clean air in such a manner that

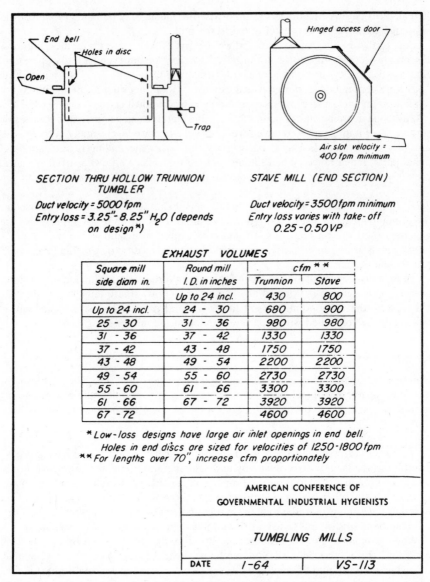

Figure 5.3. Total enclosure for a tumbling mill showing the use of either the hollow trunion or a more conventional exhaust point. (From *American Conference of Governmental Industrial Hygienists,* Committee on Industrial Ventilation, P.O. Box 16153, Lansing, MI 48901. By permission.)

dust escape is unlikely. Design of exhaust hoods for grinding equipment sometimes requires a great deal of ingenuity; the American Conference of Governmental Industrial Hygienists' *Industrial Ventilation Manual*[1] is one of the best places to find good ideas if not complete solutions. Another is the similar manual published by the National Institute for Occupational Safety and Health (NIOSH).[2]

2.3.2 Wet Grinders. When material can be ground wet, it probably should be. This usually eliminates any dust problem and even helps control noise. In addition, wet grinding is usually more efficient because grinding surfaces are kept cleaner, fines are washed away, and, if wet enough, particles have less tendency to stick together (if the mass is only damp, however, this tendency may be magnified instead of being reduced). Wet grinding also can be used for many heat-sensitive materials because the heat of attrition and friction is carried away by the grinding fluid. Finally, wet grinding almost completely eliminates the fire/explosion hazard so long as the grinding fluid is not itself flammable.

2.4 Equipment Location

Many crushing and grinding applications call for a vertical plant layout because passage of the material through equipment is by gravity. Only where the material is fine enough that work against gravity is only a small fraction of total power to the system are other arrangements likely to succeed.

Ore handling, for instance, typically proceeds from initial mining operations by truck to the primary grinder. Material passes down through that device possibly directly to the feed point of a secondary grinder or, more usually, to a belt conveyor and then to the secondary grinder. Exiting at the bottom of the secondary grinder, the material may be screened or otherwise sized or it may again be ground (crushed) before further treatment, with perhaps another ride on a conveyor. In many vertical installations, chutes and pipes are used for transfer of the material from one piece of equipment to another. Many of those pieces of equipment vibrate (grinders, screens, conveyors, feeders, weighers, etc.), and thus the chutes and pipes must have flexible connections. With greatly restricted headspace, the horizontal movement of the flexible connecter can be too large for the verticle length, and the connecter wears out rapidly. In addition to the sledge hammer for pounding on chutes, pipes, and equipment to dislodge blockages, the most-used gear in such a plant may well be duct tape. Duct tape seals do not wear well, either.

For the most part, equipment arrangement affects health hazards only when headspace becomes cramped or when the plant is designed to use only general (building-wide) ventilation through grating floors rather than local exhaust ventilation coupled with rather tight enclosures. In fact, such plants have been built with the general exhaust fans located at ground level because the dust was "heavier than air and would tend to fall." Dusty air actually travels with eddies and with general air currents because its "bulk" density is only very slightly different from that of clean air.

Using general ventilation has two main penalties. First, it forces workers to be exposed to the worst of the dust, to its highest concentrations, before that dusty air is discharged from the building. Second, it is much less efficient than local exhaust ventilation, and therefore, energy costs for moving and tempering the air are much higher than for the other method. Most always, general exhaust ventilation should be used sparingly or not at all for dust control.[3,4]

Vertical orientation of equipment can result in exposures to higher noise levels than those associated with a more horizontal layout. Vertical buildings tend to be crowded, not only because of insufficient headspace, but also because they tend to rise rather than spread out. Equipment requires floor space even in a high-rise building, and floor space is expensive, particularly when it is off the ground. When the tendency to crowd is avoided, however, a vertical arrangement of equipment can be as hazard free as a horizontal one.

3 CONVEYORS FOR BULK SOLIDS

3.1 The Process

Solid materials transferred from point to point range from relatively homogeneous substances such as limestone or grain to nonhomogeneous materials such as sugarcane, hides, and bone. Finished products are also conveyed as bulk solids. Movement of packaged materials by conveyor is excluded because that process rarely has any associated health hazards.

Bulk solid materials may be moved either continuously or discontinuously; particle sizes may range from a few tenths of a micrometer to a few meters; the amounts transferred may range from a few grams to thousands of kilograms per hour. These parameters plus properties of the material govern the choice of equipment for hazard control as well as for process needs.

Methods of moving solids will be discussed under categories of open or closed containers (discontinuous); and continuous mechanical, pneumatic, or hybrid systems. Because they are not usually associated with any distinct processing, overhead "conveyors" such as cranes and roadway and railroad equipment are omitted from this discussion.

The main health problems associated with the transport of bulk solids are dust and noise. Some kinds of conveyors are much quieter than others, but none is completely silent. Similarly, conveyors vary widely in their ability to produce dust, and none is inherently dust free. If dust is or will be a severe problem because of its effects on either equipment or people, then a technique to consider is to slurry the material in water (or other liquid) and pump the slurry rather than convey the solid. Where it can be used, this technique can eliminate the hazards of dust and result in much reduced maintenance and repair.

3.2 Discontinuous Conveyors

Open- or closed-container systems are used not only for transferring materials but also for temporary storage; in fact, the storage function or batch process flexibility may well dictate their choice. Container movement/storage is appropriate for relatively low production rates. Filling, handling, and emptying such vessels can present many sources of hazardous conditions. Bins, tubs, drums, bags, paks, wheelbarrows, and buckets are probably subject to more physical abuse than any other conveying system because they are at most only one step removed from manual handling. Costs of repair and upkeep of these containers as well as the costs of hazard control must be included in the total cost of such a system when comparisons are made with other systems.

3.2.1 Open Containers. Open transfer containers are frequently used in the first stages of the solids handling process. Open tubs and bins may be used for almost anything from small castings to phosphate ore on its way to a crusher but are rational only where contamination from external sources is not important. Historically, their use has been popular where transfers to and from the container must be made, whether by hand or with mechanical assistance.

When open containers are used, skin and eye contact with material from within the containers can be expected, and there almost always is ample opportunity for vapors and small particles to become airborne (see Sections 4 and 11). Open transfer containers are thus often a poor choice when the transferred material or a contaminant of the material (sand on castings, for instance) can be hazardous by any route of contact. Often the first attempt at reducing losses from open containers either to avoid attendant costs or to control a hazard is to close the containers—or switch to closed containers without changing much else.

3.2.2 Closed Containers. Typically, closed containers used for material transfer are exemplified by lidded garbage cans, fiber paks, or drums. A plastic bag (usually polyethylene or polyvinyl chloride) that can be "sealed" one way or another may be used. Only when such a system is outgrown is the closed dumpster or tote bin likely to be used instead.

Used carefully, closed containers can prevent eye and skin contact with transferred material almost completely and can significantly reduce the amount of material becoming airborne during movement of the container. The operative phase is "used carefully." Closed containers are subject to the same kinds of physical abuse as are open ones and quite often do not maintain seals for more than a few round trips. They are usually treated as if they were sealed, unfortunately, and consequently can actually become more hazardous and more prone to spillage than open containers.

One of the most prevalent problems of bags is breakage and subsequent spread of dust throughout a plant or warehouse. Bags are usually transported on wooden pallets. When a pallet load of bags (or other small containers) is wrapped in plastic stretch or shrink film, the problem of dust dispersion can be reduced markedly at little extra expense. This method protects bags not only against breakage but also against the weather during out-of-plant storage and/or transfer.

As material is added to a closed container, air is displaced unless the container is evacuated prior to filling. The displaced air emerges from the container laden with dust or vapor that becomes airborne in the process of loading. The great advantage of a closed over an open container here is that the airstream, if it is emitted to the plant atmosphere, is now a "point" rather than a "large area" source of dust and thus is subject to much easier control with exhaust ventilation.

Closed containers have another problem when used as direct replacements for open containers. They are much more difficult to fill and empty. This is especially important when either job is done manually and may lead to a complete lack of acceptance of the new device (in an evolving process) by workers. For this reason alone, the best replacement for open containers in many processes may be some sort of continuous system rather than a simple substitution of closed containers for open ones.

Where the material collected in the container is quite hazardous, is a by-product or scrap rather than a product, and especially where the quantities are small, closed transfer containers may be ideal. Consider lead and lead oxide scrap from a lead–acid battery manufacturing operation. Here all conditions are fulfilled. Collecting the scrap in small (to keep weight manageable) drums equipped with exhaust ventilation (see Section 11) minimizes air contamination at that point. If the drums are fitted with tight deformable covers (perhaps plastic "shower caps"), transfer to a collecting area by lift truck can take place with little or no spillage and very small amounts of airborne material.

3.3 Continuous Conveyors

Continuous mechanical conveyors are exemplified by the conveyor belt or bucket elevator. Pneumatic conveyors operate under either negative or positive pressure (or perhaps even a combination of the two). Especially for mechanical conveyors, classification of type may be on the basis of whether the solids movement is vertical or horizontal, but the direction of movement is immaterial unless a hazard is created or increased by equipment layout.

3.3.1 Mechanical Conveyors. Equipment used for mechanical conveying of bulk solids may be belts; chains or cables attached to buckets, platforms, slats, aprons, or pans; rollers; chain trolleys; draglines; redlers; screws; vibrators; chutes, spiral chutes, and roller chutes; and perhaps many more. Further, each of the main varieties may have several subvarieties developed to solve specific problems. The solutions, of course, may cause one or more further problems. Only the most important of the mechanical conveyor types will be considered here.

BELT CONVEYORS. An endless belt is looped over a driving (head) pulley and an idler (tail) pulley at the other end of its run. Material is carried on the top surface of the belt that, in turn, is supported by many idler rollers or even a continuous platform beneath the upper side and fewer idlers or another platform beneath the return (bottom) side and thus in contact with the carrying side (and subject, perhaps, to abrasion and fouling).

This kind of conveyor can discharge its contents over the end of the tail pulley and does so in most fixed installations. However, when more than one discharge point is necessary along its length, a mechanical "tripper" will be used, causing an abrupt rise and/or tilt of the belt and discharge of its contents. Belts can be made endless by clipping ends together with metal fasteners or by chemical welding and/or thermal action (vulcanizing) when they are constructed of a polymeric material. Metal clips wear more rapidly than welds and also tend to snag on and tear off at belt scrapers/cleaners. Vulcanizing, on the other hand, produces a belt that wears well and is easier to clean/scrape, but, when the belt breaks, repair can take up to 12 hr compared to a few minutes to an hour or so for clips.

Belts may be made of almost any material. Tough polymers such as rubber, Neoprene, polyvinyl chloride, butyl rubber, polyurethanes, and even fluorinated plastics tend to predominate. Whatever the polymer, fibrous reinforcement is the usual practice; the fibers can be of cotton, steel, asbestos, and the like. The only health hazards associated with conveyor belts themselves are those of pyrolysis products emitted if the belt is overheated or burned. Conveyor belts are expensive and designed for years and perhaps decades of useful life; hazardous emissions from the belt material are very rare, probably not peculiar to the belt, and impossible to design out.

If the belt is designed to change vertical direction, that direction change is usually affected by a snub roller that, as with return idlers, is in contact with the conveying surface. Snub rollers follow discharge of material from the belt and usually are protected by some sort of belt cleaner ahead of their position. Nevertheless, snub rollers still tend to pick up material from the belt surface and thus may be sources of airborne dust.

Even well-designed belt conveyor systems tend to create dust from dry materials not only at the load (feed) and discharge or transfer points, but also, perhaps, at each of the idler rollers. This dust is difficult to control without spending relatively large amounts of money on "local" (the length of the belt plus feed and discharge points) exhaust ventilation (see, for instance, Volume 1, Fig. 44.3). Even with unventilated or ventilated belt cleaners, scalpers, and scrapers, the return strand tends to dribble dust. One control is to install a catch pan beneath the return strand, providing one more piece of equipment that must be cleaned regularly and that rarely is. A "solution" to this kind of problem is to enclose the conveyor belt system entirely in a ventilated structure. This, of course, may make maintenance quite difficult and generally prevents or impedes the frequent casual inspection of the belt, idlers, and drive mechanism.

Load and discharge points for belt conveyors must be designed and constructed very carefully to provide for long life with little maintenance and also to control dusting at those points. A natural tendency is simply to construct a canopy hood over transfer points in the hope that all dust will rise to and be captured by the hood. Even when the conveyed material is hot foundry sand, the thermal air currents in a vertical direction are easily deflected by almost any horizontal or eddy air movement, and more dust is added to air outside the exhaust system. Figure 5.4 shows some of the salient features of properly designed and ventilated conveyor belt load and discharge (transfer) points.

Noise from belt conveyors is usually associated with the motor and idler rollers. Usually this noise is not sufficiently intense to be a hazard to hearing even though it may be annoying. Proper mounting of the motor to avoid vibration along with proper design of idlers and good maintenance of both will often solve or avert a noise problem. When the belt slides on a solid metal surface, changing that surface to lubricant-impregnated wood has reduced noise considerably with no reduction in useful life of either the belt or the supporting surface.

SCREW CONVEYORS. A screw conveyor is simply a helicoid inside a shell or trough. Almost any kind of material can be moved with a screw conveyor, but because of rather high power requirements and susceptibility to abrasion, runs are usually short (perhaps to 25 m or so) and either horizontal or slightly inclined, although a screw with a double-flight short pitch can be used vertically.

Because the shaft rotational velocity in a screw conveyor is low (usually 40–75 rpm), there is little tendency for dust to arise even from an open trough, and the main dusting problem is usually spillage at load points and perhaps even several discharge points. If dusting is a problem, the conveyor can almost always be totally enclosed; little or no ventilation is required.

VIBRATING CONVEYORS. A vibrating or oscillating conveyor is a relatively short (perhaps up to 100 m long) pan moved in such a way as to impart forward motion to material on its surface. Although most vibratory conveyors are horizontal, they can be operated successfully on a slope. At least one commercially available degreaser is based on a spiral vibrating conveyor that conveys metal parts from the bottom (perhaps even in the degreasing fluid) up through the vapor zone to discharge them at the top down a chute. Movement is slow enough on a vibrating conveyor (usually a few meters per minute) so that the parts emerge dry from the degreaser.

These conveyors tend to be noisy and dusty, not only at load and discharge points, but also along their length. If the material handled has any tendency to dust and if that dust may be troublesome, then some sort of exhaust ventilation will be necessary. Hoods similar to those in Fig. 5.4 can be used successfully but, because of the relatively slow material movement speed on vibrators, air volumes/velocities can be reduced significantly (50–75%) from those required for belt conveyors. As with belts, vibrating conveyors can be totally enclosed if necessary for dust control but usually a side-draft or modified canopy hood is used instead, allowing easy access to operating machinery.

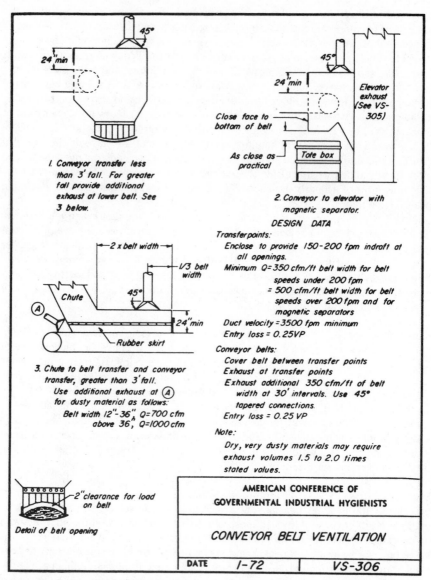

Figure 5.4. Local exhaust ventilation of conveyor belt loading, discharge, and transfer points. (From *American Conference of Governmental Industrial Hygienists,* Committee on Industrial Ventilation, P.O. Box 16153, Lansing, MI 48901. By permission.)

As with any exhaust hoods, increasing the extent of enclosure reduces the amount of air that must be moved for dust control. Exhausted air volumes should be sufficient to cause indrafts at all openings of 45–60 m/min (150–200 fpm).

Similarly, enclosure is used for noise control when necessary; sound/vibration damping material on either surface of the vibrating trough will usually reduce noise levels, but this approach is rarely practical. The usual means of controlling noise from vibrating conveyors is simply to use just about any other type if that is at all practical. No other bulk solids conveyor is so noisy.

BUCKET ELEVATORS AND SKIP HOISTS. Most skip hoists use open containers, although openness is not a criterion. A skip hoist may be used in addition to or instead of a bucket elevator to raise material from one level to another, higher one. The main hazards of skip hoists are the dust and noise that may accompany either filling or dumping or both. Because each process is intermittent, unless the hazard is great, it is usually ignored (however, see Sections 4 and 11).

For movement of parts and pieces as well as powdery, granular, or lumpy material from one level to a higher one, the bucket elevator is almost the universal choice when more or less continuous delivery is required.

Discharge may be by centrifugal motion or by a positive method wherein buckets may be more than inverted. Either kind may be filled by scooping material from a bin at the bottom of the elevator or by a feed spout either above or below the tail (bottom) wheel. Further, bucket elevators may be used on an incline rather than absolutely vertical. Each alteration, of course, affects capacity as well as cost.

Bucket elevators create perhaps more dust than any other conveying system moving the same amount of material. For this reason and to enhance safety, they are usually enclosed totally or almost totally, and the enclosure is ventilated as shown in Fig. 5.5.

OTHER MECHANICAL CONVEYORS. Conveyor types other than those already mentioned rarely cause dusting (or noise) problems except at loading and discharge points, which can be handled in a manner similar to that illustrated in Figs. 5.4 and 5.5. Continuous-flow conveyors are, by nature, totally enclosed whether used horizontally or vertically. The unique closed-belt conveyor with its zipper is essentially a closed tube throughout its length and is particularly suitable for materials that are fragile. The belt is snugged up to the material moved so that the conveyed material is not subjected to any internal movement. Power requirements are low, and this method is very versatile if somewhat expensive. Apron conveyors consist of modified buckets attached to chains. In this case, the "bucket" flanges are shaped so that they overlap flanges on preceding or following buckets to form a more or less continuous surface.

3.3.2 Pneumatic Conveyors. Ventilation engineers have long known that when dust-laden air is taken through duct work, if the air velocity is sufficiently high, there will be little or no tendency for the dust to settle in the ducts. Experience has shown, for instance, that the air velocity must be at least 1100 m/min (about 3600 fpm) in duct work serving dust control ventilation systems. In one sense, pneumatic conveying can be regarded as simply an extension of that idea.

Pneumatic conveying systems use air that must be moved with a compressor or blower. The blower may be quite noisy; in fact, it may well be the main source of noise where materials are conveyed. The usual optimum solution to this problem is to locate the blower far enough away and/or in a silenced enclosure so that noise levels in the work area associated with its operation are low.

A somewhat different way of using air to aid in the conveying of bulk solids is to fluidize the solid during the conveying step so that material flows as a fluid rather than a solid. This is done by conveyance in a double-compartment line with pressurized air in one compartment and the bulk solid in the other. The compartments are sep-

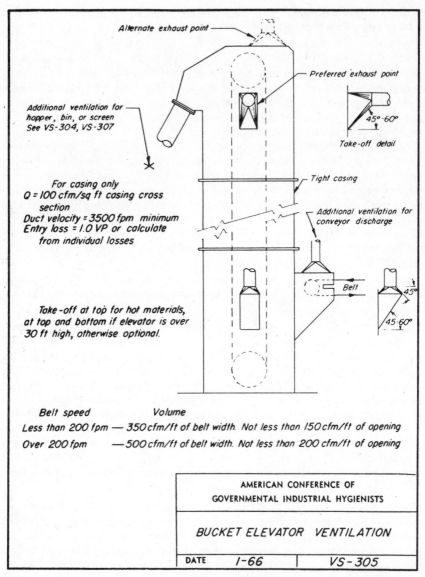

Alternate exhaust point

Preferred exhaust point

Additional ventilation for
hopper, bin, or screen
See VS-304, VS-307

Take-off detail
45°-60°

Tight casing

For casing only
Q = 100 cfm/sq ft casing cross
section
Duct velocity = 3500 fpm minimum
Entry loss = 1.0 VP or calculate
from individual losses

Additional ventilation for
conveyor discharge

Belt

45°

45°-60°

Take-off at top for hot materials,
at top and bottom if elevator is over
30 ft high, otherwise optional.

Belt speed   Volume
Less than 200 fpm — 350 cfm/ft of belt width. Not less than 150 cfm/ft of opening
Over 200 fpm   — 500 cfm/ft of belt width. Not less than 200 cfm/ft of opening

| AMERICAN CONFERENCE OF |
| GOVERNMENTAL INDUSTRIAL HYGIENISTS |
| BUCKET ELEVATOR VENTILATION |

| DATE | 1-66 | VS-305 |

Figure 5.5. Typical exhaust points for a bucket elevator (From *American Conference of Governmental Industrial Hygienists,* Committee on Industrial Ventilation, P.O. Box 16153, Lansing, MI 48901. By permission.)

arated by a perforated or porous partition; air bleeding through that partition fluidizes the solid in the other compartment.

Whether the conveying system is classified as pressure, vacuum, combination pressure–vacuum, fluidizing, or blow tank, potential inhalation problems are limited to the discharge point and to any leaks that may develop in a pressure system. Leaks are usually obvious and repaired immediately because they represent loss of material and/or power. And, because air exiting from a pneumatic system will be laden with every bit of dust it can contain, provision is almost always made to remove that dust before the air is discharged to the plant or outdoor environment. Figure 55.3, Volume 1, illustrates a process wherein an air-lift (a variety of fluidizing) conveyor is used as an integral part of a process to help effect size separation of particles.

3.4 Equipment Location

Horizontal layout of equipment and belt conveying of solids go hand in hand. Belt conveying is probably the least expensive means of moving materials from place to place within plant confines, but is associated with the most (and therefore most expensive) real estate.

Bucket elevators are, of course, associated with vertical structures and vertical plant layout. In general, the flow of material is usually from ground level to the top via bucket elevator and then by gravity through ducts, chutes, screens, blenders, dryers, and the like back to the ground level. This arrangement uses but little land, substituting the more expensive bucket elevator for the belt conveyor. However, gravity feed of bulk solids is far from being trouble free in most circumstances. One of the most ubiquitous tools to be found in many a vertical solids handling plant is the sledge hammer, used to pound on duct work and equipment to dislodge held-up solids. Not even well-built ducts, chutes, and equipment will maintain their integrity for long when they are pounded regularly with sledge hammers. Leaks with their real or potential hazards develop. Leak control may eventually cost far more than would have been spent for land and a more horizontal plant layout.

The foremost bane of free flow for solids is moisture. Many materials that flow very readily when bone dry tend to cake and resist flowing when even slightly damp. Those materials rarely cause serious trouble because that tendency is easily recognized. More troublesome are those substances that cake only when truly damp. They cause no trouble at all until the roof leaks or the weather turns unusually cold and most of the plant is outdoors, or the heating system fails; in any case, moisture (condensed or otherwise) permeates the system and disaster results. A portion of the problem may well be excessive exposure of the maintenance people to the material(s) handled by the conveyor. Past experience and pilot testing of flowability under very rigorous conditions are true prerequisites for plant design where free flow of solids is necessary.

In a vertical plant, and especially in one with open or grating floors, even a small leak in a roof can result in wetting down most duct work and, of course, the most moisture-sensitive machinery. This is far less of a hazard in a horizontal layout, but in that case there is more roof to leak and probably a greater tendency to leak because large roofs appear to develop more leaks per unit of area than small ones. Dust spilled from chutes, joints, and operating equipment can filter down from floor to floor through grillwork causing inhalation as well as skin contact and eye exposures. Using solid floors or a covering over grating floors will minimize these problems in a vertical plant.

Especially in plants where drying is done or water is given off inadvertently, duct work will at times be subject to high humidity. Even substances that show no signs of caking at a "normal" 80% relative humidity may well cake at 95 or 100%. This causes problems not only in processing equipment but also in exhaust system duct work. Under such circumstances, the only real solution appears to be to install water-flushing equipment within the duct work so that caked material can be removed easily.

Flow of solids is almost always enhanced as particle size increases. Therefore, either specifying larger particles, grinding to a larger size, or even pelletizing or prilling may solve a solids flow problem. Because moisture is such a prevalent problem in solids flow, installation of one or more dryers (see Section 12 of this chapter) should be considered in the design phase if such a problem can be anticipated. One means of maintaining dryness is a dried air or other gas blanket throughout sensitive portions of the system. This "solution" may be much more difficult to use over time than it is to plan, however, because of the tendency of all equipment to develop leaks with age.

High temperature can interfere with solids flow even when moisture is not a problem. When the material handled is itself temperature-sensitive (many low-melting compounds, for instance), it may become very sticky at only slight increases above normal operating temperatures. In such cases, cooling the process stream may well be necessary. Too much cooling may cause moisture condensation, however, with its problems. If temperature sensitivity of the material handled is known or suspected, pilot testing of conveying equipment under anticipated extreme conditions is suggested before purchase and/or installation. This is especially necessary when the material is very toxic or corrosive.

As mentioned in Section 3.1, an alternate to drying or cooling may be to slurry the solid material and pump the slurry, thus avoiding the problems. This solution should always be considered when a slurry is or can be used in further processing.

4 DUMPING SOLIDS

4.1 The Process

Material can be removed from a container a little at a time or all at once. The first of these processes requires another piece of equipment, from the hand used to get powder from a sack to a scoop or shovel used with bags and other containers, to a slide valve and perhaps a screw conveyor. When the process requires the container itself to be upended for removal of the material, the process is dumping. This usually, but not always, refers to complete emptying of a container.

Because problems differ according to the kind of container used, bags, barrels and fiber paks, and large containers are considered separately in this section. Damp or nondusty powder rarely poses inhalation problems although skin and/or eye contact still may occur and be troublesome. Dumping dusty solids, then, is the problem to be considered (see also Section 11.4).

Whenever material is added to a closed container, air is displaced. Therefore, one of the principles of dumping solids into a closed container is to remove air from that container a little faster than the dumped solid enters (Sections 3.2.2 and 7.2). In this manner air is drawn into the container along with the solid material (the same principle applies to vapor displacement when liquid is added), and no dust escapes to the general atmosphere. This principle always should be applied if practical, no matter what kind of container is being dumped.

When dumping a dusty solid, the general idea of dust control with local exhaust ventilation is to apply just enough ventilation so that excessive amounts of dust do not reach the breathing zone of the person doing the job. If too much air is withdrawn from the hood or if the takeoff point is located improperly, much of the solid may be picked up by the "vacuum cleaner" effect of the ventilation and thus not enter the process stream as intended. This admonition applies whether the air is withdrawn from the container into which the material is being dumped or to an external hood. Only rarely is an air velocity greater than 30 m/min (100 fpm) necessary close to the point of greatest dust release.

4.2 Dumping from Bags

Even though a part of a bag can be dumped, the errors involved are rarely acceptable. Instead, extensive efforts are made to utilize bags containing exactly the amount needed for a process or some exact submultiple. Thus each bag can be emptied completely as soon as it is opened. This not only avoids the problem of trying to dump exactly half a bag, for instance, but that of having a half-bag of material lying around to be spilled or contaminated. For a small price premium, bags of material in several sizes can often be obtained from the supplier, allowing almost any amount to be dumped and leaving only empty bags at the completion of dumping. Of course, if the supplier is not trustworthy and containers (or contents) must be reweighed, then many advantages of this method vanish.

One of the main problems with bags is that they can break or tear in storage. Bags are more fragile than other industrially used containers for solids and, in general, the less expensive the bag the easier it is to tear. If broken bags are a problem, then better bags, more gentle handling methods including that of wrapping pallet loads of bags in plastic shrink or stretch film, or changing to another container type are alternatives that should be considered.

When stored bags break, they contaminate not only the warehouse floor and other structural members, but also other stored bags. They may leak dust all over when being transferred from one spot to another by pallet/lift truck or other means. If material within any stored bags is more hazardous than a nuisance dust, then protecting bags, barrels, fiber paks, and the like in the warehouse area from spilled dust is practical and may be absolutely necessary. Usually a tarpaulin of cloth or plastic film will suffice if it is handled carefully in removal and is always replaced. Otherwise, more elaborate methods such as use of compartments with closed doors may be necessary.

Dumping a bag actually involves three separate actions: opening the bag, dumping it, and discarding the empty bag. Each will be considered separately.

4.2.1 Opening the Bag. If the top or bottom seam of a bag is sewn, a rip thread is often provided that with a simple action the entire end of the bag is opened completely. Finding the rip thread seems to be a problem with many people, however, and often bags furnished with such an opener are cut open instead. This is unfortunate because cut bags rarely empty completely; rip-thread-opened bags usually come closer to that ideal than others. Thus, one way of avoiding problems with material that remains in "emptied" bags is to purchase the material in rip-seam bags and insist that the bags be opened and dumped properly. Of course, even rip-seam "dustless" plastic (or plastic film lined) bags are not truly dustless and may cause problems if the solid is highly toxic and/or irritating.

When the material in the bag is hygroscopic or otherwise sensitive to air or moisture, sewn seams are rarely adequate because they are not vapor proof. Glued seams are cheaper and therefore tend to predominate even when the bags contain material insensitive to atmospheric effects. Methods for handling such bags safely are therefore necessary and have been developed.

When material within the bag is not particularly toxic or irritating, cutting the bag with a knife is the usual method of opening it. The hooked lineolum knife and the razor-blade knife are used most. A knife in which the cutting edge is easily shrouded or hidden is preferred for safety reasons.

When the whole bag is to be dumped, the best way is to make two knife cuts on a side at right angles—side to side and end to end. When the bag is then inverted over a grating, it will empty "completely" most of the time if it is shaken a little. A cut across one end, however, always leaves a pocket to catch material; that pocket is very difficult to empty completely. Bag-dumping hoods with internal bag breakers are available. The usual bag breaker either has several large sharp spikes that puncture and tear or an inverted V sharpened on its edges against which the bag is thrown. Neither device works well with all materials. Even with their use, bags rarely empty completely unless prodded and shaken by hand or by mechanical agitation of the breaker platform/grating and perhaps not even then.

4.2.2 Dumping the Bag. A properly constructed bag-dumping hood will have a grating to prevent passage of the bag but allow passage of its contents into a chute or pipe. There will be an air takeoff point within the hood, often improperly placed just above the grating where a maximum amount of the material dumped will be collected. Better designed bag-dumping hoods have internal baffles similar to those in laboratory hoods, arranged so that about half of the air is removed at the bottom of the baffle (several centimeters above the grating) and half at the top; attachment of the takeoff can then be anywhere behind the baffle. Air flow rate through the hood should be sufficient so that the average velocity at the open face is on the order of 30 m/min (100 fpm) assuming no appreciable external air disturbance; the hood should be as small as practical and yet allow easy handling of a bag without spillage outside the hood. The bottom rear baffle opening should be, as indicated, above the height of a horizontal bag and the hood should be deep enough so that this opening is several centimeters from the bag as it dumps.

If a within-hood bag breaker is to be used, the general configuration of the hood face opening should be horizontal with the grating a few centimeters (inches) below the bottom lip of the hood. If rip-seam bags are used, general configuration of the hood should be vertical with the grating perhaps 25–30 cm (10–12 in.) below the bottom lip of the hood and the opening about as high as the bag is long. Baffle openings should be at the top of the hood and at about the level of the front lip; that is, 25–30 cm above the grating.

4.2.3 Discarding the Bag. Opened bags are almost always a problem. They have little or no value and are very seldom reclaimable; they are waste and tend to be treated as such. And yet, most opened bags have some residue that if toxic or irritating may actually cause more problems than are found at any other point in this process.

Often, emptied bags are simply tossed into a waste container—a trash bin or dumpster, perhaps. When they are actually tossed, the trip through the air may cause an appreciable amount of their remaining contents to become airborne. When they are carefully placed in the waste container, pushing down the stack may accomplish much the same but closer to someone's breathing zone. There are several possible solutions to this problem, none easy to apply.

First, if empty bag contents are actually troublesome the best solution is probably to obtain the material in containers other than bags or, alternatively, in other than powder or crystal form. Barrels and fiber paks are easier than bags to empty completely (see Section 4.3); they do not collapse to expel dust-laden air. Empty barrels and fiber paks have appreciable value; they are easy to recover and reuse. They are treated more carefully than are empty bags. Larger containers share these advantages and have other advantages. And, of course, lumps are less dusty than powders; damp material is less dusty than dry; and liquid solutions or slurries are not dusty at all.

Second, if bags must be used, a part of the solution to the dust-from-empties problem is to assure that empty bags contain very little residue. The best means is to use rip-seam bags properly. Bags opened in any other way will probably continue to be troublesome.

Third, if practical, the full or emptied bags can be wetted (this technique is now often used for asbestos). When bags are emptied into water, a spray inside a dumping hood can assure that the dusting problem will be reduced considerably in its severity. Wet bags are messy to handle, however, and this method actually can trade a house-keeping problem for less air contamination.

Fourth and least desirable is to discard emptied bags carefully into a ventilated container. This "solution" may function best as an interim stopgap rather than a way of actually solving the problem; it has been applied with success only where supervision is close enough to assure that bags will be placed carefully within the ventilated container. "Administrative" controls that require close supervision rarely succeed for great lengths of time.

Personal protective devices and/or clothing are almost never a true solution to bag dumping problems. Even if the person doing the dumping is well protected (as often must be the case), nearby people rarely are. Further, airborne dust does not confine itself to the immediate vicinity of a bag-dumping station; it tends to spread and cause surface contamination and housekeeping problems even when the inhalation hazard is small.

4.3 Dumping from Barrels and Fiber Paks

To a large extent barrels have been replaced with fiber paks that are now often called barrels. These are available in many sizes and in several shapes for the same volume. For this reason, dumping facilities usually must be custom designed.

If the material to be dumped poses dusting problems, the best solution is usually to change the form of the material, if that is practical, from a powder or small crystal to something else. Materials that can be prilled successfully are much less dusty than powders. Lumps are also less dusty than powders although they do not dissolve as rapidly. Large crystals cause fewer problems than do small ones and may dissolve almost as readily. Finally, solutions and slurries cost more to ship but do not cause dust when they are added to a container.

At times, a dusting problem can be solved by changing a chemical's structure slightly. An anhydrous material, for instance, may be available only as a powder or small crystal while a hydrate or acid salt may be available as large crystals or chunks or vice versa. Elimination of a problem by substitution can be far more successful than any other solution.

The first principle of dust control when dumping dusty solids from fiber paks or barrels is the same as that for dumping bags; draw air along with the dusty solid into the container being filled. When such a "flow-along" system is used well, a good pak-dumping hood may be nothing more than a floor-level grating. With that kind of ventilation a pak can be emptied completely and then agitated a little to remove the last of the solid without causing an inhalation exposure to the worker. The emptied fiber pak is then turned upright and its cover is replaced.

When withdrawing air from the container (or chute or pipe) being filled is impractical, a dumping hood may be necessary. This hood should be sized to be larger than the fiber pak by a few centimeters in height and width and as much as 25–30 cm (10–12 in.) deeper than necessary to contain the pak completely. As with a vertical bag-dumping hood, a pak-dumping hood should have an internal baffle forming a plenum at its rear. The top baffle opening should be at the top of the hood; the bottom one as much as 25–30 cm above the grating through which the dumped solid flows. Baffle openings should allow about half of the exhaust air to enter at top and bottom; the takeoff can be anywhere within the plenum. Air velocity at the hood face should be sufficient to contain dust: 30–60 m/min (100–200 fpm) with a pak in place within the hood.

Large fiber paks are heavy, and mechanical assistance may be necessary for dumping them without undue worker stress. In most cases a mechanical dumper can be accommodated within or by a hood if the hood is designed properly for that purpose. Such facilities can automatically remove lids/covers and invert the container over the chute, all within a ventilated enclosure. Metal barrels treated in this manner have been sent directly to a barrel washer, emerging clean. Adding a mechanical dumper where a hood already exists, however, can be much more difficult. In any case, the basic principle of hood design applies—enclose the operation as much as practical.

4.4 Dumping from Large Containers

Where large containers from "tote" bins to boxcars are dumped, the facility must always be engineered for that purpose. If a dusting problem is anticipated, substitution (see the preceding two subsections) may be the best solution. When that is impossible or impractical, control by enclosure and proper application of ventilation will be necessary.

For large containers as well as small ones, the best way to control is to draw air into the system along with the solid being dumped. If air flow is adequate, no dust will escape. When the containers are large and seldom dumped, an arrangement whereby the exhaust system operates at full air flow only while a container is dumped can effect a large saving. When this is done, sufficient lag time must be built in to allow the blower to reach full speed and/or a damper to open wide before dumping proceeds.

If the dumping creates much dust, an air cleaning system will be needed in the exhaust air duct work before the blower (see Section 13). Material removed from the dust collector can then be added to the system.

Exhaust ventilation other than the flow-through or flow-along concurrent system just described will probably offer inadequate control or will be quite inefficient or both.

Tote bins and other large containers can be emptied without dumping through slide valves, belt or screw conveyors, and the like. If dumping may create a dust hazard, then another means of emptying should be considered even though one may thus trade a noise hazard, from thumpers and vibrators used to assist solids flow, for the dust hazard.

4.5   Equipment Location

There are many advantages to using gravity to assist in the flow of solids through a system, although in some cases those advantages may be more illusory than real. Nevertheless, vertical layout can, does, and probably should prevail for much solids handling.

Small containers such as bags and fiber paks can be dumped at any convenient location within a building. However, the nature of the process dictates that a receiver of some sort be beneath the dumping station, which, therefore, is usually not a ground floor. Either the receiver or the dumping station or both may be ventilated. Because the best location for exhaust blowers is usually on rooftops, receiving tanks and such are probably best located rather high in a vertical structure to avoid long duct runs, and dumping stations are located even higher.

Large containers are much more difficult and expensive to lift than small ones and this, alone, tends to keep their dumping stations near the ground. Furthermore, such facilities must be properly engineered and not added as afterthoughts; dumping is often at ground level into a receiving pit or tank, and a bucket elevator may be used to elevate the dumped solid if necessary (see Section 3.3.1).

5   BULK SOLIDS FEEDERS

5.1   The Process

A bulk solids feeder is a device to assist flow. Assistance may be necessary for a host of reasons from need for a very constant flow rate to insurance against blockage. Many solids flow freely under almost all real process conditions. For those materials, use of properly designed mass-flow bins may well obviate the need for an auxiliary feeder of any kind. For other materials that do not flow well under all possible plant conditions, that are stored in funnel-flow bins, that must be weighed, or must have a constant flow rate, feeders are desirable and perhaps necessary.

5.2   Feeders and Vibrators

Not all types of feeders can be discussed here. The most popular are gate feeders, screw feeders, belt or apron feeders, table feeders, vibrating hoppers and feeders, and star feeders. They are usually fed from bins or hoppers. Where possible, hoppers and bins should be designed for mass flow rather than funnel or cavity flow (where material flows only from directly above the outlet, creating a funnel-shaped cavity within the bulk material). In either case, flow is at times assisted by bin vibrators attached to the equipment walls. Vibrators can materially assist flow of bulk solids and are much better than blows from sledge hammers or other objects. They are noisy, however, with nearby noise levels at times exceeding 115 dBA. Hearing protection may be required even for brief exposures.

5.2.1   Gate Feeders.   Gate feeders are similar in action to gate valves in piping or blast gates in duct work. Most are simply a piece of sheet metal held in two slots, one on either side of the gate. Free-flowing solid material in a mass-flow bin can be fed at a predictable rate simply by varying the size of the opening in the hopper with a sliding gate. This simplest of feeders is often used in conjunction with a screw- or belt-type feeder as the first line of control. Unfortunately, gate feeders tend to stick and bind and therefore are much better for on/off than for flow rate control.

Gate feeders tend to accumulate powders and dusts and therefore can pose a skin contact problem. Since they are usually located under a bin or other piece of storage equipment, they may also, because of their dribbling tendencies, pose inhalation and eye contact hazards.

5.2.2 Screw Feeders. A screw feeder is a screw conveyor connected more or less directly to a source of the material to be fed, such as a hopper or bin. Usually, the hopper opening is above and along the length of the screw. To withdraw material from the hopper uniformly along the length of the opening, the screw must have a variable pitch, the helicoid being tighter near its start.

As the trough in which the screw operates can be covered along its entire length, any dusting tendency of the material fed can be suppressed by confinement. Screw feeders result in a very uniform steady flow of solid material. Especially when the material fed is reasonably uniform in size and consistency, screw feeders are usually quite trouble free.

5.2.3 Belt or Apron Feeders. The moving surface of a conveyor belt or an apron conveyor passes beneath an elongated slot opening at the bottom of a bin or hopper. For best uniform flow, the outlet opening may be tapered in both vertical and horizontal directions. Speed of the apron or belt is relatively low, but with very dusty or powdery materials some particles may become airborne. This is rarely a problem; it is usually solved not with ventilation or enclosure but by substituting a screw feeder.

The main problem with belt and apron feeders is nonuniform flow of material onto the feeder. This necessitates manual intervention and attendant skin, eye, and perhaps inhalation exposures. Any acute problems are usually solved or sidestepped through use of personal protective equipment.

5.2.4 Table Feeders. A circular flat table is rotated beneath the usually round opening of a conical hopper. Flow of material from the hopper to the table may be facilitated by means of increasing the distance between the table and hopper skirt in the direction of table travel. Material on the table is scraped off with a stationary plow.

Plowing action and resulting fall of the fed material may emit dust into the air if the material has that tendency. And, as this method of feeding works best with fine free-flowing materials, dusting may well be a problem. Here, local exhaust ventilation may be used for control because almost all dust is generated in the vicinity of the plow.

5.2.5 Vibratory Feeders. Hopper construction for use with vibratory feeders is quite similar to that used with belt and apron feeders. A tapered slot aids in the uniformity of flow. However, vibratory feeders have a gyratory motion. At the forward end of the slot the feeder is completely laden with material, and upward movement of the feeder may cause packing if there is any tendency to do so. That problem is usually solved by changing the feeder type. Until a solution is found, however, worker exposures to the material fed may be frequent and relatively severe as clogs are cleared.

Feeders that vibrate tend to emit dust into the air if the material handled is at all dusty. Because feed slots may be quite long, local exhaust ventilation is usually not chosen for control; instead, another type of feeder is used. The same consideration applies when noise must be controlled.

5.2.6 Star Feeders. The star feeder (or valve) is one of the more positive means of emptying material from a hopper. When vanes (usually four or six) are welded equidistantly to a cylindrical shaft along its length a "star" is formed, the length of which may range up to several times its "diameter." The star is rotated in a trough beneath a slot under the hopper. Each of the cavities formed by the vanes is filled successively and then discharged at the dropped side of the trough in a matter similar to that of a water wheel.

Because of their positive action in forcing material from a hopper, star feeders are often successful for substances that are too difficult to feed by any other method. Coupled with a screw conveyor (feeder) alongside the dropped side of the trough, the combination will result in very uniform feeding of many materials. Since the whole system can easily be sealed to the atmosphere, the only hazards are those associated with maintenance.

5.3 Equipment Location

Unless the feeder is a major source of dust (and it should not be), its location is almost immaterial insofar as health hazards are concerned. However, feeders of all types tend to be found beneath hoppers and bins where quarters are almost always cramped and where dusts tend to accumulate unless special provisions are made to remove them. Depending on toxicity and hazards of the dust, this, in turn, may be a severe problem for maintenance and even janitorial personnel.

Prevention being almost always the best means of control, when a problem associated with feeder location can be anticipated, it should be prevented. When the problem will be with the material handled, then specifying an enclosed screw feeder with or without a star valve may be the best means of control.

6 PUMPS AND TANKS

6.1 The Process

Tanks and pumps are available in a bewildering variety so that classification for purposes of advising on equipment selection and arrangement must, of necessity, be broad. Tanks and pumps are grouped into three classes each. For tanks, open, closed, and floating roof; for pumps, centrifugal, positive displacement, and propeller and turbine varieties.

6.2 Tanks

Tanks are used to hold or store liquids and, occasionally, wet solids. Other than the classification scheme already used, they can be subclassified as atmospheric or pressure. Atmospheric tanks have internal pressures (aside from hydrostatic pressure) at most a few pounds per square inch (psi) above atmospheric pressure; pressures greater than that call for pressure tanks. Open tanks are, by definition, atmospheric tanks, while closed tanks can be classified either way. Floating roof tanks or gas holders are a subvariety intermediate between the other two.

6.2.1 Open Tanks. Open tanks are constructed of many different materials; most common are steel, wood, glass-fiber reinforced polyester resin, and concrete. Choice of construction material has little or no influence on hazard unless the choice allows either reaction of the tank material with its contents or some other cause of rapid deterioration. If such has occurred or appears likely, then lining the tank with a nonreactive material may be an elegant solution to the problem. Tanks have been lined successfully with acid-resistant brick, plastic (polyolefin) film/sheet, asphaltic or other cement, and even more unlikely materials.

Open tanks are used to contain, store, and dispense liquids that are not harmed by contact with the atmosphere and/or its constituents, including precipitation and pollutants if out of doors. As with reactors, the liquid most often found in open tanks is water or a water solution of some sort, and the chief hazard is probably that of drowning unless the pH of the water differs from the range of 7 ± 2, in which case skin or eye contact may have serious consequences.

6.2.2 Closed Tanks. With a roof, an open atmospheric tank becomes a closed atmospheric tank. And, because the tank is closed and the liquid level within it will certainly vary, the tank must be equipped with a vent. Atmospheric tanks usually vent to the atmosphere; pressure tanks may well not. "Pumping" caused by temperature and liquid volume change may cause excessive loss of vapor through an atmospheric vent; this, in turn, may be the cause of air pollution or a hazard to those in the vicinity. Mainly because of air pollution problems, much attention has recently been given to tank vents.

When a pressure or fixed-roof tank is being filled, air saturated with vapor is displaced. One way of preventing that air from reaching the atmosphere is to vent it back to the tank being emptied. This solution is becoming almost standard for petroleum products.

Another method of preventing excessive atmospheric venting is the use of a vapor trap. The first such traps were constructed by manifolding all vents from tanks handling similar fluids and then discharging that vent to a holding tank, perhaps one with a floating roof. This system can work well especially when there are several tanks on the manifold system. When that is not practical, either absorption, adsorption, or cold trapping can be used to remove much vapor from the air before it is vented to the atmosphere. The choice of method depends on many factors; potential hazards are reduced to about the same extent by all.

Closed tanks are emptied as well as filled and thus are usually equipped with vacuum breakers to limit the amount of external pressure they must withstand. If vacuum breakers are not used, the tank must be designed and built with external (usually air) pressure in mind.

At one time or another, and possibly often, tanks must be cleaned. As with reactors (see Section 7.2.2), this is a case of entering usually closed equipment with its attendant hazards during cleaning and/or inspection. Where the tank or its contents may contain or have been contaminated with arsenic or antimony, entry may be especially hazardous. The hydrides of those substances can be formed easily upon contact of the material with even slightly acidic water. Most overexposures to arsine and/or stibine occur during tank cleaning and such overexposures are often fatal.

Leakage from storage tanks is not usually a severe problem and is only rarely a hazard to workers. Whether the tank is indoors or out, however, if leakage or overflow is possible (and it usually is), provisions should be made to handle that problem. One method is to provide each above-ground tank with its own diked area to contain large spills or overflows. Other methods may include liquid level and/or pressure sensing with alarms, remote pump shutdown, and so forth.

6.2.3 Variable-Volume Tanks. The ability to devise and construct a good sliding seal between roof and tank walls allowed the development of "floating-roof" tanks where the roof is actually supported by gas (vapor) pressure; as the pressure increases or decreases, the roof rises or falls. Under normal circumstances, a floating-roof tank discharges no vapor to the atmosphere and thus is ideal for storage of many liquids having moderate vapor pressures. Unless further protected by a fixed roof, the floating roof must be equipped with one or more drains for rain runoff to avoid overpressures.

A variant that avoids the problems inherent in the sliding-seal approach is the use of an impermeable but flexible membrane to connect the movable roof to walls. Finally, even a rigid tank can have a variable volume if it contains an internal expansion chamber constructed of some impermeable but flexible material.

Variable-volume tanks of any variety pose no new or unusual hazards to workers or to the public.

6.3 Pumps

There are probably more centrifugal pumps in use than all other types combined, but there are still jobs for which this type is not particularly suited. Nevertheless, if a centrifugal pump can be used, it probably should be. Other types in common use are positive-displacement pumps and propeller or vane-axial pumps, turbine pumps, and various kinds of ejectors. For special purposes (particularly for moving liquid metal), direct application of electromotive force to the fluid can exert a pumping action. (See Chapter 4 for additional information.)

6.3.1 Centrifugal Pumps. Characteristics of all centrifugal pumps include efficient performance under a wide variety of conditions, essentially no capacity limitations, and almost no pulsations in the fluid delivered. These pumps are not best suited to applications where the fluid viscosity is very high, the necessary discharge pressure is very high, or where the fluid is laden with erosive solids, although special designs can enable use even under such conditions.

Centrifugal pumps have rotating shafts; keeping the fluid being moved inside the pump has posed many problems that now have been solved almost completely. Where the fluid handled is very toxic, packing- or stuffing-box seals are almost never adequate because these seals must leak in order to function. Lubrication at the rotating seal is provided not by the seal itself, but by the fluid being pumped; because this is the case, some of that fluid must penetrate the seal. A local exhaust hood built around the stuffing box can reduce problems associated with leaking seals, but as these seals require more maintenance than other types, such hoods are only rarely a complete solution to this problem.

Mechanical seals for centrifugal pumps have become increasingly popular. Although their initial cost is much greater than that of conventional packing glands, well-designed and applied mechanical seals do not leak liquid (although they may leak vapor) and require much less maintenance than stuffing boxes. Mechanical seals, perhaps fitted with movable local exhaust hoods, are often the best way of handling leaks of toxic liquids that happen also to be good solvents.

A relatively recent development in the fight against pump leakage has been canned-motor pumps. These are close-coupled designs wherein both pump and motor occupy the same sealed shell. The liquid being pumped provides lubrication for the motor bearings, and no seals are required. These pumps are relatively expensive but almost maintenance free so long as the process fluid remains completely clear of solids. Canned-motor pumps do not leak as long as the shell (the "can") is intact.

Real and potential hazards of the fluid being handled should always be considerations when a pump is being chosen. Often overlooked are hazards faced by maintenance people that may be far more severe than those the operating crew encounter.

6.3.2 Positive-Displacement Pumps. When "head" (discharge pressure) requirements are severe, positive-displacement pumps are usually considered because they have essentially no head limitations. These piston, plunger, diaphragm, and rotary pumps are even more efficient than centrifugal pumps but are much more inflexible in their applications. One major disadvantage is that pressure and flow volume both pulsate unless the pulsations are efficiently damped. Another disadvantage is that the fluid pumped must be clean.

Both piston and plunger pumps have mechanical sliding seals that can and eventually do leak. Diaphragm pumps have no moving seals other than the diaphragm itself, and therefore, generally do not develop small troublesome leaks unless the diaphragm is incompatible with the process fluid. However, at some point in service the constantly flexing diaphragm must fail, in which case the leak that occurs will be a major one.

Rotary pumps come in two varieties, namely, gear and screw pumps. Neither type is suited to handling dirty liquids, and the screw pump, especially, best handles liquids that happen to be reasonably good lubricants. Both types have moving seals that are subject to leakage, although with screw pumps this is usually only a minor problem.

Another positive displacement pump variety is the fluid displacement pump called an "acid egg." This is simply a sealed container that actually may be shaped like an egg. Acid or other liquid that may be laden with solids is added to it, usually by gravity from another container. The line is sealed, and then air or another fluid not miscible with the liquid being moved is pumped into the egg, forcing the liquid out through a dip tube. Although this pump is quite inefficient and not suited to many applications, it rarely has leakage or even maintenance problems.

6.3.3 Propeller and Turbine Pumps. Pumps that act on liquids in much the same way airplane propellers do on air are called propeller or vane-axial pumps. Although this kind of a flow-through pump is not capable of high discharge pressure, it can move more fluid for its size than any other type. Because of this characteristic, a vane-axial pump is often used to provide recirculation of fluid through the pipe in which it is located at an elbow. The drive shaft must extend through the pipe wall and hence a moving seal is necessary; vane-axial pumps do leak, and this must be taken into account when their use is considered.

Turbine pumps are the kind found immersed in wells. They are usually mounted vertically and are a cross between the centrifugal and vane-axial types. Because of how they are used, leaks are rarely important. These pumps are used only on clean liquids and are capable of discharge pressures higher than those usual for centrifugal pumps. (See Chapter 4 for additional information.)

6.4 Equipment Location

Pumps and tanks go together. Most tanks are filled and emptied using pumps (usually centrifugal); hence their juxtaposition in this chapter. If a tank contains a liquid or gas that may be harmful, the associated pumps will share that hazard.

Because storage tanks filled with liquid are heavy, they tend to be located close to if not on the ground floor of buildings or else out of doors. However, if gravity flow of a liquid can be used, locating a feed tank high in the building makes sense even if the pump filling it is elsewhere. That "elsewhere" is very likely to be on the bottom floor of the building so that intake and discharge lines to and from the pump can be kept filled with liquid at all times.

Pumps are capable of developing very high discharge pressures but are quite limited in what they can accomplish on the intake side. As pressure on liquids is reduced, they tend to boil and "cavitate" inside pipelines and, worse, inside pumps. A pump that works very well with a liquid may not pump a vapor or a gassy liquid at all well. Commonly, then, pumps are located beneath feed tanks. If the fluid being pumped can be hazardous and if maintenance on the pump will be necessary (almost a forgone conclusion), special care should be taken not to locate pumps directly under tanks in a way that restricts headroom severely.

In those cases where several toxic fluids will be handled in a process, or one such fluid in several vessels, the best approach may be to locate all tanks and associated pumps in one area. Then, if necessary, leakage spots such as stuffing boxes on pumps, valves, and agitators can be equipped with local exhaust hoods. If the whole area is enclosed, local exhaust ventilation for the leakage spots becomes general exhaust ventilation for the equipment group. Although perhaps not always necessary, the general ventilation can be very welcome when pump maintenance is needed. The small penalty paid for longer pipelines to other parts of the process usually can be offset completely by the efficiency of the exhaust system or systems applied in this manner.

7 REACTORS

7.1 The Process

Reaction between one or more components of a system can occur in the gaseous, liquid, or even solid phase. Vessels designed to promote and contain reactions thus can be quite varied in design. Occasionally, reactions are spontaneous and occur upon mixing; more often, addition of heat and/or catalyst is necessary to either initiate or to speed the reaction. And, although some reactions are endothermic or isothermic, most are exothermic. A primary requirement of a reactor, then, is usually control of heat.

Sophisticated control devices may modulate flow of one or more reacting materials based on how close some desired parameter is to a set point. This is called proportional control and is the way many people seem to think a home furnace thermostat operates, or should operate. Control can also be based on a summation of deviations from a set point over the recent past (integral control) or on the rate of deviation from a set point (rate or slope control). Because of the importance of heat to reaction (a rule of thumb states that the reaction rate in organic systems doubles for each 10°C rise in temperature), the set point in reacting systems is often a temperature.

Reaction in industry can be classified as "batch" or "continuous." Because reaction vessels (reactors) vary tremendously in size and design, these factors cannot be discussed here in any detail. Instead, this section will focus on general principles of hazard control associated with reaction.

The first step in hazard analysis is to determine the exact nature of all materials in the reaction system. Next, the reaction itself is analyzed to determine what products may result. This part of the analysis should include detailed consideration of by-products and waste materials that may prove to be considerably more hazardous than either reactants or desired products. Determination of the chief source(s) of hazard will often aid materially in selecting proper reaction equipment.

7.2   Discontinuous Reactors

Almost all processes that depend on chemical reaction begin their evolution in discontinuous reactors. Furthermore, most processes begin with amounts of material much too small to be handled on a continuous basis; only growth allows development of continuous-reaction equipment.

Discontinuous or batch reactors can be either open to the atmosphere or closed. If closed, they can be operated near atmospheric pressure, under a vacuum (unusual), or sometimes under greatly increased pressure. Because the hazards of high pressure are distinct from those associated with moderate pressure, they are treated separately here.

7.2.1 Open Vessels. Open reactors are commonly used when the suspending fluid or reaction medium is water but rarely for other fluids. The reaction medium has several purposes including dilution of reactants and/or products, but one main purpose is usually temperature control. As most commercially important reactions produce or require addition of heat, a medium for heat transfer is either necessary or desirable. Because it is stable at all temperatures usually encountered, inexpensive, and readily available, water is the preferred heat transfer fluid. Thus, the material that evolves from an open reactor in the largest quantity is very likely to be water vapor or steam.

The chief hazard of an open reactor may well be that of safety. There may be, of course, an associated thermal hazard if the temperature of the fluid is well above 37°C (98.6°F). In addition, evolution of water vapor and associated surface events may cause emission of other components of the reacting mass that may pose problems for people in the vicinity. The usual means of controlling inhalation hazards associated with open reactors is simply to move the reactors outdoors and/or to close them. Even though air movement and dilution are usually greater outdoors than inside a building, that is not always the case; this "solution" of the problem rarely succeeds. Local exhaust ventilation for a large open reactor is expensive, difficult to install, and usually only partially effective. Closing the vessel, on the other hand, offers an opportunity for essentially complete control of airborne substances but also a few attendant disadvantages, the main one usually being that the reaction can no longer be observed directly with ease.

7.2.2 Closed Vessels. The main reason for closing a reaction vessel may be to keep unwanted materials out rather than to assure that desired substances are retained. Open vessels collect dust from the air, debris such as cigarette packages and lunch wrappings, and perhaps even less-desired substances at times. Closed tanks and reactors are much less likely to do so.

Another reason for closing a reactor is to allow more precise control of temperature. Closed reactors have been the subject of much experimentation and therefore may provide better stirring and mixing than open vessels. Further down on the list of advantages, but still very important, is that of hazard control.

One of the problems of adding any liquid or solid to a closed vessel is that the addition, itself, causes displacement of air from within the vessel unless the vessel is evacuated first, an unusual process step. Because the reactor is closed, air exits through any available openings, even the one used for the addition, and thus may pose a severe hazard to anyone in the vicinity of that opening (if the reactor is large enough, the main opening is usually called a manhole). The usual means of controlling this hazard is to construct and install some kind of external hood around or close to the opening to capture air leaving the vessel. Such hoods are often quite ineffective.

A better means of control is to exhaust air from the reactor through another opening so that air is drawn in through the manhole along with any material being added there. So long as the indraft is large enough (30–60 m/min, or 100–200 fpm), there will be no puffs of dust- or vapor-laden air to control. The same principle applies when material is added through a pipe or chute. If there must be an opening (for direct observation inside the reactor if for no other reason), the reactor should be directly exhausted during all additions of material.

Often a closed reactor is a designed vessel having a very limited number of access ports. The usual means of exhausting the reactor then is through the vacuum system that was designed for purposes other than handling relatively large volumes of air laden, perhaps, with vapor or, worse, dust. If the vacuum system must be used, some means of cleaning the air it handles may be necessary (see Section 13). In addition, because the volume of air drawn through the reactor by the vacuum system may not be completely adequate for control of airborne substances, a local exhaust hood at the main opening may be necessary. This should almost always be supplemental, however, and not the main means of control.

Closed reaction vessels have several kinds of seals. Such reactors are usually stirred and thus have a packing box through which the agitator shaft passes. Each pipe that enters the reactor usually does so at a flange; the manhole will have a lid or cover held to the vessel at a flange by several clamps or other fasteners. Those seals most likely to leak are the ones around rotating agitator shafts and, because they are often disturbed, at manholes. The flanges where pipes enter are much less likely to have major leaks. If vapor that may leak from the reactor can be hazardous to workers, close attention to seals is necessary, in which case temporary or permanent local exhaust ventilation is recommended at the points of most likely leakage.

There is no one exhaust hood design best for this purpose. As usual, general principles should be applied. Essentially total enclosure at the rotating shaft seal box is usually possible and even practical. Sufficient air flow should be provided so that flow through openings is in the right direction and, more important, that the largest anticipated leak can be handled. At the manhole, a specially designed hood is usually necessary because of equipment-related restrictions. The best hood is probably one that encircles the opening completely with a "slot," but that ideal can rarely be achieved even though it can be approached. This is not the area for a canopy hood of any variety; side draft of some sort is needed.

Entry of normally closed vessels poses hazards unlike most others. Reactors are usually emptied and cleaned thoroughly before being entered for manual cleaning and/or inspection. However, because of the difficulty in actually observing the interior of these vessels and having to rely on gauges and meters, procedures for entry must be developed that minimize danger.

The chief hazard is that of asphyxiation. Closed reactors may be purged with an inert gas such as argon, carbon dioxide, nitrogen, or a mixture of carbon dioxide and nitrogen produced by combustion. In some cases, bacterial action can deplete the oxygen concentration almost completely. If the oxygen-deficient gas is not replaced with air before entry, a tragedy can result. The only practical means of avoidance is to assume that all normally closed vessels are filled with an inert, toxic, or even potentially explosive gas until proven otherwise by tests independent of gauges and equipment used for normal operation. Even then, the normally closed vessel should be entered only by a person wearing a rescue harness and with at least one observer outside the vessel. If the reactor may ever be entered, the manhole should be designed so that rescue is possible (see Chapter 4 for additional information).

7.2.3 Pressure Vessels. *Pressure vessel* has many meanings and connotations. The term always implies a container that is completely free of leaks at the pressure of use. A reactor may be considered to be a pressure vessel even though the actual pressure used is modest. Although subject to many legal codes, the term *pressure vessel* has no legal definition. In practice, it is a container (not a pipe, for instance) subject to an internal pressure of a few pounds per square inch (psi) above ambient, "a few" being as little, perhaps, as 1 or 2 (approximately 7000 to 14,000 N/m^2). By that definition, most closed reactors qualify and are therefore subject to code requirements.

When pressures inside reactors may exceed some much higher limit, perhaps several hundred to several thousand psi, procedures become much more complex. Such high pressures may call for operating the reactor behind a barrier of reinforced concrete using mirrors and/or television for viewing. In these cases, the necessary remote operation may pose hazards of its own, leading to an increased likelihood of an explosion (in these cases, perhaps merely a large leak) or release of relatively large amounts of reactants. Operating procedures must be designed with that in mind; hazards are not limited to those of shrapnel or shock.

Almost all very high pressure equipment is designed for the job at hand following applicable codes and accepted engineering practice. There is very little chance for arbitrary selection.

7.3 Continuous Reactors

Reactors used in continuous processing are not normally available "off the shelf." They are almost always designed for a particular process based on processing conditions preestablished by laboratory or pilot plant work. Selection of continuous reactors in such a way as to avoid hazard thus is limited to assuring that the designers have available accurate data relating to the process and to the materials handled.

Man has never designed a failure-proof piece of equipment, even though the probability of failure may be quite low for some. For this reason, even continuous reactors should be designed in such a way that any necessary maintenance is relatively easy to accomplish with a minimum of hazard to those involved. Possible and potential failure modes should be analyzed thoroughly and means devised to minimize harmful effects should failures occur. In many modern corporations this process is a formal one that requires approval from the industrial hygiene staff at several stages from idea conception to final design. No better way of avoiding hazards has been devised.

7.4 Equipment Location

Where gravity can be used to assist the flow of liquids and solids to and from reactors, vertical orientation of the equipment is usually chosen. Because even relatively high agitation rates do not normally cause severe equipment vibration, leaks at flanges and other seals are usually slow to develop and relatively easy to repair when discovered. If leaks cannot be tolerated, welded piping and flanges can often be used. Large fluctuations in the temperature of flowing fluids can, of course, have adverse effects on piping unless such effects are considered properly in the design phase.

Because liquids and gases are so easily transferred from point to point, considerations other than hazard avoidance properly should dictate reaction and associated equipment location in most cases. Where local exhaust hoods will be used, installing them relatively close to the roof will avoid long vertical duct runs because in almost all cases the proper location for exhaust blowers is on the roof (see Section 13.3).

8 MIXERS AND BLENDERS

8.1 The Process

Mixing is a generic term referring to all combinations of the three states of matter (gas, liquid, solid), while *blending* is not used for the combining process if a gas is involved; otherwise the terms are synonyms.

Gases diffuse rapidly and only rarely is there need for special equipment for their mixing. In the same manner, mutually soluble liquids mix readily, and mixers are not usually designed for them. Mixers are designed to provide for mixing and contact between mutually insoluble liquids, between liquids and solids, and between solids. In this section, mixers for liquids, viscous materials, and solids are considered. (For additional discussion of mixers and blenders see Chapter 21.)

8.2 Mixers for Liquids

Mutually insoluble liquids are mixed for one or more of four purposes: chemical reaction, creation of permanent emulsions, separation of components in solution, and for heat transfer by direct contact. The mixing process can be either continuous or batch, giving rise to flow mixers and agitated vessel mixers.

8.2.1 Flow Mixers. The term *flow mixer* is used when at least one of the components being blended is flowing, although both may be. Almost all flow (or line) mixers are characterized by small contact times, and equipment is usually small compared to that used for batch mixing. The mixers used may be jets, injectors, orifices, valves, pumps, packed tubes, or even pipelines in turbulent flow. Only pump-agitated line mixers have moving parts at the point of contact.

The only health hazards associated with flow mixers are those occasioned by maintenance and leaks in packing glands. Leaks are important when one of the materials is both volatile and hazardous in low vapor concentrations, in which case selection of a mixer having no moving parts is recommended.

8.2.2 Agitated Vessel Mixers. Most agitated vessel mixing is a batch process, although continuous mixing is certainly possible and practical. This process is characterized by literally hundreds of designs involving a tank (usuallly baffled) and one or more mechanical agitators. Agitated vessel mixers are used for dispersing and dissolving solids in liquids as well as for liquid–liquid mixing. When the process is used for mass transfer from one phase to another, the mixing vessel may be equipped with a settling section, separated by a baffle from the mixing section, and an arrangement for decantation.

The process of mixing in agitated vessels only rarely adds additional hazards to those associated with tanks and/or reactors. Unless completely submerged (this is unusual but sometimes practical), the mechanical agitator usually has a packing gland through which a shaft passes and which may be a source of vapor leaks. Attention to maintenance and local exhaust ventilation are the usual remedies.

Unexpected hazards can occur where mixing is expected but does not happen properly. For instance, an exothermic reaction between two liquids may be controlled easily if good mixing occurs as one is added to the other in a vessel (this arrangement usually allows better heat transfer and thus control than adding the liquids simultaneously). If, however, the mixing does not occur as planned because the agitator is not activated or is broken, then reaction at the interface of the liquids can possibly generate high temperatures locally, causing "bumping" or formation of large bubbles of vapor. As the bubbles rise through the upper layer, there is an increase in agitation within the vessel, causing even more mixing and subsequent formation of more larger vapor bubbles. Under these circumstances, liquid has been "bumped" out of open manholes, closed tanks have ruptured, and explosions have occurred.

Direct observation cannot be relied on as the sole source of information concerning possible hazards of the mixing process. An interlock that will not allow liquid (or gas) flow to the vessel unless the agitator is operating is one way of preventing this problem. Flow mixers, not subject to this problem, are another means of solving problems associated with high heats of reaction and/or solution.

8.3 Mixers for Viscous Materials

Problems associated with mixing viscous materials are similar to those of mixing liquids with two exceptions. First, viscous materials require much more power input for mixing than do liquids of low viscosity and, second, heat transfer to or from the viscous material is impeded. Whether the material being mixed is a polymer formed by reaction or a paste formed by adding one or more solids to a liquid, the problems are similar.

8.3.1 Batch Mixers. The most common viscous material mixer is the "change-can" type. The container ("can") is separate from the mixer frame and is placed under or is attached to the mixer as needed. Hazards are associated with the mixing, the transfer of cans to and from the mixer frame, and with can and mixer cleaning. Operating procedures must be carefully worked out in detail to avoid problems from skin or eye contact. Change-can mixing is only rarely suitable for materials that may pose problems from vapor inhalation.

Other types include stationary-tank mixers, double-arm mixers, kneaders, helical-blade and shear-bar mixers, and intensive mixers such as the Banbury and roll mills or calenders. The most open of these, and consequently the type having the greatest potential for hazardous contact, is the roll mill. Rolls may be heated (often), cooled (unusual), or at room temperature and are often operated manually. Heating the rolls and/or the material to be mixed is common to reduce viscosity. This may induce a thermal burn hazard to the operator and/or may cause volatilization of one or more components with attendant hazards. Personal protective equipment is normally used for the heat, and any vapor problem is usually "controlled" with a canopy hood over the roll mill such as that shown in Fig. 5.6. Such a hood may well protect others, but only rarely does it offer much protection to the operator. The mill should be enclosed as much as practical (perhaps with a structure that can be removed for maintenance in a manner similar to that shown in Fig. 5.1) and then ventilated at a rate that will induce an indraft of 30–60 m/min (100–200 fpm) at all openings in addition to that exhaust requirement of any thermal head.

Most other batch mixers can be enclosed either completely or almost completely so that skin/eye contact is unlikely and any vapor problem is easily handled with more or less normal exhaust ventilation.

8.3.2 Continuous Mixers. Those continuous viscous material mixers usually encountered are single-screw or twin-screw mixers. These are similar to screw conveyors in design but are totally enclosed. The extruder commonly used for plastics is a single-screw mixer/conveyor. If the screw mixer is not totally enclosed it is called a "trough mixer" and may use either single or twin screws. This mixer may even be used where the power input is all at a pump that forces the material through a static screw-type enclosure. Having no moving parts, maintenance is nil.

For these continuous mixers, the main hazards are associated with maintenance and, occasionally, leaks. Leaks are not usually a major problem if maintenance is adequate; vapor problems are handled easily with local exhaust ventilation.

8.4 Mixers for Solids

In most cases, solid–solid mixing is more difficult to do well than even viscous material mixing. And, although much of the equipment in use is best adapted to batch operation, continuous solid–solid mixing is practical. Those kinds most commonly found are tumbling barrel mixers, double-cone and Y blenders, mullers, and ribbon blenders, but other kinds are certainly possible. As with viscous material mixing, heat transfer to or from the material being mixed can be a severe problem; power input is also necessarily great.

8.4.1 Tumbling Barrel Mixers. A barrel revolving end over end exemplifies the tumbling barrel type of mixer. The rate of rotation is important, as is the presence of baffles within the barrel or drum. In use, the barrel is sealed, but during addition or withdrawal of material the dust problem can be severe. The barrel must have at least one usually gasketed door; without good maintenance such gaskets may rapidly lose their ability to seal and the door(s) may be a difficult-to-control source of dust.

If dusting is a problem as the barrel is filled, the best solution is either to use a different kind of mixer or to exhaust air from another section so that air enters the barrel along with the feed and leaves only through the exhaust port. Because the

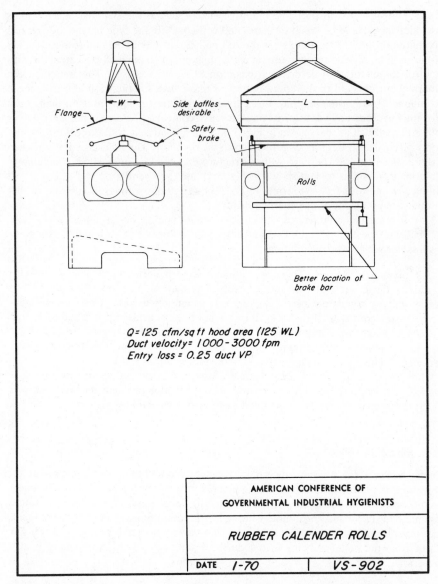

Flange

Side baffles
desirable

Safety
brake

W

L

Rolls

Better location of
brake bar

Q = 125 cfm/sq ft hood area (125 WL)
Duct velocity = 1000 - 3000 fpm
Entry loss = 0.25 duct VP

AMERICAN CONFERENCE OF
GOVERNMENTAL INDUSTRIAL HYGIENISTS

RUBBER CALENDER ROLLS

| DATE | 1-70 | VS-902 |

Figure 5.6. A canopy hood for a calender. Note that enclosure is used to as great an extent as feasible. (From *American Conference of Governmental Industrial Hygienists,* Committee on Industrial Ventilation, P.O. Box 16153, Lansing, MI 48901. By permission.)

barrel must revolve, the exhaust connection must be made and broken with each batch.

Dumping this or any other batch blender can also cause much dust to become airborne. A barrel-filling hood (see Section 11) or ventilated chute can solve that problem, but usually arrangements must be made for the hood or chute to swing out of the way when not in use.

8.4.2 Double-cone and Y Blenders. Double-cone and Y (or V) blenders are variants of the tumbling barrel and have many of the same health hazard problems and solutions.

8.4.3 Muller Mixers. A muller is a horizontal pan in which two crusher/mixer wheels revolve at either end of an axle—or the axle can be stationary (or revolve) as the pan revolves (in a direction counter to that of the axle if the axle moves—see Section 2.3.1). Although usually operated on a batch system, with two interconnecting pans each having an axle and wheels, the muller can be made continuous.

When dusting is a problem that cannot be solved by addition of water or other fluid, total enclosure is almost always the best solution.[5] The enclosure should be exhausted at a rate sufficient to handle any thermal head and to induce an indraft at all openings of the usual 30–60 m/min (100–200 fpm), as shown in Fig. 5.2.

8.4.4 Ribbon Blenders. A rotating ribbon (a helix or variant) can be used horizontally or vertically, usually alongside the wall of a cone, in a single or twin configuration to blend solids as well as viscous liquids. Dust problems at loading and discharge may occur but, because the shell is stationary, local exhaust ventilation is easy to apply.

8.5 Equipment Location

Provided that sufficient headroom is present for any necessary local exhaust system, location of mixers and blenders generally has insignificant health hazard consequences. However, especially in solids handling, gravity flow may make a vertical structure important in design and operation. This, in turn, may make installation of local exhaust ventilation difficult and expensive.

9 CENTRIFUGES

9.1 The Process

Centrifugation is a process of separating two phases that differ in density through use of centrifugal force. The terms *sedimentation centrifuge* and *centrifugal filter* are generally applied to equipment where at least one of the phases is liquid; the other may be solid or, in some circumstances, liquid. Centrifuges, therefore, find most use in separating a solid phase from a liquid one.

9.2 Batch Centrifuges

There are four types of industrial batch centrifuges, namely, basket, tubular bowl, multichamber, and disk units. Hazards of their use are associated mainly with manual or other handling of the solids; liquid discharge is through closed piping systems offering little opportunity for contact. In all varieties, solids accumulating on surfaces (perhaps covered with paper or cloth) inside the machine may be removed by hand. The removal process thus offers opportunity for skin and eye contact as well as inhalation of vapors or even dusts.

9.2.1 Basket Centrifuges. A perforated (usually metal) basket is rotated horizontally on a vertical axis. The basket is usually lined with a paper or fabric bag; slurry is added to the bag usually from a pipeline through the open lid either manually or automatically. The slurry is added until a sufficient amount of the solid phase has collected in the basket. That material is then usually washed one or more times by spraying it with "mother liquor" or another liquor as the basket turns. The machine speed is then increased to extract as much of the liquid phase as practical (usually on a timed cycle) before the cover is opened and the bag is removed by hand. Instead of removing a bag, the basket may be bottomless and the cake removed by operation of a "knife."

Opportunity for skin, eye, and vapor contact is quite great when a basket centrifuge is used even though the process may be largely automated. Personal protective devices (spectacles, face shields, gloves, aprons, boots, coveralls, etc.) are usually used but may not prevent all contact.

When vapor is a problem, the usual "solution" is to construct an exhaust hood over or around the opening through which the centrifuge is loaded and discharged. The hood interferes with operation, even perhaps with proper lid closure, and usually does not control vapor at all well. A better solution to this problem is to enlarge the liquid discharge pipeline and draw the exhaust air from the sealed receiver into which that line drains. In this manner, rotation of the basket (or "wheel" of the centrifuge) adds to air motion; air is drawn into the machine through its charge/discharge opening and, if at the proper velocity, prevents vapor escape. Air velocity should be 30–60 m/min (100–200 fpm) through the opening when the lid is wide open.

9.2.2 Tubular Bowl Centrifuges. In this equipment, a bowl a few centimeters in diameter and perhaps close to a meter long is suspended from a flexible-drive spindle. Feed enters at the bottom through a nozzle. The liquid being discharged at the top of this centrifuge is moving at a high rotational velocity and, hence, breaks up into mist and droplets (or even foam). If droplets become airborne, they may create a hazard, although in operation the unit is "sealed."

Solids are removed by hand in a manner similar to that of the basket centrifuge and with its associated problems. This equipment is used where a high centrifugal force (up to perhaps 62,000 times that of gravity) is necessary to remove small amounts of solid material from large amounts of liquid. Solids are not usually removed at frequent intervals, thus perhaps minimizing contact opportunity by "administrative" means.

9.2.3 Multichamber Centrifuges. The addition of up to six concentric tubes (tubular bowls) to the tubular bowl centrifuge increases its efficiency but adds few health hazard problems.

9.2.4 Disk Centrifuges. Turning the bowls upside down and flattening them while decreasing the space between them to as little as 0.4 mm (about 0.015 in.) so that the number can be increased to perhaps 100 or more produces the common disk centrifuge. Its hazards are similar to those of tubular bowl centrifuges.

9.3 Continuous Centrifuges

Most continuous centrifuges are used to break emulsions or to separate solids as a slurry from suspension, in either case to produce a more or less clear liquid as well. Some, however, have been designed so that nearly dry solids production is practical on an automatic basis. Especially in that case, handling of the solids after centrifugation is usually far more troublesome than during the process (see Section 3).

9.3.1 Continuous Decanters. The several varieties of continuous decanters available are equipped with centrifugation sections to separate solids from a dilute slurry and to move the solids out of the unit with an internal screw conveyor. Vapor and dust problems are handled with local exhaust hooding at the discharge point.

9.3.2 Knife-Discharge Continuous Centrifuges. These devices are very similar to the common basket units except that they are so constructed that the "basket" has no bottom. In operation, slurry is fed to the centrifuge until a sufficient amount of solids has accumulated. Rotational velocity then increases automatically to extract a maximum amount of liquid; programmed washes may follow. After the cake has been washed and "dried," rotational velocity is decreased to a few revolutions per minute and a "knife" blade is used to dislodge cake from basket walls. The cake falls through the open bottom to a bin or conveyor beneath and than a new cycle begins.

Vapor problems during centrifugation and/or washing are best handled as with the basket centrifuge, by flow-through or flow-along concurrent ventilation using the liquid discharge line and catch tank as parts of the ventilation system. If the discharged solids are troublesome, the best solution usually is complete enclosure of the area beneath the unit (as illustrated in Chapter 15).

9.3.3 Other Types. Many other types of centrifuge (especially centrifugal filters) are in common use but will not be described here. None appears to present unique hazards.

9.4 Equipment Location

When a solid is extracted from a slurry by a centrifuge (centrifugal filter), vertical equipment layout should be especially considered to facilitate subsequent handling of the solid material. This "axiom" remains true whether the operation is batch or continuous. Nevertheless, there may be other considerations.

All centrifuges vibrate; in general, those handling the largest masses of solids at the highest rotational velocities cause the most severe problems. Vibration not only increases wear rates in the equipment but also tends to loosen seals and joints and even affect whole structures and people working in those structures. This may require that centrifuges be mounted as low in a structure as is practical; ground floor concrete slabs offer the greatest resistance to induced vibration attainable in a building. Fortunately, special vibration-damping mounts are available to reduce the severity of this problem. For optimum performance, vibration dampers must be custom designed for each application.

Because of vibration, seals on centrifuges usually fail fairly rapidly, allowing leaks to occur. Where practical, welded joints should be used instead of flanges, and the areas in which centrifuges are located should be designed for spill containment, easy cleanup, and maintenance. If seals are used, they should be presumed to leak; if such leakage may be hazardous, precautions are in order. One precaution is to make certain that there is adequate room in the centrifuge area to install local exhaust ventilation at any point where a leak could occur and where dust, mist, or vapor from that point could create a problem if airborne. This precaution, incidentally, may complicate building design considerably. For reasons already discussed, centrifuges are often placed on the ground floor or at least low in a building; if the building is tall, adequate local exhaust ventilation on the lower floors can be difficult to achieve. Discharge of exhausted air at any side of a building almost guarantees recirculation of that air back into the building.

One means of resolving these conflicting requirements is to match equipment to the job rather than pressing into use whatever happens to be available. Then, the centrifuge should be mounted properly so that its vibrations are unlikely to affect either the building structure or associated piping and other equipment. Finally, to make certain that even accidental overloading will not cause a serious upset or breakdown, a vibration sensor can be used to shut down the centrifuge before any untoward event occurs. Using these precautions, centrifuges can be located anywhere in a building with the anticipation that vibration-related problems will probably be few and relatively minor in importance.

10   FILTERS

10.1   The Process

10.1.1 Definition. In filtration, particulate solid material is removed from a fluid. The term *filtration* can apply to removing solids from either gases or liquids; this section is concerned with liquids (often but not exclusively water); Section 13.5 deals with gases in general and air in particular. (This subject is further discussed in Chapter 16.)

According to *Perry's Chemical Engineer's Handbook,* filters can be classified in several ways: by driving force (gravity, pressure, vacuum), by filtration mechanism, by function, by operating cycle, and by nature of the solids. Insofar as hazard is concerned, the most important of these classifications is the operating cycle, that is, whether or not the filter is continuous. In general, batch (or intermittent) operation is associated with the greatest exposures of people to both the filtrate and the suspending medium, liquor or mother liquor.

Perhaps second in importance is function. Filtration designed to recover the solids (cake) as product often exposes operating personnel more severely than does filtration designed to clarify the liquor. Clarification is often accomplished in completely closed equipment that can be designed so most cleaning is done automatically in a backwashing cycle; infrequent maintenance then may be the only occasion for contact with either the filtered solids or the liquor.

The driving force, itself, does not usually create health hazards but may be very important nonetheless. Use of centrifugal force instead of gravity, pressure, or vacuum puts into play a whole new class of equipment: centrifuges (see Section 9).

10.1.2 The Media. Accepting the fact that almost all filtration is accomplished by the filter cake itself, that cake still must be supported by a medium that will allow rather free flow of the suspending liquid (liquor) while impeding flow of at least the larger solid particles. Filtration media play a role in determination of the hazard only under those circumstances where poor choice causes a need for excessive maintenance or results in operator exposure to any of the product components.

A separate but related part of filtration is the use of filter aids. These are materials that form a highly porous cake even when the material filtered is a slime or very compressible solid. Most commonly used are diatomaceous earth, cellulose fiber, and asbestos. Cellulose fiber—in this use at least—causes no health hazards; the same cannot be said for the other two kinds of material.

Natural diatomite is sometimes used, and its health hazard is related to the crystalline free-silica content of the respirable dust. When the material as mined is calcined to remove organic matter, some of the amorphous silica (up to 60%) is changed to the crystalline variety, cristobalite, that has about twice the ability of quartz to cause silicosis upon prolonged inhalation. Exposures most often occur when bags of the diatomaceous silica are opened and the material is added to the slurry to be filtered. Most of these exposures are intermittent and brief; when such is not the case, there may be a hazard (see Section 4). Asbestos today is used as a filter aid only when no other material can be substituted because of its well-known ability to cause cancer especially in those people who smoke cigarettes. When used as a filter aid, asbestos can almost always be handled wet (usually with the fluid being filtered), and this is probably the best way to reduce or eliminate asbestos inhalation. Local exhaust ventilation can, of course, also reduce exposures but generally is not as effective as wetting and is much more expensive. Respiratory protective devices are a poor third choice.

10.2 Batch Filters

Filtration is much more an art than a science and is quite difficult to do well. For these reasons, the filtration equipment available is quite varied in design. Very few filters have been designed with hazard control in mind, however, and this sometimes leads to unexpected problems. For hazard analysis, batch filters can be regarded as being in three categories, namely, gravity filters, pressure filters, and pressure leaf filters. Except in the laboratory, vacuum is only rarely used for batch filtration (exceptions are vacuum Nutsches and leaf filters).

10.2.1 Gravity Filters. The main use of gravity filters in the United States is for drinking water purification–clarification where the filter medium is aggregate and sand; the greatest hazard is probably that of drowning. Gravity is used as the driving force only when filtration is easy and fluid flow is rapid even when no other force is used. And, although very inexpensive to construct, gravity filters occupy large amounts of floor space and may be quite expensive to operate (in terms of worker time for operation and cleanup). They also allow large surface areas of the slurry to be exposed to air; unless the liquor is water, this may cause a vapor inhalation hazard.

Cake produced by gravity filtration must be removed from the bed manually, often with a shovel. This causes obvious contact and inhalation possibilities. The main cure is to switch to another type of filter; the switch can often be justified on an economic base alone without adding the factor of hazard.

10.2.2 Pressure Filters. Other conditions being equal, filtration rate is directly proportional to pressure at the filtering surface. Consequently, batch pressure filters have historically been the next step from gravity filters.

Pressure is applied to the liquid being filtered by some means, usually a pump, plunger, or compressed gas. Pump leaks (see Section 6.3) can be hazardous. However, hazards associated with filter use are usually greater.

Batch pressure filters may be of the Nutsche, sand bed, upflow sand bed, press, or other varieties. Of these, the most popular is probably the plate-and-frame filter press. Some of its problems and their possible solutions will be discussed as illustrative.

Filter presses operate intermittently. The plates, media, and frames are assembled, and filtration continues until the frames are full of cake as signaled (usually) by an increase of the pressure drop across the system to a preset value. The hydrostatic pressures used are high and, because the seals are not perfect, presses leak, causing a housekeeping problem even if not a hazard. When filtration is complete, the press is manually disassembled, and the frame contents are dumped into tote bins or other receivers. After a thorough washing and inspection of each filter cloth, the unit is reassembled to begin another cycle.

Skin and eye contact hazards abound during the disassembly process; the usual control is protective clothing—gloves, boots, aprons, glasses, and the like—that is impermeable to the filtrate and liquor. Such protection is rarely complete enough to prevent all contact at all times because the process is almost inherently a sloppy one.

An even more severe problem is caused when vapor from the liquor or, much more rarely, dust or vapor from the solid is an inhalation hazard. No way of absolutely preventing vapor inhalation by use of local exhaust ventilation has been devised, although a canopy or—more rarely—a side-draft hood can reduce vapor concentrations reaching breathing zones. And, because both mobility requirements and the amount of physical work necessary are relatively great with even a small press, respiratory protective devices are not accepted at all well.

Where either the liquor or the cake may be hazardous upon contact or inhalation, the filter press should be avoided despite its many processing advantages.

10.2.3 Pressure Leaf Filters. A filter leaf is a frame covered by a filter medium on both sides. Liquor is forced through the medium to the center of the leaf where it is collected and led through a nozzle that connects the leaf to a main drain. The unit consists of several leaves on a main, all in an enclosing shell. Slurry is fed to the shell, liquor drains from the main, and cake collects on surfaces of the leaves. Leaves may be circular, rectangular, or something in between and may be attached to the main at the top, bottom, or center; the shell may be a cylinder or a cone, and its axis may be horizontal or vertical; variety is great.

Because all filtration is done inside an enclosing shell, leaf filter leaks do not, of themselves, create either a housekeeping problem or a hazard. Furthermore, by backflushing with liquid, cake can be dislodged from the leaves, forming a thick slurry that can be drained from the shell with essentially no opportunity for contact with workers. If a dry cake is necessary, the leaf assembly can be removed from the shell either before or after backflushing with room temperature or heated air to dry and then dislodge the cake. This process, too, can be operated with little or no contact and inside an exhaust hood if necessary.

The main disadvantages of pressure leaf filters are the restricted cake buildup and the lower operating pressures than customary with the plate-and-frame filter press. Nevertheless, their hazard potential is far lower and such hazards as are present are far easier to control than those of the press. For these reasons, pressure leaf filters are often good substitutes for filter presses. There are other batch filter designs than leaf filters; hazards of all are similar to those of leaf filters.

10.2.4 Clarifying Filters. When the object of continuous filtration is not to produce a cake or thick slurry but, rather, a clean filtrate, the process is likely to be called clarification. And, because the amount of cake produced is usually rather small, clarifying filtration tends to be a batch operation.

Because these devices are sealed, used in line, and are opened or removed only very occasionally, hazards associated with their use are small, infrequent, and usually dealt with by personal protective devices and clothing if necessary.

10.3 Continuous Filters

The limited production capacity of batch filters has led to the development of several varieties of continuous filters. This, in turn, has reduced hazards if only because of a reduced necessity for human participation in the process. Continuous filtration can be accomplished using either pressure or vacuum techniques.

10.3.1 Pressure Filters. Most continuous-pressure filters are variants of the intermittent pressure leaf filter (Section 10.2.3) where leaves or drums are arranged in a shell and operated so that a thick slurry as well as a clarified liquor are discharged continuously. The units are sealed and cause no hazard except possibly for leaks at seals and during any necessary maintenance.

10.3.2 Vacuum Filters. Continuous-vacuum filters are more popular than continuous-pressure filters for two main reasons. First, they are about half as expensive for similar capacities; second, they produce a relatively dry cake. A characteristic that may be either an advantage or a disadvantage is that the cake is, of necessity, exposed to the air in the operating area.

Vacuum filters are constructed as internally supported drums or leaves that rotate through the slurry or as horizontally rotating pans or even belts that tilt to discharge cake. Two characteristics of vacuum filters that may cause hazards are dry cake handling and the vacuum system.

So long as cake produced by vacuum filters is on the filter, air is usually being drawn through it. At discharge, air pressure, a knife blade, or tilting may be used, each causing some material to become airborne. If this can be a hazard, then local exhaust ventilation can almost always be arranged for essentially absolute control because the discharge point or area is stationary.

In a vacuum system, the motive force for the filtration is supplied by one or more vacuum pumps. These pumps must discharge the air to the atmosphere; this discharge can be hazardous from either cake or liquor components. Assessment of vacuum pump discharge hazard is often overlooked. If that material can be hazardous, the usual remedies are to discharge either high in the air or through an air cleaning device (see Section 13).

10.4 Equipment Location

Filters operated to produce a dry cake tend to be placed high in a vertical equipment layout simply to ease solids handling after filtration. Dry filter cake is usually (not always) damp enough that only small amounts of material become airborne even during more or less violent handling. If those small amounts are troublesome, the best solution usually is to decrease handling violence first, enclose second, and ventilate if needed as the third step. With the passage of time the cake from filter discharge usually becomes much drier, increasing the intensity of this problem.

Liquor evaporates from damp cake as the cake dries. If this vapor is a problem, enclosure and ventilation are the usual solutions. As time and distance from the filter increase, vapor emission rates tend to fall and this problem becomes less acute. Aside from these considerations, location is not usually a material component of filter hazards.

11 CONTAINER FILLING

11.1 The Process

Containers to be filled may be either open to the atmosphere or closed, flexible or rigid; they may be filled with gases, liquids or slurries, or solids, and, of course, the containers may be those intended for temporary or permanent storage, shipping, or sale. This section will consider only those containers intended for shipping or sale; those used within the plant are considered in Section 6.2. The only open containers routinely used for shipping are railcars such as gondolas; they are not considered here.

11.2 Gases

Gas does not have a clear definition accepted throughout industry. A material used as a gas from a compressed-gas cylinder, for instance, may be added to that container as a (usually cold) liquid. This, in fact, is the case with almost all but the "permanent" gases such as oxygen and nitrogen and even these, of course, are sold as liquids as well as compressed gases. At the time the "gas" container is filled with a liquid, it is usually open to the atmosphere; it is sealed with a valve after filling. See Section 11.3 for liquids.

Health hazards associated with filling containers with compressed gases are, then, only those of leaks, usually at the fittings where connections are made and broken most often—those at the containers being filled. If inhalation of the gas is harmful, the filling station should be equipped with a local exhaust hood capable of capturing material emitted from small leaks. Instances of line, valve, or compressed-gas cylinder rupture with consequent release of large amounts of gas are highly unusual, and routine precautions such as local exhaust ventilation are not expected to be adequate for control in such cases.

11.3 Liquids and Slurries

Bottles, cans, cylinders, and tanks are almost always open to the atmosphere as they are filled; they are sealed later, sometimes on the order of seconds or even less. In the filling process, however, air expelled from the container as the liquid is added is presumed to be saturated with any vapor above the liquid and may also be contaminated with droplets.

If vapor from the liquid is hazardous, the first step in controlling that hazard may well be to cool the liquid. Cooling reduces vapor pressure and consequent vapor loss and for many materials is justified simply on that economic basis. Where cooling the liquid does not offer adequate control, it still may be used to reduce the extent to which ventilation must be relied upon.

When a seal can be made between the filling valve and the container being filled, it may even be possible and practical to conduct displaced vapor (and/or air plus vapor) back to the source of the liquid through a vapor line; there is no need for ventilation and no loss of vapor at all in the filling process. When this method is practical it should be adopted, because it provides essentially complete control of the vapor inhalation hazard. Where it is not practical, use of an exhaust hood at the filling station may be necessary.

Because the size of the containers to be filled with liquids varies so greatly, only a few generalizations concerning hood design are useful. First, the hood that fits tightest will control vapor release with the least air flow. For that reason, tank-filling valve nozzles have been designed with hoods incorporated so that when the nozzle enters the tank, the hood fits rather tightly over the opening and thus collects displaced vapor easily and completely with little exhaust volume. This method also prevents skin/eye contact with and/or inhalation of any mist generated in the filling process. Its disadvantage, however, is that the liquid in the tank being filled is not visible. Some method of stopping flow of the liquid automatically when the container is full must be present. As most containers are sold on the basis of weight of liquid within them, automatic cutoff at a preset weight may well be used, triggered by a scale upon which the tank or cylinder rests. Other methods are available, including the nozzle developed for filling automobile gasoline tanks with little, if any, overflow.

When liquid level in the container being filled must be observed or a tight-fitting hood cannot be used, the hood is usually constructed to be separate from the tank-filling line or nozzle. Both the nozzle and the filling port of the tank must be within the hood, of course. If a portion of the hood is constructed of transparent material, the liquid level can be observed while still providing a maximum amount of enclosure.

Other than saturated vapor displacement, the greatest hazard associated with liquid filling is that of overflows. No matter how much care is taken, overflows still occur and, in consequence, each filling station should be equipped to handle spills—or at least to minimize any accompanying hazards. Depending on circumstances and the nature of the liquid, items to consider for a filling room or area include dikes or catch pans, granular absorbent material, facilities for washdown and for making sure that chemical spills do not reach the sanitary sewer, any necessary protective equipment and clothing, "fire" and eye showers, and perhaps rags or other wipers along with containers for soiled ones.

When the containers being filled are small and numerous, automatic filling may well be done. Automatic filling machines have one thing in common: Operators of the equipment are almost always farther from the containers being filled than at manually operated stations. This, of course, may decrease hazard but does not decrease the need for process control. Local exhaust ventilation, where needed, for automatic filling stations is almost always custom designed to fit the equipment and process. The generalization that enclosure should be used to as great an extent as practical still applies, however.

In addition to vapor inhalation hazards, automatic filling of containers usually is done in conjunction with equipment that moves and perhaps sorts and orients those containers. The moving–sorting–orienting process is usually noisy. Levels on the order of 100 dBA are not at all uncommon in such plants where glass or metal containers are filled. Plastic and paper containers usually generate far less noise on contact among themselves, thus reducing overall noise levels considerably. As mentioned in Section 3.2.1, efforts at quieting conveyor systems have also been successful.

When many small containers are handled, most of the noise occurs where those containers change direction. Thus use of noise-absorbent material at such points can be somewhat effective. One of the large problems, however, is that those points where the most noise is produced are usually those that require the most attention from maintenance personnel. Noise shields/attenuators probably interfere with maintenance and consequently may not be reinstalled after being removed.

In addition to the measure already described, noise absorbers may be suspended from the ceiling, or the ceiling and walls may be constructed of noise-absorbent material. This technique can actually reduce overall noise levels in a large room by a few decibels but rarely is able to reduce noise levels for those people having the most severe exposures—the ones who work closest to the filling lines. Until such time as a method is developed to handle masses of glass and metal containers quietly, personal hearing protectors in such plants will continue to be necessary.

11.4 Solids

Filling containers with solids causes problems in proportion to the number of fine particles per unit volume of material transferred. Coarse solids produce fewer dust problems than powders unless the filling process itself is energetic enough to break up the larger particles.

Where dusting is a problem, the first step to consider is changing the material to a less dusty form. If dusting is a problem for the producer, it very probably is a problem for users who may welcome beneficial changes in the product. One such change is adding a dust-suppressing liquid. This method has worked well, for instance, for asbestos fiber where the liquid is water, and even for pentachlorophenol flakes where the liquid is a low vapor pressure oil. Even controlling humidity in the packaging area can help materially in dust suppression. In all of these cases, an important point is to use just enough liquid or humidity to suppress dust but not so much as to add materially to the product weight or to make it difficult to handle. Experimentation is usually necessary.

As mentioned in Section 4, changing the chemical nature of the product can sometimes reduce a dusting problem, as can changing a solid to a slurry or a liquid solution. Such changes are especially successful when they can be "sold" to customers as advances in technology that promise easier processing in their plants.

One of the most successful methods of altering the tendency of a material to dust is that of particle size enlargement, exemplified by prilling. Prilling is a means of making "shot" of the material by melting it and then spraying the molten material into a prilling tower. Particles fall in the air to form spheres with size ranges being controlled to rather close tolerances. During their fall, the prills are quenched to retain their spherical form and, if necessary, later coated with small amounts of an inert powder to make a free-flowing product. Prilled products are much less dusty than powders or even granular material. Urea for lawn fertilizer use is now generally prilled; prior to that step it was either very dusty or agglomerated into large lumps or both. This technique is adaptable to many materials from salts such as ammonium nitrate to viscous substances and even slurries.

Other methods of size enlargement that have proved successful in at least some applications include tableting, pressure rolling, extrusion, molding, and even agglomeration tumbling to form granules, balls, or pellets. Use of heat to produce a sinter or heat-hardened product has been successful with some inorganic materials. This means of substituting few large particles for many smaller ones has solved several dust problems.

Where substitution cannot be used or is not completely successful, local exhaust ventilation methods must be sought for control when containers are filled. The problem is mainly that of air displacement from the container being filled. The usual solution of a flow-along concurrent ventilation system is rarely available because product containers commonly have only one breach in their wall integrity, the opening through which the product is being added. Furthermore, even if there were a means of removing air from the container as it is being filled, that means would probably not be used. Containers are usually filled as completely as practical to avoid wasted space; a second opening would allow the ventilation system in too many cases to become a pneumatic conveying system, instead, as the container filled. Hence, an external exhaust hood is necessary.

Filling containers of all sizes to more or less "exact" weights has become common industry practice for a number of reasons, including that of inventory control. Exact predetermined weight is much more difficult to achieve than, for instance, either constant volume or exact (but not predetermined) weight. The advent of computers with at least terminal facilities in production plants has allowed either scheme to be used by both suppliers and customers so long as each is in agreement. When that is not the case, or where there is a need for physical as well as computerized inventory control, then the predetermined weight approach probably must continue.

11.4.1 Small-Container Filling. Several bag-filling and barrel (or fiber pak) filling hoods have been developed and standardized to the extent that their designs are published in the ACGIH *Manual of Industrial Ventilation*[1]; Figs. 5.7–5.9 are illustrative. The basic principle of each is to accomplish the job by enclosure as much as practical and yet allow access to the container being filled and then to exhaust air at a velocity great enough to capture dust or vapor as it escapes from the container. Air velocities as high as 160 m/min (500 fpm) at openings may be used, depending largely on the rate of container filling and, hence, air displacement. A few of the designs, especially for fiber pak or barrel filling, employ something that closely approaches a flow-along system (Fig. 5.9). When exact predetermined weights of containers must be achieved, a filling station similar to that in Fig. 5.10 can be used where a preweighed batch of material is discharged completely into the ultimate container every cycle. This equipment configuration requires more headroom than that usually found on a single level, but the hopper and perhaps even the feeder could, perhaps, be on the floor above the packaging station.

Bag and barrel (fiber pak) filling hoods can do an excellent job of dust control, but they do not solve all problems. First, every container-filling station must contend with product spillage either from overfilling containers (even with weight- or other-activated automatic cutoff) or simply from the filling spout dribbling or being activated at the wrong time. This problem will certainly occur despite all precautions.

Second, when the ventilation system is very effective, it will certainly pick up enough dust so that reclaiming that material should at least be considered. Close to the packaging station(s), then, should be at least some sort of roughing air cleaner such as a dry cyclone in the exhaust ventilation line. Cyclones are large (as are other types of air cleaners that might be considered for this purpose), and the space occupied by them must be planned. And, of course, provision must be made for product dust trapped in this manner to be added back to the system or separately packaged without hazard to the workers. A container-filling hood beneath the air cleaner (and sufficient headroom to allow easy container filling) may be necessary.

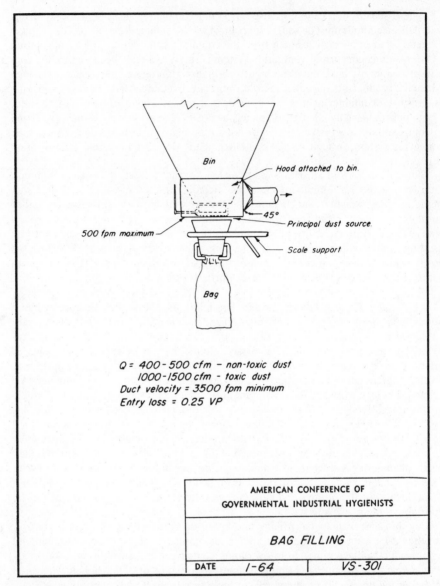

Q = 400-500 cfm — non-toxic dust
1000-1500 cfm - toxic dust
Duct velocity = 3500 fpm minimum
Entry loss = 0.25 VP

**AMERICAN CONFERENCE OF
GOVERNMENTAL INDUSTRIAL HYGIENISTS**

BAG FILLING

| DATE | *1-64* | *VS-301* |

Figure 5.7. Filling bags to a known weight. Rather than "nontoxic," "nuisance" should be used. (From *American Conference of Governmental Industrial Hygienists,* Committee on Industrial Ventilation, P.O. Box 16153, Lansing, MI 48901. By permission.)

11.4.2 Large-Container Filling. Most filling of large containers, from tank cars to closed barges to tote bins, is done out of doors in the belief that natural ventilation will solve all dust (and vapor) problems. It almost never does.

Air escaping from a container being filled with a powdered or granular solid often is laden with dust; unless collected by an exhaust system, that dust tends not only to

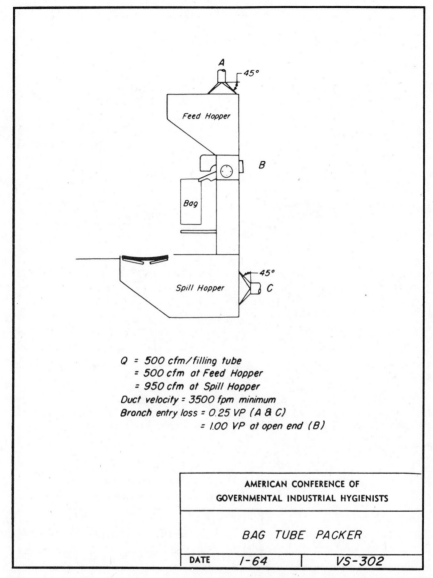

Q = 500 cfm/filling tube
= 500 cfm at Feed Hopper
= 950 cfm at Spill Hopper
Duct velocity = 3500 fpm minimum
Branch entry loss = 0.25 VP (A & C)
= 1.00 VP at open end (B)

AMERICAN CONFERENCE OF
GOVERNMENTAL INDUSTRIAL HYGIENISTS

BAG TUBE PACKER

| DATE | 1-64 | VS-302 |

Figure 5.8. Diagram of apparatus for filling tube-equipped bags and associated ventilation. (From *American Conference of Governmental Industrial Hygienists,* Committee on Industrial Ventilation, P.O. Box 16153, Lansing, MI 48901. By permission.)

migrate to breathing zones but also to settle on anything and everything in the vicinity of the operation. If the solid is at all dusty, then control with exhaust ventilation may be more necessary for filling large containers than small ones. The same principles apply except that many large containers have more than one opening (or can be made to have more than one) and, therefore, are amenable to a flow-through or flow-along

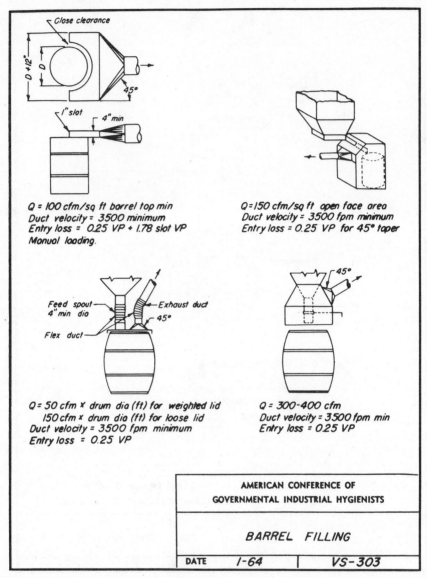

Figure 5.9. Several types of barrel or fiber-pak filling hoods. (From *American Conference of Governmental Industrial Hygienists,* Committee on Industrial Ventilation, P.O. Box 16153, Lansing, MI 48901. By permission.)

system of dust containment that should be used if practical. With sufficient air being withdrawn from the container as it is being filled, there will be essentially no dust escape; with adequate transport velocity (at least 1100 m/min, 3500 fpm) in the duct work, the dust-laden air can be delivered directly to a recovery system.

In those cases where a flow-along system cannot be used, more or less conventional barrel-filling hood designs can be modified for the large-container opening. The filling

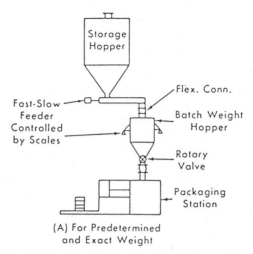

Storage
Hopper

Flex. Conn.

Fast-Slow
Feeder
Controlled
by Scales

Batch Weight
Hopper

Rotary
Valve

Packaging
Station

(A) For Predetermined
and Exact Weight

Figure 5.10. Packaging station arranged for predetermined exact weight. Note that ventilation is not shown but usually must be present. (From *Uranium Production Technology,* Van Nostrand Reinhold, New York, 1959. By permission.)

area should be enclosed as much as practical by the hood, and the air takeoff point should not be located right next to the product stream unless the solid is lumpy or otherwise contains a small percentage of fines.

11.5 Equipment Location

Most containers considered in this chapter are filled at one plant to be shipped to another point. Because transportation facilities are on the ground, or close to it, most container-filling stations should be installed near or at ground level. Exceptions to this generalization may include those where warehousing of filled containers occupies more than one floor. Even in those cases, however, gravity flow to containers is often useful (especially for solids) and for that reason alone, filling stations are usually low in a building.

Despite all precautions, spills occur at container-filling stations, which, then, should be constructed to confine spills that do occur and prevent any attendant hazards to personnel or equipment. In many plants, thorough cleanup is facilitated by a ground-floor location. Finally, filling stations are interesting to observe and are usually included in plant tours. A ground-floor location is ideal for that purpose as well as for the others.

12 FURNACES AND DRYERS

12.1 The Process

Furnaces and dryers have in common the use of heat to accomplish some usually simple task. They differ mainly in maximum temperature, which, in a dryer, is only rarely above the 300–400°C (600–800°F) range. Many dryers are unlined although they may be insulated; furnaces are lined with a firebrick usually based on silica or magnesia. Both types may be batch or continuous; most furnaces described here are batch.

In addition to producing heat (and thermal burns), these devices may increase humidity in the workplace; emit foul and/or hazardous particles, gases, and vapors; and cause even less obvious problems.

12.2 Furnaces

Furnaces have many uses, among which are smelting, melting metals, manufacturing glass, heat-treating metals, and drying high-temperature materials such as ceramics. Hazards common to all furnace operations include those associated with convective and radiant heat stress, thermal burns, inhalation of dusts and gases emitted during charging and pouring, "glassworker's" cataract, and inhalation of refractory and cement dusts during relining. Some furnaces have additional associated problems, which are also discussed.

Even well-insulated furnaces are sources of convective and radiant heat and perhaps heat stress in the workplace unless precautions are taken. Almost always, the most cost-effective control measure is infrared radiation shielding. A shield placed between the furnace wall and work area reduces—and indeed may almost eliminate—radiant heat stress while at the same time assisting in venting hot air by forming an annulus in which that air can rise without mixing with cooler air in the work area. If, then, that hot air is vented out of doors (except in the winter), a very effective system will result.

Radiant heat shielding is best done with a shiny metal surface; the least expensive source is aluminum foil or sheet. In many furnace environments, keeping the shield shiny is difficult; this problem can be reduced in two ways. First, the shield can be layered; two sheets of shiny aluminum separated by a few inches of air (the annulus again vented outdoors) makes a much better shield than one layer, even when dirty. Second, if the proper framework is used, then aluminum foil can be substituted for more rigid sheet. Since foil is very inexpensive, it can be changed often. Properly used, a radiant heat shield can virtually eliminate both convective and radiant heat as stresses in a furnace environment.

A well-constructed radiant heat shield in place a few centimeters (inches) to a meter or so (a few feet) from a furnace wall will increase the temperature of the outside furnace wall surface substantially. That, in turn, decreases the rate of heat loss through the wall insulation and hence increases fuel efficiency and decreases furnace operating costs.

No matter how much shielding is used and how well it is maintained, workers often still must view the furnace contents and/or discharges and must perform certain tasks that subject them to extremes of radiant heat. Garments made of the newer flame-resistant fabrics and coated with aluminum if necessary are available to protect against extreme temperatures and thermal burn hazards.

Especially for fuel-fired and electric furnaces, charge preheating usually is able to effect economies in operation. Much of the material used to charge furnaces in foundries today is scrap that may be contaminated with oils and possibly with metals/metalloids such as lead and arsenic. Gases emitted during preheat thus can contain metal/metalloid fume and pyrolysis products of oils of various kinds; local exhaust ventilation is the only sure control.

All furnaces used to melt metal must be charged and must deliver molten metal to another device of some sort. Both operations can be the source of large amounts of dust and fume and, consequently, attempts are usually made to control such emissions with sometimes elaborate local exhaust systems. Unfortunately, many of these seem to be designed in such a way that they are able to handle room temperature air quite well, but their capacities are much too small for the additional thermal head and induced draft associated with furnace operation. Furnaces heat air as well as the charge, and hood systems must be designed with that in mind if they are to succeed (Figs. 5.11–5.15).

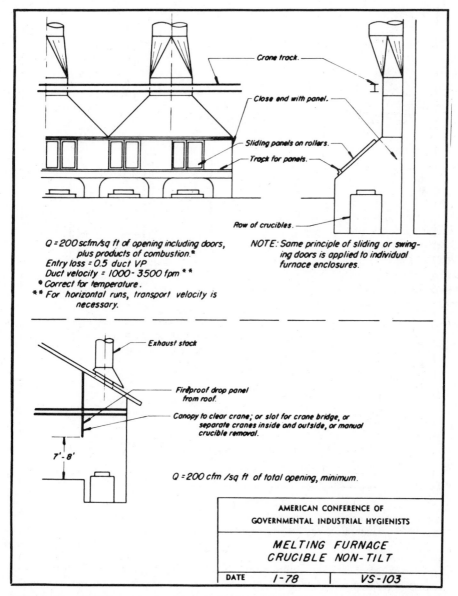

Figure 5.11. Cruicible ventilation. Crucibles may be heated by gas or electric induction; in either case allowance must be made for thermally induced draft when sizing blowers. (From *American Conference of Governmental Industrial Hygienists,* Committee on Industrial Ventilation, P.O. Box 16153, Lansing, MI 48901. By permission.)

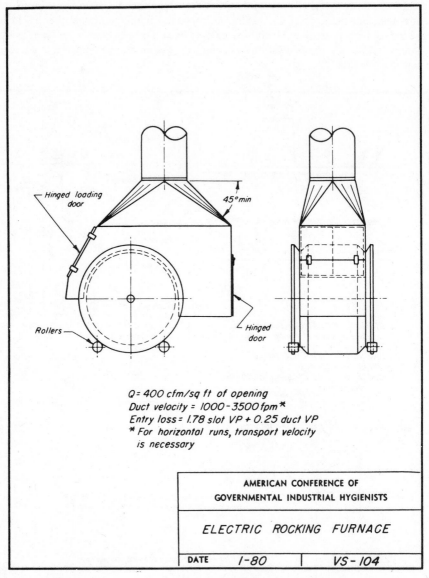

Figure 5.12. Typical ventilation of a rocking furnace that allows some control of fume during pours. (From *American Conference of Governmental Industrial Hygienists,* Committee on Industrial Ventilation, P.O. Box 16153, Lansing, MI 48901. By permission.)

Glassworker's cataract results from a denaturation of the eye lens caused by excessive absorption of infrared radiation by the iris. This process can be interrupted by viewing surfaces at red heat or hotter only through doped glass (the usual dopant is neodymium).

Furnace linings serve to protect the steel walls. Linings wear out and must be replaced. "Basic" linings are usually high in magnesia, while "acidic" linings are usually

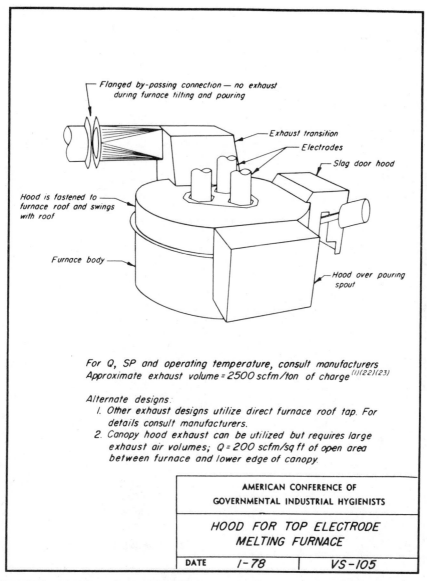

For Q, SP and operating temperature, consult manufacturers
Approximate exhaust volume = 2500 scfm/ton of charge[1][22][23]

Alternate designs:
   1. Other exhaust designs utilize direct furnace roof tap. For
   details consult manufacturers.
   2. Canopy hood exhaust can be utilized but requires large
   exhaust air volumes; Q = 200 scfm/sq ft of open area
   between furnace and lower edge of canopy.

AMERICAN CONFERENCE OF
GOVERNMENTAL INDUSTRIAL HYGIENISTS

HOOD FOR TOP ELECTRODE
MELTING FURNACE

DATE   1-78   VS-105

Figure 5.13. Electric arc furnace hood. Because of the disconnection, pouring is done with no local exhaust ventilation. (From *American Conference of Governmental Industrial Hygienists, Committee on Industrial Ventilation, P.O. Box 16153, Lansing, MI 48901. By permission.*)

high in silica. Removing old furnace lining is a hot, dusty job in most cases; the chief hazards are heat stress, thermal burns if the furnace is entered too soon after shutdown, toxic metal inclusions in the firebrick, and crystalline silica inhalation if the firebrick contains silica. In most cases the only available control of these hazards is through use of personal protective equipment coupled, perhaps, with administrative control (regulating exposure duration). Air-supplied suits or coveralls with or without a vortex

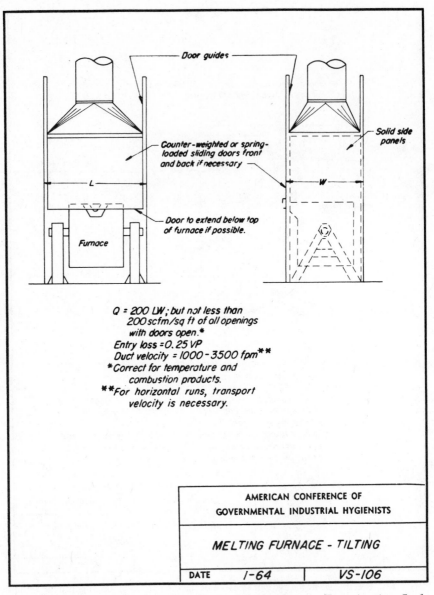

Door guides

Counter-weighted or spring-
loaded sliding doors front
and back if necessary

Solid side
panels

L

W

Door to extend below top
of furnace if possible.

Furnace

$Q = 200\ LW$; but not less than
200 scfm/sq ft of all openings
with doors open.*
Entry loss $= 0.25\ VP$
Duct velocity $= 1000 - 3500\ fpm$**
*Correct for temperature and
combustion products.
**For horizontal runs, transport
velocity is necessary.

AMERICAN CONFERENCE OF
GOVERNMENTAL INDUSTRIAL HYGIENISTS

MELTING FURNACE - TILTING

| DATE | 1-64 | VS-106 |

Figure 5.14. Canopy hood for a small gas-fired or induction furnace. (From *American Conference of Governmental Industrial Hygienists,* Committee on Industrial Ventilation, P.O. Box 16153,
Lansing, MI 48901. By permission.)

tube (to separate compressed air into warm and cool fractions) can make possible
reasonable comfort in the heat; the air can also be directed to a hood or mask if
necessary as a source of clean air for breathing.

The main hazard encountered in installing a new lining is dust inhalation, especially
when brick and other lining material is cut with a power saw. Saws used for this
purpose are often portable and seldom are equipped with effective exhaust ventilation.

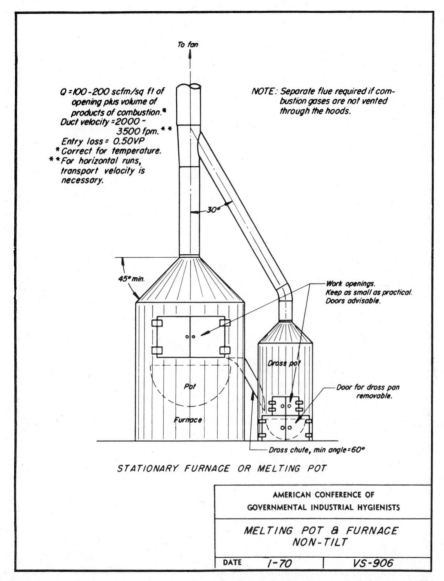

Figure 5.15. Total enclosure for a small gas-fired or induction furnace and dross pan. This arrangement is common in nonferrous foundries. (From *American Conference of Governmental Industrial Hygienists*, Committee on Industrial Ventilation, P.O. Box 16153, Lansing, MI 48901. By permission.)

Instead, an attempt may be made to control dust with a "wet table" and/or a wet saw blade; neither technique is truly effective for dust control in most situations. Exhaust ventilation such as that shown in Fig. 5.16 is usually far more effective.

12.2.1 Electric Arc Furnaces. When the furnace charge is electrically conductive, direct electric arc can be a very efficient heat source. This type of furnace is used,

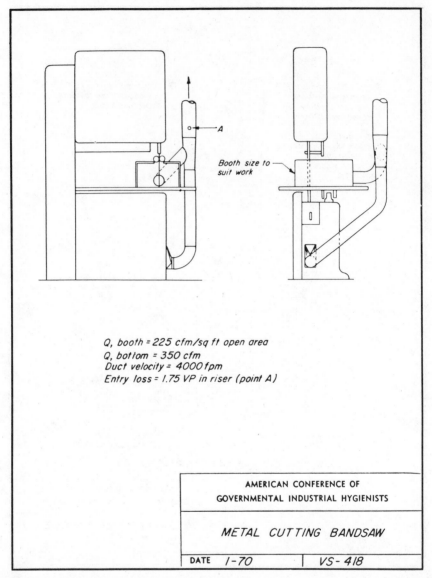

Q, booth = 225 cfm/sq ft open area
Q, bottom = 350 cfm
Duct velocity = 4000 fpm
Entry loss = 1.75 VP in riser (point A)

AMERICAN CONFERENCE OF
GOVERNMENTAL INDUSTRIAL HYGIENISTS

METAL CUTTING BANDSAW

DATE 1-70   VS- 418

Figure 5.16. Local exhaust ventilation for a band saw used to cut metal or even silica insulating brick. (From *American Conference of Governmental Industrial Hygienists,* Committee on Industrial Ventilation, P.O. Box 16153, Lansing, MI 48901. By permission.)

therefore, to melt metal alloys, some metals (those having sufficient resistance), and in the smelting of some ores, notably phosphorus. Three-phase electric current is directed through the furnace charge by three consumable graphite electrodes that, depending on furnace size, may have a diameter from a few centimeters to a few meters (a few inches to several feet).

Under arcing conditions the graphite electrode slowly reacts either with air or with oxides in the charge to produce both carbon dioxide and carbon monoxide with the latter prevailing in many circumstances. Of course, fume is also produced from metallic charges; these air contaminants and heat are thus the chief hazards of most electric arc furnaces. In addition, the arc may pose a real noise problem.

Air contaminants from electric arc furnaces are controlled either by operating the furnace under a slight negative air pressure or by hooding all leak points, or both (see Fig. 5.13). The main problem with the negative-pressure approach is that air must be drawn into the furnace interior. That air not only cools the charge but also tends to oxidize it; neither is desirable. Nevertheless, where it can be used, that is probably the most efficient system.

Some sort of hood arrangement (many of these hoods are based on proprietary designs) is usually necessary to control fume during the pour; extending the exhaust system to cover other leak points such as around the electrodes may be a relatively simple task, especially when compared to the problems associated with venting the furnace itself (see Figs. 5.12–5.14).

12.2.2 Induction Furnaces. An electrically conductive furnace charge can be made to act as an immovable (and thus "short-circuited") rotor of a motor by winding the stator in the furnace shell. A more precise way of looking at this is to regard the shell coil as the primary of a transformer with the charge being the secondary. At any rate, electrical current induced in the charge causes its temperature to rise in proportion to time and to the current flow in the furnace shell. Although less efficient than electric arc furnaces, induction furnaces offer the opportunity for more precise temperature control and far less noise.

Induction furnaces are not large fume sources except possibly at the pour, and the only gases associated with their operation are those that result from flux use. Fume is controlled by negative air pressure in the furnace during the melt and an external exhaust hood for the pour (Fig. 5.11). There is usually no need to provide large elaborate air-cleaning systems where induction furnaces are used.

12.2.3 Cupolas. A cupola is a vertical tower-type furnace. The charge—consisting of scrap metal, coke, fluxing agents (chiefly limestone) and alloying materials—is fed in at the top, usually in batches and perhaps with a skip hoist. Air is admitted at the bottom and as the coke burns both melting and smelting occur; liquid metal is drawn off at the bottom. Once used extensively in ferrous foundries, cupolas are gradually being replaced with electric furnaces because of their higher efficiency and much smaller gas/fume handling problems. Cupolas, however, are better suited than most furnaces for continuous pour operation—charging is still done batchwise.

Most cupolas have heavy firebrick linings that last for only a few "heats" before being replaced because of the very rigorous conditions within the furnace. However, it is possible to operate one of these furnaces with little or no lining by arranging for the outside surface of the shell to be covered with a film of water at all times through a series of overflow weirs. In such a furnace the "lining" actually consists of cooled, solidified iron and slag rather than firebrick and thus may last indefinitely.

Combustion takes place in the cupola, and the products of combustion can be hazardous. Those that have caused the most problems are carbon monoxide, sulfur dioxide, and even hydrogen sulfide (not everywhere in the cupola is the atmosphere oxidizing). Many of these problems are avoided by operating the cupola essentially out of doors (shielded from precipitation) where general ventilation is easier; treatment of gases and fumes emitted from the cupola is still necessary, however. Problems occur on windless days when general ventilation of the area is least effective. The best solution to those problems is probably to switch to another type furnace.

12.2.4 Other Furnaces. Many other types of furnaces are in daily operation. Their hazards are not unlike those discussed here, and the solutions are similar to those previously discussed.

12.3 Dryers

Drying can be done through the direct or indirect application of heat, in batches or continuously; the solid phase can be static, moving, or even fluidized; the potential for application of ingenuity in equipment design or application is unlimited. Regardless of equipment design, however, the problems associated with drying and the control of those problems are similar.

Where a solvent is removed in the process of drying, inhalation of that material and/or its pyrolysis products is generally the main hazard. The optimum solution is proper venting of the dryer through an exhaust system to the outdoors, perhaps through an air cleaner of some sort (Section 13.2.6). When the solid being dried is a powder, subsequent handling may result in exposures to airborne particles of that material, usually during transfer operations. Hazards associated with dumping of solids and filling containers are discussed in Sections 4 and 11.

The common tray dryer probably offers more opportunity for contact (skin, eye, and inhalation) than any other. Usually trays laden with material must be inserted into and removed from slots in the dryer by hand. Contact hazards vary with the height of the slot above the floor and the height of the operator. Mechanization of tray filling and dumping is possible and practical even if the material must be dried in thin layers.

Rotary kiln dryers usually operate continuously and often are unattended. Dust leaks from seals create most of the problems and can be minimized with good design.

Where the drying process itself creates a hazard from pyrolysis of either the solvent or the solid, a different kind of dryer is probably the best solution. Freeze drying was invented as a technique for use when the solid to be dried was very sensitive to above-ambient temperatures even though that sensitivity did not create a hazard. Nevertheless, this most gentle of drying methods is available as a potential solution to an inhalation hazard problem. For additional information, see Chapters 14, 25 and 26.

12.4 Equipment Location

Because they tend to be large and heavy, furnaces are usually located on the ground; ancillary structures can be found both above and below them. Dryers, on the other hand, are found throughout structures, and because their product is a dried solid that usually must be handled further, they are often high in a vertical structure. Location of a dryer is not usually an important determinant of its associated hazards.

Hazards associated with furnaces can be very severe. A part of their solution is, or can be, isolation. This is facilitated by interruptable transport as exemplified by a bridge crane serving one or more furnaces. The bridge crane operator can be isolated in a cab supplied with clean (and possibly cooled) air. Charge material can be delivered to the furnace by the crane and melt removed in a ladle carried by the crane. Furnace operators can be isolated as necessary in shielded (insulated) control rooms similarly supplied with cleaned and probably conditioned air. This type of operation is possible and practical only with an essentially horizontal equipment plan.

A ladle of hot metal transported by a bridge crane may become a large source of metal fume that can be handled in several ways. First, the ladle can be equipped with a removable cap. This will reduce but not eliminate fume emission. The next step might be to equip the ladle with a local exhaust ventilation system that travels along with it. Air from the hood over the ladle travels through a duct connected to a permanently installed main equipped with a traveling seal for the ladle branch. Although this method is rather expensive (it is proprietary) and does require periodic maintenance, especially for the traveling seal, it is effective.

Another method is to transport the hot metal with a rail cart. There is no weight limitation because the cart moves on rails that are on the ground. The ladle can be capped and, if necessary, the cart can be equipped with its own exhaust fan and air-cleaning system to serve as the local exhaust for the ladle.

Whether a cart or crane is used for ladle transport, the ladle should be filled under a local exhaust hood if effective fume control is to be practiced. This is often easier with a rail cart than with a bridge crane because pouring hoods tend to interfere seriously with bridge crane operation. For this reason, the rail cart approach is preferred when its lack of flexibility can be tolerated.

Interruptable transport advantages are sometimes forgone when distances are small and lined troughs are used instead. Although such troughs can be equipped with long canopy hoods to capture emitted fume, they rarely are and thus become "line" fume emitters.

13   AIR CLEANING

13.1   The Process

There is no reason why, in theory, any material cannot be removed from air. Practically, however, almost all air cleaning is done to remove particulate material; removal of vapors/gases may be very important to a particular company or process, but is of minor importance industrially compared to particulate removal. The topic of removing gases and vapors from air is treated briefly in Section 13.2.6; the remainder of this discussion is devoted to particulate material.

Air is cleaned for several reasons, some of which may operate simultaneously at any location. On a mass basis, most of the material removed from air is that which otherwise would increase air pollution outside of buildings. Perhaps next in importance are the amounts removed at intake systems serving office buildings, commercial establishments, and even industrial plants. (There are few health hazards associated with systems that clean clean air; such systems will not be discussed further here.)

Of minor importance on a mass basis, but on occasion of major importance to a process, building, plant, company, or industry is the cleaning of air contaminated by processing and then deliberately recirculating it back into the work space. This topic is discussed in Section 13.4.

This discussion, of necessity, avoids such interesting topics as how to determine what type of air cleaner is optimum for the job, how to size the equipment, test it, and so forth. Instead, the subject will be, as in the other sections, the special provisions of layout, arrangement, and selection most important in avoiding and/or minimizing hazards from operation and maintenance of the equipment itself. These subjects are often overlooked or given inadequate attention simply because under most circumstances the process would operate just as well even if no air cleaning were done. Air-cleaning equipment thus is regarded as an item of overhead expense rather than of production expense, usually installed because a rule, regulation, or law dictates its necessity (there are many exceptions, of course). A need for proper selection and location for efficient and safe operation and maintenance may thus escape the designer's full attention.

One of the more important principles in the selection and location of air-cleaning equipment is that this job should be done by someone who is adequately informed. Otherwise, the equipment may be selected on the basis of a very simple performance specification and purchased only on the basis of the lowest bid. Even though air-cleaning equipment may be simple in principle compared to much production equipment, its design and even location can greatly influence maintenance requirements and hazards as well as performance.

Much air-cleaning equipment is specified and purchased with little or no attention given to the toxicity and/or hazards of the material to be removed from the air, perhaps because the designer is not aware that such topics even should be considered. If, then, a piece of equipment is chosen that requires much manual attention and thus opportunity for contact, real hazards can result that are very difficult to reduce. People who design plants and specify equipment must have a working knowledge at least of toxicity and injury potential of substances as well as a rather thorough knowledge of the equipment (such as air cleaners) to be specified.

13.2 Equipment Types

Devices used to clean air of particulate material can be categorized in several almost mutually exclusive ways based, for instance, on the temperature or flow rate of the gas handled, size range of the particulate material, and whether the particulate material is dust, mist, or fume. Here, only a few representative types of *dust* (using the word in its most generic sense) collectors will be discussed, namely, settling chambers, cyclones, filters, electrostatic precipitators, and wet collectors.

13.2.1 Settling Chambers. Settling chambers are simply enlarged pieces of duct work in their simplest form, perhaps equipped with internal baffle or impingement plates. Their purpose is mainly to reduce the velocity of the flowing gas to allow sufficient time for the larger particles to either drop out or impinge on a plate and thus be captured. They can be operated either wet (see 13.2.5) or dry; dry is the usual state.

Efficiency of a settling chamber is increased in proportion to its volume; these devices, however, are inefficient almost by definition and tend to be quite large even so. The only good reason for using a settling chamber is a high proportion of very large particles in a gas stream that has a high concentration of particulate matter. The usual reason (read "excuse") for a settling chamber is its low first cost and low maintenance/upkeep.

Settling chambers require little maintenance if operated properly and generally pose few hazards for workers. When a settling chamber is used as a "skimmer" to recover valuable product, it is usually located quite close to a packaging station and, in fact, may actually be a part of such a station. Used only occasionally for packaging, however, the chamber's hopper/valve may not be equipped with necessary exhaust ventilation. If a particle can become airborne once, it usually can do so again with little provocation. Packaging from a settling chamber can thus be the source of major dust exposures.

13.2.2 Cyclones. Inefficient as most are, cyclones are much more efficient than settling chambers. Costing but little more in many cases, they usually are preferred for the "first cut" from a dust-laden gas stream. For the amount of dust removed, cyclones are smaller than settling chambers, but still larger than many other types of equipment.

Cyclones can be and often are operated wet, especially when product recovery is not their primary purpose. This usually means that human contact with the particulate matter is minimal; sludge is handled with a pump, and there is no need for packaging dusty material. Wet and dry cyclones have low maintenance costs, unless the particulate matter is erosive, and generally low pressure drops; high pressure drop, high-efficiency cyclones are available and useful.

The main hazards associated with cyclones have to do with packaging dusty material. These devices, as other skimmers, should be equipped with good packaging hoods if dust exposures can be a problem.

13.2.3 Filters. Filters are available in many styles and varieties; most are operated dry, but some are wetted, usually with an oil. The media may vary from woven or unwoven fabric composed of almost anything that can form a fiber to pleated or unpleated paper. Filters can have efficiencies approaching 100% even for very small particles. In general, pressure drop through them is higher than for any other type of air cleaner usually encountered. Filters can be operated and cleaned manually or automatically. They may be used to clean particulate matter from a very dusty gas stream or to remove the last traces of dust from a relatively clean gas stream. They may be operated at room temperature to a maximum of a few hundred degrees Celsius. Generalizations are difficult.

Most filters are designed to be reused (high-efficiency particulate air— HEPA—filters being exceptions in most applications, see Section 13.2.6). They must be cleaned, manually or automatically, and eventually replaced. Filters designed to be used and then discarded are found mostly in "clean air" service; they pose few, if any, health hazards.

In the form of a cylinder or bag, a filter needs only minimal support to resist the force of the gas (usually air) passing through it; most industrial air filters are in that form. With an internal framework, air can pass from the outside into the bag, an advantage in automatic cleaning; when air passes from within the bag to its outside, no internal frame is needed.

Filters are usually found in a rigid shell called a "baghouse." Bags tear, burn, become clogged, and otherwise require periodic inspection and maintenance. The method of accomplishing these tasks is fixed by the design of the structure and may involve skin and inhalation contact ranging from almost nil to gross. Design of the baghouse can be of paramount importance when the particulate material is hazardous. A few examples and precepts follow.

Automatic bag cleaning in any kind of baghouse almost always results in far less skin, eye, and inhalation contact with the dust than any kind of manual bag cleaning.

The first precept, then, when a filter system is chosen for a hazardous material, is to insist on an automatic bag-cleaning system.

Some automatic bag-cleaning systems, even within the same generic class, require far less maintenance than others and obviously are to be preferred when the dust is hazardous. For instance, in the popular pulse-jet fabric filter, increasing the size of the cage wire diameter and reducing the knuckle radius of the bent wire at the bottom of the cage will usually reduce bag wear considerably, thus resulting in fewer bag changes, less general maintenance, and perhaps far less exposure to workers. In the same kind of filter the mechanism by which the wire cage is secured to the tube sheet may work well when clean but not at all well when dirty with a sticky or abrasive dust. That, in turn, may increase significantly the time required for a bag change.

Changing bags from the top of the baghouse (so-called top-entry design) almost always results in far less exposure to the dust than does a method requiring entry through the body of the structure. Filters that require bottom entry for bag replacement and in which the cage is held in place by a rod fastened with a wing nut cause more skin, eye, and inhalation exposure than almost all others.

Many baghouse purchasers are assured by salesmen that a need for maintenance and/or bag replacement will be signaled reliably by a change in the pressure drop across the system. In practice, however, visual inspection of the bags at periodic intervals is found to be absolutely necessary when there is a real need for continual high-efficiency gas cleaning. When the system is large, visual inspection can only be achieved by entry of the baghouse. The design must accommodate such entry and resulting inspections. A good rule of the thumb is that no bag should be separated from an internal walkway by more than one, or at most, two other rows of bags.

13.2.4 Electrostatic Precipitators. Electrostatic precipitators (ESP) induce a charge on dust particles and then remove them from the gas stream by attraction to plates having the opposite charge. All large ESP units impart a negative charge to particles; most small units use less efficient positive ionization instead to avoid, so far as practical, ozone formation.

Advantages of ESP units are several. They can be constructed entirely of steel and thus resist effects of temperatures high enough to cause rapid deterioration of fabric bags. They can be very efficient for particles in the size ranges usually encountered. They have a relatively low pressure drop, require little maintenance, and are inexpensive to operate.

Their main disadvantages are large size, heavy weight, relatively high initial cost, and a need for a high-voltage power supply with its associated hazards and problems. In addition, electrostatic precipitators do not efficiently remove particles that either are very electrically conductive or are very good electrical insulators; their efficiencies are also altered by the electrical conductivity of the gas stream. Efficiency (no matter how measured) is rarely constant with an electrostatic precipitator. Electrical arcing takes place within these devices; therefore fire and explosion are possibilities when the dust contains a large flammable organic component.

Notwithstanding these disadvantages, ESP units have been used successfully to clean gas streams of particulate material ranging from fly ash to welding fume to oil mist in industrial application. Large units are usually cleaned automatically by "rappers" that vibrate dry collection plates or by being operated "wet" with flowing water or oil. Wet units produce a sludge that is usually pumped; worker exposure except during maintenance is rare. Maintenance of both wet and dry units is associated mostly with replacement of the wires used to ionize the gas and particles within it; such replacement can usually be accomplished from the top with no need to enter the device. Small electrostatic precipitators are usually cleaned manually. The process normally requires some disassembly with subsequent opportunity for contact with the dust. At such times the discharge electrodes (usually wires) are inspected and replaced, if necessary. The only other maintenance of large or small units may well be on the high-voltage power supply, unassociated with dust exposure. Where they can be used, electrostatic precipitators can greatly reduce exposures to materials cleaned from gas streams.

13.2.5 Wet Scrubbers. A settling chamber operated with a spray of water wetting baffle plates is one kind of wet scrubber. Wet scrubbers can also be tanks partially full of a liquid (usually water) through which the dirty gas is bubbled or can be towers packed with raschig rings, saddles, or more modern surface-extending shapes. Water and other fluids have been added to most types of air cleaning equipment.

Wetting the scrubber usually has a few main desired effects and perhaps a few not so desired. First, wetting usually makes the device somewhat more efficient at dust removal than dry operation. Second, wet scrubbing results in a sludge, slurry, or even a free-flowing liquid that can be handled with a pump; worker contact is this reduced. Third, wet scrubbing almost always results in some cooling and humidification of the gas scrubbed; this may be either a disadvantage (especially when it results in corrision) or an advantage. Fourth, wet scrubbing usually reduces the need for clean-out and contact associated with that kind of work; when cleaning is necessary, it can often be done with a hose.

The main disadvantage is that of sludge or slurry removal; this often creates problems easily avoided if the material is handled dry. Other disadvantages are the need for a continiuous supply of water or other wetting fluid, increase in weight of the equipment over that used dry, and so forth. When the dust is a real hazard, however, wet scrubbing should be examined thoroughly as a means of essentially avoiding contact.

13.2.6 Other Types. Many categories of dust removal systems have not been considered at all, from the mechanical centrifugal separator to the water- or steam-jet scrubber to the impingement separator. The neglected equipment has no hazards different in kind from those already discussed, and most hazards are far lower in intensity.

Recently the fluidized bed has been employed for air cleaning, not only for its high cleaning efficiency but also because the bed can be returned to process stream (see Volume 1, Chapter 4).

HEPA FILTERS. These high-efficiency particulate air-cleaning devices are constructed of pleated filter paper and are very efficient for capture of even very small particles. Most are used as throwaways, but in some cases they are adapted for reuse with a "rapping" technique similar to that used for electrostatic precipitators. Pore sizes are small in the filter media used, however, and these filters are therefore subject to clogging.

When used as disposable units, HEPA filters are often found as the last stage before discharge of the cleaned air to the environment. They remove the "traces" of contaminants remaining after pretreatment and can often stay in service for weeks or months before a change is necessary as dictated by an increasing pressure drop and reduced air flow rate.

HEPA filters are often used for particulate material that is especially hazardous to people: radioisotopes, beryllium oxide, some of the insecticides, and so forth. Especially in these cases, no attempt should be made to change the filters while the exhaust system is in operation; and people doing the job must be protected adequately. In addition, upon removal each filter should be separately packaged in an impermeable plastic bag and then disposed of properly.

VAPOR REMOVAL. A final item in the "other" category is that of cleaning air of contaminating vapor. The methods available include freeze-out, adsorption,[6] absorption, direct or catalytic combustion, and noncombustion reaction (if that may be distinguished from absorption). The only methods that pose hazards that warrant specific attention are those using reactive materials such as combustion catalysts and liquid reactants; an adsorbent saturated with a combustible vapor may be more hazardous to handle.

The most favored combustion catalysts are metal/metal oxides from the platinum group. Some of these materials are capable of causing skin and/or respiratory tract sensitization following prolonged minor or even gross acute contact. In their usual form, however, little contact is necessary.

Potassium permanganate solutions pose no particular problems although another "liquid combustion" medium, chromic acid and its derivatives, should be used only under circumstances where exposure to the material and/or to mists is controlled.

13.3   Equipment Location

Many of the problems attributed to poor hood design, inadequate air velocities or volumes, and the like are actually the result of two other causes: leaks from duct work within the building and inadvertent recirculation of contaminated air to the building interior. These will be discussed in turn.

With the possible exception of some welded material, all duct work leaks. Most welded duct work also leaks if only because a sealing weld is difficult to obtain on sheet metal. Therefore, air within duct work should be at a lower pressure than air surrounding in-plant ducts so that all leaks are into the duct. In this way, air contaminants carried within the duct work are not readded to the plant atmosphere. The only penalty of the leaks is then that of efficiency loss—a penalty exacted whether the leaks are into or out of the duct work.

Assuring that the air within inside-the-plant duct work will be at a pressure lower than atmospheric can be done only by having that duct work on the intake side of the blower serving it, never on the discharge side. The most usual means of doing this is to have the blower(s) on the roof. This is possible and practical as long as the building is low or the hoods served by the duct work are on top floors, and when no heavy air-cleaning equipment need be mounted on a poorly supported roof.

Of course, mounting blowers (and possibly air-cleaning equipment) on the roof means that people must go there to service that equipment. This may not be a particularly onerous task in some climates, but in the more northern parts of the world rooftops are quite cold and perhaps snow covered in winter. Because of that, other means have been sought to accomplish what must be done—without much success.

The first attempt is usually to mount the blower(s) just beneath the roof so that only a short section of duct is at a relatively high positive pressure. That, in turn, results in (usually but not always) only minor leaks from the duct to the indoor atmosphere. Mounting blowers and air filtration equipment just beneath the roof, however, almost always results in even more servicing difficulty than is associated with a rooftop location. When the maintenance people must find long ladders to climb for servicing, either service quality or frequency or both will usually suffer. Just above the roof is generally far better than just beneath it.

A second means of avoiding rooftop locations for blowers and ancillary equipment is to mount that equipment either just inside or outisde a building wall. Air from the system is then discharged at the same point along that wall either in a horizontal or vertical direction. If that air has been well cleaned before discharge, then probably no harm has been done by this. Much if not all of the air discharged along a building wall, however, may make its way back into the building (where it will be "recircu-lated") unless the building is much more tightly constructed than is usual for even well-made industrial buildings. This is true even for those buildings where adequate amounts of makeup air are furnished. Wind from the "wrong" direction will overcome even internal pressure created by makeup air and force contaminants through any points of leakage. Of course, many industrial buildings having exhaust ventilation are always under negative pressure, exacerbating the problem.

Finally, the blower(s) and ancillary equipment could be located beside a building to discharge air from the exhaust system through a duct (stack) that terminates high in the air, perhaps supported by the building. This method will work well to prevent inadvertent recirculation provided that the duct is relatively free of leaks, terminates high enough, and discharges straight up. "High enough" is, unfortunately, not com-pletely defined. Two rules of the thumb are: A discharge point half the building height above the roof will prevent almost all recirculation, and the discharge point should never be less than about 3 m (10 ft) above the roof. This, however, is an exceedingly complex subject and some experimentation (with the building or perhaps a wind tunnel model) may well be necessary before an optimum solution is found.

To discharge "straight up" is not at all difficult. If the blower will be operated more or less continuously, with an air velocity at discharge of at least 800 m/min (2500 fpm), no precipitation will enter the stack. The terminal velocity of the largest rain-drops is about 800 m/min.

In those cases where the blower will not be operated continuously, keeping pre-cipitation from the stack can be accomplished by use of a "concentric cylinder" weather cap that allows vertical discharge. Specifications are that it should have a diameter 2–5 cm (an inch or two) larger than the stack, should overlap the stack only enough for good support, and should have a height of about four stack diameters or somewhat more. Rain rarely falls straight down; most will strike the weather cap and run out through the annulus between the cap and the stack.

With (actually or possibly) contaminated air being discharged straight up well above the building, air for makeup can be taken from any convenient point at the side or roof with no fear of recirculation problems (unless this building is close to taller ones, in which case "well-above" applies to them, instead).

Most air-cleaning equipment in common use has been designed for outdoor lo-cations because it can be large, heavy, or both. So long as the air is actually cleaned well, its discharge can usually be at any convenient point but still should be straight up and relatively high. Even air-cleaning equipment breaks down occasionally.

13.4 Recirculation of Cleaned Air

In an attempt to save some of the energy required to heat ("temper") air, especially in northern winters, two general methods have been employed. Historically the first was use of a heat exchanger contrived so that the warm discharge air would heat the cold incoming air without mixing with it. Heat exchanger installations are expensive and rarely work as well as advertised. In the late 1970s and early 1980s the rising cost of energy caused some rather thorough investigations of deliberately recirculating cleaned air back to the work space.

Recirculation had, indeed, been practiced for many years with air contaminants that were nuisances rather than hazards—oil mist, for example. The technique was to mount an air cleaner (a small electrostatic precipitator or fan–filter combination) near and usually above the generating process. These devices cleaned the realtively clean air found at their intakes and discharged the cleaned air to the general room atmosphere. In some plants, the intake was connected to a duct that terminated close to the source of the contaminate with or without some kind of hood. Few objected to this practice because of the lack of real hazard associated with the contaminants; the air cleaners were regarded as devices to increase comfort, not to reduce hazard.

An extension of this idea was to use it where the contaminant posed a real inhalation hazard, with crystalline silica or lead, for example. Much of the research was sponsored by NIOSH, and results will be found in several publications of that institute.[2,7-12] In general, the hazard will be low provided several circumstances apply. The system must have a means of sensing the air contaminant downstream of the air cleaner in concentrations lower than the control concentration (Threshold Limit Value, permissible exposure limit, etc.). At such time as a preset limit is reached, the sensor must cause the air cleaner discharge automatically to be diverted from the workroom to the general atmosphere outside the building. And finally, air samples in the breathing zones of people subject to the recirculated air must show the absence of hazardous concentrations and/or exposures.

The presence of potentially hazardous concentrations of particulate material in duct work on the discharge side of an air cleaner is easier and cheaper to detect reliably than that of a gas or vapor. For that reason, deliberate recirculation has been applied far more often to particulate air contaminants than to gases and vapors. In addition to an in-duct detector of some sort (usually optical), workroom air is often monitored automatically or at short periodic intervals for the contaminant of most interest. Under these circumstances, this method can result in large energy savings with no concomitant increase in health hazards. (See Chapter 7.)

14 ACID WASHING AND PICKLING

14.1 The Process

Whenever a coating of any kind is to be applied to a surface, adhesion is facilitated by surface cleanliness. In plating this means that there will be absolutely nothing between the metal surface and the plating solution.

One of the first cleaning steps for several metals (particularly mild steel and beryllium–copper alloys) is pickling—treatment by acids for oxide and other scale removal. The metal is allowed to remain in a strong acid bath for a period determined by the kind of metal, thickness of scale, and so forth. The acid is usually hydrochloric (muriatic), but pickling can be done with almost any of the strong acids; cost and rapidity of action determine which is used.

Pickling is always followed by a rinse to rid the metal of excess acid and then, perhaps, by electrocleaning where an electric current, usually in an alkaline solution, is used to effect removal of remaining soil. After another rinse the metal may well be clean enough for paint or other coating, but generally more cleaning is required for electroplating.

The last step before plating the metal is an acid wash or dip to remove any remaining trace of oxide or scale (only traces or less should remain), to neutralize any remaining alkali that might be present, and to clean the metal of any other film that may remain. The treatment is with a dilute acid (hydrochloric, sulfuric, nitric, phosphoric, fluoroboric, acetic, hydrofluoric, or citric) for a brief period. In addition to (or perhaps mostly because of) final cleaning, this step also "activates" the metal surface in preparation for the plating that follows a rinse.

For some metals there may well be several other cleaning steps than those outlined here, one or more of which may use an acid. See Chapters 13 and 17 for additional information.

14.2 Batch Processes

For batch pickling or acid washing, the pieces to be cleaned are almost always lowered into an uncovered tank containing the acid solution, allowed to remain for a few seconds to a few minutes, and then transferred to a water wash tank. Reactions in the acid tank cause gas formation in may cases and, in addition, the solution is splashed and dripped as parts are lowered into and raised from the bath. Transfer of the parts to be cleaned is usually done with a bridge crane.

Agitation of the bath surface causes mist formation; this and/or gases produced in some reactions (especially NO_2 from nitric acid treatment) may pose a health hazard and also quite possibly a hazard to materials of construction in the plant. For these reasons control generally is necessary. Local exhaust ventilation is difficult to use satisfactorily because of interference with the bridge crane operation. This means, in essence, that there can be no hood or duct work above the pickling and acid wash tanks.

There have been two general approaches to this problem. In the first, the problem is simply ignored. This approach has resulted in rusting of steel building beams, nearby machinery, and even of parts produced in the plant under certain conditions.

When air space above an open tank cannot be occupied with an exhaust hood, the only real solution if a hood is necessary is to use some sort of side-draft arrangement. The ACGIH manual *Industrial Ventilation*[1] has devoted several pages to the design of side-draft hoods for open surface tanks and should be consulted for details. In addition, NIOSH has investigated in detail the use of push–pull systems for tanks and has found that such arrangements can effect excellent control of mists and vapors with considerably less exhaust than required without the push part of the system. Recent NIOSH publications on this topic are suggested reading.[13,14]

Even when a good side-draft or push–pull ventilation hood system has been devised, the problem remains how to remove the exhausted air from the tank vicinity. Bridge crane operation may dictate that absotutely no air space is available for duct work, in which case the only solution is to go down instead of up. This, in turn, means that the building where the pickling and acid washing is done usually must be equipped with a basement through which duct work can be directed; basements are difficult to retrofit. A far less satisfactory means of finding room for duct work is to raise the tanks and install duct work beneath them. This "solution" may cause as many problems as it solves.

One step toward a solution is to cover acid wash and pickle tanks when they are not in use. Unfortunately, such covers may do little good because mist and vapor leave the tanks mostly during use, not while the liquid surface is quiescent.

The usual attempt at retrofitting a solution is to install hoods on tanks perceived to be the main sources (often without measurement to quantify the problem) and run undersized duct work wherever possible (along floors, walls, etc.) to accomplish partial even if not complete control. The best solution may well be to construct a new facility based on proper design.

14.3   Continuous Processes

Many modern electroplating operations are conducted in a series of tanks served by an automatic cycling system arranged to dip the parts to be plated into tanks successively. One or more of these tanks will almost certainly be an acid wash (pickling is unlikely in such a system). Local exhaust ventilation may be necessary as outlined in Section 15.3.

Pickling and acid washing pose special problems when the metal to be pickled or washed is in the form of a continuous sheet or roll. Not only is the bridge crane approach still necessary to transfer rools, but at each bath, arrangements must be made to unwind the roll at the proper rate, lead the metal through the bath (and usually a following rinse), and then dry and wind it again. A necessity for these operations puts even more constraints on any possible local exhaust ventilation system than those associated with batch operation. Retrofitting any local exhaust may be nearly impossible, not simply impratical.

Neither the kind of hazard nor its magnitude is likely to be altered significantly by changing from batch to continuous operation or vice versa; the amount of acid mist and/or vapor produced is a direct function of the area of metal surface exposed to the bath, the amount of scale on the surface, and duration of exposure to the acid.

14.4   Equipment Location

Large tanks full of acid for washing and/or pickling are heavy and for that reason alone tend to be on the "ground" floor. Necessity for service by a bridge crane may convert a tendency to an absolute need. Local exhaust ventilating hoods at such tanks must be attached to duct work and, for many reasons, the best place for that is beneath the tanks but not on the operating floor. Trenches might well suffice for the duct work except that the interior of ducts becomes contaminated, making routine cleaning necessary. Furthermore, duct work deteriorates with time. These considerations suffice to indicate that a basement where people can stand and walk upright is very desirable.

Horizontal process equipment layout with exhaust ventilation duct work in a basement thus can accommodate most pickling and acid wash facilities nicely. Any other configuration is likely to be more troublesome and, in design stages, should be compared with that indicated here before a final decision is made.

15 ELECTROWINNING AND ELECTROPLATING

15.1 The Process

Electrowinning and electroplating are processes that use a flow of electric current to produce chemical reactions. In electroplating the object is to cause metal atoms to precipitate ("plate") onto an object to be plated, the cathode of an electrochemical cell. Electrowinning uses a somewhat similar process (but usually far different equipment) to produce a more or less pure element (often but not necessarily a metal) from a chemical compound of that element. In both cases direct electrical current is used at quite low voltages but often rather high current densities.

15.2 Electrowinning

Many metals and a few nonmetals are or can be produced by electrolysis of a compound, usually a salt. Metals that can be electrically "won" include aluminum, calcium, cerium, cesium, copper, gold, lithium, magnesium, manganese, nickel, potassium, rubidium, silver, sodium, strontium, tantalum, thallium, thorium, tin, uranium, and zinc in addition to all of the rare-earth metals; commercially, electrowinning is used for alumium, copper, magnesium, and zinc. Nonmetals that are or can be produced electrically include the halogens (fluorine, chlorine, bromine, and iodine) and hydrogen; commercially, electrowinning is used mainly for chlorine (that, in turn, is used to produce the remaining halogens) and fluorine.

When aluminum and magnesium are electrowon, the processing is anhydrous, the electrical current being passed through a molten fused salt or mixture. Electrolysis of a water solution is commercially important in the electrowinning of only copper and zinc. Other metals that may be or are produced in this manner include cesium, gold, indium, manganese, nickel, rubidium, silver, thallium, tin, and uranium.

15.2.1 Fused-Salt Methods. Aluminum is produced exclusively by fused-salt electrolysis; magnesium is produced either by that method or one called silicothermic. For aluminum, bauxite is beneficiated so as to be essentially pure Al_2O_3 and then dissolved in a mixture of metal fluorides for electrolysis. Differences in processing technique are related to the electrodes used and their configuration. When the electrodes are prebaked, exposures to coal-tar pitch-related material driven off from the hot electrodes are less in the aluminum reduction rooms than when the Soderberg process is used. In the Soderberg method, precast but unbaked electrode material is added as electrodes are consumed; heat of the process bakes them, causing graphitization, but in the process volatile materials are given off from the coal-tar pitch component of the green electrodes.

Soderberg process electrodes can be used in either a vertical or horizontal configuration. Limited data suggest that concentrations of the coal-tar pitch volatile material is highest with the horizontal electrodes. In all three methods, exposures to fluoride salts are easier to control with ventilation and judicious operation than the pitch-related materials.

In some aluminum processing, molten metal from the electrowinning operation is further processed by bubbling through it a reactive gas such as chlorine to further remove impurities. Exposures to the gas can be significant and are best controlled by process design in which the gas is substantially consumed by the molten metal. Local exhaust ventilation is often difficult to install after the fact, however, because of a lack of room. In those cases, control has been attempted with general ventilation, sometimes using what is almost "plug flow" with limited success. For additional information see Chapter 4 of Volume 1.

Electrowinning of magnesium is done from electrolysis of $MgCl_2$ that, in turn, is produced from $Mg(OH)_2$ precipitated from sea water with $Ca(OH)_2$. In addition to the magnesium, chlorine is also produced to be recycled and react with the $Mg(OH)_2$ to form the $MgCl_2$. Molten magnesium reacts violently with air; oxygen and air must thus be excluded from the process, which also serves to assure that the chlorine produced will be captured efficiently. Fluxes used to exclude oxygen from the molten metal during the casting of ingots, however, may cause exposure to sulfur dioxide and/or fluorides.

15.2.2 Water Solution Methods. Several metals are or can be prepared in essentially pure form by electrolysis of water solutions; greatest commercial importance is associated with production of copper and zinc.

When a water solution of any salt is electrolyzed, there are at least two kinds of reactions that occur; the desired one in which metal is precipitated at the cathode and an undesired one in which the water is dissociated into oxygen and hydrogen. Electrolysis of water not only wastes electrical current but also causes gas formation; gas bubbles break the surface of the liquid causing mist formation that constitutes the chief hazard associated with most electrolysis processes.

Copper is plated from a solution of cooper sulfate ($CuSO_4$) kept acidic with sulfuric acid (H_2SO_4). The $CuSO_4$ is made by reaction of relatively pure blister copper with sulfuric acid. Zinc is usually plated from a basic solution (acidic solutions can be used but are more suited to electroplating) made by leaching a relatively pure ZnO with a NaOH solution. In both cases electrolysis is conducted at relatively high current densities of, perhaps, a few hundred amperes per square foot. Electrode efficiencies are very high for copper electrolysis; less so for zinc; gassing is therefore less serious for the copper electrowinning process than for zinc. In either case, particles of mist have good "warning properties" (irritation of the upper respiratory tract). Operating conditions that result in excessive mist formation are wasteful of electrical current and therefore are not usual.

Ventilation of electrolytic cells is somewhat difficult because of the necessary presence of bridge cranes to handle the metal produced in the process and the many electrical cables in the cell room. Nevertheless, when there is a problem with mists, local exhaust ventilation[2] is by far the most certain method of solving it, usually by means of a side-draft hood or even a push–pull system (see Section 13.2).

Mist generation at the surface of electrolytic solutions can be suppressed by addition of surface-active agents to reduce surface tension and by floating material on the surface such as plastic chips or balls that cause much of the mist formed to condense and run back into the bath. Neither of these solutions is generally practiced in electrowinning of copper and zinc but could probably be used if necessary. For additional information see Volume 1, Chapters 16 and 33.

15.2.3 Nonmetals. Electrowinning for nonmetals is practiced chiefly with fluorine and chlorine. Chlorine is of far greater commercial importance; this discussion is restricted to it.

Almost all chlorine is produced by the electrolysis of a solution of sodium chloride called brine. In that process, chlorine is formed at the anode and sodium hydroxide at the cathode. Diaphragm cells separate anodes from cathodes by means of a permeable wall usually formed from asbestos fiber; sodium hydroxide from the cathode area is kept from migrating toward the anode by hydrostatic pressure across the diaphragm. Diaphragm cells have had commercial success in several modifications (Allen-Moore, Dow, Gibbs, Hooker, LeSueur, Nelson, Townsend, Vorce, etc.), each differing in important aspects from the others; they have in common a diaphragm protecting the anode area from the remainder of the cell. In all, essentially pure chlorine gas is evolved at the anode; liquid in the cell accumulates hydroxyl ions to the point where half the salt has been transformed into caustic (sodium hydroxide).

The mercury cell uses liquid mercury as the cathode. Sodium forming at the cathode amalgamates with the mercury to be recovered later by reaction of the mercury with water to form sodium hydroxide and essentially pure mercury that is recycled.

Hazards associated with chlorine (and caustic) production are several. In an effort to reduce the hazards associated with handling asbestos in the formation of diaphragms, several other materials have been used, including glass and ceramic fiber and even plastics. Nevertheless, asbestos appears to have more of the required properties than any of its potential substitutes and therefore its use continues.

Cell diaphragms do not last forever; they must be removed, reformed, and replaced at fairly regular intervals that depend on the type of cell in use. Probably the single most important method of limiting exposure to asbestos during this process is to make certain that the asbestos, whether in the old diaphragm or the new fiber, is always wet or at least damp. In this manner, dust formation and consequent fiber inhalation are inhibited almost completely. The used diaphragm material should be securely packaged (usually in sealed plastic bags) for proper disposal.

The main hazard associated with mercury cells is the mercury itself. Hazards of mercury inhalation have long been recognized, and processing facilities have been designed to minimize them. In fact, in the cell room and in the caustic formation room measures adequate to prevent irritating concentrations of chlorine and sodium hydroxide in the air will almost always be adequate to assure less-than-hazardous mercury concentrations.

In cell rooms one of the main concerns is to collect every bit of chlorine gas that forms. To this end, electrolytic cells are always equipped with a chlorine collection system operated at a slight negative pressure with respect to the atmosphere. When an upset forces that system to a positive pressure, the gas leaks out into the cell room air. Also, cells must be opened for maintenance occasionally, and if some chlorine remains or the collection line is not blanked properly, some gas will escape to the general atmosphere. The result of these circumstances is that most chlorine cell rooms are characterized by a faint odor of chlorine most of the time and by higher chlorine concentrations some of the time.

In diaphragm cells, hydrogen is formed at the cathode and chlorine at the anode. At the cathode the breaking of hydrogen bubbles, if there is liquid at the cathode, causes mist formation that consists of varying amounts of sodium chloride and sodium hydroxide in water (that usually evaporates, leaving the solid behind). Therefore, in many chlorine cell rooms the occasional presence of chlorine in the air is impressed on a background of particulate material containing caustic. This does not occur in mercury cell rooms; the problems associated with caustic-containing particulate material are reserved for the caustic recovery (production) area.

Chlorine is quite corrosive to most materials of contruction (even concrete), especially when it is wet or damp. Because of this, complete containment of chlorine in cell rooms is usually not regarded as practical. Keeping the collection system at negative pressure is essential, but if the differential is too great, leaks allow too much air to enter; the pressure differential is therefore usually kept rather small, thus allowing occasional upsets. Where the next leak will occur cannot be predicted well; therefore, chlorine cell areas usually are designed for rather large amounts of general ventilation.

Rather than attempt to provide leak-free cells and collection systems or local exhaust ventilation at the most leak-prone points, many chlorine cell buildings are equipped with more or less elaborate general ventilation systems. These may supply tempered (in northern areas) or untempered air if the need for a supply system has been recognized, but always exhaust air from the building at one or more points. An attempt may be made to induce plug flow by causing air to enter at one end or side of the building and flow with little or no distrubance to the opposite side where it is exhausted. This kind of system may offer more efficient dilution within the building than that provided by the usual roof exhausters but seldom allows complete abandonment of respirators.

General ventilation of any type has two other distinct disadvantages particularly in chlorine cell rooms. First is that chlorine can and does react with steel construction materials, causing them to corrode much more rapidly than in other atmospheres. General building ventilation, of course, acts to supply fresh chlorine to the corroding metal at all times. The concentrations need not be above those tolerated easily by humans to be quite destructive. Protective coatings can deter but seldom if ever prevent this attack. Second is the air pollution problem. General ventilation systems of necessity move relatively large amounts of air having relatively low concentrations of contaminants. If air pollution regulations were to limit emissions on a mass per unit time basis, control would be much more difficult and expensive than for any local exhaust system.

Other protective equipment normally used in chlorine cells consists of rubber boots and gloves, safety glasses, and a rubber or plastic apron. Hazards of skin and eye contact are considerable and thus, fire and eye showers should be easily available to all. For additional information see Volume 1, Chapter 14.

15.3 Electroplating

The process of electroplating is disccussed rather thoroughly in Chapter 13 and will not be repeated here except in summary.

The chief hazards in electroplating are those of mists from the solutions used (chromic acid is one of the main problems) or the materials handled in making up the solutions—various salts of hydrocyanic acid in particular.

Hazards associated with mist formation are best controlled with local exhaust ventilation. Usually the optimum hood in this process is the side-draft variety, perhaps with a push system on the opposite side of the tank (see Section 13.2). Push–pull tank ventilation can work very well to control mists, but if used incorrectly can actually compound problems. Careful attention to operating parameters, interlocks, and maintenance is almost always necessary.

Electroplating solutions are usually at least somewhat corrosive to the usual materials of construction and, in addition, the material collected by the ventilation system is a wet mist. Wet mists tend to collect in duct work and thereby cause other airborne material to be trapped even when the transport velocity is high enough to allow little or no dropout under other circumstances. Hence ventilation systems for plating tanks must be constructed of material inert to this environment or heavily protected with such a coating inside and out, and they must be cleaned at far shorter intervals than those normal for other systems. Need for cleaning can be signaled by measurement of either static pressure, air velocity, or simply passage of time. Ventilation systems for plating operations should be designed with the need for periodic cleaning in mind. One common method is to permanently install a water flush system in the duct work.

Many plating operations require the use of one or more cyanides. If so, then one the salts usually on hand is sodium cyanide. This white, crystalline material may be supplied in the form of briquettes to make it easier to handle and to reduce the surface area per unit weight. Carbon dioxide in the air is acidic enough to cause HCN formation by reaction with NaCN; this gives NaCN its typical bitter almonds odor. The reaction is, of course, inhibited by reducing available surface area.

All cyanide salts will release HCN in contact with any acidic material. Most electroplating requires one or more acids either for the plating solution(s) or for neutralization of alkaline solutions used either in washing or plating or both. Acids and cyanides must be stored in such a way that there is absolutely no chance for accidental contact. This means that one kind of material must not be stored on a platform over the other; that one kind of material not be stored in a compartment next to the other, and so forth. When cyanide salts contact a strong acid, HCN formation can be incredibly rapid; concentrations easily attainable in the open air can be rapidly fatal. There must be no chance for cyanide solutions to become mixed with acid solutions through equipment or floor drains.

Wherever cyanides are handled, one of the precautions that can be taken is to have on hand and easily available one or more "antidote" kits. These kits contain "pearls" (small glass vials wrapped in gauze) of amyl or butyl nitrite along with hypodermic syringes, needles, and injectable sodium nitrite and sodium thiosulfate. Directions for proper use are included with the kits. The main problem with these kits is to keep them easily available for use in an emergency (they can be and have been lifesaving) in such a way that the pearls, syringes, and needles are not easily stolen. These kits should only be used under medical supervision.

For additional information, see Chapter 13.

15.4   Equipment Location

Neither electrowinning nor electroplating is well suited to vertical plant layout and as a result, most such operations are located in mainly horizontal structures. Chlorine cells, especially, may be constructed with rather high ceilings so as to maximize the air volume available for dilution. Electrolysis cells, whether anhydrous or not, are usually heavy and, because electrical currents are high but voltages low (in this business, 12 is a high voltage), electrical cable is also heavy. Cell covers, electrodes, and even cells and disphragms tend to be large enough to require the presence of lifting aids, usually bridge cranes, in production facilities. All of these constraints almost dictate that electrowinning and and electroplating be done mainly in ground-floor locations.

Many, but not all, electrochemical operating facilities require local exhaust ventilation systems, and these are usually easiest to install and operate properly when the blower(s) can be on the roof and not separated from the main duct work by a large distance. This constraint may well cause the facility not only to be located on the ground floor but to be of the single-story variety.

Other than these considerations, there appear to be no other pressing health hazard control reasons why one type of structure or process location should be preferred over another.

REFERENCES

1. *Industrial Ventilation: A Manual of Recommended Practice*, latest edition, American Conference of Governmental Industrial Hygienists, Committee on Industrial Ventilation, P.O. Box 16153, Lansing, MI 48902.

2. *Recommended Industrial Ventilation Guidelines*, National Institute for Occupational Safety and Health, NTIS No. PB-266-227, 1976.

3. J. Goldfield, Contaminant Concentration Reduction: General Ventilation Versus Local Exhaust Ventilation, *Am. Ind. Hyg. Assoc. J.* **41,** 812–818 (1980).

4. H. D. Goodfellow and M. Bender, Design Considerations for Fume Hoods for Process Plants, *Am. Ind. Hyg. Assoc. J.* **41,** 473–484 (1980).

5. R. J. Heinsohn, D. Johnson, and J. W. Davis, Grinding Booth for Large Castings, *Am. Ind. Hyg. Assoc. J.* **43,** 587–595 (1982).

6. A. Turk, L. K. Otis, A. E. Steinberg, and T. B. Wolf, Assessing the Performance of Activated Carbon in the Indoor Environment, *Am. Ind. Hyg. Assoc. J.* **45,** 714–718 (1984).

7. R. T. Hughes and A. Amendola, Recirculating Exhaust Air, *Plant Eng.* March 18, 1982.

8. *Recirculation of Exhaust Air*, National Institute for Occupational Safety and Health (unavailable from NIOSH) NIOSH Publication No. 76–186, NTIS No. PB-267-396, 1976.

9. *The Recirculation of Exhaust Air . . . Symposium Proceedings*, National Institute for Occupational Safety and Health (available from NIOSH) NIOSH Publication No. 78–141, GPO No. 017-033-00303, 1978.

10. *Recommended Approach to Recirculation of Exhaust Air*, National Institute for Occupational Safety and Health (unavailable from NIOSH) NIOSH Publication No. 78–124, GPO No. 017-033-00281-9, 1978.

11. *Validation of a Recommended Approach to Recirculation of Industrial Exhaust Air, Volume I (Spring Grinding, Chrome Plating, Dry Cleaning, Welding and Vapor Degreasing Operations)*, (unavailable from NIOSH) NIOSH Publication No. 79–143A, GPO No. 017-033-00350-5, NTIS No. PB-80-161-474, 1979.

12. *Validation of a Recommended Approach to Recirculation of Industrial Exhaust Air, Volume II (Lead Battery, Woodworking, Metal Grinding and Enamel Blending Operations)*, National Institute for Occupational Safety and Health (unavailable from NIOSH) NIOSH Publication No. 79–143B, GPO No. 017-033-00351-5, NTIS No. PB-80-161-482, 1979.

13. D. J. Huebner and R. T. Hughes, Development of Push–Pull Ventilation, *Am. Ind. Hyg. Assoc. J.* **46,** 262–267 (1985).

14. *Development of Design Criteria for Exhaust Systems for Open Surface Tanks*, National Institute for Occupational Safety and Health (unavailable from NIOSH) NIOSH Publication No. 75–108, NTIS No. PB-274-222, 1975.

GENERAL REFERENCES

Burgress, W. A., *Recognition of Health Hazards in Industry*, New York: Wiley, 1981.

Clayton, G. D., and F. E. Clayton (eds.), *Patty's Industrial Hygiene and Toxicology*, 3rd rev. ed., Vol. 1., New York: Wiley, 1978.

Kent, J. A., *Riegel's Handbook of Industrial Chemistry*, 7th ed., New York: Van Nostrand Reinhold, 1974.

Perry, R. H., and Cecil H. Chilton (eds.), *Chemical Engineer's Handbook*, 5th ed., New York: McGraw-Hill, 1973.

CHAPTER SIX

Building Types

Knowlton J. Caplan

1   GENERAL CONSIDERATIONS

1.1   Migration of Contaminated Air

While there are some types of processes that almost demand a certain style of building, most industrial processes can be accommodated in any one of several types. The nature of the building has a strong influence on the migration of contaminated air in the workplace, and on the control of such migration when necessary. If the industrial process is one of the majority of kinds that can be adapted to different building styles, the requirements for control of general ventilation and migration of contaminants should perhaps take precedence over minor advantages and disadvantages relative to the process itself. Obviously, the lower the threshold limit value (TLV) for the contaminants in question, or the more difficult it is to achieve really effective local control of emission sources, the more important this factor becomes.

It is axiomatic that perfect containment of hazardous materials by the process equipment or by local control ventilation is impossible. There will be some emission into the general room air. The general ventilation requirements depend on the relationship between the amount of such emissions, the allowable concentration of the contaminant, and the specific relationships of where workers are located relative to other building spaces. There are so many possible combinations of these parameters that only generalities can be discussed in this chapter.

In practical terms, there is always some general air movement in any closed space. The ventilation engineer's definition of "still air," for example, is actually an air movement in the range of 15–20 fpm, (0.07–0.10 m/s) usually random in direction. This air movement is caused by local thermal sources such as people, lights, machinery; solar radiation on sides and roof of the building, wind pressures; and in most factories, air currents created by exhaust ventilation systems and by replacement air.

Very small forces are sufficient to create the air movement involved. The building type and design has an influence on whether these random migrations can be controlled, and how difficult and expensive such control may be. Let us postulate, for example, the presence of a single relatively significant heat source, such as a drying oven, which emits small amounts of contaminant that escape local control, in one corner of a large flat building. A significant thermal plume will rise from this heat source, which will then distribute itself along and under the ceiling. The direction of that distribution may change from time to time depending on sunshine and wind pressures. The contaminant carried by this air current will show up in unexpected places where there is no local source of air contamination. Even the tracing of such air currents is a difficult task, especially since they are variable. In general there would be only two ways to control that migration, either by full-height partitions or by hanging baffles in the roof spaces accompanied with significant amounts of exhaust ventilation; both solutions are expensive. Other types of building design, planning for isolation of this source, would be more effective and less expensive.

1.2 Background Contamination of Room Air

The relatively small amount of contamination that would migrate throughout a building from a "well-controlled" process should be well below the TLV for the contaminant and therefore assumed to be of little concern. Further examination, however, shows that this assumption may not be valid. This may best be explained by example. Assume a process for which the contaminant has a TLV of 10 ppm that is well controlled, and the exposure of the worker at or near that process is only 5 ppm. This obviously is significantly below the exposure limit and would be regarded as a good situation. However, there is also the implication that the general air migrating away from that process is contaminated to the level of 5 ppm. Let us assume that the plume of contaminated air migrates across the building under the ceiling and reappears in the working area near the floor on the far side of the building somewhat diluted, with a concentration of 2 ppm. These air currents, being fairly large in magnitude and of very low velocity, do not enjoy a rapid rate of dilution by mixing with room air. Other well-controlled sources may add to the burden, also.

At the remote impact point on the other side of the building, workers are also involved with the same material with the same 10-ppm control limit, but the migrating contaminant results in a background concentration of 2 ppm. In order for the worker exposure to remain below 10 ppm, the exposure due to the imperfect control of this second process must be limited to 8 ppm or less. This then makes the control requirement of the second process in effect 20% more stringent and more difficult than it would be if the background contamination were not present.

1.3   Reentry

Reentry is the term used for the phenomenon of the return into the building, through replacement air intakes, open doors and windows, and the like, of contaminated air that has been discharged through some exhaust system. This phenomenon is the cause of many unnecessary problems, especially for air contaminants of low TLV or objectionable odor characteristics, and its solution is relatively straightforward although not necessarily inexpensive. The cost of preventing reentry is influenced not only by the process but also by the type of building design and furthermore by the design of other nearby buildings.

The basic nature of the reentry phenomenon is illustrated by Fig. 6.1. The simplest case is a flat building on a flat plain with no nearby obstructions. The wind flowing over the building follows streamlines as indicated in the figure. The streamlines closer to the building are brought back down to the ground on the downwind side, and in addition create eddies in the lee of the building itself or of any significant obstructions on the roof, including the leading edge of the roof. In order for reentry to be prevented, the contaminated air must be discharged at a height such that under most wind conditions it is outside of the envelope of streamlines that create downwind eddies. The only other alternative is to install air-cleaning equipment in the discharge gas, which is usually quite expensive for the very low concentrations and relatively large air volumes involved. The contaminant concentrations of concern in reentry problems are typically well below any emission standards promulgated by the Environmental Protection Agency (EPA), and otherwise would not require a gas-cleaning device.

It should be noted that fugitive emissions—those which leak out or are discharged through natural nonmechanical ventilation devices such as roof monitors, or through general ventilation roof fans, also are ideally placed to cause a reentry problem.

Attempts to arrange air discharge and intake points to "take advantage of the prevailing wind" are, in general, futile. There are very few locations where the prevailing wind does in fact prevail for a sufficiently large percentage of the time to represent a satisfactory solution to reentry. In most locations, the wind will blow from any major point of the compass 10% or more of the time. Further, most buildings being square or rectangular, fine gradations of wind direction will not be significant in determining whether there is reentry. The configuration of the building—tall versus flat—has a strong influence on the parameters causing reentry and on the difficulty and expense of preventing reentry.

1.4   Stack and Discharge Characteristics

The behavior of the contaminated air plume discharge from a stack is complex. The major parameters are the stack discharge volume, the stack discharge velocity, temperature, and meteorological factors such as turbulence, wind speed, and the like. The complexities are such that most significant environmental air pollution problems

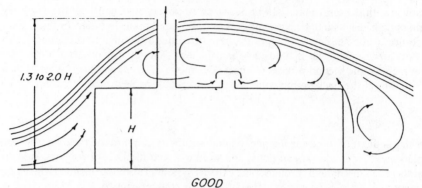

GOOD
High discharge stack relative to building height,
air inlet on roof.

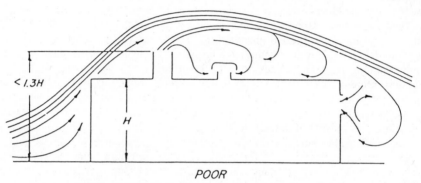

POOR
Low discharge stack relative to building height,
air inlet on roof and wall.

These guidelines apply only to the simple case of a low building
without surrounding obstructions on reasonably level terrain.

Figure 6.1. Illustration of wind streamlines and eddy flow around a building. (Courtesy *Industrial Ventilation—A Manual of Recommended Practice,* American Conference of Governmental Industrial Hygienists. From J. H. Clark, *Air Flow Around Buildings,* May, 1967, Heating, Piping, and Air Conditioning, Cleveland, OH, by permission.)

require computer modeling or wind tunnel testing, both of which are significantly expensive, to compute downwind concentrations. Further, these models usually are not effective close to the stack, and are really not applicable to the reentry problem.

The U.S. Environmental Protection Agency has promulgated a set of rules for "good engineering practice" (GEP) stack height. The purpose of these rules from the EPA point of view was to prohibit "credit" for stack heights greater than GEP in air pollution modeling computations. However, the same set of rules can be used in reverse, that is, to determine what stack height is advisable in order to minimize reentry:

$$H_{GEP} = H_S + L_S \quad \text{(minimum } H_{GEP} = 65 \text{ ft)}$$

where H_{GEP} = stack height

  H_s = height of tallest nearby structure

  L_s = lesser of the height or length of nearby structure

All heights are measured from grade. Many contaminated air discharges are of such large volume and low concentration that a real stack is not required. A stub stack may be sufficient in many cases. For this type of reentry situation, data is furnished by Wilson in the *ASHRAE Handbook*.

It "goes without saying" but must be said anyway, that stack discharges provided with a weather cap or any other device that directs the discharge downward, such as an elbow on the end of a duct or a horizontally discharging blower, represent poor practice with regard to reentry.

2 TALL BUILDINGS

2.1 Advantages of Tall Buildings

Depending on the type of process, tall buildings can offer many advantages in contaminant control. They make possible an arrangement of equipment that minimizes contaminant emission points or makes their control easier; they provide a structure that encourages the thermal lift of contaminants, especially if the structure can be partitioned as discussed later; the building itself provides a structure constituting in effect a large part of a stack in case a moderately tall GEP stack is required; the horizontal transport of bulk materials and the resulting multitude of contaminant control points is minimized; the building walls furnish extensive areas and support for duct work for either exhaust or replacement air, which is especially advantageous if the "inside out" building skin design is adopted. Dust-gathering horizontal duct runs inside the plant are minimized; and it is possible to isolate or segregate process elements in a different fashion than in a flat building.

2.2 Process Characteristics Amenable to Tall Buildings

Tall buildings will be advantageous from both a process and a contaminant control point of view for processes that have the general characteristics of (a) movement of bulk materials, granular or powdery solids; liquids; and gases; and (b) can utilize gravity for movement of these materials from one process step to the next. Care must be taken to allow adequate difference in elevation between major process units for intermediate machinery such as feeder valves, control valves, and flexible connections. A frequent example of lack of vertical space is the short length of flexible connectors between process equipment that vibrates or oscillates. That short length introduces excessive strains in the materials of the flexible connector, resulting in short life, leakage, and high maintenance. In many situations it is advisable, for contaminant control, to install intermediate, small, airtight hoppers that function as a dead-end stop for high-velocity contaminant-bearing drafts, and vertical space is needed for these.

In general, any process for which the raw materials can be pneumatically or mechanically conveyed, hoisted, pumped, or blown to a higher elevation, and then can be moved from one process step to the next by gravity, will also enjoy significant advantages in terms of the effectiveness and ease of providing good general ventilation and minimizing migration of contaminants.

2.3   Utilization of Thermal Head

In terms of natural draft ventilation, a tall, open building is similar to a stack. The thermal head or the "stack draft" is a product of the building height and the difference in density between the air inside and outside the building. A tall building without openings on the vertical surfaces (above doorways at grade) will provide effective upward natural ventilation of contaminated air from warm or hot sources. Further, since the horizontal dimensions are presumably correspondingly smaller than in a flat building, there is less opportunity for horizontal migration and downdrafts that carry the contaminated air back to the first floor. This behavior is highly advantageous in the event that the process requires the presence of workers predominantly in the lower levels of the building. The tall process building can be rather easily segregated by vertical non-load-bearing partitions for isolation of equipment that needs only infrequent inspection or analysis. An example of such an arrangement is shown in Fig. 6.2.

2.4   Access for Duct Work

In a tall building configuration, the horizontal distance between contaminant-generating process points and the outside walls is shorter, and the walls furnish convenient support for both vertical and horizontal duct runs. Large duct work may conveniently be run outside the building, thus conserving building volume and space potentially useful for process equipment.

2.5   Maintenance Access

Maintenance access for heavy equipment is obviously more difficult in a multistoried tall building than in a single-story flat building. However, if previously planned, in an example such as shown in Fig. 6.2, difficulty of maintenance access is minimized. An alternative method, for equipment of intermediate size, is the installation of a freight elevator.

It is obvious that industrial construction in the past few decades has increasingly avoided the use of elevators. Perhaps the main reason for this is the general trend toward single-story flat buildings. On a rational basis, it may be appropriate to revive the use of well-designed high-quality elevators if the process otherwise dictates a tall building. When one considers the technology and investment in high-speed elevators for the ever increasing number of tall office buildings, it would seem obvious that a comparable application of technology and investment to the industrial elevator should result in a reliable and safe method for handling of heavy equipment for maintenance.

When bulk materials are suitable for such handling, bucket elevators can solve the material handling problem; believe it or not, bucket elevator casings can be made leakproof. Enclosed, ventilated skip hoists may solve the hoisting problem for coarse materials.

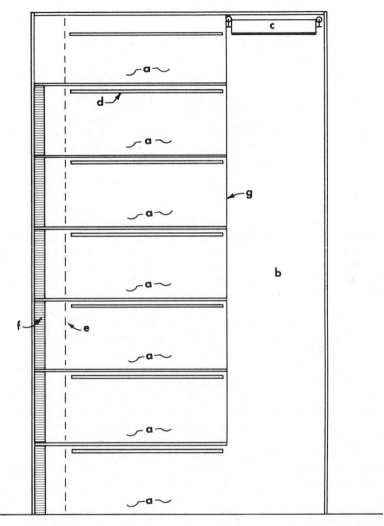

Figure 6.2. Tall building concept for hazardous gravity-flow materials. Structure deleted for clarity. (*a*) Process equipment area; (*b*) maintenance crane bay, not necessarily inside building; (*c*) bridge crane with monorail; (*d*) extension monorail at each level; (*e*) controlled environment isolation area for workers if required; (*f*) stairs/elevator; (*g*) partition panels, removable or with doors.

2.6 Flooring

The process equipment in a vertical array may require personnel catwalks or a number of floors at different levels. If catwalks are adequate, they usually will be of open grating; and whether the catwalk is of open grating or solid decking is of little importance to the vertical movement of ventilation air. If complete flooring is to be utilized, then whether the floor is solid or open has a very significant effect.

For processes handling dry, dusty powders, solid flooring is recommended. If the flooring is open grating, any material that leaks in the upper reaches of the building can cascade through the entire height of the building and contaminate all the lower floors. If the air contaminants are vapors or gases, open flooring presents the advantage of providing general ventilation for the entire space, and any leaking contaminant will probably be carried upward in the general air movement. If both particulate and gaseous contaminants are present, solid flooring is indicated.

Solid flooring makes the general ventilation of the building both more difficult and more costly, but also more controllable and effective. Each floor can be controlled as a separate entity without being influenced by floors above and below. This arrangement is recommended for substances with a very low TLV.

2.7 GEP Stack Provisions

Consideration of the GEP stack formula and rules (Section 1.4) indicates that for a tall building, the stack extension above the building is proportionally much less than for a low flat building. Wilson's formulas also support this conclusion. Thus a tall building represents an opportunity to have a stack discharge at a considerable additional height above grade without constructing the stack itself for such a height, that is, the building height counts toward the stack height.

If the initial planning of the building provides for such stacks, the increased cost of the building structure is moderate. If such GEP stacks need to be added later as a retrofit, and the building structure is inadequate, the cost is high.

Of perhaps greater importance is the height of nearby buildings, either existing or planned. A low building in the lee of a tall building presents a difficult problem. In such a configuration, any reasonable stack discharge height for the low building will be inadequate to prevent reentry when the wind is blowing from the high building to the low building. Conversely, if the wind is blowing from the low building to the high building, contaminated air discharged from the low building will impact all air intakes for the high building. If the low building can instead be constructed as a tall building, these effects can be minimized at reasonable expense. The other alternative is to have extraordinarily effective air- and gas-cleaning equipment on all discharges from the low building plus adequate height for discharges from the tall building.

2.8 Disadvantages of Tall Buildings

2.8.1 General Comments. A tall building is obviously inefficient for some industrial processes. An example would be a plant fabricating artifacts, such as parts or objects that cannot be handled by gravity flow but instead are moved from operation to operation by carts, bins, or horizontal conveyors. A tall building also would be inefficient for various electronic manufacturing operations where the close proximity of extensive air-handling and conditioning equipment, together with the need for flexibility in air handling to accommodate frequent and rapid process changes, is important. For that situation, the air-handling equipment should be on a floor immediately above the operating floor and in close proximity. Putting that array in a tall building would almost automatically double the height of the building. If the demands of the process strongly favor a single-story building, the advantages of the tall building diminish in relative importance and the process requirements have priority.

2.8.2 Personnel Movement and Egress. A tall building has disadvantages for labor-intensive operations. Personnel working on the upper levels of the tall building, whether infrequently or permanently stationed there, must have easy and rapid egress. In addition to safety considerations, there is the normal personnel movement involving lunch hours, breaks, direct-contact communications, and the like, all of which are more difficult and time consuming in a tall building. Here again, if the tall building is obviously advantageous for other reasons, another look should be taken at personnel elevators.

2.8.3 Poor Roof Access. A tall building obviously has a much smaller roof area for roof-mounted equipment, and access to the roof is more difficult, especially if heavy equipment is involved.

3 FLAT BUILDINGS

Pertinent characteristics of flat buildings are diametrically opposite to those for tall buildings in almost all respects. It is more difficult to prevent migration of contaminated air horizontally through the building, especially if hot processes and significant numbers of thermal plumes are involved. More partitions or roof-hung dividing baffles are required. The free unobstructed horizontal movement of materials or personnel, a feature for which a flat building may have been selected in the first place, will be interrupted by such partitions. If there is a large number of contaminant-producing operations that require local exhaust ventilation, and if that exhausted air requires cleaning before discharge to the atmosphere, long runs of horizontal exhaust duct work may be required in order to bring the contaminated air to one, or a few, large air-cleaning units. If the contaminant is particulate, these long horizontal runs must be designed at high velocity, with correspondingly high suction pressures; the combination of these two factors requires greater structural strength of such duct work and correspondingly increased cost. When extensive air cleaning is required, it is usually regarded as preferable to have a smaller number of larger units than a large number of small units, although this is not always the case.

On the other hand, the vertical distance to the roof is obviously shorter, there is a much larger roof area for location of ventilation equipment, and access to the roof is easier and less expensive.

If extensive contamination control is required, and if the requisite equipment is large and heavy, the popular low-cost "packaged" steel buildings are not appropriate for the application in spite of the low base cost. The structure of such buildings is of minimal design, to support the building walls and roof only for the minimum necessary loading requirements of wind and weather. Such structural design will not support additional equipment, not even extensive duct runs that could become filled with particulate. The costs of accommodating such structural problems in a building of this type frequently can override the initial cost advantage.

One accommodation for the disadvantages of the packaged building design is to arrange the process layout so that those operations requiring extensive ventilation control are grouped together and located along outside walls. This accommodation may result in layouts that are not advantageous for movement of materials or personnel.

3.1 Amenable Process Types

The large flat building design is advantageous for processes that handle a large number of small parts of artifacts that cannot be transported in bulk or by gravity, and for industries that are labor intensive so that large numbers of people must be accommodated. For many industries with such characteristics, the number of operations requiring extensive contaminant control is frequently rather small, and appropriate layout and partitioning can accommodate the exposure control problems. Perhaps one of the most common exceptions to this generality is the industry that has a large number of noisy operations. A heavy-duty metal fabricating shop, for example, has numerous and scattered noisy operations. Part of the flat building advantage would be lost by the incorporation of numerous partitions, with necessary sound treatment, to keep the building from becoming one large echo chamber.

3.2 Thermal Head Characteristics

The behavior of thermal plumes from hot processes in a large flat building is difficult to control. The rising (and presumably contaminated) plume of warm air will migrate horizontally underneath the roof in an uncontrolled and difficult to control pattern, showing up as background contamination in unexpected places, as described earlier in this chapter. The layer of contaminated air under the roof interferes with what otherwise would be convenient location of replacement air units and/or heat conservation methods for recycling that warm air down to the working level.

The only dependable and predictable means of controlling this random horizontal air movement is the construction of partition walls. A less expensive alternative, but somewhat less dependable and efficient, is a system of hanging light partitions or baffles from the roof down to the bottom of the roof trusses. For large or high-temperature heat sources, the low roof may lead to excessively hot working conditions for personnel on the floor, and this problem is made worse by the installation of partitions or baffles to control horizontal air movement. This situation in turn leads to the installation of large volumes of "roof fan" general ventilation exhaust, which then leads to the need for large volumes of expensive replacement air. The management of significant thermal plumes is, as seen in the previous section, easier and less expensive with the tall building design.

One seldom seen solution for this type of problem is a manufacturing building with a combination structure, that is, one tall section for the hot processes, and the remainder of the building a flat structure for the other parts of the process. A single unit combining two types of building construction is no doubt more expensive than a single type; but one suspects that the failure to arrive at that design is, more often than not, an accidental result of "selecting the type of building for the project" that overlooks the possibility of such tailor-made combinations.

3.3 Roof Access

Access for duct work to and from the roof is obviously easy, and to and from the sidewalls is difficult unless the operations to be controlled are close to the sidewalls. If local exhaust ventilation is required for large portions of the process, it may not be practical or advantageous to attempt to locate so much of the operation along the sides. The result is a need for long duct runs, above the roof or inside the building, if a relatively small number of large-capacity air-cleaning units is to be used. If the contaminant to be collected is particulate, some minimum transport velocity is required, which in turn leads to higher static suctions in the duct work and heavier duct work construction to furnish adequate strength to resist those pressures. In addition, the possibility of solids settling in the duct work results in a potential structural load that can be very significant. If the duct work is inside the building, and if the contaminant of concern is particulate matter, the top of the ducts represent collecting surfaces that add to the housekeeping problem.

An alternative is to install a large number of smaller air-cleaning devices on the roof, thus minimizing the horizontal duct runs. Depending on the characteristics of the process, a large number of small units may have production advantages in that an equipment outage represents shutdown of a smaller part of the production operations. For dust collectors, the collected dust must be returned to process or packaged. The large number of small units is advantageous only if the units can be located for direct gravity return to process. For most other considerations, installation of a large number of small units is more expensive in terms of initial cost and maintenance cost.

Obviously either alternative requires structural strength of the building far exceeding that available from standard packaged light construction steel buildings. The choice of these alternatives involves other considerations (see Section 3.5).

3.4 Segregation Characteristics

If the number of operations requiring segregation for good contaminant control or noise control is small, the use of partition walls is, typically, not excessively burdensome on the flow of materials or people. The detail design and integrity of such partitions varies with the TLV level of the contaminant to be controlled. If the TLV is very low, with very tight control required, the partition walls should be designed in many respects as though they were fire-rated walls, even though such a rating may not be required. The wall may need to be extended clear to the roof, and all openings including pipe penetrations well sealed. Light paneling with crackage between panels may be unsatisfactory. For control to lower contamination levels, open doorways may not be permissible, and in some circumstances an airlock or vestibule arrangement is indicated.

Partition walls severely inhibit the use of bridge cranes for movement of material. One solution that allows the use of bridge cranes, if they are "absolutely" necessary, is to provide a transfer arrangement for moving the load from one side of the partition to the other. This typically would consist of some kind of a car or cart, perhaps on rails or tracks, which moves the load from one side of the wall to the other through the doorway; the load is placed on the car by separate bridge cranes on both sides of the wall.

All considerations regarding the integrity of the segregation arrangement become increasingly important as the TLV level of control is lowered. For very low allowable contaminant concentrations, and if use of a low flat building is otherwise indicated, the proper execution of the principles of segregation justifies considerable planning and expenditure.

3.5 GEP Stack Provisions—Multiple Stacks

For operations requiring a large amount of local exhaust ventilation, the low flat building presents a difficult problem if a large number of small exhaust systems is chosen. Unless the cleaning of the exhaust air is exceptionally thorough (and therefore exceptionally expensive) these stacks must project to a sufficiently high elevation to minimize the contaminant reentry problem (see Section 1). For a multiplicity of such systems, the end result is a "forest of stacks" protruding above the top of the building. This situation is almost always offensive to the architectural aesthetics involved; depending on the location of the plant, the "forest of stacks" may also represent a difficult public relations situation. There are only two alternatives, both of which are quite expensive.

One alternative is to clean the exhaust air so that reentry into the building does not constitute a problem. Air-cleaning equipment adequate for this purpose is usually much more efficient and costly than would be required by air pollution regulations concerning the general ambient atmosphere. Further, depending on the nature of the potential hazard, requirements include not only exceptionally high efficiency but exceptionally high reliability; in other words, if a short-term or intermittent dosing of replacement air with the contaminant may have immediate serious effects, extraordinary steps must be taken to avoid it.

A point made previously deserves repeating here: If contaminated air is discharged into the building envelope where turbulent eddies occur, there is no location in the immediate vicinity of the building that constitutes a reliable source for the intake of clean air.

The second alternative to the forest of stacks is the single large stack. The proper design of such systems is frequently complicated, and beyond the scope of this general discussion of building types. However, an additional induced draft fan is usually required for the collection flue into which the large number of small systems discharge; and of course then there must be the foundation and necessary structure for a single large-diameter tall stack. Tall stacks are frequently found offensive by the architectural community. However, they are probably less offensive to the neighbors in terms of public relations.

The reader is warned that the modern practice of architecture uniformly and strenuously avoids visible stacks. The reader is also warned that compliance with this architectural criterion can result in extraordinarily expensive solutions to problems that would be less severe if appropriate stacks were used. One alternative is the architectural masking of the stack, which considerably improves its appearance and minimizes the public relations impact, while still providing the necessary function.

3.6 Ceiling Height

This section has discussed large flat buildings, without comment on the criteria for ceiling height. Even if the large flat building is otherwise acceptable, there is some minimum requirement for ceiling height if contamination-producing processes are involved. Many buildings currently constructed for general-purpose light-industry occupancy have clear space, floor to bottom of the roof steel, of 12 ft (3.7 m). In the opinion of the author, this is completely inadequate for any operation that requires contamination control. The clear head room should be at least 16 ft (5 m), and 20 ft (6 m) is preferable.

4   MILL-TYPE BUILDINGS

For purposes of this discussion, a mill-type building will be loosely defined as a large, fairly open building that is of significant size both horizontally and vertically. The vertical clear height from floor to roof steel may range from 25 to 50 feet (7.5–15 m). The name with which these buildings are dubbed no doubt derives from the classical structure of the steel mill or smelter.

While such buildings may have very significant amounts of exhaust ventilation or process ventilation, the building itself is typically served only by natural ventilation. The building will usually have a roof ridge ventilator or monitor, large areas of opening in the sidewalls, large doors permitting entry of large haulage equipment, and the like. Thus the building in general is not designed to be ventilated in any planned and controlled fashion. This type of construction provides weather cover for the equipment, process materials, and personnel, but does not provide a controlled environment.

If the operations conducted in this type of building require strict contamination control, the solutions are difficult, expensive, and only partially successful. For new construction, significant departure from traditional methods is indicated, including improved layout and arrangement of processes and material handling equipment. Because specific solutions to these problems will depend highly on the nature of the process, only a few generalities, as food for thought, will be offered here.

Figure 6.3 is an illustration of a recently constructed mill-type building for casting of toxic metals. It is small, as mill buildings go, about 200 ft (60 m) by 50 ft (15 m), by 35 ft (10 m) high, and is provided with a bridge crane. The building columns are sheathed on the outside with sheet metal and on the inside with insulated panels. The roof trusses are above a false ceiling. Dust control duct work, furnace exhaust, and utility lines run in a large central trench. Most of the exhausted air is cleaned for recirculation with two stages of backup filters and is returned to the plant along one sidewall through low-velocity distribution panels. Temperature-controlled dampers mix outside air with recirculated air in cool weather, and supplemental heating coils are provided. The equipment layout is such that the workers perform most of their tasks in the area between the air supply plenums and the contaminant-generating operations. The metal dust and fume is controlled to less than 50 μg/m³.

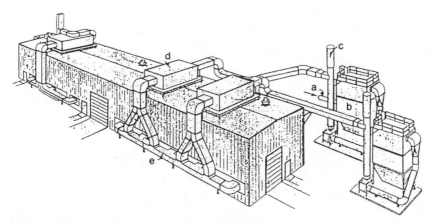

Figure 6.3. Mill-type building with controlled environment. (*a*) inlet ducts from trench in building to fabric filters; (*b*) fabric filters; (*c*) bypass stack; (*d*) backup filters, heating coils, dampers, outside air inlet, blower; (*e*) supply air distribution duct.

4.1 Functional Purpose and Characteristics

The functions to be accommodated by mill-type buildings involve several basic process and equipment characteristics. The process usually involves handling large amounts of process material, high tonnage per day flow rates, and/or massive objects. Very high temperatures—red-hot solid to molten metal temperatures—are usually involved in large fractions of the process. Movement of materials and machinery is traditionally by bridge crane and belt conveyor.

The present state of technology cannot accomplish adequate control of contaminant emissions from large, hot objects or large molten-metal ladles being moved by bridge crane. One method of control that satisfies requirements for outdoor air pollution is building containment. This consists of encapsulating the entire building, withdrawing contaminated air from the roof ridge line, and processing it through an air-cleaning device of adequate efficiency to satisfy air pollution regulations. This solution to the air pollution problem almost invariably results in degradation of working conditions inside the building. The basic reason for this poor result is the extraordinarily high air volume that would need to be handled and treated if adequate general ventilation for worker health and comfort were to be achieved inside the building. The cost is so high as to be prohibitive. Such solutions, forced by government regulations, improve one environmental situation at terrific cost both in money and resources, and in the degradation of the inside environment.

While various combinations of partial solutions and compromises may satisfy problems of retrofitting existing plants, a significantly different approach, if at all possible, would be advantageous for any new construction. The new approach obviously would involve some process changes, rearrangement of equipment with relation to preceding and following process steps, reexamination of material handling methods, and so on.

4.2 Achieving Segregation—Material Handling Problems

In order to achieve segregation of operations where necessary, a combination of approaches would need to be creatively applied to the specific process requirements. Typical segregation by partition walls would vary from difficult to impractical to impossible, depending on the specifics of the process. Therefore a combination of new material handling techniques, or possibly the revival of old techniques, would be indicated in combination with segregation where practical.

For example, the "tall building" concept to minimize the use of belt conveyors would be applicable in some parts of some such processes. Although the process equipment is large and massive, it could be (and in some cases has been) installed at a sufficiently high elevation so that the bulk material handling following the operation is by gravity. Belt conveyors are commonly regarded as the most practical and least expensive way of moving large amounts of bulk materials. However, if strict dust control is required, the cost of the belt conveying system plus the cost of the dust control may change the economic balance of choices. For many contaminants, adequate control of conveyor belts requires complete enclosure of the belt for its entire length, with corresponding increased maintenance and operational problems. If the bulk materials were to be hoisted vertically, other means of handling might be made practical, such means varying from large containers to staged bucket elevators, and perhaps other mechanisms.

Transport of ladles of molten materials giving off toxic fumes is currently impossible to control except by building encapsulation. Process changes, or rearrangement of equipment allowing gravity flow through pipelines or open troughs, would be a solution to some of the problems. Large vehicles on tracks would be a helpful change in other situations. Basic process modifications would offer potential solutions, but are mentioned here only as a future possibility and obviously would vary drastically from one process or one industry to another.

4.3 Controlled Environment for Workers

In many operations carried out in mill-type buildings, the workers can and do spend a significant portion of their work day at a given station or area. For those workers, an effective although partial solution to the problem of reducing their exposure is to provide a controlled environment at that work station to compensate for the inability to control completely the sources of air contamination.

If a completely enclosed room is practical, that room can be pressurized with clean air so that the contaminant concentration inside the room is well below the allowable limit. Ventilation design for such enclosures would be different depending on whether the contaminant of concern is particulate or gas.

If the contaminant is in the gas or vapor phase, the room can be pressurized with air inducted from a clean air source, thus avoiding the problem of cleaning the intake air. For most vapors and gases, adequate cleaning would require activated carbon, an expensive installation either in terms of installation or alternatively in frequent changing of carbon filters. If the contaminant of concern is particulate, the maintenance of clean conditions inside the control booth is more difficult because of the contamination brought into the room on work clothing. However, when in specific circumstances it may be difficult to find a location for the induction of suitably clean air, adequate cleaning of the air by filtration is usually practical.

The control booth or isolation area usually requires air conditioning. The application of ordinary comfort air-conditioning designs to these booths does not effectively pressurize them against the infiltration of contaminated air because the percentage of outside air handled in such systems is so low as to preclude ensuring an outflow of air through the crackage in the room and the opening and closing of the door by its occupants.

Two general solutions are available. The most attractive, if space permits, is a strenuous attempt to make the room airtight, and provide an air lock vestibule so that the exfiltration load of an open doorway is not encountered. If the contaminant is particulate, provisions should be made, at minimum, for cleaning work shoes and leaving heavily contaminated work clothing outside the room. The more stringent the necesary control, the more stringent the requirements for clothes changing and the like become, in order to avoid contaminating the room by personnel traffic.

An alternative method is more expensive in first cost but more effective and has the advantage of not inhibiting personal movement in and out of the controlled area. Such free movement enhances, rather than inhibits, personal contact communication between workers and supervisors. This method provides a custom-designed air-conditioning system for the booth. It constitutes handling much larger volumes of air in order to provide a controlled downflow through a perforated ceiling to the floor, thus ensuring that contaminated shoes and floors do not result in air contamination. Contamination from work clothing is generally released below the breathing zone and is removed downward. If an air lock is not possible, the percentage of outside air must be increased to the point where a definite outflow of at least 50 and preferably 100 fpm (30 m/min) is achieved through the open doorway. The conditioning of the outside air represents an unusually high air-conditioning load and requires a design different from conventional systems in order to prevent freezing of coils.

In the event that the operations dictate that the workers involved should not be enclosed and separated to that degree from their work, the concept of a "clean-air island" may be practical in some circumstances. The clean-air island consists of a supply plenum furnishing clean air, usually not air conditioned but tempered in cold weather, over the area that the workers would consider to be the work station. The clean-air plenum must be relatively low, no more than 7–8 ft (2.5 m) above the floor and should introduce air through a perforated plate or filters to achieve even distribution at a low velocity, typically 100 cfm/ft^2 (30 m^3/m^2 − min).

5 OPEN-AIR PROCESSES

Open-air processes are briefly discussed in this chapter only for completeness. Such processes are discussed more fully in Chapter 4.

Most open-air processes are, almost by definition, closed processes in themselves. Petroleum refineries, or some chemical plants, represent typical examples. The exposures to the operating personnel are lower than they would be if the process were enclosed in a building, but the popular conception that the exposures are nil is not necessarily correct. Being in the open, the natural dilution of contaminants escaping from pump seals, valve packings, and the like, is obviously much more rapidly diluted than it would be if inside a building. However, the dilution is neither complete nor instantaneous. The exposure to the worker obviously depends on whether he or she is upwind or downwind to the source, and the weather conditions.

Most modern outdoor plants have a multiplicity of "control rooms" where the workers can spend much of their time in a controlled environment. Some of the same principles discussed in Section 4.3 apply to such control rooms.

Reduction of worker exposure during normal operations in outdoor plants essentially consists of reducing the leaks and, at selected operations, the installation of local control ventilation or improved process controls. As a generality, no doubt with many specific exceptions, it could be stated that if a process itself is amenable to being out in the open and unhoused, the resulting worker exposure will be easier to control than if the process were enclosed in a building. This generality does not include the control of fugitive emissions from outdoor handling of large quantities of raw materials such as some types of ore concentrates and intermediates.

ACKNOWLEDGMENT

Mr. H. L. Rowland, Director of Engineering and Maintenance, St. Joe Lead Company, furnished valuable commentary on the substance of this chapter that is gratefully acknowledged.

GENERAL REFERENCES

ASHRAE Handbook, "Fundamentals," American Society of Heating, Refrigerating, and Air Conditioning Engineers, Atlanta, 1988.

CHAPTER SEVEN

Building General Ventilation

Knowlton J. Caplan

1 AIR FLOW PATTERNS

1.1 General Characteristics of Air Flow Patterns

"Still air," in the vernacular of the ventilation engineer, has a velocity random in direction at a speed of 15–20 ft (0.07–0.10 m/s) per minute. This definition arises from the fact that in any ordinary building, with no mechanical ventilation and all doors and windows closed, that type of air movement will be found. It is caused by small temperature differences created, for example, by sunshine on walls and roof, by minor heat sources within the building such as people and lighting, and by wind pressure, Infiltration and exfiltration of air through cracks in the building structure and around doors and windows add to the air movement created by thermal differentials and wind pressures. In an industrial building, the random air motion is typically higher due to higher intensity heat loads from motors or processes. Thus "control" over general air movement cannot be established unless the control velocity is significantly greater than 20 ft (6 m) per minute.

1.1.1 Energy Conservation and "Tight" Buildings. Buildings constructed before the onset of the energy conservation campaign in the mid-1970s experienced a significant general ventilation rate merely by the process of infiltration and exfiltration of air through cracks in the structure itself, around doors and windows, and so on. For office buildings and multistory buildings in general the ventilation rate was 0.5–1.5 air changes per hour due to this cause. For most industrial buildings, excepting mill-type buildings, the air change rate due to infiltration/exfiltration was 2–3 air changes per hour. For most types of occupancy this ventilation air needs to be heated, cooled, or otherwise conditioned. These ventilation rates constitute a considerable energy cost that was quickly targeted as an opportunity for conservation. Building codes, ASHRAE standards, and similar guidelines that required introduction of outside air by mechanical ventilation were quickly revised to indicate a lower outside air requirement. Building designers made strenuous efforts to minimize the cracks in the building construction and to improve the insulation value of the outer shell. Most, if not all, codes requiring lesser amounts of outside air ventilation carried an exemption for industrial operations where noxious materials might be present.

The greatly decreased ventilation rate due to the drastic reduction in outside air supply created problems unforeseen by many. Very small sources of airborne contamination emanating from modern synthetic building materials, furnishings, office machinery, smoking, and even the people themselves, created a new fad called "indoor air pollution." (It is interesting to note that, historically, many building codes specifying percentages of outdoor air in the ventilation system were based on the control of body odors, in the days before the American people were quite so thoroughly washed as we are today). Nevertheless, it is true that the accumulation of these minor air contaminants, perhaps accompanied by the lack of that indefinable feeling of clean, fresh outdoor air, has resulted in a number of complaints and psychosomatic illnesses. Health risk from formaldehyde and carbon monoxide can be significant, especially in residences. The formaldehyde problem is caused primarily by improperly applied foamed-in-place urea–formaldehyde insulation or by extensive use of improperly cured particle board. Carbon monoxide has always been a problem, primarily caused by incorrect venting of flue gases, but it is made worse by "tight" buildings.

Although true health risk may be quite minimal (except for formaldehyde and carbon monoxide), there is at least a subjective problem of nonnegligible proportions.

Much research is being conducted on the various possible sources of trace air contamination. One suspects that, when all the research is done and all the factors evaluated, a somewhat increased ventilation rate may be, overall, the least costly correction of the problem.

As already noted, industrial operations are usually exempt from code requirements minimizing the outside air ventilation. Insulating the outside walls of a factory building in a cold climate, or an air-conditioned building in a warm climate, may appear to be cost effective. However, it is the opinion of the author that extraordinary attempts to make the factory building airtight in terms of minimizing crackage are not, in most circumstances, cost effective. The energy cost of the relatively large amount of ventilation required for other reasons is such that the energy cost of a modest infiltration/exfiltration rate of perhaps 0.5 air change per hour is relatively inconsequential.

The large amount of air required to dilute contaminants is not appreciated by many. Take, for example, a workroom 100 × 50 ft (30 × 15 m) with 12-ft (3.5-m) ceilings, and assume an outdoor air ventilation rate of 0.5 air change per hour. The dilution rate would be 30,000 ft³/hr, or 500 cfm (14 m³/min). Assume a solvent, such as methanol, is being used. The threshold limit value (TLV) for methanol is 200 ppm, a relatively high—that is, not stringent—value. Evaporation of 9.75 oz (290 cm³) of methanol per hour (a little more than a glassful) if evenly mixed with the dilution air, would be enough to bring the concentration to 200 ppm. For a similar substance with a TLV of 1 ppm, about 0.049 oz (1.5 cm³) would be sufficient to reach the TLV.

1.1.2 Thermal Air Currents. Most industrial processes have at least a few local high-temperature sources of thermal air currents. The convective and radiant heat loss from these sources creates an upward moving current of air that rises to the ceiling, and then migrates in an uncontrolled manner to other portions of the building. These air currents may also contain small concentrations of contaminant perhaps generated by the same process, which has escaped the imperfect local control. Even though the degree of contamination may be small, it nevertheless will add to the background contamination of the air in other portions of the plant and may be significant if additional contamination is generated in those areas.

The nature of these air currents, the significance of the background contamination, and the ways of controlling this migration of contaminant are discussed at length in Chapter 6 with respect to the characteristics of tall buildings or flat buildings.

1.2   Air Jets

Where significant amounts of contaminant control local exhaust ventilation are employed, it is necessary to replace the exhausted air with adequate supply air, suitably treated. Air issuing from supply openings creates an air jet of magnitude greater than usually appreciated, and is a major source of air movement in the industrial setting.

Air issuing from a supply opening at high velocity squirts through the room air almost like water from a hose except that it is invisible. The jet, after passing through a transition zone, expands at a solid angle varying from 17° to 22° and entrains and mixes with room air on all sides of the jet. The centerline velocity of the jet diminishes slowly, so that at a distance of 10 "diameters" from the supply opening the centerline velocity is still about 10% of the nominal velocity at the supply opening. This phenomenon is sometimes called the "throw" of the jet. Regardless of the shape of the supply opening, beyond the transition zone the jet assumes a round cross section. It is not possible to control the shape of the jet at significant distances from the supply opening by changes in the shape of the supply opening. The behavior of the jet can be modified by the use of supply diffusers such as commonly seen in office and public building air-conditioning systems. The elements of simple jet behavior are shown in Fig. 7.1 and further details are elucidated in *Fan Engineering,* the *ASHRAE Handbook,* and "Heating and Cooling for Man in Industry" (see the General References section at the end of this chapter). Although the characteristics of air jets are fairly well understood by ventilation engineers, they are frequently misapplied in the industrial setting where contaminant control is important. Air jet behavior can be deleterious in two respects: in the creation of high velocities at local exhaust hoods and by mixing the room air.

In ordinary heating and air conditioning the throw phenomenon of the air jet is deliberately used to mix the air in the room. It is the proper application of this technique that avoids uncomfortable temperature differentials between the heads and the feet of the occupants that would otherwise occur due to temperature stratification. There is a major difference, however, in the desired application of general air movement in offices or residences and that in industrial operations. The air flow rate in office air conditioning is very small in proportion to the room volume, as compared with ventilation of the factory. Use of the air jet throw to induce mixing in the office is necessary to prevent stratification. However, the air flow rate through the industrial shop, especially where contaminant control is of significance, will be so high that temperature stratification and discomfort from that cause is of no consequence; and mixing in general is not desired.

Local exhaust hoods control contamination at the source by partially enclosing the source and exhausting air from the hoods so that the velocity at the open "face" of the hood is sufficient to contain and control the contaminated air. Other types of local exhaust hoods do not enclose the source, but create a capture velocity at some point in space, external to the hood, sufficient to capture the contaminated air. Effective face and capture velocities are frequently in the range of 50–100 fpm (15–30 m/min), and only in the most extreme of circumstances do they get as high as 500 fpm (150 m/min). Obviously the designer of the local exhaust hood attempts to secure effective capture with as low a face or capture velocity as practical, in order to reduce the capital and energy cost of the local exhaust system.

It has been determined from research on laboratory fume hoods that a room air current more than about one-half of the face velocity of the hood is sufficient to

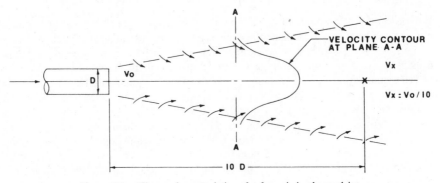

Figure 7.1. Throw characteristics of a free air isothermal jet.

degrade the capture efficiency of the hood. By analogy, approximately the same proportion should apply to most exhaust hoods. Thus an air jet from a supply opening may be deleterious even at a velocity of 50–100 fpm (0.25–0.50 m/s) if it still maintains that velocity by the time is has arrived at an exhaust hood designed for local contaminant control.

At times the controlled motion of large-scale, low-velocity air movement is part of the scheme for controlling the contaminated air, and in that case mixing of the clean air jet with the otherwise contaminated room air is specifically undesirable. This is especially true in those work environments where it is impractical to control the entire space to safe levels of contaminant concentration, and clean air islands at normal work stations are provided.

A dichotomy of cross purposes arises when high-velocity air jets are desired for alleviating a heat stress situation. One of the methods for alleviating heat stress is to direct a high-velocity jet of relatively cool air at the worker. The air supply may be obtained directly from the relatively cooler outside air, or alternatively the outside air may be further cooled by evaporative cooling (see Section 2.6). The problem arises when a given task involves both heat stress and air contaminants. Then one of the most effective methods of combating heat stress, spot cooling, directly conflicts with the need to minimize cross drafts to permit effective capture of contaminant by the exhaust system. Such situations require some other method of overcoming the heat stress, or truly creative engineering in revising the process, and are not usually considered to be part of the general building ventilation scheme. The solutions vary widely according to the particular circumstances.

2 REPLACEMENT AIR

2.1 Terminology and Definition

Until recently, the term *makeup air* was rather loosely used and could be taken to mean either the outside air induced into a building ventilation system to replace that air that was exhausted or exfiltrated in general building ventilation and air conditioning; or, it could mean the air blown into a building to replace that air exhausted by contaminant control exhaust systems. Recently the taxing authorities have preempted the use of the term *makeup air* to mean that part of the permanent building ventilation system not specifically related to replacing exhaust system air, and require that makeup air equipment can be depreciated only at the rate specified for the building air conditioning. Thus separate systems installed and intended for replacement of contamination control exhaust air have been renamed in order to avoid improper inclusion in the wrong category for tax consideration. Such systems are now called *replacement air*.

In an ordinary building, it is obvious that if a cubic foot of air is removed, a cubic foot of air must flow into that space since the space cannot be literally evacuated (more than a fraction of an inch water gauge). The air flowing into the space to occupy the void created by air exhausted from the space is an inescapable phenomenon, whether or not controlled. If the air enters in an uncontrolled fashion, a variety of deleterious effects can and usually do occur. Consequently replacement air is defined as that air that is introduced into a space in a controlled fashion, suitably conditioned, to replace the air removed from that space by contaminant control exhaust systems.

2.2 Distribution of Heat Load

To gain an understanding of the benefits of well-designed replacement air systems, it is worthwhile to contemplate the situation prevailing in a typical large industrial plant with respect to various heating loads and reasonable comfort conditions. Let us consider a large, single-story manufacturing plant, hundreds of feet in each horizontal dimension and perhaps 25 ft high. Various heat-producing operations and workers are scattered throughout the area and an adequate replacement air system has not been provided. A significant amount of exhaust ventilation is required at several places in the building for contamination control. Powered roof ventilators are also installed to remove hot air in warm weather and to remove the contaminated air that accumulates in the ceiling space. The situation is shown schematically in Fig. 7.2.

On a cold winter day, cold air enters through the building sidewalls through cracks, open windows or doors, and so on, to replace the air exhausted from within the building. Those exhaust fans that have a low-pressure characteristic may be withdrawing less than the rated volume, but nevertheless they are still removing some air and creating a slight suction in the building. The cold air flowing into the building around the entire perimeter affects the workers near the perimeter who perceive that they are cold. Demands are made for additional heating, and in response unit heaters are installed inside the building around the perimeter.

Meanwhile, in the interior spaces of the building, heat is being generated by motors, machinery, and ovens as well as by lighting and people. As a matter of fact, the interior space of a large factory building needs no heating even in cold weather, and usually needs some heat removal. The unit heaters installed around the perimeter are only partially effective in alleviating the complaints of workers there, because they are still subjected to cold drafts, and the unit heaters are doing their job a bit too late. However, the air flowing from the perimeter into the interior spaces of the building has been

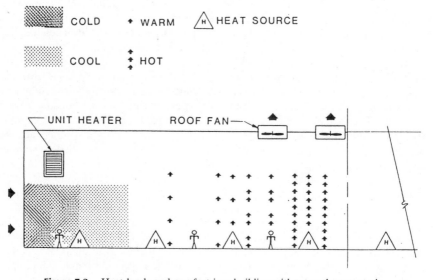

Figure 7.2. Heat loads and comfort in a building without replacement air.

heated more than previously. Since what is needed in the interior is heat removal, not heat supply, complaints of excessively warm conditions arise from those workers. The typical response to that complaint will be to install additional powered roof ventilators to more adequately exhaust the overheated air. Now the building is more "air starved" than previously, and a new wave of complaints of being too cold originates with those people working around the perimeter. And so it continues, sometimes until water pipes around the perimeter freeze in cold weather even when the plant is in operation.

One energy conservation method offsetting this scenario is to install circulating fans that take in the hot air under the roof and discharge it at lower, working levels in the building. This solution may be appropriate if there is little or no air contaminant involved. However, most contaminant producing sources cannot be brought under perfect control by the local exhaust ventilation, and the air underneath the roof will be higher in contaminant concentration as well as in temperature. It may not be appropriate to recirculate that air down to the worker level without cleaning it.

The entire fiasco can be avoided by the supply of adequate replacement air. For the attainment of an acceptable level of warmth in the winter, the heat is going to be supplied in any event; energy is conserved if it is supplied in a controlled fashion.

2.3 The Economy of Replacement Air

For the attainment of a given set of temperature conditions in the industrial plant, provision of properly designed replacement air systems will result in less energy consumption than can be attained in the absence of such systems. Another way of phrasing it is that less energy will be wasted if the energy is properly applied.

Kenneth E. Robinson, retired "home office" ventilation engineer for a large manufacturing company, tells a story that is, unfortunately, not "legally" documented. Nevertheless it rings true. One large establishment he visited was undergoing expansion, with increasing demands for space heating as well as for process heating. The plant used steam heat but did not have replacement air systems for space heating. A shortage of steam was projected in the near future, and an additional steam generator was being budgeted. The plant management was persuaded to try the installation of a large replacement air system, and the results were favorably received. One advantage was that such systems could be installed stepwise, not requiring a major single installation such as a steam generator. After two years of installing replacement air systems, the plant canceled the budget request for an additional steam generator.

Other less obvious economies are also thus revealed. The systems can be installed stepwise without major production interferences. The same system can be used in hot weather to alleviate heat stress or otherwise improve comfort. Better working conditions are achieved than would, in reality, be achieved without the replacement air. If air distribution is proper, exhaust hoods and enclosures will be more effective and/ or will require lower capture velocities and thus be less costly.

In the absence of adequate replacement air, the following undesired effects can and often do occur.

2.3.1 Exhaust Hoods. Exhaust hoods operate improperly for two reasons. First, the negative pressure created in the building increases the differential pressure against which the exhaust fan must move air. If it is a low-pressure-type propellor or tube-axial fan such as commonly found on spray booths, the air flow may be inadequate.

Additionally, strong cross drafts drawn in through openings in the building walls, doors, windows, and so on, may create drafts that interfere with the proper capture of hoods.

2.3.2 Cross Drafts. Strong cross drafts through windows and doors may disperse contaminated air from one section of the building into another and may interfere with the proper operation of some types of process equipment such as solvent degreasers. Settled dust, in extreme cases even the dust on the floor, may be dislodged and redispersed throughout the workroom.

2.3.3 Natural-Draft Stacks and Flues. Moderate negative pressures in the building can result in backdrafting of flues. This may cause a lethal health hazard from the release of carbon monoxide in the workroom. Burner life may be shortened by the flame being drawn back over the burner. Difficulty will be experienced in maintaining pilot lights; temperature controls may operate improperly; corrosion may occur in stacks and heat exchangers due to condensation of water vapor in the flue gases.

2.3.4 Differential Pressures. In the extreme, differential pressure on doors may be sufficiently high to make them difficult to open or shut, even to the degree of constituting a personal safety hazard. Roof skylights or roof panels may prematurely collapse under the combined load of the negative pressure and snow load or settled dust load.

2.4 The "Air Change" Fallacy

Basing the capacity of general building ventilation systems on "air changes per hour" is a popular method. The reasoning involved, however, is usually fallacious. The fallacy is obvious; the air change method bases the ventilation rate on the size of the building rather than on the size of the problem.

   Circumstances frequently arise where the preferred method of controlling an air contaminant is of general (or "dilution") ventilation, rather than by local exhaust capture. Examples of these circumstances would include a large number of very minor sources scattered throughout a room, such as use of minute amounts of solvent in electronic manufacturing; or a source that is in motion, so that local exhaust cannot be applied. There may also be other special circumstances where it is impractical, if not literally impossible, to apply the principles of local exhaust ventilation. In such circumstances, it is appropriate to utilize general ventilation. The principle is to reduce the concentration of contaminant by diluting it with large volumes of clean air, preferably in a controlled fashion, so that the dilution is adequate before it enters a worker's breathing zone. Control of direction of air flow is important in this regard. The required amount of dilution ventilation can be estimated in a number of ways: from a knowledge of the generation rate of the contaminant; from prior experience; by measuring the existing contaminant concentration and the existing ventilation rate and extrapolating to the desired conditions; or by other methods that will occur to the investigator faced with this problem. The size of the room or the size of the building for which this dilution ventilation is required is of minor consequence.

   The point is easily illustrated. Suppose sources exist for which a dilution ventilation rate of 10,000 cfm (280 m³/min) is required. If these sources were located in a room 100 × 50 × 20 ft high (30 × 15 × 6 m high), with a cubage of 100,000 ft³ (2800 m³), the air change rate would be 6 per hour. If the same sources were located in a space

150 × 100 × 25 ft high (45 × 30 × 8 m high) with a cubage of 375,000 ft³ (11,000 m³), the air change rate would be 1.6 per hour. There obviously is no direct relationship between the air change per hour rate and the degree of control of the hazard. As a matter of fact, the situation with the lower air changes per hour may be somewhat better, merely due to the larger building volume and (presumably) wider spacing of individual sources.

"Air changes per hour" has a few advantages. It is simple, easily understood by all whether technically trained or not, and the computation is easy. An air change basis may indeed be appropriate for circumstances that are reasonably standard. Schoolrooms, for example, are seldom built with a ceiling lower than 8 ft (2.5 m) or higher than 12 (3.5 m); there is a limit to how tightly the students can be packed into the room, and there must be aisle space and a pathway to the teacher's desk. Furthermore, no serious health hazard is involved, at least in the nature of toxic air contaminants. Therefore, an air change basis for the ventilation of schoolrooms in general may be appropriate. The same rationale may be applied to other circumstances of standardized occupancy, including some few industrial operations that are reasonably "standard" from one plant to another.

For the sophisticated ventilation engineer or hygienist, however, the air change method is a method of last resort, when no more rational basis can be derived.

2.5 Evaporative Cooling for Heat Stress

Evaporative cooling is a process where air is cooled by evaporating water into it. This may be accomplished in an ordinary spray chamber, by wetted filter pads, and similar mechanisms. The temperature of the air is reduced and the humidity of the air increased. In desert climates the outdoor air is normally of very low humidity, and evaporative cooling serves adequately for comfort air conditioning. In more humid climates such as prevail in most of the United States, less cooling is achieved (because the entering air is more humid to start with), and the air produced is of such high humidity as to be "uncomfortable." Thus evaporative cooling is in a state of poor repute, and justifiably so, for comfort air conditioning in nondesert climates. However, the same rationale does not apply to evaporative cooling for relief of heat-stress conditions, even in nondesert climates. The principles are the same, but the absolute values are different, and the results are different. In alleviating heat-stress conditions, we no longer are concerned primarily with achieving optimum comfort conditions, although assuredly a less heat-stressful situation is more comfortable than a higher heat-stress situation.

Under heat-stress conditions the air temperature and/or the radiant heat load is sufficiently high that the importance of humidity is reduced. In ordinary midwestern climates the warmest summer days do not coincide with the highest ambient humidity conditions, so that a surprising reduction in air temperature can be achieved by evaporative cooling. Further, the incoming air to the evaporative cooler will be at a temperature sufficiently above that of the typical water supply so that some conductive cooling also takes place by contact with the cooler water, in addition to the cooling caused by evaporation of water into the air.

The quantitative aspects of this situation are adequately outlined by Caplan (see the General References section). Furthermore, the proof of the application is apparent in the many situations in which it has been applied, although there has been little publicity accompanying it, perhaps because it is so simple and inexpensive.

2.6 Auxiliary Air at Local Exhaust Hoods

The volume of replacement air that is completely treated to suit the occupancy specifications may be reduced by the provision of auxiliary air supply, less completely conditioned, at or near exhaust hoods. The principle involved is that such auxiliary air furnishes some of the air entering the exhaust hood and that the auxiliary air does not pass over the worker (or perhaps sensitive equipment) on its way from the supply opening to the exhaust hood. Various trade terms are applied to this basic principle. *Push–pull* is a term frequently applied to the ventilation of large, open-surface tanks. *Compensating hoods* is a term used for some foundry shake-out ventilation systems. *Auxiliary air* is a term used for laboratory fume hoods that are so arranged.

Although it is true that the introduction of auxiliary air at the hood reduces the requirements for replacement air, such systems are not without cost. There is typically some energy savings for heating or otherwise treating the auxiliary air but an additional and separate system must be installed. *Utmost caution* must be applied to such installations. The most common failing is an excess of auxiliary air supply—that is, too much push—so that the contaminant is blown past the hood opening rather than entering it.

For fully air-conditioned applications such as research laboratories, even more caution is advised. Under summer conditions, when the partially treated auxiliary air is warmer than the fully air-conditioned occupied space, the warmer air coming from the auxiliary system may not be completely captured by the exhaust hood, and the condition of the general space will be downgraded proportionally.

3 NATURAL VERSUS MECHANICAL VENTILATION

3.1 Natural Ventilation

Ventilating a factory building entirely by natural ventilation is apparently a custom of the past. The concept depends on large openings at or near grade level, and large-capacity natural-draft roof ventilators. For such an arrangement to produce much in the way of ventilation rate, there must be high heat-producing sources in the building to create a thermal head. Many such sources also generate significant amounts of air contamination and fugitive emissions. The ventilation rate is of course subject to all the vagaries of wind and tmeperature. In cold weather the temperature is essentially that of the outdoors, except that the warming effect of sunshine is absent. Local radiant heat sources will offset this. If local exhaust ventilation is applied to some of the operations in such a building, on windy days the high-velocity draft will spoil the capture effect of the hoods and disseminate the contaminant horizontally in the building and the working area. If the fugitive emissions escaping the roof ventilator are in excess of outdoor air pollution standards, they must be captured and the air cleaned or else the sources controlled. Most existing old buildings that were initially designed on this principle have been modified at least to the extent that high-power roof ventilator fans have been installed.

As a generality, one might speculate that such a building and its ventilation would be appropriate for a process that could take place out of doors except for the need for protection from rain or snow. In that case the building logically should consist of a rain shelter roof only, and the process equipment designed as an outdoor plant.

3.2 Modes of Natural Replacement or Natural Exhaust

In seeking to avoid the cost of air-handling units and duct work distribution systems, two types of building general ventilation have been conceived.

3.2.1 Building under Negative Pressure. One system utilizes mechanical exhaust ventilation, both in local exhaust hoods and in roof ventilators or wall fans, but "economizes" on replacement air. Since replacement air will enter the building anyway, deliberate openings with controllable louvres are placed at locations judged to be strategic. The slight suction in the building draws air in through such openings. The adjustable louvres are intended to direct the air flow so that it does not create uncomfortable drafts on workers in cold weather, and invariably those louvres are built so they may be completely closed. The theory is that some openings can be completely closed while others are open, thus directing the general nature of the air flow in the building, compensating for winds, and so on. The facts are, however, that in cold weather all the louvres will be closed. As we have seen previously, however, all air exhausted must be replaced, and it enters in a less controlled fashion through accidental openings in the building and the exhaust volumes are (probably) correspondingly reduced due to the increased resistance of smaller openings.

An additional difficulty is caused by the fact that the air entering the building through such an opening is in effect issuing from a pressure jet, as illustrated in Fig. 7.1. The air on the upstream side of the opening—that is, outside the building—is at a pressure higher than that inside the building; therefore the behavior of the air flow inside the building is that of air issuing from a pressure jet. As described previously, this jet of air will throw for considerable distances. For example, air entering a wall opening 5 × 10 ft (1.5 × 3 m) in cross-sectional area at a velocity of 500 fpm (150 m/min) will still have a centerline velocity of approximately 115 fpm (2.5 m/s) 100 ft (30 m) from the opening. This high-velocity draft may well upset the capture velocity of local exhaust hoods. If such a jet encounters a contaminant-laden thermal plume where the plume tends to carry the contaminant up and out of the roof ventilators, the jet will disturb that plume and blow the contaminant horizontally into the work area.

Any apparatus for cleaning, heating, or cooling the incoming air imposes enough resistance to be significant. Then the building would need to be constructed to withstand such suction, and still be liveable and operable, and to overcome all the objections to the lack of replacement air described in Section 2.4.

While this mode of natural replacement air cannot be completely rejected as impossible, it is usually quite unsuccessful in terms of controlling air contaminants and providing an otherwise acceptable work environment. Its use is, obviously, limited to warm climates.

3.2.2 Building under Positive Pressure. The other scheme is the converse of the situation described in Section 3.2.1; that is, supply air is blown into the building, heated or otherwise treated as necessary, and air is allowed to escape the building by opening strategically placed louvres in wall openings. The theory is that the parts of the building that are cold will vary, depending on natural forces such as wind direction. The flexibility afforded by opening and closing various wall openings then permits openings to be created on the cold side of the building, thus permitting the warmed air to generally flow in that direction.

A frequent characteristic of this system is that there is little or no distribution duct work furnished on the supply air units. Instead, large volumes of air are dumped into the building at a few locations. Here again, the throw from these air jets may disturb whatever local exhaust hoods are utilized. Opportunities for energy conservation or energy recovery are virtually nonexistent.

This system of mechanical supply and "natural" exhaust is, however, at least workable for some kinds of manufacturing. If the process is one in which only small amounts of contaminant are generated at widely spaced locations in the building; and those few local exhaust hoods can be placed out of harm's way, not in the throw of the air jet from a supply air unit; and the contaminants generated are of low toxicity, such systems are acceptable.

3.2.3 Wind Pressures and Velocities. Both the natural exhaust ventilation and the "natural" replacement schemes suffer from the vagaries of the wind. A wind speed of 1 mph is equal to an air velocity of 88 fpm (0.5 m/s). On a calm day, wind speeds are in the range 1–3 mph, and winds in the range of 5–10 mph are very common in most parts of the country. Thus if a "natural ventilation" opening is facing into the wind, it will frequently be subjected to opposing wind velocities of 440–880 fpm (2–4.5 m/s). The presumed control over air currents leaving or entering the building by adjusting louvres is thus frequently largely negated.

4   AIR CONDITIONING

4.1   Technical and Popular Definitions

Technically, the term *air-conditioning* means the treatment of air to control one or more of the properties of temperature, humidity, or cleanliness. The popular definition implies the cooling, and perhaps dehumidification, of air to achieve comfort conditions. The popular definition is so popular that even the technicians usually use the term in the same sense. In some industrial operations, the temperature, humidity, or cleanliness of the air is dictated by considerations of the process or the product, and such conditions may not be optimum for human comfort.

4.1.1 Cost Factors. Even comfort air conditioning, as employed in an office building, for example, is expensive. In order to understand the cost implications of industrial air conditioning, the characteristics of comfort air conditioning and industrial air conditioning need to be compared.

Most comfort air-conditioning systems do not involve humidity control in the literal sense. For a properly designed system, some dehumidification, and therefore improved comfort, is achieved by the design of the cooling coils but without specific control features. In a typical comfort system, almost all of the air being exhausted or "returned" to the air-handling unit from the occupied space is passed through the cooling coils and recirculated back to the occupied space. Only a very small amount of outside air is introduced, typically less than 10% of the volume being circulated. The air passing through the cooling coil will issue at a temperature of 55–60°F (18°C) and be resupplied to the occupied space. The actual surface temperature of the coil is much colder, perhaps 35–40°F (4°C). The portion of the air flow that constitutes the slow-moving laminar-flow layer of air immediately adjacent to the metal will be cooled much more than the average temperature of the air, to approximately 40°F (4°C). That portion of the air will be dehumidified by condensation at that low temperature. This partial dehumidification of the air, without specific controls, will result in adequate comfort conditions in the well-designed system.

For a cost comparison, let us assume an ordinary office occupancy with an inside design criteria of 75°F (24°C) and 50% relative humidity (RH), operating on a typical summer day where the outside conditions are 95°F (35°C) and 80% RH; and that the air passing through the cooling coil is 90% recirculated air at the design conditions plus 10% outside air. The mixture passing through the coil will be at 80°F (27°C) and 50% RH. It is this mixture that will be cooled to 55°F (13°C) and 50% RH. A relatively small flow of this cool air is introduced to each occupied room in sufficient quantity to remove the heat generated by personnel, lights, and office machinery so that the design condition of 75°F (24°C) and 50% RH is reached. For a moderate sized air-conditioning system in the capacity range of 20–50 tons (1-ton refrigeration equals 12,000 Btu per hour), the capital cost of the refrigeration equipment and cooling equipment will be about $2000 per 1000 cfm (28 m³/min). The energy cost of the cooling and dehumidification will be 56,000 Btu per hour (4.7 tons) per 1000 cfm. The actual annual energy cost will be significantly less than equivalent to the per-hour cost at maximum load because obviously the system is not working at maximum load for much of the time.

For the air-conditioned factory, the cost is much higher. This arises from the fact that in the factory that will be a significant number of local exhaust hoods, and the air from such exhaust is normally not recirculated. Only the surplus supply air, over and above that required to satisfy the exhaust, will be recirculated; and in some circumstances, even that air is not recirculated. The system handles much larger volumes of air in the first place than would be required in an office system; and equally or even more important, perhaps as much as 50% and sometimes as much as 100% of that air is outside air, not recirculated air. The outside air under summer design conditions will be of much higher temperature and humidity than the recirculated air in a comfort system. On the other hand, because such large volumes of air are handled, it need not be chilled to a temperature of 55°F (13°C) in most circumstances.

Taking an example similar to the comfort air-conditioning situation, assume an outside condition of 95°F (35°C) and 80% RH, a design condition in the building of 75°F (24°C) and 50% RH, and assume the supply air need only be cooled to 65°F (18°C). For a 50-ton system, the equipment cost per 1000 cfm (28 m³/min) would be about $4000, and the conditioning energy cost per 1000 cfm (18 m³/min) at design conditions would be 142,000 Btu/hr or 11.8 tons. Comparable values for conventional air conditioning in the previous example were $2000 equipment and 56,000 Btu/hr.

If humidity control is also required for process or product reasons, the capital cost and energy cost are further increased. Humidity control is generally obtained under summer conditions by overcooling the air to condense more moisture from it, and then reheating it to the design conditions for supply to the room. An alternative method is to remove moisture from the airstream by various absorption techniques. Obviously, either process requires energy.

There may be some industrial operations where little or no contaminant is involved. In such cases the air conditioning, absent any process or product requirements, would be similar to comfort air conditioning in cost.

The point of all this is to illustrate that conditioning outside air is very expensive. This then leads to the conclusion that extraordinary efforts and expenditures to control sources of airborne contamination, to utilize local exhaust hoods that minimize the required exhaust air volume, and to institute and enforce excellent work practices are all important in reducing air-conditioning cost.

4.1.2 Recirculation of Conditioned Air. Because factory conditioned air is so expensive, great economies can be achieved by recirculating it, thus reducing the amount of outside air that must be treated. A basic criterion for such recirculation is that the return air to the cooling unit must be low in contamination level; and if it is not, it must be suitably treated to remove the contaminant in a manner that is adequately efficient and reliable for considerations of health and safety. If return air is clean enough to recirculate without further cleaning, no particular cost is incurred. However, if high-efficiency air cleaning is required, this has an impact on the air-conditioning cost in addition to the cleaning cost.

The methods of air cleaning applicable to this situation are, generically, filtration, wet scrubbing or wet absorption columns, activated-charcoal adsorbers, and electrostatic precipitators. The selection of equipment would depend on the type of air contaminant, whether particulate or gaseous, and its other properties. However, each type of equipment has serious cost implications of one kind or another. Part of the cost is that pumping air through a resistance consumes energy, and that energy typically appears as an increase in temperature of the air. The temperature increase occurs at the blower as the air is "compressed." For every 1-in. (2.5-cm) differential total pressure through flow elements of the system, the air temperature will increase 0.47°F— that is, about half a degree. This adds to the heat that must be removed in the cooling coil.

Each class of air-cleaning equipment has the following impact on the air-conditioning load.

(a) Filtration of air to remove particulate imposes a significant resistance. If, for example, a filtration recirculation system consisted of a fabric filter to clean the air of a significant dust load, followed by a roughing filter and a HEPA (High Efficiency Particulate Filter) filter as redundant backup in case of failure of the fabric filter, the total pressure drop through that system including duct work and damper losses would be about 8 in. (20 cm) water gauge, resulting in an increase of 4°F in air temperature. This is a significant added energy cost for the air cooling.

(b) A wet scrubber or a wet absorber has a pressure drop typically on the order of 2–3 in. (5–7 cm) water gauge. For safety, two scrubbers in series would frequently be required, thus doubling the pressure drop. Further, the passage of the air through the scrubbers would increase the humidity almost to saturation, imposing a further air-conditioning load for dehumidification.

(c) Activated-charcoal filter beds have a significant pressure drop. The thin-bed devices, configured something like an ordinary air filter, have a limited adsorption capacity and must be changed frequently. There is no way to predict when such a charcoal bed is approaching breakthrough. One of the safe ways to install such a system is to install two sets of charcoal filters in series, with some kind of a sensitive (and probably expensive and hard to maintain) monitoring instrument between them, so that the instrument would indicate breakthrough on the first bed while the second bed is still functional. Then the filters would be changed, moving the second filter up to first place and installing new filters in the second section. The charcoal filters themselves are expensive. As waste, depending on the contaminant, they may be classified as "hazardous waste," with further attendant cost. Deep-bed regenerative carbon-bed adsorbers are much more expensive and have a much higher pressure drop than the thin charcoal filter configuration.

(d) Low-voltage electrostatic precipitators in themselves have a low pressure drop and a reasonably low power demand for the precipitation function. However, unless the plates of the precipitator are coated with adhesive, flakes of agglomerated collected dust blow off of the plates and a secondary filter is required anyway to achieve high efficiency. Thus the pressure drop includes the resistance of at least one filter section. High-quality and heavy-duty low-voltage precipitators will have an automatic system for washing the plates and recoating them periodically; the added cost of this mechanism, plus the necessity to furnish hot water for washing and drainage, diminishes the cost advantage of this type of equipment. The least expensive type of low-voltage precipitator, with dry plates and after-filter or provisions for manual cleaning only, is not, in the author's opinion, suitable for recirculation of any toxic contaminant.

High-voltage (Cottrell) precipitators are not usable in this application because they generate excessive amounts of ozone.

4.2 Operational Characteristics of Air-Conditioned Industries

Industrial air conditioning, as we have seen, is so expensive that very few factories are air conditioned for personnel comfort only. Requirements of the process or the product are usually the reason for industrial air conditioning. The major categories of industries so treated are the electronics industry, some pharmaceuticals and specialty chemicals, laboratories, and some ultra-high-precision machining operations. The electronics industry and the laboratories probably represent by far the greatest numbers of air-conditioned industrial facilities. The electronics industry is labor intensive and involves a large number of toxic chemicals, usually in fairly small quantities. Many clean rooms are used to avoid particulate contamination of the product. At least currently, the industry is also characterized by the necessity for making very rapid changes in production facilities in order to stay competitive.

These factors combine to make life difficult for the air-conditioning engineer. The occupancy of a workroom can be doubled almost overnight by moving the workbenches closer together and hiring more people. Such a change significantly increases the cooling load for the area and also may increase the burden of air contamination from a number of small sources. Changing the air-conditioning system, particularly the duct work, is an especially upsetting procedure in most electronic plants because of the contamination that is spread by that work; a virtual shutdown of a complete section of the building may be required while it is going on. This then leads to around-the-clock and premium-time work for the construction contractor in order to minimize the length of shutdown. On the other hand, it surely must be difficult to get approval for designing the air conditioning for a new building so that the capacity is twice that required for the initial occupancy!

Another characteristic of electronics plants is that they are frequently built in light-industry/office parks or semiresidential areas. Buildings can be designed to be esthetically pleasing. As a result, discharge stacks are not as high as they should be to prevent reentry of contaminated air into air-conditioning intakes, and retrofit installation of tall stacks is regarded as esthetically "verboten." Installing air-cleaning equipment for the large air flow containing low concentrations of contaminant is very expensive, and unless the equipment is placed inside the building, it too will incur esthetic disapproval.

One solution for this dilemma, at least for the types of electronics manufacture that have high cleanliness requirements, would be to take two drastic steps. One would be to build the air-conditioning systems on the floor below or above the operating floor, with the two floors separated by solid concrete with a large number of precast openings. This arrangement would permit changes in the air conditioning to take place with minimal upset in terms of particulate contamination to the operating floor. It would also almost double the height of the building since every such operating floor would require an "equipment room" type of additional floor to go with it. The second major relief for the problems of the industry would be to erect, as part of the intial construction, an "architectural mask" structure that would in fact contain the exhaust gas discharge stacks raised to a level high enough to avoid reentry.

While the initial cost of such construction would be much higher than ordinary, the cost of making changes both in terms of lost production, spoiled product, and direct cost, would probably pay for the initial cost increment several times before the end of the useful life of the building.

5   RECIRCULATION OF CLEANED AIR

5.1   Problems of Definition

Any discussion of the propriety of recirculating air, cleaned of toxic contaminants, back to the workroom involves the concepts of dilution ventilation, local exhaust ventilation, and recirculation. Local exhaust ventilation involves the concept of enclosing or partially enclosing the source, and sucking the contaminated air away before it enters the general workroom and before the worker is exposed. A related form of local exhaust is where the source is not enclosed, but the suction opening is placed very close to it so that it captures the contaminated air.

The concept of general dilution ventilation implies that the contaminated air is not captured, but instead is diluted with clean air so that the concentration is below the applicable TLV before any worker is exposed.

The concept of recirculation implies that contaminated exhaust air is cleaned sufficiently to be safe, and deliberately returned to to the workroom as supply air.

Although these concepts are simple in themselves, attempts to define them rigorously lead to endless and usually fruitless debate. There are a number of circumstances that are accepted as good practice which that examined closely violate the "principles" classical in the field of industrial hygiene. The following sections attempt to shed some light on this question.

5.1.1 Local versus General (Dilution) Ventilation. Practitioners of industrial hygiene almost uniformly insist that local ventilation is preferable to general or dilution ventilation. As a generality, that policy position is correct. Again as a generality, the local exhaust system results in handling less air, makes the application of an air-cleaning device more economic, thereby reducing any insult to the outside environment and the cost of replacement air. Further, it exercises better control in that the contaminant is removed from the work space before it is breathed by the worker, thus usually providing better protection. These generalities break down, however, if an attempt is made to develop rigorous definitions.

Recirculation is, in concept, the deliberate return of exhaust air to the work space through a supply air system. Implied is the need for some air-cleaning device so that the air is sufficiently clean to be so recirculated.

5.1.2 Local and Incidental Dilution or Recirculation. Problems with rigidly defining these concepts can best be illustrated be example. The first example would involve a moving source of particulate contamination or a source of contamination that is so located that it is impractical to run duct work from it to an outdoor discharge. In either situation, it may be practical to install a small fabric filter on the moving machine or on the source in the inaccessible location. In that case, the cleaned air emanating from the fabric filter would be in fact "recirculated." The policy stance of some hygienists and the rules of some regulatory bodies would prohibit this. If the fabric filter therefore is not installed, is not the contaminated air being "recirculated" without using a fan? Or is it "dilution"?

In some circumstances, local exhaust ventilation is partially applicable, but complete control by that technique is literally impossible. An example is the cutting house in smokeless powder manufacture. In this situation, the operator must stand, an appreciable portion of the time, between the cart carrying and feeding the strands to the cutting machine, and the cutting machine itself. Both the cart and the machine are major sources. The cutting machine is amenable to local exhaust ventilation. The cart is not. One solution to the problem consisted of applying local exhaust ventilation to the cutting machine, and establishing dilution ventilation in a controlled flow pattern so that the general flow of air was from the operator toward the cart and then to a general ventilation exhaust duct. In this case, is not "dilution" preferable to local exhaust?

In most industrial operations, not *all* sources of contamination, no matter how minor, are controlled. Those minor sources not controlled by local exhaust are therefore, by default, being controlled by dilution ventilation. The air drawn through the workplace by the local exhaust system provides dilution ventilation, whether or not that is intended.

Some operations involving the application of small amounts of toxic solvents to small workpieces can generate significant exposures in the breathing zone of the worker actually doing the task. Successful and inexpensive control of this operation in the proper circumstances constitutes an exhaust hood with a low-power fan but no duct work, discharging into the room. This hood adequately protects the worker doing the task, and solvent concentrations throughout the room are well below the applicable TLV. This is a flagrant case of dilution and recirculation both, but is it poor policy? Should it be contrary to regulation?

5.1.3 Sources Amenable to Local Dilution. For some kinds of tasks, deliberate dispersion of the contaminant into the room air, with adequate general ventilation to insure dilution and removal, is the only practical solution. An example involves the welding of very large flat sheets of steel. The structures are so large that the welder must kneel or sit on the plate to make the weld. The natural posture of the welder results in leaning over the weld to some degree, with the breathing zone then being near to or within the rising plume of fume from the welding source. The sheets are large, the work is conducted throughout the shop, and overhead cranes are necessary for moving the work. The classical flexible-duct exhaust hood, which needs to be placed close to the work, is throughly impractical. In a situation such as this, small supply air jets may be utilized to create a strong horizontal air current between the weld and the welder's breathing zone. The welding fume is thus blown horizontally, diluted and mixed with the room air deliberately. If the general ventilation of the building is adequate, the allowable concentrations for welding fume will not be exceeded at the welder's breathing zone. This then would be a thoroughly satisfactory solution, inexpensive, easy to operate, and safe.

In an operation with a considerable amount of local exhaust hooding, the air flowing into the exhaust hoods is being replaced in the room. Hopefully, it is being supplied by a replacement air system in a controlled fashion. But even if it is not, all the air exhausted through the exhaust hoods also flows through the room as replacement air. Does that not also constitute dilution ventilation as far as the rest of the activities in the room are concerned?

These examples should illustrate the basic futility of tightly drawn legalistic or technical definitions of these ventilation characteristics. It is the concepts, and the proper application thereof, that are important.

5.2 Benefits of Recirculating Cleaned Air

The major benefit of recirculating cleaned air is energy conservation. In the parts of the United States where there is a significant heating season the energy costs per thousand cubic feet (28 m³) per minute capacity will be in the range of 65,000–90,000 Btu/hr at design conditions and typically from 75 to 180 million Btu per heating season. If the exhaust air can be adequately cleaned for recirculation, its heat content is almost 100% conserved. The power required to pass the air through adequate cleaning devices is only incrementally larger than would be required to supply the same volume through a replacement air system, and that energy appears as heat in the air. For a situation that is relatively simple—that is, if the contaminant is particulate and there are no complications either due to the "difficult" nature of the paritculate or complicating factors of contaminant gases—the capital investment in the recirculation system is quickly recovered. The simple payout period of the investment, neglecting factors of interest, taxes, and the like, varies from one to two years under such circumstances. If the contaminant is gaseous in nature or there are any complications, the payout period is typically longer.

The air-cleaning equipment used in a recirculating system obviously must be adequately efficient so that the cleaned air is of satisfactory quality. In addition, from a health protection point of view, the reliability of the system must be considerably improved over that considered customary. Better design, better maintenance, and high-quality equipment will assist in this. However, it is necessary to provide a monitoring instrument on the clean side of the equipment to detect any breakdown or leakage of significance and to shut down the system or by bypass the "cleaned" air to atmosphere, sound an alarm, and initiate whatever appropriate action is indicated for the particular situation. Alternatively, a redundant, backup air-cleaning device may be provided if some way can be devised to indicate when the primary device has failed.

For particulate material of significant toxicity, the collection efficiency of the typical fabric filter may be somewhat marginal. In that event the primary cleaner, that is, the fabric filter, should be followed with a secondary cleaning device such as a HEPA filter or a 90%-efficient media filter, the choice depending on the safe concentration of contaminant. Such a secondary filter provides a backstop that prevents contamination being reintroduced to the plant in the event of sudden failure of one or more filter elements in the fabric filter. It also serves as a monitoring device, conceptually "sampling" the entire airstream rather than a portion of it. Such a filter has no moving parts; if properly installed, it maintains its efficiency, and indicates significant leakage of the primary filter by a rapid increase in the pressure drop through the secondary. Normal leakage through the primary will require periodic replacement of the secondary in any event.

The heating effect of passing the recirculated air through the filters means that the air being returned to the building, except for heat losses in outdoor duct work, will be 4–5°F (2°C) warmer than the exhaust air, and ironically energy conservation is frequently greater than desired. The interior of a large factory building will require some cooling, not heating, in all except the most severe winter weather. Therefore such systems are usually provided with a thermostatically controlled damper system that bleeds in colder outisde air and dumps some of the exhaust air under most operating conditions. the same or an additional set of dampers can provide for bypass to the atmosphere in the event of failure of cleaning equipment or to change the secondary air filters without shutting down the system.

It has been the historical stance of most industrial hygienists and regulatory authorities to prohibit the use of recirculated air if the contaminants involved are considered toxic. Under the historical circumstances that policy stance was correct. Most air-cleaning equipment is not part of the production components in the plant and usually receives less than adequate maintenance. Even though the rated efficiency of the equipment would be sufficient to assure health protection, it frequently does not operate at rated efficiency.

There have been no real breakthroughs in the technology of air cleaning that have changed the picture with respect to the advisability of recirculation. The shift to favor recirculation is largely economic. In the days of cheap energy, it was less expensive to dump the exhaust outside and replace it with suitably conditioned air than to safely recirculate it. As energy becomes relatively more expensive, the investment in better engineering, more and better equipment, and adequate maintenance is justified by the energy cost savings. In addition to the monetary considerations, there have been (any may again be) regulatory or supply restrictions on availability of heating energy.

In addition to the economic benefit, there is another important benefit to be anticipated. If the exhaust air is not to be recirculated, and as replacement air gets more and more expensive, it is predictable that the design or (operating) air volume of exhaust hoods will be cut back to the minimum in order to conserve energy. Even in today's environmentally conscious climate, all too often the exhaust systems do not perform as adequately as would be desired in the first place. If exhaust air volumes are pinched, it is predictable that worker exposures to contaminants will be increased. If recirculation is permitted, there is less cost incentive to cut back on adequate exhaust volumes.

5.2.1 Total Contaminants to be Considered. In most industrial environments, the air contamination frequently involves more than one component that could be a health risk above its safe concentration. Usually, one of the air contaminants predominates and the others are relatively minor. Thus the design of the exhaust system focuses on the predominant contaminant and rightly assumes that if that contaminant is controlled to a safe level, all the others will be also. Thay may not be so in the case of the recirculating system, particulary if the air-cleaning apparatus is capable of removing one component but not the other. For example, if the major contaminant is particulate, a filtration system would be applied; but if gaseous contaminants are also present, they will not be removed from the air by a filtration system and could build up to hazardous concentrations during continued recirculation. A common example is that of welding fume. In most situations, it is the particulate material in the fume that is of primary concern and therefore, if welding fume exhaust is to be recirculated, a particulate filter would be installed. The arc welding process produces other harmful contaminants in minor amounts, for example, nitrogen oxides, ozone, and if a CO_2-shielded arc, carbon monoxide. These secondary contaminants are all gases. With nonrecirculating exhaust, if the particulate fume component is controlled, the gaseous contaminants will be of no concern. However, in a recirculating system those gaseous components could build up to harmful concentrations.

All exhaust systems should be designed to avoid mixing incompatible chemical contaminants. Reaction products that may be explosive, combustible, corrosive, or create sticky deposits or "unfilterable" aerosols should be avoided. This factor is especially important in a recirculating system.

In the design of a recirculating system, all contaminants must be considered. That consideration may dispose of the question of any potential hazard, but the process of consideration must take place to ensure that such will be the case. If there are secondary contaminants that would not be removed in the proposed air-cleaning system and could become too high in concentration, one simple expedient is to indulge in a small percentage of "blowdown" from the system, discharging some of the contaminated gas and replacing it with clean outside air. A second alternative, usually more expensive and complicated and therefore not utilized, would be the installation of two or more different types of air-cleaning devices so as to clean the air of all the comtaminants of concern.

5.3 Particulates versus Gases or Vapors—State of the Art

At the state of the art, recirculation is indicated more clearly for particulate contamination than it is for gases or vapors. The available technology and equipment permits the removal of particulate to be generally of higher efficiency and reliability than the removal of gases. The safe or redundant-safe design is also more straightforward for particulate materials. Several suitable monitoring instruments for particulate contamination are available, and though they are delicate and require frequent calibration and maintenance, they can be kept adequate. Alternatively, redundant or secondary filters can be installed that are less expensive and more reliable than the primary filter, although not capable of handling heavy dust loads. If the system is functioning properly, the secondary filter does not experience heavy dust loads; if a heavy dust load does occur, that means the primary filter has failed and the secondary filter is performing its basic function.

Gases and vapors are removed from an airstream primarily only by wet absorption towers or by beds of activated carbon, although other and usually more expensive methods can be used. There is no secondary or backup air-cleaning equipment that is significantly less expensive than the primary equipment; so backup redundancy requires, in effect, double the primary investment in terms of equipment. The equipment itself is somewhat more complex to keep in proper operating condition than is a particulate filtration system. Wet absorbers, for example, involve maintaining proper flows of solution, maintaining the chemical makeup of the solution, preventing the plugging of the sprays, and so on. Adequate monitoring devices for the cleaned air are available for some kinds of gases and vapors, but for others may be unduly complex and subject to failure.

Beds of activated carbon vary in their effectiveness in removing gases, but for the gases with favorable adsorption characteristics, such equipment functions well. Carbon beds consisting of thin layers of a configuration similar to ordinary air-conditioning air filters are usually not sufficiently reliable, and do not have sufficient adsorption capacity, for application to a system recirculating materials of significant toxicity. Deep-bed carbon filters are much more expensive and that alone would rule out the application, on an economic basis, for most recirculating systems.

5.4 Monitors versus Backup Filters

Assuming that the primary air-cleaning device in a recirculation system has adequate efficiency, the safe operation of the system can be assured either by the use of an instrument to monitor the cleaned air, or a backup/redundant air-cleaning system. Of the two choices, at the current state of the art, the choice of the backup system, for filtration of dry particulate matter at least, would seem to be the best choice.

Particulate monitoring instrumentation has been developed to meet EPA specifications of sensitivity and reliability and on this score these instruments would be adequate for the task. However, the concentration ranges of particulate that are permissible for stack discharge relative to outdoor air pollution considerations are typically 100-fold greater than would be tolerable for recirculation to the workplace. A recirculation monitor would need to be 100 times more sensitive than the stack emission monitors. This represents a significant change in the technology, reliability, and cost. Frequent and highly skilled maintenance attention is required for such instruments to compensate for zero drift, loss of calibration, or just plain malfunction. In most factories, that level of skill and dedication may not be already available, and the cost of adding this capability to the cost of the hardware itself should be considered.

As already described, for particulate filtration systems a secondary tier of equipment is available that is reasonable in cost. The secondary filter not only serves as a "monitor" in that its pressure drop will increase if leakage is occurring through the primary, but it also functions as a backstop in the event of more catastrophic failure of the primary filter. Primary filter failure could consist of split seams in bags, bags coming detached from thimbles or frames, and similar relatively sudden occurrences of large leaks. The backstop capability of the secondary filter is attractive for preventing dosing of the workroom with a high concentration of particulate, even if only for a short time. In addition, the secondary filter system further cleans the air of particulate that would normally penetrate the primary filter even when it is operating correctly. This achieves a greater margin of safety during normal operating conditions, and reduces the background level in the workplace so that the control of individual sources need not be quite as perfect as if a high background were present.

For equipment removing gaseous and vapor air contamination, the same principles apply, but the application of them is different. The backstop or redundant equipment is more expensive, and the monitoring instrumentation may be, relatively speaking, simpler and more reliable, if needed an appropriate instrument is available for the contaminant of concern.

5.5 Industrial Hygiene Basis of Recirculation

It is obvious that unless the contaminant that is being recirculated is of very low concentration, that is, a small fraction of the applicable TLV, some consideration is necessary of the impact of the recirculation on the worker exposure in the workplace. Obviously there will be a quantitative relationship between the amount of contaminant released into the exhaust air, the efficiency of the air-cleaning device, the concentration of the contaminant in the recirculated air, the contaminant in the room air not captured by the exhaust or other control systems, and the amount of other ventilation supply air that is not recirculated air. That relationship is represented by Eq. (7.1):

$$C_R = \tfrac{1}{2}(\text{TLV} - C_0) \times \frac{Q_T}{Q_R} \times \frac{1}{K} \qquad (7.1)$$

where C_R = concentration of contaminant in exit air from the collector before mixing, any consistent units

Q_T = total ventilation flow through affected space, cfm

Q_R = recirculated air flow, cfm

K = an "effectiveness of mixing" factor, usually varying from 3 to 10, with 3 = good mixing

TLV = threshold limit value of contaminant

C_0 = concentration of contaminant in worker's breathing zone with local exhaust discharged outside

The mixing factor is important in locating the recirculated air return supply openings. Unless the recirculated air is well below the TLV, the supply opening should be located remote from any work station so that there is adequate opportunity for mixing with the room air, which is presumably below the TLV. A further consideration is that in some circumstances, the air in the upper spaces of the building will be more contaminated than that determined as the value of C_0 in the work area. If that is the case the recirculated supply air should be not introduced in such spaces because it will entrain and mix with that more contaminated air on its way to the work areas. Note that in Eq. (7.1) the assumption is made (although not stated) that supply air to the workplace other than from the recirculation system contains essentially zero level of contaminants.

Figure 7.3 illustrates the results obtained by application of that equation to an assumed set of circumstances and for different values of C_0. For this example, the TLV was assumed to be 100, the mixing factor 3, and various values of C_0 ranging from 20 to 80 were taken. Note that even if the work situation is such that one is just slightly under the TLV, $C_0 = 80$, under reasonable circumstances the recirculated air could contain concentrations higher than 60% of the TLV.

The advantage of Eq. (7.1) is its simplicity and ease of application, although some common sense is required. The disadvantage of the equation is that it requires a knowledge of the value of C_0. For an existing situation that is to be retrofitted with recirculation capabilities, C_0 is directly measurable. For a new facility, C_0 may be estimated accurately only if there is in existence a situation closely similar to the projected one, where C_0 can be measured. If neither of those circumstances exist, there is no rigorous nor even approximate method of computing C_0 that has been validated. Some practitioners of the art may be able to come surprisingly close with

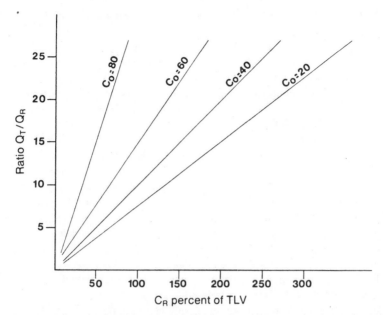

Figure 7.3. Concentration of contaminant in recirculated air (C_R) versus ratio of total ventilation to recirculated ventilation.

such estimates, but the process cannot be codified so as to be generally applicable by the designer.

Further, there are several general considerations that should be taken into account by the designer contemplating a recirculating system. The following list is adapted from the ACGIH *Ventilation Manual:*

- Air contaminated with certain substances should not be recirculated under any conditions. Substances that can cause permanent damage or significant physiological harm from a short overexposure should be considered in this category.
- It is usually considered necessary to provide general ventilation air in addition to that recirculated so that there is, in effect, continuous dilution of contaminants not cleaned from recirculated air, or escaping local control. If it is proposed that *all* the supply ventilation air be recirculated air, care must be taken to evaluate *all* the possible contaminants, not just the major contaminants normally of concern. A fabric/filter–ultra-high-efficiency air filter combination, for example, although providing essentially 100% collection of particulate, could allow gases and vapors to build up in the course of operations from insignificant to hazardous concentrations.
- Recirculating systems should, whenever practicable, be designed to bypass to the outdoors, rather than to recirculate, when weather conditions permit. If a system is intended to conserve heat in winter months and if adequate window and door openings permit sufficient replacement air when open, the system can discharge outdoors in warm weather. In other situations where the work space is conditioned or where mechanically supplied replacement is required at all times, such

partial to full bypass operation can be achieved by a thermostat-controlled damper system.

- Wet collectors also act as humidifiers. Recirculation of humid air from such equipment will usually cause discomfort and condensation problems. Auxiliary ventilation or some other means must be used to prevent excess humidity.
- It may be expected that there will be a variation with time in the exit concentration of typical collectors. Design data and testing programs should consider all operational time periods.
- The layout and design of the recirculation duct work should provide adequate mixing with other supply air and avoid uncomfortable drafts on workers or air currents that would upset the capture velocity of local exhaust hoods.
- A secondary air-cleaning system is preferable to a monitoring device because it is usually more reliable and requires a less sophisticated degree of maintenance.
- Odors or nuisance value of contaminants should be considered as well as the official TLV values. In some areas, adequately cleaned recirculated air, provided by a system with safeguards, may be of better quality than the ambient outside air available for replacement air supply.
- Routine testing, maintenancce procedures, and records should be developed for the recirculating systems.
- Periodic testing of the workroom air should be provided.
- An appropriate sign should be displayed in a prominant place reading as follows:

CAUTION

AIR CONTAINING HAZARDOUS SUBSTANCES IS BEING CLEANED TO A SAFE LEVEL IN THIS EQUIPMENT AND RETURNED TO THE BUILDING. SIGNALS OR ALARMS INDICATE MALFUNCTIONS AND MUST RECEIVE IMMEDIATE ATTENTION: STOP RECIRCULATION, DISCHARGE THE AIR OUTSIDE, OR STOP THE PROCESS IMMEDIATELY

5.5.1 Source Strength Estimates. At the behest of the National Institute for Occupational Safety and Health (NIOSH), several research projects have been performed to refine and quantify more closely the parameters necessary to rigorously define the performance of a recirculation system (see General Reference section at the end of this chapter). Unfortunately this rigorous approach requires the quantification of parameters for which, in general, quantification is almost literally impossible. One version of such equations requires knowledge of the generation rate of contaminant from the subject operation(s) and the subsequent quantification of the fraction of that emission source that is captured by the local exhaust system. Data on generation rate of contaminant does not exist for most sources. Another parameter requiring quantification is the fraction of the breathing zone air that comes "directly" from the return airstream. Estimation of this is theoretically possible but a great deal of detail work would be involved.

The other version of this approach involves the knowledge of the breathing zone concentrations in general plant areas "not under the direct influence of strong flow fields induced by local exhaust hoods" (commonly called "general air" samples) and also the breathing zone concentration in plant areas that are under the direct influence of strong flow fields. This is essentially the same information, broken down into a large number of discrete portions, required in Eq. (7.1).

Typical equations representing this approach are not reproduced here since it would serve no purpose but to indicate the complexity thereof. The set of equations required in the more simple models occupies a whole page.

It has been our observation, in one or two attempts to apply this approach, that most of the parameters in the page-long set of equations drop out, in any practical sense; and one is almost, but not quite, reduced to the same basic approach as represented in Eq. (7.1).

The more complex method does have the advantage of presenting to the designer, who may be competent at manipulating mathmatical equations but not particularly competent or experienced in ventilation for health hazard control, an approach that is rigorous and that will prevent him from overlooking some aspect that may be important. The disadvantages are that it is extremely tedious, gathering the data represents almost a research project, and crucial parameters as already described are usually not quantifiable; and certainly no attempt should be made to estimate them by the amateur.

An interesting sidelight on the studies on recirculation air is the attempts to rigorously quantify the reliability of the mechanical systems involved. The reliability estimating techniques, or at least some of them, are taken from the aerospace industry and their application is interesting. Reliability, that is, "mean time before failure," has been estimated for all components except the sensor of the monitoring instrument. The sensor involved is typically the least reliable of all components in the entire system. It needs to be checked and/or maintained almost weekly. At the present state of the art, if an instrument monitor is to be used, it would seem appropriate to have reliability requirements for that portion of the system. For the rest of the system, well-engineered and quality component systems and machinery are "state of the art." They have been developed over a long time and are apparently satisfactory for most applications.

In considerations of reliability, the matter of complete power failure to the whole system has been ignored, and it is our experience that in the typical manufacturing plant, this may occur several times a year—much more frequently than does failure of various components.

Similar reliability criteria have not been proposed for ordinary industrial exhaust and replacement air systems where recirculation is not encountered. The protection of the workers' health would be equally as adversely affected by component failure in these systems as it would be in the recirculation system, with one exception: the effectiveness of the air-cleaning device that removes the contaminant from recirculation. A better approach, in our opinion, would be to require adequate safeguards for that component of the system and to provide shutdown or bypass facilities since it is known that failure will occur sooner or later. With few exceptions, it will prove to be more workable and much less costly to provide procedures for coping with failure of the system (especially since the bypass/shutdown must be provided anyway) than it would be to indulge in extensive failure analysis and tight component specification in that regard.

6 EMERGENCY VENTILATION

In general, there are two kinds of "emergency ventilation" situations, accidental ruptures and failure of the ventilation system itself.

6.1 Accidental Rupture

One type of emergency situation can arise in a plant where the hazardous materials are completely or almost completely enclosed in the process equipment. This situation is typified by a chemical plant with pressure vessels, reactors, pumps, piping, valving, and so on. In the absence of an emergency situation, some contaminant may be emitted from leaking pump packing, valve packing, or the like; although for extremely hazardous materials, even this can be locally controlled or prevented.

Appropriate plant layout with regard to hazards, one of the major subjects of this entire book, can minimize these problems. In this type of plant, the emergency ventilation would be called into action in the event of an accident, such as the rupture of a pipeline, massive or complete failure of valve or pump packing, or accidental spills. For those situations, the emergency ventilation typically consists of wall-mounted or roof-mounted propeller fans, moving large volumes of air at low pressure, and providing enough ventilation to clear out the building, or at least some portions of it, so that rescue or repair work can take place.

6.2 Ventilation Failure

Need for another kind of emergency ventilation occurs when the normal ventilation sytem itself fails, and the consequences either to the workers or to the process can be immediately serious. The immediately serious effects could be overexposure to the workers for a short time, which might have serious health effects; or possibly a disruption to the process that could lead to damage to equipment, damage to product, or could create conditions leading to fire or explosion.

Emergency ventilation in some cases may also involve general ventilation fans to move air through the place at a high rate as was previously described, but in addition will require tailor-made provisions for the specific events that may occur.

One good method of coping with such situations is to provide installed spares. Use of installed spares is usually limited to items such as fans, blowers, pumps, and their respective motors, on the basis that the "spared" equipment is necessary to keep the air moving or process safe. Sometimes the instrumentation is made redundant, and if not should always be fail-safe. The typical installed spare is connected to power and piping, but the power is off and the spare is valved off from the system when not in use. Putting the spare into action requires specific actions, either manual or automated (another reliability item) to start the motor and switch the valves so that the spare comes into operation promptly enough to prevent whatever problem was anticipated.

Another scheme is to install two smaller blowers instead of one larger one, each blower handling of the rated load. If one blower fails, the odds are very small that the other blower would fail at the same time (except during power failure) and the air-handling system will operate at somewhat more than 50% of normal capacity. Even in this circumstance, the dampers or "valves" usually must be closed to prevent bypassing of the system into the operating blower through the inoperative blower.

Instrument engineers will be familiar with the problems of fail-safe design. If the instrument fails, or motive power to it fails, the instrument will "fail" on or off, in one direction or the other, being actuated typically by springs. The problem frequently is to determine which consequences of the control action, going on or off, represent the most serious failure. Either condition is out of control, and whatever is being controlled will also be out of control. This is a consideration of detailed design, and the right answer is specific to each problem. It must be kept in mind by management, however, that simply specifying fail-safe is not enough.

Another fail-safe application is represented by the shear pin in a drive shaft, or by some other mechanical component that is designed to fail first, so that more destructive results do not occur. At least in the case of the shear pin, this is a self-defeating approach. Once a given shear pin has failed more than two or three times, one can be assured that its replacement will be of greater strength, and therefore more difficult to shear than was called for in design. A more appropriate approach, difficult to achieve in all designs and impossible in many, is to design the machinery so that if some part of it overloads or jams, the motor is stalled and the power drops out without anything being broken. It will be somewhat easier to prevent the installation of a more powerful motor than it is to prevent an improper shear pin replacement.

6.3 Power Failure

Power failure to a plant, or to a section of a plant, is not uncommon. As a matter of good plant design, the process/production machinery and the general production system will be designed with this eventuality in mind. The problem arises, what to do about the ventilation equipment when the power fails?

Even though materials involved may be quite hazardous, it is usually not necessary to provide emergency power for all the installed ventilation to prevent hazardous conditions. For example, if dust control equipment is installed on packaging machine and the power fails, there is no need to continue the dust collection function at the packaging machine because it is no longer making any dust.

Appropriate criteria for dealing with failure of the main power supply are difficult to generalize. Obviously there must be sufficient emergency lighting so that personnel are not trapped in the dark. Whether or not certain ventilation equipment needs to come on automatic alternate power source or automatic emergency power is a question that can be answered only by analysis of the process, and the anticipated eventualities of power failure.

Emergency power supply to the industrial plant requires a supply of power to certain equipment from a source other than that normally used. The appropriate arrangements will vary with the specific circumstances. One common provision is an engine-generator set that comes on automatically within less than a minute of the failure of the initial power. The capacity of such emergency generators is usually much less that the power consumption of the plant when normally operating, and the available power must be utilized wisely and sparingly. As with all such automatic safety equipment, the automatic system should be tested periodically to make sure that all the necessary components function when needed.

6.4 Other Considerations

6.4.1 Location of Manual Controls. The location of manual controls to activate emergency ventilation provisions is a specific design problem for a specific set of circumstances. Only a few principles can be outlined here.

In many cases it is wise to have redundant controls. For example, if there is more than one egress route from a section of plant where failure of ventilation could be serious, there should be large palm switches at each egress so that workers can easily and quickly activate the system as they leave the premises. A similar emergency switch might be installed at local work stations, for those types of situations where the worker at the location would be the first one to observe signs of trouble.

Room air monitors, which continuously sample and the test the room air for the presence of a specific contaminant, can be used to activate emergency ventilation automatically. These monitors require frequent maintenance and calibration, as described previously. This is easy to say and more difficult and costly to execute. However, it is not just a platitude; if the monitoring instrument is to be depended upon for protection of the worker health, or—even more serious—the prevention of a fatality, it *must* be properly maintained and always in operation. Is is not uncommon to find such an instrument, guilty of giving frequent false alarms, completely turned off or unplugged as the solution to the problems. Further, maintenance instructions, necessary spare parts, and the like, must be available and usable on all operating shifts; maintenace instructions that are locked up in the foreman's filing cabinet are not very useful.

6.4.2 Estimation of Ventilation Rates. If emergency ventilation of the "air the place out" type is contemplated, there must be some basis of estimating its capacity. Here again, the popular "air change" basis is likely to be fallacious (see Section 2.5).

Utilizing the geometry of the work space to be ventilated is one approach. For a long narrow space, for example, providing an air movement from one end to the other by installing exhaust fans in one short wall may be the most effective approach. In such a circumstance, it may be possible to provide an average air velocity, from one end to the other, of 25–50 fpm (0.13–0.25 m/s) and still maintain a practical situation with respect to the number of fans and their locations. A tall, narrow building may be subject to the same treatment by installing roof fans.

Another approach is to estimate the emission rate from some anticipated accident such as breakage of a pipeline, major failure of a valve stem, and so on. If the pipeline is carrying a gas, the estimate of the emission rate should be fairly straightforward. If it is carrying hazardous liquid, estimating the generation rate of its vapor is more complex, but at least an order of magnitude estimate can be made. Once that estimate is made, the ventilation rate can be calculated to provide dilution ventilation, and also if possible some semicontrolled direction of air flow, so that extreme hazardous concentrations are minimized.

The situation is somewhat self-limiting. For a small room, an accidental break or spill could release enough contaminant to cause hazardous concentrations in a very short time. In that situation, a very high ventilation rate on an emergency basis is usually practical. Such situations have been observed where the ventilation rate on an air change basis turns out to be 30 or even 60 air changes per hour.

As the affected work space gets larger, however, the application of such high air change rates becomes impractical. On the other hand, a given accidental emission will not fill up the entire space with high concentrations nearly as rapidly as the same emission would in a small space. Therefore, if emergency ventilation can be strategically located in the regions where such eventualities are most likely to occur, a large volume of exhaust from that region will keep the rest of the building reasonably clear.

In all such emergency ventilation situations, the criterion for adequate ventilation is different from the TLV or the OSHA permissible exposure limit (PEL). There are two stages to this more lenient criterion. One is the short-term exposure limit (STEL) provided in the ACGIH TLV booklet. These STEL values are usually three times the normal TLV unless more specific information is available for the given contaminant. The STEL is a good target for emergency ventilation design *if* it can be achieved.

Another criterion for emergency design is the "immediately dangerous to life and health" level (IDLH), as provided for some chemicals in the NIOSH/OSHA *Pocket Guide to Chemical Hazards*. Obviously, the emergency situation should control the contaminant to well below the IDLH level if at all possible.

In both of these criteria, it is assumed that rescue/maintenance work will be conducted by personnel suitably equipped with personal protective apparatus.

6.4.3 Escape Gear. Unless the emergency ventilation provisions and egress routes can virtually guarantee the escape of workers in the event of accident, the provision of appropriate escape apparatus is indicated. Such apparatus varies from conventional gas masks to self-contained breathing apparatus (the "air pack") to small respirators, and even to small oxygen-supply apparatus. These types of equipment all vary in their convenience, availability, and the length of time required to don them.

Conventional gas masks and self-contained breathing apparatus are most reliable, but are typically kept in cases strategically located throughout the premises. They must be periodically inspected, maintained, and immediately replaced after use. The speed with which then can be donned is important. The untrained, unpracticed person, putting on self-contained breathing apparatus for the first time, must first read the directions. The minimum time interval for amateurs is about two minutes to read enough of the instructions to know what to do, to get the apparatus on, and be able to breathe again. Not many people can hold their breath for two minutes, especially in an emergency situation. On the other hand, once donned, the self-contained breathing apparatus provides virtually 100% protection for 20 min or more.

The gas mask, easier to don and requiring less training, will filter out of the air various contaminants for which it is designed but does not furnish oxygen or clean breathing air. At the other hand of the spectrum are emergency devices so small and light that they can be worn on the worker's belt at all times. These vary from special emergency respirators (not rated by NIOSH) to small emergency packs providing oxygen for up to 5 min.

Another consideration in the selection of such equipment is whether or not the worker's eyes will require protection on an emergency basis from a given contaminant.

The choice of appropriate protective gear should be made only after a number of factors have been weighed: the ease and speed of use, the number of individuals that can be supplied, and the integrity of the protection afforded.

6.4.4 Training. All types of escape gear require training to some degree. The larger and more reliable types require actual drills, performed fairly frequently, perhaps in conjunction with the drill for escaping the premises (the "fire drill"). These drills should include all actions that should be taken in the event of an emergency evacuation. The more convenient and less cumbersome equipment also requires some practice, as well as inspection and maintenance, which may be more difficult to conduct if the number of such units is large.

In any event, in really serious emergency situations, lack of training and drill will result in unfortunate consequences.

ACKNOWLEDGMENT

Mr H. L. Rowland, Director of Engineering and Maintenance, St. Joe Lead Company, furnished valuable commentary on the substance of this chapter that is gratefully acknowledged.

GENERAL REFERENCES

ASHRAE Handbook of Fundamentals, latest edition, American Society of Heating, Refrigerating and Air Conditioning Engineers, New York.

Caplan, K. J. "Heat Stress Measurements," *Heating, Piping and Air Conditioning*, February 1980.

Fan Enginering, 7th ed., Buffalo: Buffalo Forge Company, 1970.

Guidelines for the Recirculation of Industrial Exhaust, Arthur D. Little Co., National Institute of Occupational Safety and Health, Draft Report, July 1977.

Heating and Cooling for Man in Industry, American Industrial Hygiene Association, Akron, OH, 1970.

Industrial Ventilation—A Manual of Recommended Practice, latest edition, American Conference of Government Industrial Hygienists.

A Recommended Approach to Recirculation of Exhaust, Arthur D. Little Co., National Institute of Occupational Safety and Health Research Report, January 1978.

CHAPTER EIGHT

Noise Control

Carl D. Bohl
Paul A. Kannapell

1 DESIGN CRITERIA

There are many considerations when choosing criteria for the design of a new plant or the improvement of an existing manufacturing location. The intended use must be considered. Will it be a manufacturing area, an office, a conference room, or a customer service area? The exposure of employees to noise raises questions of worker's compensation for noise-induced loss of hearing and of compliance with state and federal regulations such as the Occupational Safety and Health Administration (OSHA). There are concerns for the external environment. It must be realized that neighbors do not wish to be unnecessarily or continually disturbed from normal activities. An outdoor paging system may disturb sleep at night. Continuous noise levels may not be of concern during the day, but may be considered a major irritant at night. Both the number of times a day people will be disturbed and the hours of the day when disturbances will occur must be determined. The natural noisiness or quietness of equipment and the cost of reducing the noise are important factors.

1.1 Worker Health and Safety

1.1.1 History. The payment of workers' compensation claims in the states of New York, Wisconsin, and Missouri during the 1950s was the stimulus for industry to start efforts in controlling noise. During the 1960s many states added noise-induced loss of hearing to their list of compensable illnesses. By the late sixties most states granted awards for loss of hearing and recognized that there was some safe level for noise exposure.

The federal government took its first steps toward control of worker exposure to noise in 1969. This was carried out under its authority granted by the Walsh–Healey Public Contracts Act of 1936. These regulations applied to those manufacturers who sold materials to the government. Then late in 1970 Congress passed the Williams–Steiger Act and OSHA was born. The noise regulation specified by OSHA was the same as that set forth in Walsh–Healey. Only OSHA regulations apply to all non-governmental employers in the United States. OSHA amended the noise regulation with a Hearing Conservation Amendment (HCA) in 1983.

1.1.2 OSHA dBA 85 or 90. Currently there are two levels of noise that require action on the part of the employer. One is for engineering controls of noise. The other is for compliance with the HCA.

When an employee has an average daily exposure (called a time-weighted average, or TWA) exceeding 90 dBA, feasible administrative or engineering controls shall be used to reduce exposure. The definition of feasible and the amount of reduction—not necessarily to 90 dBA—have been argued in court since the first citation was given an employer by OSHA. A broad and general attitude is that if the TWA exceeds 90 dBA, then an employer shall reduce the noise exposure to the extent feasible. Certainly this does not mean every work area above 90 dBA must be controlled to less than 90 dBA. If an employee spends only 30 min a day in a work area with a noise level of 95 dBA and the rest of the work day in work areas with less than 90 dBA, then his or her TWA is much less than the allowable 90 dBA. Thus, no corrective action is needed. For the determination of the need for engineering or administrative control, only noise levels that exceed 90 dBA must be considered. If reduction to a TWA of less than 90 dBA is not achievable, the use of personal hearing protectors is prescribed.

The Hearing Conservation Amendment (HCA) of OSHA has an entirely different criterion for noise measurement than the portion dealing with engineering and administrative controls. The HCA requires the measurement and integration into the TWA of all noise between 80 and 130 dBA. This includes steady-state, intermittent, and impulsive noise. There can be problems in accurately accessing the TWA to determine OSHA compliance in noise fields with significant impulsive noise. Problems can arise, when using integrating sound level meters and dosimeters, in assessing the TWA of impulsive noise fields. These problems are caused by the 5-dBA doubling rule of OSHA. The TWA noise level is overestimated because of the response of the averaging circuit in these meters. This is true even when ANSI meters are used. The details of these problems are beyond the scope of this chapter. When it has been determined that the TWA exceeds 85 dBA, the employer is required to have a hearing conservation program. The legal details are given in the *Federal Register* for March 8, 1983.

Therefore, two criteria must be considered when designing or retrofitting workplaces to comply with federal regulations.

- An administrative and engineering control requirement to meet 90 dBA TWA when measurements start at 90 dBA.
- A hearing conservation program when exposures equal or exceed 85 dBA TWA with measurements starting at 80 dBA.

1.1.3 Hierarchy of Control Measures. The regulations of OSHA have a hierarchy of noise control measures implicitly built into them. Paragraph B orders feasible engineering controls first, administrative controls second, and as a last resort the use of personal hearing protectors. The ordering of control measures without a specific definition of feasibility has resulted in extensive judicial review of many citations issued by OSHA. OSHA-accepted interpretation of these regulations is stated in *Field Operations Manual.* The publication of this manual and its acceptance by employers has led to a significant reduction in the number of noise citations issued by OSHA and has almost eliminated judicial review of noise citations.

The guidelines generally stated are:

- In work situations where employees have a full-shift time-weighted average exposure of 100 dBA or greater, all feasible engineering and administrative controls shall be used.

    When the worker's full-shift time-weighted average exposure falls between 90 and 100 dBA, the employer is still required to use all feasible engineering and administrative controls, unless the employer can show that it is cost effective to use personal hearing protectors.

    Engineering controls include any physical or chemical change to the workers' environment, equipment, process, or raw materials that reduces the level of noise to which the worker is exposed. Engineering controls are the preferred method of control. If engineering controls are not feasible, then administrative controls are used, that is, changes in work and worker schedules that reduce the amount of time employees are exposed to noise.

• The use of personal hearing protectors (ear muff, ear plugs, semiaural devices) is allowed in two situations by the OSHA regulations. Those situations are when feasible engineering or administrative control cannot reduce worker full-shift noise exposure to less than 90 dBA and when an employee is exposed between 85 and 90 dBA on a full-shift basis and is found to be suffering a STS when examined for purpose of the Hearing Conservation Amendment.

The effectiveness of personal hearing devices as used by workers is difficult to measure. OSHA has described a number of evaluation schemes in Appendix B of the Hearing Conservation Amendment. It is a good reference. However, the current issue of the OSHA *Field Operations Manual* recommends reducing that value by one-half.

1.1.4 State Programs. Many states have contractual agreements with OSHA to administer safety and health programs. These agreements require the states to have regulations as effective or more stringent than the federal regulation. In these states, it is necessary to determine how compliance with the regulations can be achieved. Some states, both with and without federal contractual agreements, have consulting programs. These programs usually include "inspections" where violations of health and safety standards are pointed out and advice is given on methods of compliance. These inspections are usually performed under the condition that OSHA will not be informed of the findings.

1.1.5 Workers' Compensation. The payment of claims for noise-induced loss of hearing under state WC (workers' compensation) laws is of concern to employers. Two states have recently changed their WC regulations denying compensation for loss of hearing when the employer has an enforced program for the use of personal hearing protectors. These enforced programs require reprimands and other disciplinary action of an offending employee. If an employer carries out this program, the offending employee is not entitled to compensation. The success of this defense has not been tested.

1.1.6 Good Practice and Corporate Policy. Many corporations and corporate health officials have requirements that differ from federal and state regulations. There are many reasons for these differences of opinion. Regulations are written at a specific time, and knowledge gained later never influences that regulation. However, an employer should use all the information available to protect the health and safety of employees. A source of reviewed data is the threshold limit value (TLV) of the American Conference of Governmental Industrial Hygienists. These values are reviewed annually and reflect more up-to-date knowledge and data interpretation than older government regulations.

1.2 Worker Performance in Manufacturing Areas

1.2.1 Communications. Everyone knows that noise interferes with communications. The extent of this interference is seldom realized. Only when a costly error has been found that is explained by "I didn't understand the instructions," do we begin to think of the cost of poor communications. A good measuring scheme to determine over what distances speech can be clearly understood is the speech interference level (SIL). It has been used successfully since about 1950. One word of caution: An older work force, or one that has been exposed to noise with a resultant loss of hearing, needs quieter work environments than that recommended by the SIL. People who are hearing impaired have disproportionate difficulty in understanding speech.

Telephone communications present similar problems in noisy areas. A telephone supplier may be able to give recommendations and suggestions.

1.2.2 Other Factors. The U.S. Environmental Protection Agency (EPA) has monitored the world literature for physiological effects other than loss of hearing from noise exposure. The EPA has reported irritation, loss of sense of balance, increased blood pressure, reduced efficiency, and dilation of the pupils of the eye as being caused by exposure to high noise levels.

1.3 Worker Performance in Nonmanufacturing Areas

1.3.1 Noise Criteria for Room Use. Noise criteria (NC) curves for room use were developed by Beranek in 1957. This series of octave band curves indicates maximum sound level by octave for the intended use of a room. The designer of a private office or executive conference room, for example, would choose the appropriate NC value and design a room in which the noise is less than the indicated value in each octave.

1.3.2 Open Spaces. In business today it is fashionable to have open space, no walls, and the minimum of partitions. There are some advantages to this type of design. However, there are also many disadvantages, the most obvious being lack of privacy. Cross talk or noise from other personnel doing their jobs can cause embarrassment and misinterpretation due to the work environment. Reasons for embarrassment are left to the reader's imagination. An illustration of misinterpretation is a centralized order-processing station where several employees take telephone orders for the company's products. As business improves and telephone orders increase, the employees must talk louder and louder to make themselves heard by the customers or other employees with whom they are conferring. These sounds increase in volume until the background noise levels are interpreted by the calling customers as a great party.

It should be understood that certain noisy operations are to be segregated from the general work environment. Billing machines and high-speed printers are two illustrations. Each piece of equipment should be given an individual evaluation for the noise level it generates.

1.4 Communities Regulated by State and Local Laws

1.4.1 Federal EPA Guidelines. In 1975 the EPA published a model community noise ordinance (EPA 550/9-76-003). This recommendation was adopted by some communities and a few states. All modified the model to some extent. The federal government at this time does not have authority to regulate community noise. The EPA Office of Noise Control and Abatement has recently been discontinued, and its efforts at regulating product noise are at a standstill.

1.4.2 State Regulations. Some states have enacted noise regulations that vary in form and enforcement penalties. Generally, the rules are based on dBA or octave band measurements, slow or fast response time, and time-integrated or single-event intensity levels. In addition to these parameters, the allowable emitted noise levels are often related to the use or zone of the neighboring property. Before a new plant is designed, expansion is considered, or retrofit is planned, consultation with state authorities is in order.

1.4.3 Local City–county Ordinances. These rules are even more diverse than state regulations. Enforcement can be either more stringent or nonexistent, depending on the community and age of the law. A good reference source of local ordinances has not been compiled. The only source of this information is city and county officials.

Some communities have nuisance ordinances. The test of conformance is whether or not the noise environment generated by the plant disturbs the common man. If a group of neighbors is disturbed and takes the prescribed legal action, some corrective action usually adversely affecting the operation of the new noise source will be required. Truly the test of being a good neighbor and having good neighbors is adherence to a nuisance ordinance.

1.4.4 45-dBA Average Guidelines. Most communities with noise control programs have some type of averaging scheme in the evaluation of neighborhood noise. These programs generally have a level close to 45 dBA, even though that level may be based on an octave band scheme. Other schemes use a 45-dBA day–night evaluation. This procedure allows a little more than 45 dBA during working hours (7 AM–10 PM) and a little less than 45 dBA during the sleeping hours. This type of evaluation often takes special noise-measuring equipment. It is important to determine what type of analysis local officials expect with these types of ordinances.

1.5 Community Guidelines in Nonregulated Areas

1.5.1 45 dBA. A good starting point for environmental design if nothing else is known is 45 dBA. This level will not generally disturb people. Rural areas, where there are no major highways and no commercial businesses, are areas where designing to less than 45 dBA is advisable because these communities may have noise levels less than 35 dBA.

1.5.2 Existing Noise Levels. Another approach that may save money in design and equipment cost is to survey the neighborhood to determine existing noise levels. If this is not possible, refer to an EPA publication on neighborhood noise levels, *Community Noise* (NTID300.3, December 31, 1971). Match the operations of the new location to a similar one reported by the EPA. Using this comparison, an idea of the noise level to use as a design criterion can be gained.

1.5.3 Increasing the Environmental Noise Levels. The installation of a new noise source into the environment will occasionally create a situation where the neighborhood noise level will have to be increased. Then the question arises about the consequences of such an increase. Some rules of thumb, which first appeared in a book by L. L. Beranek, *Noise Reduction* (1960), will be valuable.

- If there is no increase in the noise level, there will not be any complaints.
- If the increase in the noise level is 5 dBA, occasional complaints will be received.
- An increase of 10 dBA will bring strong and widespread complaints including letters and threats of community action.
- Increases above 15 dBA will probably bring some type of legal action to force reduction of the new noise level.

2 DESIGN GUIDELINES

2.1 General Comments

When noise control is introduced during the plant design stage, it is usually more effective and less expensive than treatments that are instituted after the plant is started up. One need only look at, study, and listen to those well-designed units of equipment with built-in noise controls to fully appreciate what conceivable actions would have been required later had not noise controls been incorporated in the original design. Certain noise problems are technologically and/or economically impossible to solve later, which at times results in impractical, costly Band-Aid solutions. On the other hand, some noise problems can be solved by retrofitting.

To initiate noise control in the design stage before there is a noise problem obviously requires management understanding and support.

Noise should be one of the many considerations before a site for a new plant is selected. The economics and available noise control technology for adjacent neighborhood requirements should be evaluated for the sites under consideration. An industrial site that permits 75 dBA at the new plant property line requires considerably less noise controls than one adjacent to a quiet community that requires only 45 dBA at that property line.

It is very important for all key personnel to agree in writing on all measurable noise level criteria at the outset of any project. This group usually consists of project, environmental, medical, noise control, and plant personnel.

The noise control engineer should become involved very early in the project development stage before all plant designs are chiseled in stone. The knowledgeable engineer can evaluate and recommend alternate, economical, and quieter routes, designs, and equipment by recognizing the potential noise problems. For example, is a quieter alternate process available? Can manufacturing buildings be used as noise barriers between noisy rotary-positive-type blowers or cooling towers and the community? Can noisy sources such as cooling tower louvres be oriented to minimize noise intrusion? Can cooling towers be operated at half-speed to obtain inexpensive noise reduction during the critical nighttime hours? Can a quiet underwater granulator be used for sizing polymers instead of a very noisy conventional dicer or should one merely provide a quiet control room for the dicer operator and hope for control of the operation from there.

Prime variables for noise control by the engineering routes are:

- Feasibility, technological, and/or economical
- Effects on efficiency of operations, maintenance, and energy usage

2.2 Definitive Design

Definitive noise control engineering efforts should be undertaken as soon as layout drawings and equipment specifications become available.

The present, representative, ambient noise levels should be obtained at the planned site at this stage.

Next the noise levels of proposed standard equipment should be estimated. These noise levels should be combined to determine the levels in the in-plant operating areas where personnel will be exposed. If the predicted noise levels are higher than the criteria limit, the problem equipment should be identified and listed. Noise control of these units should be planned.

The preceding in-plant procedure should be repeated for estimating property line noise levels. All new facilities should be closely scrutinized for potential community problems. These could include some seemingly innocuous sources such as plant public-address systems or thin PVC walls for manufacturing buildings. Any possible noise source may be important when very low property line dBA values must be met.

The next step is to determine the maximum acceptable noise level for each equipment unit identified and listed as a potential noise problem. The maximum acceptable noise levels should be specified to the vendors and made a part of the purchase specification package. (For more detailed noise level specifications, see Section 2.3).

Equipment quieter than the standard equipment or equipment with added-on noise controls will be available from vendors in certain cases. If not, noise control treatments will be required by the company noise control engineer. However, the vendor should be required to furnish noise level data to provide an engineering basis for these in-house treatments.

At some point during project development the noise control engineer should provide an estimate of the total cost of noise control for inclusion in the project cost estimate.

As the project progresses, the noise control engineer should update the initial estimates of work area and property line noise. Continuous updating and review should be done until noise is under control on paper.

After equipment has been installed, unforeseen noise problems may become evident during mechanical checkout and project start-up. However, one is cautioned to avoid overreacting to some very temporary noise problems during initial start-up and before the process is balanced out at specified design rates and conditions. For example, water chillers using centrifugal compressors will be noisier when operated at less than design capacity. Centrifugal fans initially operated at other than design air flows and pressures will be temporary noise problems. However, for clearly identified unexpected noise problems corrective action should be prompt. While hearing protection can be provided to plant personnel until these in-plant problems are resolved, apprehensive residents living near the new plant will complain right after or even during their first sleep-disturbed night.

After the plant is started up and is operating at specified conditions and rates, the noise control engineer should make a noise level survey to ensure that the project criteria are met. This is valuable documented information for the plant, OSHA, and community relations files. Any future noise problems due to a lack of adequate plant maintenance, for example, can be uncovered and highlighted with noise level comparisons to the plant when new. This post start-up information is very valuable to the noise control engineer because it permits a critique and improvement of performance and techniques by comparing actual noise levels to the predicted noise levels.

2.3 Noise Level Specifications

Most of the manufacturing corporations and engineering firms have their own printed forms with detailed instructions for obtaining guaranteed or certified noise level data from vendors. Rather than attempt to develop and exhibit the universal forms here, some recommendations in handling specifications are given instead.

The noise control engineer should be very familiar with all the details of the many applicable industry or national standards or test codes for determining and presenting noise data. All these codes and standards provide a uniform means of comparing and evaluating vendor quotes; they permit translation of noise data into the purchaser's acoustical environment. For example, centrifugal fan vendors provide sound power levels in octave bands per AMCA 300-67, "Test Code for Sound Rating Air Moving Devices." Vendors of portable air compressors supply sound pressure level data in octave bands or dBA sound level data per ANSI S 5.1-1971, "Test Code for the Measurement of Sound from Pneumatic Equipment." If there are no applicable codes at present for the equipment being purchased, the purchaser and vendor should mutually agree on some specific detailed methods of measurement such as "free field conditions, one meter from the equipment."

Demanding guaranteed noise levels from vendors, when accompanied by a premium to guarantee, certainly conveys to the vendor that the purchaser is very serious about his noise control. Guarantees are very difficult to enforce rigidly in most cases, but this is not to say they should not be accepted when given freely, of course. The difficulties lie in significant differences in acoustical environments between the purchaser's installation and the vendor's test facilities. This is why it is preferable to insist that the vendor shall certify noise data per the measurement code, standard, or other method to the purchaser. However, if one wishes to pursue guarantees, the cost of guarantees should be clearly identified and evaluated. In a number of cases the vendor's proposed charge to the buyer to guarantee noise levels was in the thousands of dollars range for only a few dBA less than noise levels of standard equipment. Of course these should be summarily rejected in favor of in-house treatments. One way to avoid guarantees and to assure that the design is right to begin with is for a knowledgeable noise control engineer to provide noise control design specifications such as silencers and acoustic insulation on a certain type of equipment. For example one may select a specific silencer model as part of a rotary-positive-type blower package. Some very difficult noise control design problems occasionally crop up for newly developed equipment. Rather than try to solve these by demanding vendor guarantees, it is best to work with a knowledgeable vendor toward solving the problem on a best effort basis.

The problem of handling noise level specifications can be very complex for very large plants with thousands of units of new equipment. There are several ways to simplify the problem. First, there is no point in writing noise level specifications for equipment that must be treated by the noise control engineer. Such equipment may be newly developed large centrifugal compressor trains in refineries; there can be no proven standard treatments by vendors for this inherently noisy equipment. Second, there is no point in cluttering up the files with specifications for every piece of new equipment. Classes of small-horsepower and/or slow-speed equipment such as electric motors, for example, can be exempted. Certainly there is no point in paying for guarantees for noise levels of this equipment. Third, the specifications for a large plant such as a petrochemical facility with several thousand pieces of equipment can be simplified. If the design requirement is 90 dBA max, for all areas, a single number for *all* equipment (such as 85 dBA max, at one meter, free-field conditions) is far superior to attempting to write thousands of different noise level specifications. This is a very practical compromise with strict acoustics technology, but, if the plant is well laid out and well spaced, combining noise from adjacent equipment and/or reflecting surfaces will produce a 90-dBA plant. This technique has been demonstrated. Incidentally, an outdoor situation such as this permits a good opportunity to enforce vendor guarantees or challenge the certified data. Compare this with other reverberant indoor plants with crowded equipment, near field effects, reflecting walls, standing waves, and so on.

Observing preshipment tests in the vendors' facilities to determine specification compliance, if it makes good sense to do so, pays dividends. In one instance the vendor's noise level data indicated that additonal secondary mufflers at a cost of $100,000 would be needed for the cracking heaters (furnaces) to meet the design criteria. This cost and the quality of available data clearly indicated that additional representative single-burner tests should be run at the vendor's factory with the participation of the purchaser's noise control engineer. Extrapolation of this data showed that expensive secondary mufflers would not be required and this was later confirmed after plant start-up. On the other hand, running preshipment noise tests on a large centrifugal compressor train makes no sense because actual operating conditions cannot be approached due to problems with valving, speed, driver, gas composition, and temperature.

Even using the foregoing recommendations it should be abundantly clear that the mere initial writing down of equipment noise level specifications and formats for such equipment will not guarantee that the design criteria will be met. The process of obtaining reliable data is a continuous one requiring dogged persistence and follow-up by the noise control engineer who won't take no for an answer. Questionable data that makes no sense must be challenged. Recently one highly reliable and reputable vendor stated that the equipment sound power level would be so many decibels at a 50-ft (15-m) distance from the equipment.

The noise control engineer should fully evaluate any special designs and acoustical treatments proposed by vendors to meet the noise level requirements. The vendor should state the effects of these treatments on operation, maintenance, and equipment life. The noise control engineer must be kept apprised of any project design and equipment changes before it is too late because not every engineering discipline or project manager can fully recognize the impact of design changes on noise control. Thus a close working relationship with the multidiscipline engineering personnel in the design of new facilities is a must. Likewise, a professional, close working relationship with vendors is recommended for effective noise control rather than an adversary approach.

When a vendor cannot meet the noise level requirements, alternate vendors should be sought; if this fails, the engineer takes over.

3   NOISE CONTROL EXAMPLES

3.1   Air-Cooled Heat Exchangers

The predominant noise generation mechanism for either induced-draft or forced-draft units is vortex shedding at the trailing edge of the fan blade. Blade loading is the second most important factor. A typical spectrum for an axial-flow fan is broadband as based on octave band measurements, with maximum dB levels usually below 500 Hz. Pure tones that could be a nighttime community problem are rarely present.

Air-cooled heat exchangers must be designed "right the first time" to meet both operating performance and noise level requirements. There are no simple Band-Aid corrective treatments once the units have been installed, only very expensive modifications. Thus it is recommended that accurate (± 3 dB) noise level predictive equations that quantify the important parameters, and that are available in noise control engineering literature, be used in the design stage for in-plant and property line noise level assessment.

Here is a general guideline: Noise radiated at 3 ft (1 m) below either a forced-draft or an induced-draft air-cooled heat exchanger will be below 90 dBA if the fan blade tip speeds are limited to 10-,000 ft/min (3000 m/min).

Other design stage considerations are:

- Increasing the number of blades to allow the tip speed to be reduced while providing the same level of service:
  - Increasing the number of blades from four to six reduces noise 3–5 dB.
  - Increasing the number of blades from four to eight reduces noise 7–8 dB.
- Maintaining a given air flow by increasing the blade chord length (blade width) and decreasing the blade tip speed, thereby reducing noise.

Thus to make any of these modifications after plant start-up requires highly objectionable temporary curtailment of operations or additions of more heat exchangers plus mechanical modifications cost.

Air-cooled heat exchangers with improved blade design have been built and demonstrated that are 30-dB quieter than standard units. Thus, energy conservation should be a consideration when evaluating the details of vendor proposals.

3.2 Centrifugal Fans

The most common source of air noise in fans is turbulence, but closely related to turbulence is vortex shedding, which produces noise as tight eddies of air leave the fan tip blade. Blade frequency noise originates as pressure pulsations.

A question sometimes asked is "Why don't we buy *quiet* fans for a change and avoid all these fan problems such as expensive silencers that consume energy, acoustically treated duct work, fan enclosures, and the supporting structures?" The noise level of fans is only one consideration of many factors such as initial cost, the service, efficiency, reliability, maintainability, and so on. For example, any large radial blade centrifugal fan, such as one rated at 40,000 cfm (1133 m³/min) and 58 in. (147 cm) water static pressure, is extremely noisy with prominent dB sound peaks at its blade frequency and higher harmonics. But if one wishes to move moist dust-laden erosive air at a given cfm against a given differential pressure, then this is the type of fan to use. The curved blades of other quieter designs would clog or erode and would very quickly fail to achieve the primary purpose, that is, to reliably move the required cfm of air. In short, noise is a by-product of all fans (like horsepower), and noise must be dealt with and treated by the noise control engineer working closely with HVAC engineers in the design stage of the air-handling system.

Fan selection for any given system is in the province of HVAC engineers, and this is a case where the noise control engineer can clearly prepare noise control design specifications to make sure the design is right to begin with, as referred to in Section 2.3. Let us take an example where the HVAC engineer and the noise control engineer work together to design an outdoor fan system with an open fan suction to meet the noise level criteria near the fan while achieving maximum fan system efficiency. With the necessary fan data inputs from the HVAC engineer, the noise control engineer can specify a prefabricated silencer that according to ASTM Designation E 477-73 has the required dynamic insertion loss for reverse flow in dB for each octave band center frequency at a given face velocity and at a given static pressure drop. The point is simply this: Neither the fan vendor not the silencer vendor can be asked to take the responsibility for the noise control design at minimum energy requirement for the system; the vendors can only provide valuable assistance.

Additional recommendations are as follows:

- To prevent community complaints, fans with prominent pure tones such as vane-axial and raidal blade fans should be acoustically treated, usually with silencers, so that their pure-tone dB levels are 5–10 dB below the property line or community ambient level. Also if their intakes or discharges are untreated, closely grouped clusters of any such identical fans develop beats when their rpms are only very slightly different. Surrounding exterior plant surfaces change the location of variable-length standing waves. Pure-tone sound fluctuates by as much as 6 dB depending on the phase angle when any two fans instantaneously run at precisely the same rpm and frequency. All this variable output of pure-tone sounds 24 hr a day, 7 days a week is most annoying to the nearby community, as bitter experience has shown with five such 100,000 cfm (2830 m^3/min) untreated vane-axial fan systems.

- When selecting fan silencers, the economics of low -pressure-drop slightly higher initial cost silencers with the required dynamic insertion loss should be thoroughly evaluated. Energy-saving prompt payouts for fans in continuous service are the rule rather than the exception. In addition, pressure drop should be further minimized by using appropriate guidelines for installing silencers relative to other system components such as elbows.

- For airstreams contaminated with particulate matter, recent developments by various silencer vendors have led to new silencer designs to overcome these previous problems.

- Acoustic enclosures are not always needed for fan casings as casing noise is roughly 20 dBA less than noise inside the fan. Some noise will leak where the drive shaft enters the casing as will noise from thin flexible duct work isolators such as canvas. The latter problem can be overcome by using limp high sound transmission loss material such as leaded vinyl or the equivalent.

3.3 Compressors

Like centrifugal fans, centrifugal compressor broadband noise is produced by turbulence allied with vortex shedding from the blades. Either blade-passing frequency or blade-rate frequency can be the major noise source in centrifugal compressors. Blasde-rate frequency differs from blade-passing frequency because it includes the noise pulse when rotating blades pass stationary or diffuser vanes. However, the blade-passage frequency is usually the major noise problem.

The API's Subcommittee on Mechanical Equipment had a task group make a study of a group of acoustically untreated centrifugal compressors about 1973 to determine if impeller Mach number and horsepower could be correlated with sound level. No statistically valid correlation was found, but it was concluded that any acoustically untreated centrifugal compressor over 500 hp would be a potential noise problem, that is, probably over 90 dBA.

If a centrifugal compressor has a very heavy casing as most do, generally most of the noise is radiated from the connected gas process piping and from the driver. The first consideration for noise control then is inlet and outlet silencers bolted directly to the compressor and silencers between stages, if any. Pressure drop through the silencers is certainly an energy consideration, but this can be largely overcome by using low pressure drop straight through dissipative silencers. In addition, erosion protection for the absorption material plus an all-welded construction should be considered for inlet silencers. But even with this, a question at this point is: "What happens to the compressor if an inlet silencer comes apart?" This is a question often asked by concerned plant personnel who have seen all the internals blown out of high-pressure steam let-down silencers that have been operated well above their original design capacity.

If silencers are ruled out for whatever reasons, a well-designed acoustic insulation system is an expensive alternate effective treatment for the compressor piping. Reductions of 15–20 dBA have been demonstrated by this route.

When the compressor casing is too noisy, a commercially available, well-constructed, gas-tight, modular, 4-in.-thick enclosure works very well—30–40 dBA reductions have been demonstrated. Special provisions must be made for adequate cooling for running and for emergency shutdown conditions. Maintenance access must be provided. However, if the gases being compressed are flammable, such an enclosure or building is definitely not recommended because of the additional potential explosion hazard created by the enclosure. For nonflammable gas compression, oil consoles are also potential fire hazards, and it is recommended that they be installed outside of the enclosure where fires are easier to extinguish. Direct acoustic insulation, which is good heat insulation, on a compressor casing is usually satisfactory, so long as the main bearings are not insulated, but replacement of the insulation after compressor repairs is one more maintenance problem. Sheet metal roofs over compressors raise the noise level by a few dBA.

Reduction of high-speed drive shaft coupling noise should be worked out with the vendor. Some couplings can be partially enclosed acoustically without overheating problems. A very large steam turbine drive may require a partial enclosure over the hot section that can be provided by the turbine vendor.

It is usually a complete waste of time and money to take noise level readings on a large centrifugal compressor for process gas service in the manufacturer's shop simply because it is impossible to duplicate or even approach plant operating conditions. In one case, a very large, new-technology centrifugal compressor was being developed and tested in the manufacturer's shop by using temporary energy sources that included a steam source and flow, a modified jet aircraft engine, and another combustion source and flow. The real question there was: "What compressor noise?"

Several instances were noted in the previously described API study in which compressors in resonance with the connected piping were up to 15 dBA noisier than their nonresonant counterparts. Compressor system noise levels can be amplified by resonance of blade-passing frequencies and response frequencies in the connected piping, but piping resonance can be avoided in the design stage.

Reciprocating compressors in general are not noise problems. Design-stage analog analysis has become a standard practice to ensure that dangerous low-frequency pulsation buildup will not occur.

Acoustically untreated screw compressors are real screamers at about 115 dBA. In one case, three 1500-hp rotary screw compressors would be installed in a row; two would run continuously with an installed spare. Reactive silencers with their shells externally acoustically insulated were custom sized and provided by the silencer vendor. The silencers were bolted directly to the compressor suction and discharges. The compressor casings were so heavy it was decided that no further treatment would be needed. The compressors were mounted on and bolted directly to seismic blocks. After start-up and achievement of design rate and conditions, all the compressors ran under 90 dBA at 3 ft (1 m) with no further treatment required.

3.4 Control Valves

Conventional valves controlling high-pressure-drop gas processes contribute substantially to radiated noise levels from the valve body, from connected piping, and from certain equipment connected to the piping. Aerodynamic noise is the result of Reynolds stresses and shear forces that are properties of turbulent flow; turbulent flow noise is more common in valves controlling gases than in valves controlling liquids.

Two approaches are used to treat control valve noise: source treatment and path treatment. Quiet-design control valves are available that reduce or eliminate the noise that would otherwise be generated by a conventional valve. Path treatment involves heavier scheduled pipe and/or acoustic insulation and possibly silencers after noise has been generated by the valve. While source treatment with a quiet-design control valve is the preferred route, it is not always economical or even physically possible to do so. In those cases path treatment is the only reasonable method available, but effectiveness will extend only as far as the treatment is extended. There are other cases due to application requirements where quiet control valves alone will not provide sufficient noise reduction and acoustic insulation is needed to provide additional attenuation. Thus, in each case specific application requirements and alternatives should be thoroughly analyzed.

Here are a few additional guidelines:

- Quiet-design control valves are available whose noise levels are 20–30 dBA below those of valves selected only on the basis of capacity and control application requirements.
- Pipe wall attenuation varies with pipe diameter and the pipe schedule. Tables are available from vendors to quantify the amount. But for one example for 8-in. (20-cm) steel pipe, schedule 80 pipe will attenuate noise about 5 dBA mopre than schedule 40 pipe.
- Conventional thermal piping insulation will attenuate noise only a few dBA per inch (2.5 cm) thickness of insulation.
- Acoustic insulation attenuation is additive to piping schedule attenuation shown in item 2, but in most cases acoustical insulation has an economic advantage over increasing the piping schedule. Properly designed and installed acoustic insulation will provide 20–25 dBA noise level reduction.
- It is to be noted that in path treatments of conventional line-of-sight control valves such as ball valves, the noise level from the upstream piping will be about the same as the downstream piping.

3.5   Cooling Towers

Cooling tower noise is rarely an in-plant problem, but it is often a community problem. Similar to air-cooled heat exchangers, cooling towers should be designed "right the first time" to avoid retrofitting.

Recommended noise control considerations in the planning stage are:

- Locate the cooling tower as far away from the property line as is feasible.
- Limit the blade tip velocity to under 10,000 ft/min (3000 m/min).
- Orient the louvred side away from the nearby community. The noise level from the louvred side is about 5 dBA higher than that from closed side at 50 ft (15 m) horizontal distance.
- Use a manufacturing building as a noise barrier to the community.
- Determine if the tower fan can be operated at half speed during the critical nighttime hours. This will reduce the fan noise level by about 15 dBA while maintaining roughly two-thirds cooling tower capacity.

In writing noise level specifications the purchaser should specify the requirements at a 50-ft (15-m) distance and not some near field distance like 5 ft (1.5 m). Vendors usually provide octave band and dBA data at both 5-ft and 50-ft distances from each side and vertically from the stack for induced-draft towers.

In the event that community complaints are traced to induced-draft cooling towers, the noise control engineer investigating the complaint should first determine the source of the problem for the particular cooling tower and environmental conditions. Strong evidence was recently presented in the engineering literature indicating that noise from the louvred face (falling water plus a portion of the fan noise) dominates noise from the tower stack.

3.6   Diesel Engines (Stationary)

Fortunately large diesel engines used as emergency generators are not continuous sources of noise. For noise control an acoustic enclosure is required together with a silencer for the diesel exhaust. Operating controls can be mounted outside the enclosure.

3.7   Electric Motors

There have been recent, excellent design improvements in electric motors. Energy conservation payout is obtained with the new high-efficiency motors plus quiet-design built-in features. Noise level improvements, for example, are compared as follows for 1800-rpm, 100-hp, TEFC motors:

Motor Description	PWL, dBA
Standard	90
Standard with unidirectional fan	85
Quiet design	78
High-efficiency plus quiet design	78

In addition, recent tests per IEEE standard 85 witnessed on a high-efficiency WP-

II, 1500-hp, 1800-rpm motor with a quiet-design fan averaged only 81-dBA sound level at one meter for a no-load condition.

Cooling fans are generally the major noise source; for standard motors up to 200-hp noise levels are about 3 dBA higher for a loaded motor than an unloaded one. Commercially available silencers are available to reduce the noise levels of existing motors by about 6 dBA when fan noise is the problem. For more stringent requirements an innovative commercially available spiral design silencer can reduce fan noise by about 20 dBA.

3.8 Flare Towers

As soon as the plant main flare goes into operation, the highly visible light attracts the attention of nearby residents. This, the cause and effect is obvious to the listener. Noise from this necessary operation is caused by combustion and steam injection for smoke control. Multiport rather than single-port steam injection nozzles provide substantial noise reduction. New burner designs reduce combustion noise. As shown in the noise control engineering literature, unsteady combustion noise from periodic gas surging can be overcome by eliminating seal drum sloshing with perforated baffles in the seal drum.

The purchaser should contact vendors for recent noise control developments.

3.9 Furnaces

Furnaces are major potential sources of in-plant noise and noise transmissions to the adjacent community. Furnace noise is caused by complex combinations of noise-producing mechanisms that vary with furnace design and fueling.

In the design stage of natural-draft furnaces it is recommended that quiet multiport burner nozzles and acoustic plenums be evaluated. A well-designed plenum controls both the low-frequency combustion noise from the secondary air register and the high-frequency noise from the primary air intake. Accurate analytical procedures have been developed and described in the noise control engineering literature for practical plenum designs to achieve insertion losses up to 30 dBA. In one instance these mathematical procedures were even used to design effective plenums for older gas-fired boilers in a power house to resolve personnel grievances over excessive noise and personal safety due to burner flashback. The emergency problem arose when the power plant switched from fuel oil to natural gas.

Innovative burner silencers and plenums are commercially available. Considerable cost is involved in the noise control design of a large number of furnaces. Hence, representative tests are recommended on an individual burner at the vendor's facilities to ensure success of the final design.

In one instance a community complaint two miles from a plant was traced with great difficulty to noise from a furnace stack. A stack splitter silencer was designed and installed, the top of the stack was beveled off from the complainant area and toward an industrial area, and burner block modifications were later made to minimize excessive vortexing and turbulence within the furnace. All this successfully resolved the problem.

3.10 Gear Units

The noise level of gear units is dependent on many variables such as gear type, precision of the gear sets, loading, speed, backlash, housing, and others.

For a rule of thumb, AGMA 12–15 high-grade gear units above 100 hp will generate noise levels in excess of 90 dBA at 3 ft (1 m). In-house designed acoustic enclosures can be provided if needed for large gear trains provided that ample oil-cooling capacity utilizing external heat exchangers is also added.

Small-horsepower gear units of lesser quality can be quite noisy. In certain applications quiet nylon gears may provide a solution.

3.11 Grade Level Flares

Grade level flare towers are about 10 dBA quieter than elevated flares of comparable capacity. The tower wall not only reduces the noise level, but it also reduces light emission that may also be objectionable. The grade level flare should be located as far as possible from the property line and the adjacent community. In one case this was not economically feasible. This tower was then designed and later demonstrated to meet 45 dBA maximum at a horizontal distance of 500 ft (150 m). Multiple low-pressure burners were used together with an acoustical labyrinth for the natural-draft combustion air. The tower walls were lined with thermal and acoustic absorption material.

3.12 Pelletizers, Dicers, and Granulators

Steady improvements have been made and continue to be made in noise control of pelletizing and dicing operations as a result of development work by equipment vendors and by users' plant studies and engineering designs. Therefore, what is described here is the result of an overall team effort. Three successful case histories are provided in this section. They are not intended to be all-inclusive, and they are not the only methods to control noise. Rather, they illustrate some of the improvements using classical noise control methods.

- Noise reduction at the source
- Blocking the noise path from the source
- Blocking personnel from the noise path

Pelletizers are used to cut continuous strands of polymers. Dicers cut a continuous sheet of product into cubes or octahedrons. Underwater granulators cut strands of product. The term *granulator* is also used for (dry) scrap granulation.

Using a real-time analyzer, a pelletizer manufacturer measured decibel (dB) sound level peaks at the pelletizer rotor blade frequency (fundamental) and higher harmonics. In the test facilities when a 60 helical blade pelletizer was run at 530 rpm, dB peaks were detected at 530 Hz, 1060 Hz, 1590 Hz, and higher. The dB level at the fundamental is generally higher than the dB levels of the harmonics. Therefore, to reduce the dBA level of a pelletizer alone, it follows that the key is to reduce both the dB amplitude and the frequency of the fundamental cutting noise generating mechanism. To reduce the dB amplitude, sharp helical blades are used. Specifically, to take advantage of dBA weighting by a net shift of the fundamental blade frequency downward, more blades are added (perhaps to a total of 70), more strands are added, and the rpm of the cutting rotor is significantly reduced. Pelletizer structure-borne noise is reduced in design by isolating the cutting chamber from the rest of the pelletizer. Effective silencers can be built into pelletizers as an integral part. The heretofore noisy hydraulic pump and pump drive motor are submerged in the hydraulic oil reservoir to reduce their noise contributions or the hydraulic unit be installed in a nearby unattended area.

In summary, depending on the product being cut and other factors, present-day designs of pelletizers have reached or approached 90 dBA. But the technology to significantly reduce conventional dicer noise, especially when cutting extremely hard product sheets, is not at all this advanced despite some all-out efforts.

Ahead of pelletizers or dicers, air knives are used to blow water off strands or sheets emerging from cooling baths. Noise generated by air knives and strands is produced by high slot velocity induced turbulence plus vortex shedding from the strands. This noise spectrum is typically high frequency, and the noise level increases with the number of strands, but generally air knife noise is less for drying sheet than for strands, as would be expected. Noise around air knives can be unnecessarily increased by using more air than is needed. In addition, noise from pelletizer air knives has been reduced by 10 dBA by using a partial acoustic enclosure together with a remote exhaust fan. The enclosure also minimizes the slipping hazard of water and pellets on the operating floor.

Products from pelletizers or dicers are screened and the screening noise from particle impacts typically contains high-frequency components.

3.12.1 Case History: Noise Reduction at the Source. A full-scale production model underwater granulator was measured at an average of 85 dBA around the cutting device and its drive. The electric motor was the predominant noise source. This compares very favorably with a conventional unsilenced dicer, which ran an average 103 dBA at the operator's position for the same product and production rate. However, the underwater granulator requires considerable auxiliary equipment that is not required by conventional dicers. Many other questions must be answered before anyone can give "blanket approval" to the underwater granulator for any one product or product line. These questions include, but are not limited to, finished product form acceptability, process technology, economic feasibility, and available space for all the equipment and auxiliaries required. In one case product quality was improved with an underwater granulator. Newer improved designs of underwater pelletizers are available.

3.12.2 Case History: Blocking the Noise Path from the Source. While the previously discussed underwater granulator reduces noise at the source, in this case history the noise paths from conventional dicers and air knives are successfully blocked from operating personnel. Two lines of water baths required for cooling the continuous sheets extend through an acoustically designed concrete-block dicer room to the air knives and dicers inside. Lintels in the dicer room openings above the water baths provide water traps. By acoustic impedance mismatch between water and ambient air, the noise from the dicer room is sealed from the operating area outside the dicer room. To avoid "telegraphing" noise, the continuous sheet must be kept submerged throughout the water baths. The plant reported the noise in the unattended dicer room at 115 dBA and outside the room at 87 dBA or a minimum of 28 dBA reduction. The 87 dBA was due to other equipment.

After completing the continuous sheet feed all the way through the water bath outside the dicer room, the sheet is fed by hand to another operator who is in the dicer room only long enough to pass the sheet through the air knife to the dicer feed rolls and to make sure the operation is started and running properly. Start-up is easy to do as three workers start the unit and two then return to their normal duties. There is an acoustically rated door between the two water baths, along with acoustically rated observation windows above the water baths. Double-leaf acoustic doors on the opposite wall permit maintenance access.   In another plant, almost the same net result has been accomplished by stopping the water baths short of the unattended dicer room. Here the air knives are outside the dicer room, and noise levels run about 92 dBA in the unattended air knife area. Noise from the dicer room at the small openings for the sheet feed is controlled with in-house designed silencers mounted at the wall exterior.

3.12.3 Case History: Blocking Personnel from the Noise Path. Noise control has been achieved in several plant locations by providing comfortable acoustically designed control rooms for the operating personnel. Reductions of noise from the conventional pelletizer operating areas up to 42 dBA have been realized, thereby also permitting good speech intelligibility.

Such control rooms are of value only if they are strategically located to permit good visual observations and control of the operations. For this type of operation, however, the disadvantages of the control room as the sole means of noise control are obvious.

3.12.4 Scrap Granulators. These units are very noisy. Some vendors can provide custom-built enclosures; otherwise in-house built enclosures should be provided. Depending on the size and type of scrap being fed, noise emanating from the scrap feed inlets and outlets may be best controlled using enclosed screw or pneumatic conveyors. Belt conveyors enclosed with removable, acoustically treated tunnels are another consideration.

3.13 Piping Noise

Most of the piping noise we hear is generated by connected equipment such as compressors or noisy fluid-flow control valves. Flow noise per se through piping will usually not exceed 85–90 dBA as long as the following rule-of-thumb maximum velocities are not exceeded.

For gases and two-phase flow:

$$V = 100 \sqrt{\text{specific volume}}$$

For liquids:

$$V = 30 \text{ ft/sec (9 m/sec)}$$

where V is in ft/sec and specific volume is in ft^3/lb.

3.14 Pumps

Liquid, centrifugal pumps as a general rule are not noise problems. In some existing installations noise from older electric motors usually predominates the pump–motor package. Centrifugal pumps are usually quieter than positive-displacement pumps. Special high-pressure, high-rpm pumps will exceed 90 dBA at a 3-ft (1-m) distance.

3.15 Rotary Positive Blowers

Sometimes called lobe-type blowers, these units are intensely noisy due to turbulence in delivering gas with a pulsating flow. Both suction and discharge reactive and/or combination reactive/absorptive silencers are usually required. Even with silencers, an acoustic enclosure for the blower–silencer package may also be required depending on the noise level criteria.

Some blower packages are mounted on hollow steel bases with all components rigidly bolted together. While this is a low-cost installation, the entire package is an efficient noise radiator. Some improvement in noise level can be achieved by filling the steel base with nonshrink grout to reduce vibration and structure-borne noise.

3.16 Steam Ejectors

Uninsulated high-capacity steam ejectors and certain associated discharge piping are sources of high noise levels consisting of predominantly high-frequency noise. In one case these ejectors were in-plant problems at 100 dBA in the operating area and a community problem across a river from the plant. There were two adjacent steam ejector systems; the ejector discharge piping diameter for one system was 30 in. (75 cm), the other 18 in.(45 cm). Each system discharged through thin-wall expansion joints, 90° elbows, and conical pipe expanders to surface condensers.

Noise was reduced in the operating area to 89–90 dBA, and the community complaints were resolved with the addition of acoustic insulation. The expansion joints were suspected of being the primary problem source, but the piping configurations could not be ruled out because of the extremely high steam flow rates.

The insulation included the ejector diffuser bodies up to the surface condenser flanges. A high sound transmission loss mastic was used for irregular surfaces such as the special acoustic jacket design to permit travel of the expansion joints.

3.17 Vents

Control systems that must be vented to the atmosphere are usually very noisy because of the high pressure drops and excessively high exit velocities. Use of a quiet-design vent valve with a downstream vent diffuser (tube with multiple ports) to divide the total pressure drop quiets both the control valve and the vent. A well-designed combination system can reduce the overall system noise by 40 dBA.

Where quiet-design vent valves cannot be used, combination absorptive–dissipative or absorptive silencers are used to control the noise downstream from the relief valve. Silencer inlet diffusers not only reduce the gas velocity by breaking it into lower velocity jets but also convert the noise spectrum to high-frequency noise where absorption material in the silencer is more effective.

Here are a few additional points:

* As a rule of thumb the exit velocity of the silenced gas stream should not exceed 250 ft/sec (76 m/sec) to avoid regeneration of noise due to turbulent mixing with atmospheric air.

- Evaluation of vendor proposals can be simplified sound level requirements of a fixed conservative angle and distance like 10 ft (3 m) from the outlet.
- Some process equipment can be operated well beyond its design capacity, but this is not true of vent silencers; plant personnel have found some empty shells when inspecting their well-designed silencers to determine why they were ineffective.

4 LAYOUT

4.1 General Comments

When the site has been selected, the obvious major considerations in new plant layouts are operating efficiency and high productivity, ease of maintenance, hazard control, ventilation, lighting, personnel safety, housekeeping, economics, and future expansions of facilities.

4.2 Layout for In-Plant Noise Level Criteria

In the design phase there may be opportunities to locate relatively unattended units of noisy equipment in a separate room. For instance in one case several rotary positive blowers were located in a single concrete-block room instead of spreading them around the populated indoor operating area. This room required hearing protectors for those entering. In other instances continuously operating air compressors without noise treatments were located in the center of a busy machine shop, near operators in a manufacturing area, near mechanics work areas in a power plant. Of course none of this was necessary; remote locations were available at the outset at nominal initial incremental cost. In some cases there may be doubts as to whether or not to provide enclosures for new equipment. If a decision is to be deferred until after start-up, sufficient space should be provided for this possibility. But in one case, when this decision was deferred for large air compressor casings, speed reducers, and motors, the multiple enclosure penetrations for piping, duct work, and conduit were not laid out for this, thus requiring multiple cuts and seals in the modular enclosure.

4.3 Layout for Property Line Criteria

In order to meet strignet plant property line criteria such as 45 dBA or less, the following considerations are recommended.

The first is obvious to everyone; locate the equipment as far from the property line as possible to take advantage of inexpensive distance decay of the noise. Admittedly, this is not always economical to do, and the move may create new problems, but it is always worth a hard look.

Orienting any openings that are noise sources 180° away is best; 90° is next best. Typical noise sources that have been so oriented are maintenance doorways for utilities buildings containing boilers, air compressors, and water chillers. Ventilating openings fall in this category as do open intakes or discharges of centrifugal fans with silencers. Louvres for cooling towers are potential problems.

Another consideration is the use of buildings as barriers provided that acoustical requirements are met. In the design stage three 95-dBA rotary positive blowers were relocated with no difficulty from the property line side of a manufacturing building to the plant side. In another case the front page of the local newspaper decried all the noise complaints from residents, including aldermen, who lived several hundred feet from two cooling towers. The severity of the complaints and the resulting poor public relations were greatly amplified by the aldermen. All this was most embarrassing to the plant management who demanded an immediate solution to the problem. Expensive, high-energy consumption silencers were being considered when the corporation for other reasons sold the plant—and its noise problem. The cooling towers could have been installed initially on another side of the plant facing the railroad tracks at modest cost.

Incidental shielding is another form of noise barrier attenuation that benefits the plant noise control. Ground-level rows of similar equipment, such as pumps and motors in a refinery that are perpendicular to the property line, will attenuate broadband noise by something on the order of 5 dBA due to small barrier effects. This of course does not apply to large equipment such as centrifugal compressors mounted on a pedestal.

Perhaps the subject of planting trees and shrubs around the plant property line deserves a few words. These will help eventually if the residents consider the plant to be unsightly or if light from the plant is bothersome. But they will only help noise control by a few dBA at most even if a fairly dense forest of large trees is planted.

5 BUILDINGS

The general comments under Section 4 apply here.

5.1 Room Absorption Material

The use of sound absorption materials on walls and ceilings (such as tiles) or unit absorbers such as cylinders and blankets can achieve indoor noise reduction provided that a careful acoustics analysis of each case is made first.

First and foremost, an operator or mechanic within several feet of noisy equipment will benefit by only a few dBA with the addition of considerable absorption material to the large reverberent manufacturing room or machine shop. If the same operator or mechanic moves 25 ft (8 m) away from the machine the absorption material will reduce the noise from this machine significantly—on an order of magnitude of 10 dBA. Other personnel at this distance will also benefit provided they are not also operating noisy machinery. Thus the noise control engineer considering the use of absorption material should first determine exactly what objectives are to be accomplished and where.

After the ascoustics are settled, there are other important considerations.

- What is the fire rating of the absorption materials? What are the products of their incomplete combustion? Sound absorbing foams can produce toxic gases under conditions of incomplete combustion.
- Will any oily mists or release of inflammables in the room condense and collect in the absorption materials or on their thin protective membranes resulting in a (flash) fire hazard in time? Will suspended unit absorbers create a sprinkler and/ or lighting system problem?

- Will collections of dusty manufactured product or other dusts pose a problem? Clogging the absorption materials with dusts (or paint) will render them ineffective. Is there opportunity for contamination of products such as medicinals, pharmaceuticals, or flavorings?

If these are not problems to the noise control engineers' application and if the acoustics provide exactly what is wanted, a viable solution is at hand. Furthermore, on the more positive side, certain types of wall construction such as commercially available slotted concrete blocks have high sound absorption coefficients. They are effective through the design of resonant cavities, which behave similarly to Helmholtz resonators over a range of frequencies. These have been used in product-sensitive industry.

5.2 Enclosures

5.2.1 General Comments. Enclosures can be an effective means of containing and blocking equipment noise from the rest of the operations. Likewise, when designed as control rooms, they block equipment noise from personnel. When and when not to use enclosures depends on a thorough evaluation of each individual case. Equipment enclosures are used when there is no known technology to treat the noise source(s) of certain equipment. Personnel enclosures can effectively reduce the operator's noise exposure if in fact the high noise level equipment can be controlled from them.

5.2.2 Guidelines

- The precautions discussed in Section 5.2 also apply here. In addition any equipment within an enclosure that can release inflammable gases or vapors will convert the enclosures to a bomb.
- To be effective, the enclosure should be airtight. For the moment, think of sound waves as a gas under pressure; if gas could leak in or out so will noise.
- Adequate ventilation with acoustically treated ducts, duct elbows, and fans must be provided. At first glance this may appear to contradict the previous statement.
- The enclosure components, like doors and windows, should have sound transmission loss ratings as close as possible to the wall and ceiling materials.
- Penetrations of the enclosure with piping, duct work, and conduit should be effectively sealed. The sealant should also be flexible, like sponge neoprene, to uncouple these penetrants from the enclosure walls. Likewise, to further minimize structure-borne noise, the enclosure base should be uncoupled from enclosed machinery vibrations.
- Maintenance considerations may indicate the choice of commercially available modular panels that can be disassembled and reassembled. Large acoustically rated maintenance doors for ease of removal of machinery are another consideration.
- Personnel doors should permit rapid egress and access; adequate lighting should be provided.

5.2.3 Modular Panels. These are available from many vendors and about any size enclosure can be constructed from them. Their sound transmission loss is high. They are nominally constructed of perforated steel plate on the inside, about 4 in. (10 cm) of acoustic absorption material that can be protected, and a solid steel plate on the outside. Acoustically rated doors, windows, ventilation systems, floor vibration isolation, and lighting basically complete the vendor's package.

5.2.4 Other Materials. Other materials for in-house construction of industrial equipment enclosures include coarse, unpainted, hollow core, dense concrete blocks, or steel or aluminum sheets. A concrete slab ceiling with acoustic ceiling tiles may be used with concrete block walls. Various densities and thicknesses of mineral wool or fibrous glass may be affixed to the sheet metal with expanded metal. The absorption material may be further supported and protected with no more than 1–2 mils thickness of polyethylene sheets between the expanded metal and the absorption material. Any materials selections should be based on the equipment's octave band analysis.

If this section appears too sparse to the reader, an entire book has been written on materials for noise control (see first reference under General References).

5.2.5 Partial or Open Enclosures. These are useful for equipment only if the room acoustics have first been carefully evaluated. For a personnel shelter enclosure indoors where diffuse high-level noise is present, they are generally useless. Even well-designed commercially available open telephone booths in plants with diffuse high-level noise are only marginally acceptable.

Small in-house constructed three-sided enclosures with a solid top may be effective in certain cases if they are lined with absorption materials. For instance, they can reduce noise from an open suction centrifugal fan driven by an electric motor. Fan intake air enters the partial enclosure on the open side next to the motor end. A fan suction silencer will accomplish the same thing of course.

5.3 Building Partitions

A case history will be used to show how a plant noise problem was solved with floor-to-ceiling partitions. An indoor plant manufacturing area 200 by 200 ft (61 by 61 m) was divided as follows: a 200 by 100 ft (61 by 30 m) rectangular area was used in the manufacturing process. The remaining rectangular area was used for physical testing of the products. Nine employees per shift in the physical testing area were required to wear hearing protection as they were exposed to 90–93-dBA noise that emanated from the production area. Techniques underway to reduce noise at the machinery sources in the production area would be very difficult to accomplish even in the long term. The decision was then made to build a 200-ft-long (61-m) center partition separating the production and physical testing areas. This partition would extend from the floor to the ceiling, 27 ft (8 m) high, to reduce the noise in the physical testing area to 85 dBA max. or less. However, production requirements would not permit any delays in the flows of material between the two areas through any built-in bottlenecks in the doorways in the center partition. Acoustical vestibules or labyrinths at the doorways could not be tolerated, but two sets of quick opening sliding doors would be satisfactory from the production efficiency standpoint. The two sets of quick opening doors for 7 by 8 ft (2 by 2.4 m) doorways would be located 100 ft (30 m) apart in the new partition. The question then was how much noise would be radiated into the physical testing area when these doors are open. Would the open doorways negate the value of the center partition as some of the physical testing equipment be only 15 ft (4.5 m) from a doorway?

Calculations were then made showing that the noise radiated from open doorways would decay to 85 dBA at a distance of 15 ft (4.5 m) in this particular case. These calculations were proven correct after the wall was built. Thus the design objective of 85 dBA max. at 15 ft from the open doorways was achieved without the need for installing bottlenecking acoustical vestibules or labyrinths. Sound transmission loss through the center partition weas high enough, of course, so that noise radiating from the wall itself was not a factor.

The partition was constructed of 8-in (20-cm) concrete blocks to a height of 8 ft (2.5 m). From 8 ft to the ceiling a gyp board, resilient channel system lined with fibrous glass, was used. Concrete blocks were selected in this case to minimize partition maintenance from operations.

5.4   Indoor Barriers

These are of little value for noise controls unless the room in which they are used contains considerable absorption materials on the walls and ceiling. Without this, sound reflects back to personnel on the "quiet" side of the barrier from the room corners, walls, and ceiling.

5.5   Building Openings

Open windows and doorways can radiate objectionable noise outdoors from high noise level operations within. It is simple to calculate the effect of these openings at any distance using the area of the opening as the initial area of radiation.

5.6   Building Exterior Walls

Building exterior walls especially if they must be constructed of light-gauge PVC panels or equivalent can also be objectionable noise radiators to the property line. The noise control engineering literature shows how to determine exterior noise levels at any distance.

5.7   Building Ventillation and Exhaust

Requirements for silencers and/or other acoustic treatments of duct work should be integrated into the fan selections in the design stage to meet indoor and outdoor noise level criteria.

GENERAL REFERENCES

Compendium of Materials for Noise Control, NIOSH Publication No. 80–116, Washington, D.C., U.S. Government Printing Office.

Handbook of Noise Measurement, 9th ed., Gen Rad, Corporate Communications Dept., Concord, MA.

Industrial Noise Control Manual, rev. ed., NIOSH Publication No. 79–117; Washington, D.C., U.S., Government Printing Office.

Sound and Vibration (magazine), Bay Village, OH.

Noise & Hearing Conservation, 4th ed., American Industrial Hygiene Association, Akron, OH.

CHAPTER NINE

Building Features for Hazard Control

Timothy J. Colliton

1 FEATURES FOR ROUTINE CLEANING

1.1 General Requirements

Virtually all industrial processes produce some degree of airborne emission. Particulate emission, such as dust or fume, deposits on the interior building surfaces and eventually must be cleaned, at least for housekeeping purposes. Even dust, which is coarse and settles rapidly, generates fine airborne particles when disturbed. The large particles have fine particles adhering to them that are easily dislodged. The air drag of a falling large particle carries some fine particles into the settled pile. When disturbed, such as sifting off a roof girder, the obviously falling coarse material generates fine airborne particulate that may be significant.

Solids on the floor, even if very coarse granular material, will be ground to fine dust by wheeled traffic. Many vehicles, especially fork-lift trucks, discharge a high-velocity air current from the cooling fan that further disperses the dust generated by the wheels.

For "harmless" dusts, heavy deposits of dust on the floor, say $\frac{1}{4}$ in. or more, will create a nuisance. Floors, at least in areas of significant traffic, should be kept clean. Periodic (e.g., annual) cleaning of overhead structures will usually suffice for nuisance dusts [threshold limit value (TLV) of 10 mg/m³]. Many dusts termed *nuisance* from a health hazard point of view are flammable or explosive. Better housekeeping is needed for such materials.

For dusts of health significance, cleanliness of surfaces becomes more important. If the TLV is 0.1 mg/m³ or lower, the most complete and thorough treatment practical, in terms of building features, is frequently required. In the TLV range between 0.1 and 10 mg/m³, intermediate treatment and application of good judgment is required. An example of a judgmental consideration is the physical properties of the settled dust. Some dusts cling tightly to surfaces and are not easily dislodged, and some are the reverse, for example.

Scrupulous cleanliness of vertical surfaces, such as walls, is not usually required unless:

- The TLV is very low, say 0.01 mg/m³ or less
- The dust is radioactive, so that thin films on vertical surfaces "shine" at the occupants
- The dust emits a volatile material, such as an absorbed or adsorbed reagent, which has a low TLV

Accordingly, a prime consideration is to minimize horizontal surfaces. For example, in buildings with interior horizontal channel girts between the columns, it is more desirable to install the girt with the web surface vertical and the flanges toward the inside surface of the building siding. Another example is to install, where permissible, angle iron structural members with the legs pointing downward at an angle of 45° from vertical. Other, more effective techniques, are discussed later.

Where background contamination limits are stringent, a slanted surface can be installed on structural members to prevent or at least minimize deposition. The angle of the slanted surface should be greater than the angle of repose of the contaminant (if known) or as great as possible. This can be light sheet metal, tack welded in place, or other materials, depending on environmental conditions.

Where structural trusses are used for roof support, the multitude and variety of collection surfaces makes these approaches impractical. A method that has been proven successful is to enclose, or "box in," the truss using vertical panels on both sides. Light-gauge sheet metal, tack welded or bolted in place, or other fire-resistant materials are useful for this purpose. These concerns can be largely eliminated by using self-supported roof designs such as prestressed concrete.

1.2 Vacuum Cleaning

Since it is virtually impossible to eliminate all interior building surfaces where accumulation can occur (especially floor and process equipment surfaces), it is necessary to clean the surfaces on a routine basis. Vacuum cleaning is one of the two preferred methods; wet washdown is the other.

1.2.1 Central Systems. Central vacuum systems consist of a stationary, motor-driven vacuum producer that draws air through one or two filtration devices arranged in series. A network of rigid tubing, which extends throughout the building to be served, is connected to the filter inlet. Inlets to the tubing network are strategically placed at various locations where cleaning will be required. A flexible, portable hose is connected to an inlet to the tubing system to provide suction for the cleaning operation.

The selection procedure for major system components is well known and will not be reiterated here.

A most important aspect of central system design is to provide an acceptable piping layout so that the convenience of the personnel assigned the vacuum cleaning task is maximized. The acceptability and continued use of the vacuum cleaning and, hence, background contaminant concentration depends on this aspect being well thought out prior to installation. The importance of this concentration cannot be overemphasized. Since routine vacuum cleaning is labor intensive, if it is too time consuming and/or inconvenient, it will not be done. It follows that the capital investment will have been wasted and the environmental goals not achieved.

The basic layout design can be made assuming a radius of use around each inlet of 20 ft. (6 m). Other inlets can be added to this basic layout to account for interior partitions, elevation variations, and machinery layout. The maximum hose length should be limited to 25 ft (7.5 m) as longer hose lengths increase the difficulty of handling them and the performance of the cleaning task.

Special tools and/or extension wands should be provided to facilitate cleaning dust accumulations from overhead pipes, and the like, that may not be amenable to the "slanted surface" treatment described earlier.

Caution should be exercised when allowing a vacuum system equipment vendor to design the piping layout. All too often the vendor cannot be cognizant of all the needs of the plant and may have an incentive to minimize the cost of the piping systems. A common consequence is that there are not enough inlets, and they are not located where they are needed most, which engenders an employee attitude that the system is not usable. For more information, see Volume 1, Chapter 9.

1.2.2 Portable Machines. Many manufacturers of industrial vacuum cleaning equipment manufacture portable machines that are complete with suction source and filter on a wheeled base. These types of machines are suitable for serving small areas of a plant, but they can be difficult to move to the cleanig site due to obstructions from walls, machinery, and the like. Other factors that should be considered include lack of ready availability at more than one work station or, for spill cleanups, the need for nearby electrical power (usually three phase), and the difficulty of unloading collected material. Therefore, care should be exercised in using these types of machines on a widespread basis. Such machines can be used successfully in light-duty applications such as laboratories, pilot plant, or yard spills of toxic or valuable material.

The usual filtration device supplied on portable machines is usually sufficient for contaminants of low toxicity. However, in those areas where highly toxic dusts are to be cleaned, the exhaust of the primary filtration device should be provided with a high-efficiency backup filter to preclude the potential for recirculation of the contaminant back to the room. Portable machines with this feature built in are commercially available. An alternative to providing a high-efficiency filter is to connect the exhaust from the vacuum producer the plant's local exhaust ventilation system. However, this will usually drastically limit the usage area or require appropriate branch ducting from the ventilation system.

The same restrictions regarding hose lengths and tool selection apply to portable systems as to central systems.

1.3 Interior Washdown

Interior washing presents a possible alternative to vacuum cleaning, especially for processes that involve solutions and/or slurries. In general, washing can be performed at a much faster rate than vacuum cleaning, thus reducing labor costs. Building design to accommodate this cleaning method are discussed.

1.3.1 Electrical Considerations. As is obvious, the electrical service should be impervious to water leakage from hose streams. This would require, at the minimum, NEMA Type 4 watertight electrical enclosures and all wiring in rigid metal conduit with threaded joints. The electric motors in the plant should be totally enclosed or totally enclosed fan-cooled types.

Similarly, lighting fixtures, electrical service outlets, and other stationary electrical equipment require suitable protection from water intrusion.

1.3.2 Floor Slope. Flooring should be sloped at least $\frac{1}{4}$ in. (0.6 cm) per 10 ft (3 m) of run toward a sump or other water collection device. Floor contours or channels around floor obstructions should be provided to prevent puddling. The cost for proper design provisions for drainage will be more than offset by the continuing labor costs required to "move" a standing puddle out of a low spot and toward the water collection area.

1.3.3 Liquid Collection. The wash water will usually have to be transported to a water treatment system or, if permissible, at least to the sanitary sewer system. Since most industrial particulate contaminants are largely insoluble in water, the primary concern is to maintain adequate flow velocity to preclude settling and plugging of the transport system. Some of these methods are:

- Lighter particulate of small particle size can economically be handled using standard floor drains and piping sloped to a sump tank. Since particulate settling in a sump is inevitable during periods of nonuse, it is advisable to provide a motor-driven agitator in the sump, electrically interlocked with the sump pump motor to maintain the particles in suspension during pumping. Pumping rate and pipe size from the sump should be chosen to prevent settling and plugging in the pipe between the sump and the ultimate destination. Frequently, piping is sized for a transport velocity of up to 10 ft (3 m) per second.

- At the other extreme, heavy particles of relatively large size may necessitate installation of in-floor drainage trenches with water sprays tht scour the bottom of the trench to maintain particle transport. The trenches slope to a sump tank with interlocked agitation. If very large particles, not suitable for pumping, will be encountered routinely, a "drag flite" or other type of conveyor to remove the large particles from the tank bottom should be provided.

1.3.4 Other Considerations. Other aspects of the washdown approach that should be considered are briefly described:

- Is water, with or without detergent additives, a sufficient cleaning medium? If detergent and/or pH adjustment of the wash water is needed to perform adequately, the cost, nature of dispensing, and potential water treatment aspects should be addressed.
- Will heating of the water be required?
- Will the water clean effectively at normal supply pressure or is it necessary to install high-pressure capability.
- The susceptability of the process equipment to corrosion and/or other damage should be considered.
- Natually, water washdown is not advisable for processes involving high temperatures, molten materials, or if freezing temperatures are encountered during the winter months.

1.4 Wall Surfaces

If periodic cleaning of walls is required, wall surfaces must be compatible and amenable to cleaning.

1.4.1 Treatment of Conventional Walls. Most industrial plants are constructed with walls of sheet metal siding, concrete block, precast concrete slabs, or architectural panels. The interior wall surfaces should be smooth and resistant to environmental attack from indoor processes or contaminants. The following details the treatment of each of these three wall types:

- Sheet metal siding: Most types of siding available provide a ready made, smooth inside surface with tight joints or seams. Choosing a siding that is prepainted on the inside surface enhances these features. The only treatment that may be necessary is to apply a seal coat of paint to fill unavoidable surface cracks and, if necessary, to preclude environmental attack.
- Concrete-block wall: The rough nature of this type of surface can result in a difficult or incomplete vacuum cleaning job. The porous nature of the block does not lend itself well to water cleaning. To combat these concerns, one obvious part of the solution is to paint the interior surfaces with a high-quality enamel or urethane coating. Also, the masonry contractor should be instructed not to point the mortar but rather to strike it flat with the wall surface and as smooth as practical.
- Precast concrete slabs: This method of construction is economical and presents

an aesthetically pleasing exterior facade. The interior surface is inherently smooth with only minor indentations. Seams between slabs are vertical and do not present a location for dust deposition. Painting of the inside surface is advisable and would further minimize the need for routine cleaning, but it is not normally required.

1.4.2 Acoustic Considerations. The wall surfaces described will result in an acoustically "hard" space that increases the difficulty of noise control. In general, smooth surfaces for ease of cleaning are not compatible with noise control objectives.

Acoustically absorptive panels that can be fastened to the ceiling or walls are available from many manufacturers. Any such panel chosen should, to the maximum extent, retain the cleanability of the interior surfaces. Usually the panel types that meet this objective are made of open-cell polyurethane foam, fibrous glass wool, or mineral wool encased in a completely sealed bag made of Mylar or Tedlar or other thin (1–2 mills) material. Assistance in selection of the specific material and the quantity required to achieve a certain noise control objective should be sought from a qualified acoustical consultant. In general, for most commercially available materials a cost-effective end point is reached when approximately 50% of the interior surface of the space is treated.

1.5   Floor Surfaces

1.5.1 Process Considerations. For most processes, the ideal floor surface is a poured concrete slab that has been trowelled to a fair degree of smoothness. The strength (thickness) of the floors is naturally dictated by the load requirements for the machinery, products, and mode of material handling. For most manufacturing processes, the concrete floor is the most economical and durable choice.

In those processes that involve handling and spilling of high-temperature molten materials or caustic materials, a concrete floor will quickly be damaged, consequently increasing cleaning difficulties. Wheeled-vehicle traffic over the damaged areas can, in some cases, cause significant airborne concentrations of contaminant.

One approach that has been successful in industries that routinely handle molten metals is to embed or anchor steel plates into the concrete floor where the majority of high-temperature spills are expected. Depending on the load-carrying duty required, the plates can range in thickness form $\frac{1}{4}$ in. (6 cm) to over 1 in. (2.5 cm). They provide an impervious, usually nonstick, surface for many high-melting-point metals, slags, and dross.

The plates (and the concrete, for that matter) are most easily damaged by falling objects, though in absence of such extraordinary events, they will eventually wear out and require replacement.

To prevent deterioration of concrete floors from caustic or other reactive material spills, an appropriate seal coating of the concrete is required.

1.5.2 Floor Maintenance. Usual practice in most of industry is to routinely sweep floors and aisleways using wet or dry techniques. Where large areas with reasonable access are involved, usually this is done using some type of powered floor cleaning machine. These machines generally incorporate a scrubbing action along with a suction system to pick up the loosened material plus the cleaning solution. Use of these types of machines (especially when done dry) is not recommended unless the exhaust airstream from the suction producer is provided with a high-efficiency backup filter. Without a backup filter, the usual primary filter of these machines allows enough dust penetration so that the cleaning machine exhaust can be a very significant source of airborne contamination. Machines that use a wet cleaning method are less prone to this reentrainment problem, the extent of which depends on the characteristics of the contaminant.

Effective cleaning of floor surfaces using automatic machines is, of course, dependent on the smoothness of the floor surface. Large cracks or other sharp indentations will tend to trap and hold materials. Therefore, maintenance of the basic floor structure is essential to effective cleanup of the floor surface.

2 FEATURES FOR EMERGENCY STIUATIONS

2.1 Liquid Release

Inadvertent or emergency release of hazardous liquids requires features for containment and cleanup of the liquid. Flammable, reactive, or liquids with significant vapor pressure require special considerations.

2.1.1 Washdown Provisions. Most liquids can be diluted and cleaned up by effective water washdown. Depending on the building layout and the nature of the liquid, this can be accomplished by a hard pipied floor spraying system or by manually washing the floor. An example is the case of an acid spill. This is not appropriate for flammable or volatile organic liquids.

Since the spilled liquid will most likely end up on the floor, the need for watertight electrical service is minimized.

2.1.2 Floor Slope. If water washdown is the chosen method of cleanup, the floor from the expected point of accidental release to an appropriate sump or holding area should be sloped at least $\frac{1}{4}$ in. (0.6 cm) in 10 ft (3 m) to minimize puddling. The sump is necessary to comply with local water pollution regulations. In the case of very large areas for wash down, strategically placed floor trenches or drains may be necessary.

2.1.3 Emergency Dilution Ventilation. Emergency ventilation provisions should be made in areas where a significant liquid spill could result in life-threatening vapor concentrations. A most common approach is to provide propeller fan(s) in walls or ceilings that can be activated by palm buttons located at the entrances/exits to the area. Dangerous concentrations should not be allowed in adjacent occupied areas.

In the case of flammable materials, the ventilation volume should be sufficient to dilute the vapor concentration to less than 20% of the lower explosive limit under some assumed conditions of size of the liquid spill and evaporation rate. In the case of toxic materials, the ventilation should be sufficient to reduce concentrations to the threshold limit value or a multiple thereof that is safe for short exposures. Adequate openings for uncontaminated replacement air to the area should also be provided. In some cases this may have to be done using motorized dampers interlocked with the emergency ventilation fan.

2.1.4 Floor Material. The floor material should be chosen so that it does not react at a significant rate with the spilled material. This is necessary to prevent deterioration of the floor (and hamper subsequent cleanup) or to prevent formation of toxic or hazardous reaction products. A trowelled concrete floor is applicable in most situations. For some substances, it may be necessary to provide a topcoat of a polyurethane or other sealer. Such sealers, however, usually will not withstand high-temperature spills.

2.1.5 Diking. For potentially large spills, it is advisable to construct a dike around the area to contain the spilled material. The dike should be sized to accommodate the volume of the anticipated spill plus 10%. Other provisions, such as emergency ventilation, should also be addressed. Most materials can be recovered for reuse from the diked area.

2.2 Gaseous Release

Emergency release of gases can, depending on the gas, quickly present a life-threatening condition. This is especially true if the gas has no warning properties.

2.2.1 Emergency Ventilation. Ventilation equipment similar to that discussed for liquid releases, should be provided wherever there is a potential for a massive release of toxic gases. In the case of gases with warning properties that can be perceived at concentrations well below the hazardous concentration, the emergency ventilation system can be manually activated. For gases with no warning properties, such as carbon monoxide, an automatic system that would be activated by a sensing instrument is necessary. In addition to turning on the emergency ventilation, the system should also produce an audible warning alarm so that occupants can vacate the area safely. Thse instruments require frequent attention relative to calibration, maintenance, and the like. A manual alarm activation capability should always be included as part of an automatic system. The concentration at which the sensing device activates the emergency ventilation should be at the threshold limit value or a multiple thereof to allow safe egress of the occupants.

2.2.2 Emergency Egress. Because of the rapidity with an emergency situation can occur in the event of a release of a toxic gas, egress routes should be planned so that personnel can travel in a direction away from the gas source to an exit (preferably outside) where there is no contamination. This can be accomplished only by adequate planning during the layout of the process in relation to the building.

2.3 Solid Spill

The accidental spill of solids is usually easier to deal with than gaseous or liquid materials. The exception to this general rule is if the solid material has a substance adsorbed on it that has a significant vapor pressure thay may cause a hazardous exposure situation.

2.3.1 Vacuum Cleaning. Cleanup of spilled solids using a central or portable vacuum cleaning system is most applicable to granular materials of smaller particle size. An example of an application would be in a warehousing operation where broken bags of material are inevitable and, in the case of hazardous materials, should be cleaned up promptly to minimize airborne dust generation due to vehicular traffic. For potentially large spills, the central vacuum system is preferred because of its greater dust-holding capacity, and it is generally more powerful. Vacuum cleaning can remove material from cracks in floors and around equipment most effectively.

2.3.2 Wet Cleanup. In general the provisions for Section 1.3 are also applicable to wash down techniques for spills of solid material. Caution should be exercised in choosing this method of cleanup in that, especially with powdered materials, the impact of the first spray of water can disturb the pile of material sufficiently to create a significant concentration of airborne dust. It may be more advisable to gently wet the pile of material to a consistency of "mud" and remove it using a shovel or other similar tool. The remaining area where the spill is located would then have to be washed down using water spray or scrubbed by hand.

Naturally, concerns about how the material behaves when it is wetted need to be taken into account. This method may be more applicable to those materials that dissolve readily in water and, perhaps, can be easily extracted from the solution for reuse.

2.3.3 Coarse Materials. Materials over about $\frac{1}{2}$ in. (1.3 cm) in particle size present special handling problems. Conventionally designed vacuum cleaning systems are generally not suited for pickup of large chunks of material due to the potential for plugging the system.

A potential method is to gently wet the material with water or another appropriate solvent to minimize dusting. The remaining mud can be flushed to a sump, cleaned up by hand (small amounts), or cleaned up by broom and shovel. Proper respiratory protection should also be used if appropriate.

3   STRUCTURAL FEATURES

3.1   Basic Structure

The structural aspects of a new building can be modified to minimize those features in conventional buildings that make hazard control difficult and time consuming.

3.1.1 External Structural Members. A building design with the building skin on the inside of the structural members will minimize the places for contaminants to collect and greatly simplify any routine cleaning that may be necessary. This concept is essentially an "inside-out" building wall. For this construction, the building skin (actually the inside walls) can be prefabricated architectural panels that are bolted to the building structure. In climates that experience significant snowfall, it may be necessary to install a standard roof deck on the outside of roof structural members to carry the snow load adequately. Also, for energy conservation, the space between the roof deck and the interior skin can be insulated if desired.

3.1.2 External Utility Runs. Since cleaning of overhead pipes, ducts, and the like is at best difficult, it may be desirable to install these services on the outside of the building with penetrations to the equipment served as necessary. This can be done with not only the usual utilities (gas, water, and electrical) but can also be done for exhaust ventilation duct work, replacement air duct work, vacuum system piping, and any other utilities unique to the operations performed. Naturally, proper attention is required for weather protection, heat loss from heated air ducts, and freezing protection as appropriate to the climate.

3.1.3 Design Roof Loading. It is usually economical in the long run to design a new building such that it can carry a roof load of 10–25% or more than that specified in the local building code. During the life of the building and the process, it is a common experience that the only available space for a new piece of auxiliary equipment (such as a dust collector) is "on the roof." The predicament arises when it is realized that the building roof was designed to meet the minimum specified by the building code and cannot accept additional loading safely without reinforcement of the building structure. Also, equipment is routinely installed inside of the building and supported from the roof structure.

Since the future uses of the building and/or the process evolution are normally unknown at the time of building design, it is prudent to overdesign for added roof loads to some degree. The sight increase in building cost will be well worth the prevention of a future building failure and the consequent liability potential as well as process interruption.

3.2 Interior Features

It is recognized that it is not always possible to completely eliminate internal runs of utilities or duct work. In these cases, the following suggestions are offered:

3.2.1 Pits or Trenches. Pits or trenches can be used with great success to eliminate utility and duct work runs that would otherwise run overhead and require routine cleaning. The major pitfall in designing pits or trenches is that they eventually evolve to being "too small." As the process evolves and utility needs change (or regulations change), the pit or trench gets increasingly crowded to the point where maintenance access is extremely difficult. Therefore, it is recommended that the size of the pit or trench be chosen during the initial design stages to be $1\frac{1}{2}$ to 2 times the minimum size required. The practical limit of this oversizing is, of course, the size and maneuverability of the pit or trench cover. The chosen design load for the trench cover is dependent on the process for which the building is designed and the type of wheeled traffic anticipated.

It should be kept in mind that when piping gaseous or liquid fuels in a pit or trench, there should be adequate and continuously operating ventilation provided to prevent creation of a flammable fuel–air mixture due to a piping leak. Openings for adequate replacement air to the ventilated area should be provided at the end of the pit or trench opposite from the ventilation fan.

CHAPTER TEN

Maintenance

Charles H. Stevens

1   GENERAL COMMENTS

Maintenance operations are big business; estimates are that approximately $1.5 trillion are spent annually on maintenance. Numerous reference books are available on the various aspects of maintenance—as related to production, to organizational structures in industrial operations, to labor problems and solutions, and to modern diagnostic and maintenance tools available.

However, maintenance as related to the industrial hygienist or the environmentally

oriented design or plant engineer is the subject of this chapter. The following aspects
of plant maintenance are considered to be of most interest to these persons:

- Health, safety, and environmental regulations
- Health protection for maintenance employees
- Respiratory protection
- Eye protection
- Hearing protection
- Head and face protection
- Gloves
- Safety hats
- Safety clothing
- Health protection
- Access for maintenance
- Decontamination of equipment
- Record handling (directed toward environmental control equipment)
- Health safety, and environmental regulations
- Interrelationship between plant layout, equipment selection, and the maintenance
  engineer

2 HEALTH, SAFETY, AND ENVIRONMENTAL REGULATIONS

Although the primary purpose of an industrial organization is to turn out quality
products safely and at a profit, the production process must comply with internal
company standards and the multiplicity of government regulations that affect every
industry. Health, safety, and environmental regulations are frequently foremost and
crucial among these.

It is reasonable to assume that when a plant or production line was originally built,
it was designed to comply with then-existing government regulations. These regulations
are a major concern for management, which is responsible for developing a systematic
way to keep up with changes and additions to the regulations and incorporating them
into the program.

Policy in complying with regulations must be explicit and understood by all concerned. For instance, if breakdowns occur in production and environmental control
equipment malfunctions at the same time, which gets priority attention? Will the
production line start up before all industrial hygiene and environmental equipment is
functioning properly? The right answer is no. A line should not be restarted until that
equipment is working correctly. The maintenance manager must at all times take into
account compliance with regulations by all employees. The plant manager has overall
responsibility for making it clear that everyone is responsible for avoiding violations
of safety and environmental regulations. For instance, a well-conceived series of exhaust hoods designed for proper air flow can easily become ineffective if fan belts are
permitted to slip or bag collector back pressure is allowed to build up. It is the duty
of whoever becomes aware of such a problem to report it or take appropriate action.

Compliance with governmental regulations, though, is only one facet of the picture.
Management must, per se, accept the responsibility of protecting the employee from

job-related health hazards and thus, "good engineering pratices" must be overriding. For additional information, see Chapter 29.

The maintenance manager should have a voice in planning of and design for environmental controls and should work with others to establish maintenance programs and assign responsibility for checking system performance.

We have all seen production lines (or portions of them) installed to increase production or manufacture new products. Many of us are concerned with the design of these new facilities and the tremendous effort needed to assure successful plant start-up with minimum downtime. It is hoped that serious consideration is always given to the proper maintenance as well as operation of the new production line. Maybe even more important is the performance of the necessary preventive and routine maintenance once the design engineer's assignment is completed. The concern that engineering itelf is only a part (although a very important part) of the production line's operation is the focus of Chapter 11. Since much of the follow-up of engineering design requires well-planned maintenance, many of the ideas introduced in Chapter 11 and those discussed in this chapter complement each other.

3 HEALTH PROTECTION FOR MAINTENANCE EMPLOYEES

Health protection for maintenance workers is often more difficult than for production employees. Maintenance workers cannot predict the next work assignment or location. They frequently must, on short notice, make quick and efficient repairs in cramped quarters that are hot, dusty, damp, or otherwise uncomfortable and unhealthy, and that production workers normally do not have to enter. Access to areas where maintenance work is required will be discussed later; however, careful evaluation of the hazards and potential hazards involved in maintenance work must be made by supervisors, foremen, and, most importantly, by the worker who must enter a hazardous area.

Training of maintenance workers to recognize hazards and potential is an essential function, one to be conducted on a regular and organized basis. Such training may be done primarily and formally by industrial hygiene and/or safety personnel at the plant or corporate level; however, maintenance foremen and supervisors must be capable of training their people to assure safe performance of work at all times.

Equipment of many types is available on the market to provide for the workers' health and safety. The two major types involve respiratory protection and physical hazards protection.

3.1 Respiratory Protection

Protection from harmful dusts, fumes, vapors, and gases is accomplished by wearing respirators of the type approved for the particular contaminant. Respirators range from simple nontoxic particle masks to air-supplied full-face masks, which may also be part of a complete protective suit.

3.2 Eye Protection

Many companies have compulsory programs requiring every employee and every visitor in a plant to wear safety glasses, usually with sideshields. Lenses may be safety glass or plastic composition.

Maintenance workers sometimes require additional protection, and may wear goggles, face masks, or welding shields for certain tasks.

3.3 Hearing Protection

Ear muffs or ear plugs for hearing protection are frequently required in or near noisy areas. They must be carefully selected to meet the noise situation and to fit the wearer and should be used consistently because of the cumulative effect of noise on human ears.

3.4 Head and Face Protection

Safety hats and caps provide protection from falling objects and from bumps and bruises that occur when working in close or confined areas. These head protectors can be designed to incorporate eye and hearing protective equipment. Materials from which thay are made are chosen to give maximum protection with minimum weight. Headliners to protect against winter cold and sweatbands to absorb perspiration in summer are often used with safety hats or caps.

3.5 Gloves

Many different materials are used to manufacture gloves worn by maintenance workers for protection against cuts, bruises, punctures, abrasion, and adverse effects of chemicals, such as acids, caustics, solvents, oils, greases, and alcohols. Gloves may be made of cotton, vinyl, polyetheylene, rubber (or compositions that have similar properties at less cost), leather, polyester blends, or canvas. Some uses require specialized composition materials.

Gloves may be lined with cotton for comfort and to minimize perspiration. Good "feel" through the gloves is important for repair work. Heat protection may require special materials. Asbestos was formerly used for gloves and pads to handle hot objects and surfaces but has lost favor because of the potential hazard from release of asbestos fibers to the atmosphere. It is most important that gloves be selected in advance and available for protection against the agent in question.

3.6 Safety Clothing

Workers may require overall clothing to protect them from excessive heat or cold, and from skin absorption of liquids, vapors, or gases. It is well known that dermatitis is the most common industrial disease; barrier creams and/or protective clothing, including gloves, are often used to shield the worker against skin problems.

Safety clothing ranges from aprons and sleeves to complete air-supplied suits. Aprons, sleeves, and complete acid suits to protect against spills and splashes are often made from polyvinyl chloride (PVC), although other materials are used for certain chemicals. Paint spray hoods are common in most plants, because paint products applied by spraying are required for many types of manufactured products. Paint spray hoods are frequently supplied with breathing air lines and are generally made from various types of cloth.

Rainwear, head covering, and safety footwear are standard items in most maintenance departments. Less common but frequently required are air-supplied suits, often made of PVC, which come in various conformations. They may include provisions for self-contained breathing apparatus and may include boots, gloves, and viewing hoods.

4 ACCESS FOR MAINTENANCE

Most serious employee health hazards are caused by inhalation of toxic materials—dusts, fumes, mists, and gases. To capture and remove these contaminants, reducing their concentration in the breathing atmosphere to levels that are no longer hazardous, is usually accomplished by installation of well-designed hoods or enclosures. To maintain a safe and healthy plant environment for production workers these devices must be readily accessible for maintenace. Of equal importance is easy access to the hooded or enclosed production equipment.

Proper engineering and design of hoods and enclosures and their appropriate exhaust systems are essential to maintain air levels below the threshold limits. Special provisions may be required to supply makeup air (see Chapter 7). Adequate and quick access to the hoods or enclosures for maintenance has to be part of the original design.

4.1 Design of Hoods and Enclosures

Where possible, it is desirable to design a hood or enclosure to remain in place while normal maintenance operations are performed. For example, lubrication of equipment in an enclosure can frequently be done from the outside through properly designed fittings and lubricating tubing. Belt changes should be possible from the outside of an enclosure.

On the other hand, replacement of major pieces of production equipment may require removal of all or part of a hood or enclosure. Proper design, therefore, will not only allow for routine repair and maintenance, but will provide for removing portions or all of the enclosure in a manner that will permit reassembly with no loss of integrity or function. Examples of typical configurations are shown in Figs. 10.1. and 10.2.

Is is most important to avoid, if possible, removable panels that are held in place by a multiplicity of nuts and bolts. Maintenance must, perforce, remove these to gain access. However, in the pressure of other maintenance needs, all the nuts and bolts seldom get replaced—in deed, some are lost and who has the time to go back to the tool room to get replacements? Large and preferably "quick opening" fasteners are needed. For example, the design may permit the top to be lifted off an enclosure, and provide sliding doors for access to belt-driven fans.

Industrial hygienists often recommend hoods or enclosures around small furnaces, such as induction furnaces. A wide door, or whole side of an enclosure, which can be slid aside on a track to permit furnace relining or other repairs, makes such furnaces accessible. If a hood or enclosure has not been designed for convenient access, too often a cutting torch is used to remove a panel the first time service to the equipment is required, and the enclosure never fits correctly again.

In designing enclosures and hoods for easy access, it is well to remember that the need for respiratory and physical hazards protection may require suiting and air packs

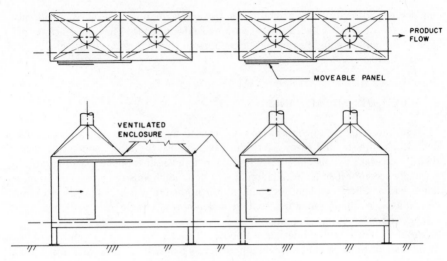

Figure 10.1. Quick access opening.

that make a maintenance person "look like an astronaut." More space is needed for such a repair person to maneuver than for an unencumbered person in a less hazardous atmosphere.

For an existing enclosure, current health and safety standards that require breathing apparatus, special clothing, and protective gear may require enlargement of access openings or partial diamantling of the enclosure. It is obviously unacceptable to require a maintenance person to enter a space where one may be overcome by residual chemicals, lack of oxygen, or hazardous fumes. A "Work Permit" should be required and strictly enforced for entry into confined spaces or on critical equipment involving the safety of operations as well as maintenance. For additional information, see Chapter 3, Section 5.6, and Chapter 26, Section 5.3.

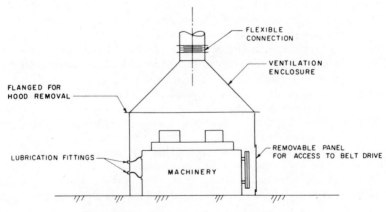

Figure 10.2. Access opening—removable panel.

For enclosures that are entered frequently for maintenance of the equipment, permanently installed lighting is helpful and a time saver. Ladders, steps, and platforms or gratings may pay for themselves in safety and convenience over the life of these enclosures. Supplied-air lines inside the enclosure that can be connected to breathing supply equipment, and water and drain lines for washing down the enclosure from the inside are time and cost savers. For instance, work in a dust collector (which is bound to be dusty inside) is easier if compressed breathing air lines are available inside the unit.

Proper water and drain lines inside an enclosure, which make it possible to wash down the space before anyone enters it, are sometimes both desirable and feasible, particularly if the material in the hood or enclosure is hazardous.

Vapor-proof and/or explosion-proof lighting and electrical outlets should be used in certain atmospheres.

5 DECONTAMINATION

Decontamination of equipment or machinery that produces hazardous dusts, fumes, or mists is often required before maintenance people can perform scheduled maintenance or make repairs following breakdowns. Sometimes such cleaning and decontamination can be performed without moving the equipment. In other instances, a machine may have to be moved to one of the plant's shops for service. Other large and complex machines sometimes must be transported to the manufacturer's home plant or some other outside repair shop. In these instances, when highly hazardous or toxic substances are present, the equipment may have to be sealed in plastic sheeting and transported to a cleanup location for thorough decontamination before it can be repaired.

The type and amount of contamination can run the gamut from minor irritants such as dust, grease, and organics to highly hazardous dusts or severely corrosive or flammable materials. The severity of the problem varies with the amount and type of contamination. Small amounts of dust on a machine to be repaired in a beryllium or lead production plant may require a major cleaning effort, while considerably more dust on equipment in a cement plant can be handled more easily because the hazard is much less.

Any maintenance program should address these issues. The industrial hygienist or safety specialist should develop rules or guidelines to protect maintenance workers from potentially hazardous exposures during equipment decontamination. These rules must be enforced by foremen and supervisors. A background of exposure data should be obtained and developed for this purpose, and the rules and guidelines regularly updated as conditions and hazards change. Protective clothing, barrier creams, respirators, and other equipment must be available, with clear directions for their use. When in doubt, err on the side of safety.

Plant personnel should be familiar with the degree of toxicity of items in various stages of production. The Occupational Safety and Health Administration's (OSHA's) Harzard Communication regulations require that information on all toxic and hazardous material be made available to every employee. This is accomplished through proper labeling and availability of material safety data sheets at or near the place of use of hazardous materials and the required initial and annual training in hazardous materials safety. These regulations have improved plant workers' knowledge of such issues.

However, potential hazards from new products or production operations need to be identified promptly, by the industrial hygienist, for production workers and particularly for maintenance people who do decontamination.

5.1 On-Site Decontamination

Decontamination and cleaning of equipment or machinery in place is often done in confined areas, frequently with limited or difficult access. Some of the problems discussed in the previous section apply. For example, a plating tank containing a hazardous plating solution may require inspection for adequacy of the tank lining. Maintenance personnel may be expected to wash down the tank thoroughly from the outside, then enter and clean the tank from the inside before a supervisor or specialist can make a thorough check of the tank lining. Needless to say, the maintenance crew must wear and use the proper protective equipment for this task. Another example might be machinery that must be repaired or modified by a specialist from the manufacturer's plant. If the machine is contaminated, it will have to be cleaned and decontaminated by maintenance staff before the specialist can work on it.

Where applicable, temporary exhaust ventilation may be used during cleanup. For large installations, temporary platforms, safety lines, and other safety devices may be required, in addition to clothing, respirators, gloves and the like.

Vacuum cleaning of surfaces, with portable units or using a central vacuum cleaning system, is often useful for dusty conditions. In other instances, water or steam cleaning may be the method of choice.

The key to safety in all such operations is advance planning, so that methods and equipment are available as needed. The documented planning must be complete, must be available to, and must be used by the maintenance personnel who do the work.

5.2 Maintenance Shop Decontamination

Often machinery and equipment repairs can be done best in the plant maintenance shop. For such equipment, a preliminary cleaning should be done, by vacuum or water if practical, before the piece of equipment is removed from the production line.

Then, depending on the type of contamination, handling in the shop and during repair should be done with care. It may be desirable or necessary to set aside a temporary "quarantine" zone for this work, with ventilation provided, a vacuum system available, and drains in the floor to allow a thorough cleaning of the equipment before the repair or maintenance work is begun.

When hazardous materials are handled frequently, a permanent decontamination room or booth is generally provided to clean equipment before maintenance work is done. This is an area specially designed with adequate ventilation, easily cleaned walls, a sloped floor, drains, and adequate lighting. If frequent welding is required, a movable, flexible welding exhaust system should be installed.

5.3 Off-Site Decontamination

Equipment or machinery that cannot be properly repaired by plant maintenance personnel, equipment incorporating specialized materials, or highly sophisticated instrumentation or controls may have to be sent to a off-site repair shop that specializes in hazardous materials decontamination and disposal. Such materials include PCBs, dioxin, some of the metals, and radioactive materials. Tools, electronic equipment, and various types of instrumentation can be cleaned for reuse. Contaminated materials such as disposable protective clothing, rags, gloves, hose, and other materials used in contaminated environments sometimes can be cleaned and reused; in some instances, however, it is more practical to dispose of them as hazardous waste.

It is especially important that equipment sent outside be thoroughly decontaminated and checked carefully for cleanliness. How clean is "clean?" This is a difficult question to answer. Few standards exist to identify how clean equipment or machinery should be before it leaves the plant. Therefore it is the responsibility of management to develop a standard by tapping the combined experience of technical, legal, and management personnel.

The usual way of measuring surface cleanliness is to take wipe samples of surfaces that have been cleaned. These samples are analyzed in a laboratory and compared to a "safe" level of cleanliness, such as a given number of milligrams per square foot of surface cleaned. The results are compared to the standard limit set by the procedure described in the previous paragraph. If the reading is over the limit, the machine should be cleaned again until new wipe samples meet the standard.

The potential for occupational illness claims from the outside repair service and related adverse litigation against the company should not be underestimated. Wipe samples should be taken in an approved manner and analyzed in a well-established analytical laboratory, preferably an outside laboratory if time permits. Records of wipe samples should be kept in permanent files. A form letter should notify the outside repair service that the equipment to be repaired was used in a contaminated work area, but has been throughly cleaned and checked with wipe samples. The contaminant or contaminants should be named.

Some decontamination firms bring a specialized van and operators to a plant and will clean materials on the spot and return them. These vans contain their own pollution controls.

The costs and effectiveness of this type of service as well as the reputation of the firm should be carefully checked before engaging them. Potential liability to a cleaning firm or its employees is certainly a main consideration. Do not risk incurring third-party liability to the outside service or liability because of improperly discarded hazardous wastes generated by equipment decontamination. The legal aspects of using outside services of this type should be thoroughly investigated.

Planning for decontamination that involves hazardous materals should result in clearly defined rules and regulations, specifying protective and cleaning equipment to be used, and also specifying procedures to use and standards to be met. These rules and regulations should be concise and easily understood by the people who will do the work; foremen and supervisors should make sure that regulations are understood and followed to the letter. Some disposal items may have to be decontaminated before they can be sent to landfills or for incineration. Other items can be placed in drums, plastic bags, or other containers for safe disposal, with proper labeling and instructions to avoid any undesirable incidents.

6 RECORD HANDLING

It appears that virtually all of man's endeavors today generate enormous amounts of paper and records. Storage and retrieval of this information becomes a prodigious task, yet for management of a maintenance department, access to complete records is essential. Computerizing maintenance recordkeeping can help this situation under some circumstances, but whether manual or computerized, maintenance records and written rules, regulations, and planning forecasts are essential. Some of the maintenance functions that involve recordkeeping are listed. Adaptation to a specific situation may require more or fewer records.

6.1   Basic Systems

- Cost records and cost control data leading to reduction in maintenance costs
- Maintenance authorization forms
- Job descriptions
- Written procedures for maintenance management systems
- Periodic review and appraisal forms

6.2   Work Order System

- Outline of how they are originated, who authorized them, how priorities are determined
- Maintenance log sheets
- Time and job cards
- Work order processing and data storage
- Coding of maintenance operations and functions

6.3   Standardized Maintenance Operations

- Records of typical, repetitive work (e.g., lubrication of a specific type of bearing)
- Labor time data to be used for estimating time needed for similar work in the future
- Standard "details"—outlines of repairs or maintenance often developed by equipment manufacturers that can help on relatively routine work

6.4   Maintenance Planning

- Recordkeeping to plan maintenance budget (e.g., yearly for a 5-year period)
- Development of a long-range planning document outlining overall maintenance program with anticipated labor needs, equipment requirements, overall organization, and, of course, dollars needed for each part of the plan
- Day-to-day management of backlog, using simple forms

6.5   Schedule Sheets

- Daily and/or weekly task assignments for individuals
- Critical and PERT scheduling to be integrated into overall scheduling

6.6 Computerized Reporting

- Programming EDP requirements
- Input and output data identification
- Variance reporting
- Computerized control of spare parts and maintenance materials
- Computer-assisted design of maintenance sytems

6.7 Preventive Maintenance

- Outlines of procedure
- Periodic (annual) review of data inspection reports
- Machinery and equipment repair records
- Preventive maintenance repair records
- Preventive maintenance work orders
- Preventive maintenance scheduling forms
- Report of equipment and machinery breakdowns or failures
- Diagnostic maintenance data

And on and on. It sometimes seems that the number of forms to be filled out, records to be kept, and reports to be submitted in a sizable maintenance operation is limited only by the imagination of the people involved. Old forms are revised or eliminated, but new ones seem to spring up to take their places. It requires constant vigilance to keep the forms and the system lean and useful; but the results of such careful management are worth the trouble.

6.8 Record Handling for Environmental Control Equipment

The plant engineer, industrial hygienist, or environmental specialist most interested in the information presented in this book is often more involved in installation of new equipment and the recordkeeping related to that equipment than to general maintenance management and records. Therefore this aspect of record handling is reviewed in some detail.

Whenever new equipment is installed, either by an outside contractor or by the plant staff, operation and maintenance (O&M) manuals or other vendor instructions deserve careful study. Based on the equipment manufacturer's instructions, the following data need to be kept, in writing, and developed or incorporated into books or manuals for maintenance department use.

6.8.1 Warranty by Manufacturer. How long the equipment is under warranty and what maintenance work can or should be done by plant personnel without voiding the warranty must be determined. If the vendor is obligated to send service personnel during the warranty period, this should be noted with relevant addresses and telephone numbers, including night numbers if the equipment is operated on two or three shifts. The expiration date of the warranty should be recorded, and the decision made as to whether the maintenance department will fully take over all repair and maintenance functions or whether the vendor's service personnel will continue to be called. Under special circumstances, production, quality control, research and development, or other departments may take over responsibility for the equipment.

6.8.2 Safety Instructions. These instructions relate not only to protection of maintenance personnel, but to avoiding damage to equipment. For example, lockout procedures for electrical equipment, used almost universally in manufacturing plants, must be identified, understood by everyone concerned, and enforced. It should not be possible for anyone else to start equipment being worked on by a maintenance person unless that individual knows about it and has approved it. Of course, similar safety problems can develop with various mechanical drives, such as a steam turbine driving dryer rolls through shafts and belts. For additional informaiton, see Chapter 3, Section 5.6.

Another type of safety instructions for equipment protection relates to high operating temperatures. For example, when some equipment is shut down for repair, air, water, or oil cooling flow must continue for a specific length of time (spelled out in writing) to avoid equipment damage. Instructions should specify whether the continuation of coolant flow is handled manually or automatically by instrumentation, and how this is done. It is assumed that the production operator has been trained in how to shut down the equipment properly and in the right sequence.

Specific safety rules for persons and equipment should be a part of the instruction book or manual. It is critical that the use of safety equipment be mandatory, and that enforcement procedures, with penalties for safety violations, be clearly spelled out.

Depending on the breathing hazard, the instruction should require the appropriate type of respiratory protection—self-contained breathing apparatus, air-supplied masks, or full- or half-face respirators. Consultation with the industrial hygienist should determine which type is necessary, if any.

Hearing protection may also need to be part of the instructions. Although the equipment to be repaired is usually shut down, nearby equipment may generate severe noise overexposure. Safety instructions should also specify other safety equipment as needed such as goggles, welding shields, or other eye protection; aprons; and barrier creams. Again, consultation with the plant health and safety professional is needed.

6.8.3 Specific Maintenance Jobs. As each new piece of production equipment is installed, specific preventive and routine maintenance chores should be identified, performed at the time if necessary, and scheduled for the future. Manufactuer's instructions call for what should be done to assure maximum equipment performance, and how often each task has to be done. These needs for each piece of equipment, when applied to the overall production line, describe the total maintenance needs; incorporated into the maintenance plan with specifics on how they are to be accomplished, these are essential parts of a total maintenance program.

Items such as lubrication, welding repairs, belt tension checks, surface inspections for onset of corrosion, checks of possible excessive vibration, and other investigations on a regular basis form another part of the fabric of a maintenance program.

Obviously, maintenance workers must service many pieces of equipment. A laborer may be assigned routine lubrication for many production machines and parts, but individual lubrication jobs are based on the needs of each machine, its bearings, and other moving parts. A worker may use a stroboscope to check pulley or cylinder speeds, or other rotating parts during a routine tour of duty. Manufacturer's instructions for maintenance are necessarily general. Specific types and frequencies of maintenance service should be developed for each plant, based on local conditions and types of production, taking into account the type of workers, overall plant cleanliness requirements, and availability of equipment for maintenance.

6.8.4 Troubleshooting. If a machine or an entire production line is not performing properly, typically the production supervisor will call for maintenance service. These calls for help may be for routine service or may require detailed investigation to determine what needs to be done.

Equipment manufacturers or production system vendors usually include trouble-shooting sections in their manuals. Reference to those sections can often save a call to the vendor's service department and a long production delay. A typical trouble-shooting sheet for a fan is shown in Table 10.1. Please note that material references in Table 10.1 relate to further detail not identified in this chapter for sake of brevity. Note that the first item of the troubleshooting list shows, as one solution to the problem, selection of a larger fan. This remedy is not usually the province of the maintenance group and may have to be brought to the attention of the design engineer. (Of course, of the maintenance supervisor selected the fan, it becomes the supervisor's respon-sibility.) Good cooperation between design engineer and maintenance personnel is critical; and in many instances, especially in larger firms, the design engineer is avail-able during the start-up of new equipment or production lines.

If in-house personnel feel that outside service is needed, vendors are usually asked to supply experts on service of the equipment involved. In special cases, particularly when problems involve an entire system, a consulting engineering firm may be called in to review the operation and make recommendations.

It is important for the maintenance manager to train his personnel to become experienced troubleshooters. They should be trained and encouraged to think for themselves, to reason out logically why the production line is not working properly, what needs to be done and why, and who to contact for help. A Band-Aid approach or baling wire fix only leads to further and more serious breakdowns. Training may require attendance of one or several people at a formal seminar on equipment or systems that are new and familiar. Informal assignments to study available literature on equipment may be made and followed up by the supervisor.

Good troubleshooting helps to keep a production line operating; poor trouble-shooting can be costly in downtime and production loss.

6.8.5 Maintenance Instruction Book. A maintenance instruction book for an entire sytsem, such as an air pollution control installation or a wastewater treatment plant, is useful for a maintenance department. It will include lubrication requirements, equip-ment data, parts lists, and related sections. A typical equipment lubrication schedule (Table 10.2) and a maintenance data sheet (Table 10.3) are shown and are usually part of a maintenance instruction book.

7 INTERRELATIONSHIP BETWEEN PLANT LAYOUT, EQUIPMENT SELECTION, AND THE MAINTENANCE ENGINEER

7.1 Maintenance Engineer

As indicated earlier in this chapter and in Section 2 of Chapter 11, one of the major functions of the maintenance group is to work closely with the design engineer to assure, to the greatest degree possible, that plant layout and equipment selected allow

Table 10.1. Troubleshooting Fan Problems and Possible Solutions

Symptom	Cause	Solution
Capacity or pressure below rating	Total resistance of system higher than designed for	Select larger fan
	Speed too slow	Check drive system
	Dampers or variable inlet vanes not properly adjusted	Reset
	Poor fan inlet or outlet conditions	Increase speed, provide turning vanes or baffles in duct work
	Air leaks in system	Repair
	Damaged wheel	Repair or replace per Sections 3.6, 5.1.3
	Rotation direction incorrect	Reverse electrically
	Wheel mounted backward on shaft	Correct per Sections 5.1.2, 5.1.3
	Bearings, coupling wheel	Realign per Sections 3.13.2, 3.13.3, and 5.1.2
	Unstable foundation	See Section 1.6
	Foreign material in fan causing unbalance	Clean per Section 5.1
Vibration and noise	Worn bearings	Replace per Section 3.8
	Damaged wheel or motor	See Sections 5.1, 5.1.2, and 5.1.3
	Broken or loose bolts or set screws	Tighten or replace
	Bent shaft	Replace
	Worn coupling	Replace
	Fan wheel or driver unbalanced	See Section 5.1.2
	Magnetic hum due to electrical input	Check input line voltage for transients or high levels
	Fan delivering more than rated	Reduce speed or close dampers
	Loose dampers of VIVs	Tighten or replace
	Speed too high or fan rotating in wrong direction	Reduce speed, check for electrical
		Reinstall wheel
	Wheel has shifted on shaft because of balancing procedure	Reposition, tighten, and rebalance per Section 5.1.2
	Too much grease in ball or roller bearings	Clean and regrease per Section 5.2.1
	Poor alignment	Realign
	Damaged wheel or driver	Repair or replace
	Bent shaft	Replace shaft
Overheated bearing	Abnormal end thrust	Reselect bearing
	Dirt in bearings	Clean bearing per Section 5.2
	Excessive belt tension	Realign per Section 3.13.2
	Unbalanced rotor	Rebalance per Section 5.1.2
	Excessive fan temperature	Recheck system
	Lack of cooling water or air to sleeve bearing	Recheck system

Table 10.1. (*Continued*)

Symptom	Cause	Solution
	Speed too high	Recheck driver
	Specific gravity or density of gas greater than design rating	Check system
	Discharging over capacity because system resistance is less than designed for	Reduce speed or close dampers
Driver overloaded	Packing too tight or defective (fans with stuffing box only)	See Section 3.12
	Rotation direction wrong	See Section 2.2
	Shaft bent	Replace
	Poor driver alignment	Realign
	Wheel rubbing or binding on inlet bell	Reinstall wheel, or realign inlet bell
	Bearings improperly lubricated	Clean and relubricate per Section 5.2.1
	Motor wired wrong	Rewire
	Moved gas temperature below design level	Check system

proper, cost-effective maintenance. Considerations in plant layout that concern the maintenance manager are numerous. Some examples follow.

7.1.1 Access. Layout will undoubtedly include aisles of sufficient width to permit vehicle and personnel traffic to transfer raw materials and semifinished or finished product. However, the following questions should also be considered in developing a plant layout. Is there sufficcient access to permit maintenance, removal, and replacement of equipment. Are cranes or hoists provided for maintenance work or, if not, is there room to bring in a movable hoist? Does the layout accommodate regular, routine maintenance such as lubrication or drive belt changes?

7.1.2 Housekeeping. Regular periodic cleaning of a production facility is important in most plants but is critical in operations involving chemicals, food, or toxic materials where regular cleanup has to be integrated with production processes. Layout must provide opportunities for easy convenient cleaning of equipment and its supports, building structures, and floor areas. Provision must be available for washup, including properly pitched floors and floor drains, adequate space between equipment and the floor, and generally good access for required cleaning. If dry materials are processed, a central vacuum cleaning system should be considered in plant layout.

Housekeeping also involves orderliness. Layout of a plant should include adequate provision for raw materials and in-process materials storage in a neat and orderly manner.

7.1.3 Plant Services. Plant layout should incorporate plant services needed for maintenance. Examples include steam lines for steam-cleaning equipment, properly designed floor drains systems for washdown of plant and machinery, and material han-

Table 10.2. Wastewater Treatment Plant Maintenance Equipment Lubrication Schedule

Item	Description	Location	Type Lube	Frequency and Miscellaneous
B-1	Air blower—Schwitzer	Gear box, Fill plug	Esso American S-1, 20 oil, or Shell Rotella 20-x-100 20	Change at 500 hr, fill to middle of sight glass 30 oz 1000 hr or 3 months
P-1A	Main sump pumps—	Grease fittings #77	Grease moisture-resistant #2 ball-bearing grease	
P-1-B	Worthington	A, B, C, shown on Figure 2		
P-2-A	Caustic metering pump	Oil sump	Gulf multipurpose gear lubricant EP-90, antifoam agent or W & T U-18443	Change every 3 months; check dipstick to maintain marked level motor bearing oil-less— See mfgr. data
P-2B	Foxcroft—W & T			
P-3A	Surge tank sump	Standard pump lubricating	Has sealed shaft bearings	Thrust bearing, self-motor—see nameplate
P-3B	Johnston 20 SPB			
P-3C,D	Neutralized waste pump— Worthington	Same as P-1		

P-5A,B	Polyelectrolyte metering—W & T pumps	Oil Sump	Gulf multipurpose Gear Lube EP-90, anti-foam or W & T #U-18443	Change every 3 months; check dipstick to maintain marked level—Motor bearings—Oil-less. See mfgr. data.
P-6A,B	Caustic cutback metering pump W & T	Same as P-5A		
P-7A	Sump Pump—Sump #2 Worthington	Same as P-1		
P-8A	Sludge pump—Air diaphragm Wilden	Lubricator, Fill plug	SAE-10 Lube oil	Add oil as required; check level weekly
P-9	Place wash supply pump—sump type Worthington	Same as P-1		
M-1A,B	Mixer—polyelectrolyte tank Eastern RG	Gear box pregreased at factory for life; see motor mfgr. for motor lube	Gear-box when rebuilt—Texaco Marfax EP-1, 1 pt. grease	
M-2A,B	Caustic mixing tank—Eastern	Same	Same	
M-3A,B	Surge tank mixer—Eastern	Same	Same	
	Clarifier rotating mechanism -do-	Gear box casing	Lithium #2	Change every 2 yr
		Main spur gear 1.5 gal capacity	Winter—AMGA #2 (SAE 2D) Summer—AMGA #4 (SAE 40)	Change every 6 months

Table 10.3. Maintenance Data Sheet

PUMP

Make _____	Supplier _____
Agent _____	Size and Type _____
Serial _____	Model _____
Date Installed _____	Instruction Manual _____

Pump Rating

Capacity _____ USgpm	Shaft speed _____ rpm
Head _____ ft	Shut-off head _____ ft
Number of stages _____	Direction of rotation _____
Impeller diameter _____	Performance curve number _____

Stuffing Box/Mechanical Seal

Size: OD _____ ID _____

Packing manufacturer _____

Number of packing rings _____

Size of rings _____

Manufacturer of seal _____

Model number _____ Type _____

Serial number _____ Size _____

Item	Material Specification	Part Number
Casing		
Bowl assembly		
Impellers		
Impeller rings		
Casing rings		
Shaft		
Shaft sleeves		
Stuffing box bushing		
Gland		
Impeller nut		
Gasket material		
Shaft bearings		
Coupling		
Item	Material Specification	Part Number

DRIVER

Make _____	Supplier _____
Agent _____	Size and type _____
Serial _____	Model _____
Date installed _____	Instruction manual _____
hp _____ rpm _____	Frame _____
Volts _____ Phases _____	Cycles _____
Number of cylinders _____	Bore and stroke _____

For details of nomenclature see Hydraulic Institute Standards.

MAINTENANCE RECORD

Date inspected _____

Inspected by _____

Repairs _____
_____

Repaired by _____

New parts installed _____
_____

Cost of repairs _____
_____

Table 10.3. (*Continued*)

Down time	_____
	_____
Remarks	_____
	_____
	_____
	_____
	_____

dling equipment such as hoists and job cranes, and access aisles for fork or lift trucks, not only for movement of production materials but also for movement of equipment.

The design engineer should review the plant layout with the maintenance manager to discuss the number and location of electrical substations and transformers, boiler room size and location, numbers and types of air compressors, condensate return systems, and the like to allow for cost-effective maintenance after the plant or addition is built. The ability of the plant's sanitary, industrial waste, and storm water systems to handle increased or changed loadings from the plant should be related not only to total waste quantities but also to plant layout and the ability of individual branches or parts of the system to perform. Gravity flow of existing systems should be reviewed before adding to their load. For example, the ability to maintain chemical treatment and sludge removal portions of treatment facilities must be checked.

7.2 Equipment Selection

New equipment proposed by the design engineer should be carefully reviewed to assure long-time, trouble-free performance. Who is more qualified for such a review than the expert who maintains the plant in operation and who knows what tyes of machinery perform day in and day out?

As new equipment is developed by the research and development (R&D) or product development groups, the opinions of experienced maintenance engineers and managers should be sought. Sometimes new equipment looks great on the drawing board but presents problems in the plant. The help of people who have learned by experience what works and what does not is invaluable.

Many more aspects of maintenance could be discussed; however, these seven items are probably of most interest to the health and safety professional. For further study, the following are some general references.

GENERAL REFERENCES

L. R. Higgins and L. C. Morrow (eds.), *Maintenance Engineering Handbook,* 3rd ed., New York: McGraw-Hill.

H. F. Lund (ed.), *Industrial Pollution Control Handbook,* New York: McGraw-Hill, 1971. (See especially Chapter 11, Dust Collectors; and Chapter 22, General Design Considerations.)

G. Salvendy (ed.), *Handbook of Industrial Engineering,* New York: Wiley, 1982. (See especially Chapter 8, Quality Assurance; Chapter 10, Facilities Design; and Chapter 11, Planning and Control.)

Magazines and Journals: *Plant Engineering,* Cahners Publishing Company, Inc., Barrington, IL; *Manufacturing Engineering,* Society of Manufacturing Engineers, Dearborn, MI; *Maintenance Technology,* Applied Technology Publications, Inc., Barrington, IL.

CHAPTER ELEVEN

Engineering Is Necessary But Not Sufficient

Charles H. Stevens

1 GENERAL COMMENTS

Previous volumes and chapters of this book have described the engineering requirements for design of a variety of equipment needed to control the potential hazards to human health and the environment that can arise from production operations. Good, sound engineering of such equipment is absolutely essential and basic to a well-run plant. But engineering is not in itself sufficient to protect the health of employees, the surrounding community, and the environment. Only commitment by all the people concerned, from directors and top management to the man or woman on the shop floor, will make the engineering expertise effective.

2 MANAGEMENT BACKUP

The passage of environmental and health legislation such as the Clean Water Act, the Clean Air Act, the Resource Conservation and Recovery Act, the Toxic Substances Control Act, CERCLA (the Superfund Act), and the Occupational Safety and Health Act by Congress, plus similar laws by states, has laid a groundwork for health and

273

safety protection unrivaled in our history. Managements of industrial companies have accepted that health and safety factors are crucial to their planning and thinking. Very few projects are designed and engineered today that do not include built-in environmental controls to protect workers and plant neighbors from health hazards, present and potential. However, no matter how well an engineering control system has been designed, no matter how much money has been spent in building it, a continuing management commitment must be made to its operation and maintenance.

As knowledge of the harmful effects of certain substances and actions on the environment has increased, society has accepted, and expressed in the form of laws, the conviction that we must be careful not to harm our people and our environment.

A first aspect of management commitment, or community commitment for that matter, is adequate, continuing funding. Many desirable things in our society have fallen flat because ongoing monies were not appropriated. One of our major national problems is to find the funding to meet the requirements of the law and, sometimes stricter than the law, our consciences.

Part of the engineering job of designing a dust collection facility or a wastewater treatment plant is the development of the necessary program to keep the system operating without unscheduled shutdowns. It is essential to provide adequate funding for a budget to maintain a piece of equipment or a whole plant. Management needs an informed estimate of the funds needed to maintain equipment, the capital dollars for updating, and the training dollars for operators and maintenance people. The designer, production personnel and maintenance personnel should all have a hand in developing such maintenance budgets, including the outlay for supplies, spare parts, personnel costs, and other costs.

For example, what happens if the ventilation system or a wastewater treatment plant malfunctions and insufficient air flow occurs in that plant or the contaminants are discharged to the river? The author believes it is absolutely necessary that the particular polluting operation be shut down should such problems occur—unless backup protection in the form of redundant systems can take over. Is this attitude realistic? It is the only realistic approach in the late 1980s and hereafter. Serious industrial environmental problems—Chernobyl, Times Beach, and Bhopal come immediately to mind—have aroused adverse public opinions about the industries involved, in an atmosphere of tremendous environmental concern. This widely held and intense public concern has required drastic changes in the thinking of business owners and managers.

Top management must be able to forecast the possibility and therefore prevent the actuality of events that can seriously hurt or even bankrupt their businesses. Environmental occurrences such as those already mentioned above, as well as the Three Mile Island nuclear power plant accident, the total or near bankruptcies of several asbestos manufacturing companies, and hazardous waste problems at Love Canal, New York, present only the tip of the iceberg. This type of serious problem can, often inadvertently, reduce the environmental protection of workers and the surrounding environment sufficiently to do significant harm to the people and the firms involved.

Thus, although in the past it seemed uneconomical to shut down a production operation to avoid pollution of a stream, the air, or the workers' atmosphere, in reality, we know now that it is only good business to set up both emergency and preventive programs and to give these strong backing. Most large corporations have done or are doing just that through deliberation and planning.

To carry out such a mandate of top management and to provide proper health and environmental protection at all times, plant managers must first accept the responsibility, then assign it to both maintenance and production managers. These managers in turn need to place the same priority on maintaining and operating the environmental control equipment that has traditionally been placed on keeping production flowing.

The Occupational Safety and Health Administration (OSHA) Hazard Communications program, and state laws such as New Jersey's and Pennsylvania's right-to-know laws, have aided in this process by developing a greater awareness among employees of some of the potential hazards, and by inventory and labeling of hazardous materials. However, some questions and potential problems need the more specialized knowledge of industrial hygienists and safety specialists.

For instance, while adequate ventilation may normally control a problem, under certain circumstances some contaminants can be absorbed by or through the skin; provision should be made for those circumstances. Or if two contaminants are generated at the same location, what happens? Will there be a chemical reaction that can liberate a toxic gas or some other form of hazardous material? Or possibly generate an explosion?

It is important that such possibilities be considered in advance and provided against—and that emergency plans be ready if such events occur. The rules, the knowledge, and the equipment for preventing as well as for correcting such events must be in place and kept up to date. And potentially dangerous processes should be monitored for problems. Title III of the Superfund Amendment and Reauthorization Act of 1986 deals with this matter.

3 MAINTENANCE SUPPORT

Once management backing has been assured, it is the responsibility of the maintenance group to implement and safeguard the maintenance portion of the design engineer's well-planned ideas. Efforts needed to assure continued satisfactory operation of health and environmental equipment include the following:

3.1 Air Flow Checks

Design of ventilation and makeup air systems is reviewed in Chapters 6 and 7. Air flow measurement is an important industrial hygiene aspect of most plant operation and is, therefore, covered in detail. However, the most effective ventilation design—good hood design, well-balanced exhaust duct systems, and properly selected makeup air distributed to various sections of the plant—becomes virtually useless if it is not operated and maintained properly to assure sufficient air flow and air quality at all times.

Certain air flows are needed to maintain contaminant levels below the permissible exposure limit (PEL). Therefore the following minimum program is required:

- During initial start-up of the exhaust system, balance the system to obtain desired air flows at each hood or enclosure. If dampers are used—and they should be—they should be set to give the desired air flows, and the setting should be marked on the duct! It should not be necessary to change damper settings unless the system is redesigned or changed because of removal or addition of equipment requiring ventilation. (Unfortunately, additional ducts are often carelessly added to an existing system, rendering it ineffective.)

- Install manometers across filters, bag collectors, packed towers, demisters, carbon adsorbers, or other contaminant collection devices in the system. A range of maximum and minimum pressure drops across the filter, dust collector bags, packing, and the like should be determined (probably by the design engineer or the equipment supplier). When the maximum acceptable level of pressure differential across the dust or fume cleaning device has been reached, the filter should be cleaned or changed; the dust collector bags shaken or replaced with new ones if needed (they should be replaced only after routine bag shaking or repressuring fails to correct the problem); plugged tower packing removed, cleaned, and replaced; or other flow obstructions removed.

- It may be necessary to clean ducts occasionally, although this is rarely necessary with properly designed exhaust systems.
- Air flows at individual hoods should be determined by checking the "hood suction." This requires a static pressure check at a small hole drilled in the exhaust duct just after the hood or enclosure. Hood face velocities should occasionally be checked against design values.
- Frequencies of air flow checks depend on how critical the ventilation system is to the maintenance of satisfactory in-plant air levels. A typical checking system might involve keeping records of manometer pressures across collectors daily and hood suctions weekly. Monthly face velocities may be indicated; records should be kept of readings noted.
- There is not much point in maintaining records if they are not reviewed on a regular basis and action taken accordingly. A production or maintenance supervisor should be responsible for the proper operation of exhaust and makeup air systems and for starting troubleshooting and corrective action promptly as needed to assure workers' health and protection at all times.
- Makeup air equipment is not as vital to the health protection of the worker as are the exhaust systems and dust, fume, or vapor collectors, but should be checked regularly. Filter changes should be a regular part of the maintenance program along with normal checks of the heating equipment (the makeup air may not be heated in southern climates, although it may be cooled in some situations.)

3.2 Exhaust System Operation and Maintenance

3.2.1 Exhaust Fans. The heart of any exhaust system is the fan. The design engineer has selected the type and size of fan to give certain air flows at different pressure drops in the system. Once the system has been checked out and the fan speed or rpm has been determined to be correct, there is not much more that must be done to the fan. Fans often run for many years without maintenance except lubrication of bearings or a periodic bearing change. (See Chapter 10 for discussion on maintenance requirements for fans.) The writer has seen fans that are 50 years old and still performing in various industries, although obviously at a lesser fan efficiency than the modern fan. The one item in a fan that needs frequent checking is the belt that connects the drive motor with the fan. Belts tend to stretch over a period of time causing slippage; it is important that belt tension be checked monthly (or bimonthly) by determining the fan speed either with a tachometer or, if available, with stroboscopic equipment. Proper design of exhaust systems is of great importance since it is essential to have sufficient air flow at various hoods and enclosures at all times.

3.2.2 Dampers. Changes to any system are likely to occur sooner or later due to machine rearrangements, hood modifications, or for other "normal" reasons. Dampers in the exhaust system then become very helpful. For this reason, most systems should be designed with duct dampers. When the exhaust system is started up, it has to be balanced to produce the designed air flows in the ducts and at the various hoods or enclosures. This can be achieved with the use of dampers. After design flows are achieved, industrial hygiene samples, face velocities, discomfort of workers near hoods, or some other reason may dictate further modifications of the system. However, after satisfactory balance has been achieved, the dampers should be locked in place by tack welding. The setting should be marked on the duct, and the dampers should then be left alone, if equipment changes occur later that require exhaust system changes, the tack welds can be broken loose and the system rebalanced. The dampers can then be tack-welded again.

It should be remembered that dust collector back pressure varies within a certain range because of dust buildup on bags over time; also, fan belts may slip and drop air flow in the enclosure or hood. This should be corrected by proper bag shaking and regular partial bag replacements. Fan belts should be tightened or replaced as needed. Leave the dampers alone.

There are situations, such as exhaust systems handling explosive dusts, where the use of dampers is not feasible. It is important to design the exhaust system in such situations with special care and to reevaluate the system when changes are made to the production line that require exhaust system modifications.

3.2.3 Automatic Controls. The next step beyond regular scheduled checking of the proper operation of equipment is the installation of automatic controls that signal operators or shut off equipment when certain conditions occur, such as a pressure drop or too high a concentration of a contaminant in treatment system effluent. The cost of such controls can often be at least partially balanced by the decreased number of manual checks required.

In some instances the automatic control can economically and efficiently change the operation of a piece of equipment. An example is a device that monitors the level of solvent vapor coming off a carbon adsorption unit and automatically switches the flow of exhaust air coming from a paint booth to a second carbon unit, and at the same time starts the steam purging process in the chamber that could no longer adsorb the solvent.

For safety reasons, such automatic controls are especially useful in avoiding emergencies that could be dangerous. Automatic shutdown of equipment combined with flashing lights and a sounding klaxon get immediate attention to emergency situations.

3.3 Miscellaneous Items

3.3.1 Manometers. In calculations of pressure losses in the exhaust system, the contaminant collection device often adds a significant portion of the pressure loss in the system. Bag collectors can range in pressure drop from 3–6 in. (7–15 cm) in shaker-type collectors to 8–12 in. (20–30 cm) in reverse air collectors. Pressure drops across various types of scrubbers can vary significantly also. The device that measures pressure drops across the collection device is usually the manometer. It can be installed at the contaminant collection device and read there or piped to an operating panel to make the reading more centralized. However, it has been found that manometers and manometer tubing will plug. Too often readings are faithfully recorded day after day, even though the reading does not change. Obviously, when this occurs, the line has been plugged and should be cleaned promptly. A portable U-tube manometer can be used to check a manometer quickly and easily.

3.3.2 Pilot Lights. Pilot lights are often used to show whether equipment is or is not operating. In an exhaust system, the fan is usually provided with a pilot light. Pilot lights are also used on shaker motors, on reverse air blowers on bag collectors, and on pumps on scrubbers. Although it is always possible that a pilot light could indicate that a fan is energized electrically when the fan is not producing air flow, this is a highly unlikely occurrence. Belts may slip but relatively seldom will the entire set break completely. Normally it is no problem for an operator to know whether there is air flow into the hood or enclosure, but it is very difficult for the operator to know whether the air flow is sufficient to provide safe working conditions. In another example, a pilot light may indicate that the sprays on a scrubber are in operation when actually they are not. Additional instrumentations could be used to assure that not only is the pump motor energized, but the pump is pumping.

3.3.3 Instrumentation. It is possible to provide instrumentation for practically every possible situation that can be imagined. However, from a practical viewpoint, over-instrumentation should be avoided. Instrumentation design can be fun, and it would seem that the more instruments in use, the better the system; however, instruments will fail at times, and they require maintenance. The minimum amount of instrumentation that will do the job is recommended.

3.3.4 Automation and Instrumentation. While the principle that the simpler a system can be, the fewer problems it will have, is generally a sound approach, the increasing automation of some types of production may require increased instrumentation, automatic recording of data, and control of operations. As a general principle, an instrument as close as possible to the action one is trying to monitor is the one to use. To refer to the previous example, instrumentation that would record actual flows from scrubber pumps would be more desirable than a pilot light that tells that the electric drive motor is electrified or that the pump is rotating. If one had such an instrument the other two would not be needed.

Automation of instrumentation systems may be a necessity in some types of automated production processes. The complications of checking and maintaining the more complex instruments, and possibly of bringing them together for on-line, real-time reading and recording may be a price worth paying. The human overseer of the instrumentation may be concerned with seeing that the instrumentation system is working, not with reading the measurements and controlling the production line. The instruments will do that automatically.

4   MOTIVATION AND TRAINING OF PEOPLE

Engineering is never sufficient in itself without people, no matter how simple or how automated a system may be. While the mechanics of designing plants and exhaust systems and providing for health protection and instrumentation are basic, operational success still depends on people who have the motivation and interest to do the job. It is therefore essential to determine what kind of personnel should be selected for operation and maintenance of dust-collecting systems, wastewater treatment plants, solid waste disposal programs, and the like. These people need to be motivated by their supervisors to feel that their environmental control equipment is important, and therefore that they and their jobs are important. Lip service is not enough. Management must convince the maintenance and operations people, with actions and funding support, of that importance. The thrust of this book is that all workers, including maintenance workers, are entitled to and must have health protection. Furthermore, the boundaries of concern do not stop at the shop wall or fence. Everyone in the plant, management and workers alike, must realize the importance of being a good neighbor. Everyone connected with the plant has a degree of responsibility for seeing that the plant does not discharge any toxic or annoying substances to the surrounding air, water, or soil.

Much of the development and implementation of health protection programs is the responsibility of the industrial hygienist. Some industrial hygiene groups are placed under the direction of the company medical director; other management systems have the industrial hygienists and the safety people both reporting to the industrial relations department. In still others, industrial hygiene is part of the engineering department. Wherever the industrial hygiene group appears on the company organization chart, the single most important factor in their successful operation is the support of top management. Training and instructing people to be concerned about health protection is essential.

One of the steps toward accomplishing this goal is to insist on the discipline of good housekeeping. As a general principle, the person that generates contamination is the one to clean up the "mess." However, this is not always feasible or even possible when a production worker has to keep up with a running production line and a steady flow of product. One alternative is to provide a cleanup crew for production areas.

This principle may be worked out in various combinations. For example, a production worker may be responsible for cleaning up within a certain number of feet of the work station. Area beyond that distance is the responsibility of the maintenance department. Considerations might include the following: Is a washdown at the end of a shift sufficient to keep the area clean or is vacuum cleaning at various times during the day required? Are the drains kept open so that contaminated wastewater does not remain in the area? Has provision been made to treat contaminated wastewater properly before it is discharged to the environment?

Achieving good housekeeping is not easy. Often, a detailed cleaning program must be developed. Who is responsible for cleaning the tops of equipment, motors, shelves, whatever? The best way to clean up is to avoid the generation of contamination in the first place, but this is not always possible. Absolute cleanliness may not be necessary to maintain safe air levels in the plant, except when handling highly hazardous material. In that instance, close to absolute cleanliness may be required.

Cleanup crews are usually part of the maintenance function and are used to clean up equipment before maintenance or repair work, as well as for general cleaning, which is not the responsibility of production workers.

5 CONCLUSION

Many environmental and health protection systems are engineered with great care and cost. But no matter how well designed these systems may be, their proper operation depends on competent, motivated people. Engineering is sufficient only when supplemented by people with the imagination and concern to assure that the engineered systems will work at all times.

CHAPTER TWELVE

Assay and Quality Control

Paul F. Woolrich

1 INTRODUCTION

Quality may be defined as the composite of those characteristics that differentiate individual units of a product and have significance in determining the degree of acceptability of that unit by the buyer/user. Quality may be considered as a specification or set of specifications that are to be met within given tolerances acceptable to the buyer while minimizing costs for the vendor. Quality must also be controlled for raw materials and supplies, labor, and machines, plus management functions such as budgeting, inventory, transportation, and the like. Quality control techniques should be applied to the complete manufacturing and marketing enterprise to obtain as efficient an operation as possible.

As the subject of quality control becomes more of a science and less of an art, less reliance is placed on individual personal judgment and more on physical or chemical tests treated statistically. While the maintenance of quality should be the concern of every individual employed in the processing, manufacturing, and handling of products, responsibility for the control of quality must be delegated to an individual or a department in order to ensure the consistent production of satisfactory products at minimum costs. A well-functioning quality control organization will contribute to the reduction of rejects, maintenance of uniform quality, increased customer satisfaction, and employee morale, while at the same time minimize costs.

This chapter will cover some general principles of quality control; the responsibilities and organizations of quality control and its relationship with other groups within the organization; sampling and inspection; and quality control assay including precautionary measures required in handling certain high-risk materials.

2 FUNCTIONS OF QUALITY CONTROL

2.1 Establishment of Specifications

Specifications for raw materials, supplies, in-plant processes, containers, and the finished product must be available so that everyone involved can appreciate exactly what is wanted. Such specifications should be established with the assistance of sales and production personnel since they are in a position to know customer requirements and production capabilities.

2.2 Development of Test Procedures

Quality levels and production variables must be tested on some scale. Hence, it is up to the quality control department to find, or to develop for any specific purpose, a means of measuring every quality attribute and production variable of importance at every step in the process from the raw material until the product is consumed. These tests may be developed for specific purposes with the aid of research and development personnel, or they may be adopted in whole or in part from existing standards developed by trade association or government agencies.

2.3 Development of Sampling Schedules

Since 100% inspection is rarely feasible or desirable, it is the function of the quality control department to establish efficient procecures for handling samples and for determining the number of units and frequency of sampling so that quality may be evaluated with maximum reliability at minimum cost.

2.4 Recording and Reporting

The quality control department is responsible for setting up the necessary forms so that results may be recorded easily and transmitted promptly to the proper personnel (usually production) who are in position to take action when and as soon as necessary.

2.5 Troubleshooting

When a situation is found to be out of control, the quality control department should see to it that the situation is corrected immediately.

2.6 Special Problems

When special problems arise in any part of the organization (e.g., customer complaints, poor raw materials, or equipment or personnel problems), the quality control department should serve as the channel of communications among departments to facilitate solution of the problem.

2.7 Training of Personnel

Line personnel may or may not be under the direct supervision of, and responsible to, a central quality control department. However, line personnel should be instructed by the quality control department in the sampling, testing, and reporting procedures. Another educational function of the quality control department, together with the support of top management, is to promote a spirit of quality mindedness among all personnel.

3 QUALITY CONTROL RELATIONSHIPS WITH OTHER DEPARTMENTS AND BUYERS

Figure 12.1 depicts the interrelationships of the quality control department with other units of the organization.

3.1 Relation to Management

Organizationally, there should be a direct line from the quality control department to top management. Quality control reports provide management not only with information on whether the production operation is in control, but with the basic information for decision making on such basic problems as inventory, pricing, and budget policies.

3.2 Relation to Sales and Purchasing

Since the sales department is the primary contact between the processor and customer, it is the salesperson who is in the position to appreciate most fully exactly what the buyer is looking for in the product. A product can be marketed successfully only if it meets the customer's requirements, *not* the manufacturer's opinion of what these requirements should be.

The relationship with a purchasing department is quite similar. In this case, however, the organization itself is now the buyer. Thus, specifications for all raw materials and supplies must be established by the quality control department for use of the purchasing department or in the acceptance inspection of incoming materials.

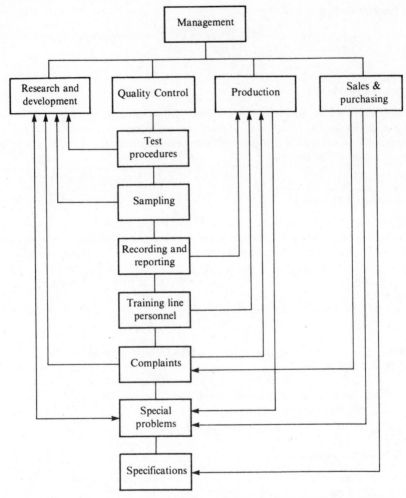

Figure 12.1. Functions of quality control and its relationship to other departments.

3.3 Relation to Research and Development

Quality control provides the continuing assessment of a current operation, while research and development is a search for something new or different. Quality control may indicate where research and development is needed. Research and development, in developing a new product, process, or equipment, may elicit a change in quality control procedures.

3.4 Relation to Production

The responsibility of the production department is to produce maximum quantities in terms of product yield and plant capacity; the responsibility of the quality control department is to ensure that this objective is achieved without impairment to quality and with a maximum of profit.

It follows that a central quality control department should be independent of the production department, and would, therefore, imply that the quality control inspector on the production line, as well as in the laboratory, should be directed by, and report to, the central quality control authority. There are exceptions to this:

- For quality control to be more effective, the line inspector, a member of the quality control group, should be able to report directly to the production employee, who can take remedial steps immediately.
- In many cases, it may be impractical to have anyone but the production employee handle quality control, as in the case of an automatic machine operator, whose major responsibility is simply to see that the equipment is operating properly. The nature of the tests for evaluating performance and the method of recording and evaluating the data should be established by quality control, however.

3.5   Relation to Buyer

The quality control system, as well as the production system, is properly geared to the buyer's specifications. The buyer is not necessarily the ultimate consumer, but he may be a broker, wholesaler, distributor, or another manufacturer using the particular product as a raw material. Thus, a quality control cycle should begin and end with the customer's specifications.

4   SAMPLING—INSPECTION AND ASSAY

Once buyer and vendor specifications are clearly understood and defined, it is the function of quality control to establish means for measuring these quality attributes.

The next step is to set up instructions and procedures by which the characteristics of interest to the customer may be measured. These methods should be as precise and accurate as possible, while at the same time as rapid, simple, and inexpensive as possible. Objective methods should be used where available. Once the methods have been selected, a sampling schedule must be worked out that will provide maximum information at minimum cost, and control stations set up either in the plant or in a central laboratory.

4.1   Sampling Inspection

Sampling inspection is a significant part of the quality control operation. Rapid statistical procedures are utilized to derive a sampling schedule that will ensure the drawing of samples that are adequately representative for inspection. Too frequently, results of inspection are taken at face value, and it is assumed that test values are identical with true average values of the lot tested when, in fact, inspection results become more representative as the frequency and size of sampling is increased. An optimum inspection procedure should provide an estimate of lot quality at the minimum acceptable level of accuracy at the least cost. This optimum procedure may be selected only by the use of certain statistical tools.

4.1.1 Influencing Factors. Factors that influence the sampling procedure are:

- Purpose for which the inspection is made
  Accept or reject
  Evaluate average quality
  Determine uniformity
- Nature of the material
  Homogeneity
  Size
  History of material
  Cost of the material
- Nature of test methods
  Importance of test
  Destructive or nondestructive to valuable material sample
  Time and equipment utilization
- Nature of lots being sampled
  Size
  Sublots
  Loading (pallet, drum, tankcar, etc.)

4.1.2 Continuous Sampling. In acceptance inspection situations, the buyer ordinarily does the sampling and testing in order to determine whether to accept, reject, or establish the price for the lot of material offered to him. In the case of continuous sampling, it is usually the manufacturer, who is the prospective vendor, who does the testing for the purpose of determining whether the product offered for sale will be acceptable. Only in the case where the manufacturer wishes to verify the quality of a lot already accepted is the manufacturer in a position where continuous sampling can be used. An adequate continuous sampling inspection, therefore, will maintain uniformity and minimize rejects. It should also make possible the complete labeling and packaging of the finished product directly as it leaves the production line, thus avoiding the necessity of storing and retesting before the proper labels may be attached.

5 MEASUREMENT AND CONTROL

Acceptance sampling and testing of the raw product should provide a basis for accepting or rejecting the material as well as information on how to blend, screen, and sort the material. Information obtained during this part of the operation should in turn indicate whatever modification may be required in the filling, processing, or packaging procedures.

5.1 Quality Attributes

The quality attributes referred to as *sensory* may readily be classified in accordance with the human senses by which they are perceived: sight, touch, taste, and smell. These subjective evaluations are accomplished whenever anyone uses a product.

Objective measurement is free of the human element. However, unless certain principles are adhered to, major errors may be committed that could exceed any error resulting from human evaluation. Objective measurements are utilized for those "hidden" attributes including nutritive value, harmless adulteration, toxicity, biological agents, flammability, reactivity, and incompatibility.

The hidden characteristics of quality are those the consumer cannot evaluate with the senses, and yet are of real importance to health and economic welfare.

5.2 Manual Testing

Subjective testing has a useful role in quality control, but since the human instrument is used, it suffers from the human limitations of being influenced by environmental conditions, state of health of the tester, lack of absolute point of reference, tendency for comparative rather than absolute evaluations, and personal bias.

When an attribute of quality is clearly defined, and the physical, chemical, anatomical, or physiological basis clearly understood, it is possible and desirable to utilize an instrument that is capable of measuring the same attribute directly. The material tested may be organic or inorganic, live or inert, hazardous, safe, or even beneficial.

Two principles are involved in any type of measurement—precision and accuracy. *Precision* is the ability to duplicate results, with one instrument or with different individual units of the same instrument, while *accuracy* refers to the degree of correlation with primary standards or human evaluation.

5.3 Instrumental Procedures and Subjective Testing

An instrumental procedure may serve as the precursor to a more elaborate instrument that not only measures the quality level, but also maintains production within prescribed quality control limits by automatic sorting or adjustment. An instrumental procedure may therefore be considered as a first step in the automation of a process.

An instrument can be made merely to measure a particular quality attribute. In such instances, the value obtained may be utilized to ascertain the quality level of the lot, or if obtained on the production line, it may be used as an indicator for adjustment.

6 INDUSTRIAL HYGIENE ASPECTS OF QUALITY CONTROL

Quality control personnel may be faced with potential exposures to materials before any other plant personnel are exposed, and all too frequently industrial hygiene awareness is lacking among these personnel. The industrial hygiene aspects of assay and quality control require employee awareness of toxicity and biological risks, corrosivity, flammability, reactivity, and incompatibility. Instrumental procedures are virtually required, and in this area skill and sophistication are necessary.

6.1 Quality Assay of Toxic Materials

A large number of common substances are acute respiratory risks and should not be sampled in unventilated areas without respiratory protective equipment. It is also notable that simple asphyxiants such as nitrogen, methane, ethane, or hydrogen can dangerously reduce the oxygen content of air and, except for nitrogen, can create explosive mixtures.

6.1.1 Reaction Kettle and Process Line Sampling. Quality control personnel in chemical plants are sometimes exposed to toxic concentrations of gases or vapors during sampling from reaction kettles or process lines. It has been the practice in many operations to sample through the kettle manhole or large opening in the reaction vessel, which permits the escape of considerable gas volumes.

The use of kettle manhole ventilation will provide satisfactory control during sampling periods. If the manhole is being used only for sampling and not additions of raw materials or for process observation, it is not necessary to use the large volume of exhaust air required for manhole control. There are several methods by which sampling may be performed with a minimum of exposure:

- Sampling may be accomplished by the use of vacuum. By this method, the reaction vessel can be kept closed and the only exposure that occurs will be when the sample is drained into a container. This exposure is of short duration and will not ordinarily constitute a health problem unless extremely toxic materials are encountered.
- Sampling may also be accomplished by the provision of a valve on the reaction kettle or in process pipe lines. This method is not always satisfactory, especially when a series of samples must be taken, as its difficult to flush lines properly and obtain representative samples.
- It is not usually necessary to obtain large volumes of material for sampling. Samples may be taken through small valved nozzles in the top of the reaction kettle or by a chain and sample container. The sample nozzle need only be opened during the sampling period. It is usually unnecessary to provide exhaust ventilation for this operation. Slot ventilation utilizing a minimum air volume, however, will provide satisfactory control if necessary.
- If desired, samples may be taken through a small hand hole in the reaction vessel.

For additional information, see Chapter 4.

These methods of sampling pertain to reactions carried out under negative atmospheric or slight positive pressures. For sampling reactions carried out under higher pressures, an adaptation of the vacuum sampling method may be used. In this case, the pressure in the reaction vessel will provide the energy necessary to elevate the liquid to be sampled.

The sampling lines utilized in vacuum and pressure sampling should be of small diameter so that a representative sample may be obtained with a minimum of flushing of the lines.

6.2 Control Procedures for Biological Agents

Control procedures must be comprehensive in that infestations may begin with the raw product before arrival at the plant and continue through each step of the production line until all the product is used. Thus the quality control department may have to determine the extent of such contamination, and control measures at several inspection areas may have to be exercised.

Contamination by biological agents in the food, pharmaceutical, and cosmetics industries is considered a serious defect by the Food and Drug Administration who allow zero or near zero tolerances for such contamination. The presence of live pathogenic organisms is a potential health risk that is strictly controlled by regulatory agencies.

6.3 Special Considerations

6.3.1 Flammability. Flammability determination is needed for obvious reasons. The unrestricted volume and container size of flammable materials presents the potential for intense or prolonged fires that can be prevented with prudent storage practices. The Occupational Safety and Health Administration (OSHA) regulations address the types and sizes of containers permissible. National Fire Protection Association (NFPA) standards concentrate on the amount of solvents that may be stored outside of approved flammable-liquid storage rooms or cabinets. Recommended storage limits according to NFPA guidelines depend on a facility's size and available fire protection equipment.

Other considerations in the handling of flammable materials incude ensuring that accidental contact with strong oxidizing agents such as chromic acid, permanganates, chlorates, perchlorates, and peroxides is not possible; and that sources of ignition are excluded.

6.3.2 Explosive and Shock-Sensitive Materials. Organic peroxides are among the most hazardous substances normally handled in industry. As a class, they are low-power explosives, hazardous because of their extreme sensitivity to shock, sparks, to other forms of accidental ignition. Certain types of compounds from peroxides: ethers, aldehydes, and chemicals containing benzylic hydrogen atoms or allylic (CH_2=$CHCH_2R$) structures.

Peroxides form by the reaction of a peroxidizable compound with molecular oxygen through a process of autoxidation or peroxidation. The reactions are highly dependent on chemical structure. For instance, most ethers containing an alkyl group bonded to an oxygen atom tend to peroxidize readily and present a significant risk. In contrast, ethers having an aromatic group bonded to an oxygen atom do not generally peroxidize and can be handled more routinely.

If a bottle of isopropyl ether, for example, is found on a little used storage shelf, the ether may have evaporated, leaving some visible white crystals at the bottom of the bottle and around the area of the screwcap. It is quite dangerous to attempt removal of the cap to confirm the presence of peroxide since such action might provide the friction or mechanical shock necessary to cause detonation.

Ethers must never be distilled unless known to be free of peroxides. Containers of diethyl or diisopropyl ether should be dated when they are opened. If peroxides are found, the material should be decontaminated or destroyed.

Appropriate action to prevent injuries from peroxides in ethers depend on adequate labeling and inventory procedures, personal protective equipment, and suitable disposal methods. Table 12.1 is a partial listing of peroxide-forming chemicals.

6.3.3 Incompatible Materials. Certain combinations of chemicals are remarkably explosive, poisonous, or present a high risk in handling in some other way. Table 12.2 may serve to alert the quality control personnel during conduct of their sampling and analysis work as well as production workers during theirs.

Table 12.1

Peroxide-Forming Chemicals	
*Acetyl	p-Diethoxybenzene
*Allyl ether	1,2-Diethoxyethane
Allyl ethyl ether	*Diethoxymethane
*Allyl phenyl ether	2,2-Diethoxypropane
*iso-amyl benzyl ether	*Diethyl ether
n-Amyl ether	Diethyl ethoxymethylenemalonate
p-n-Amyloxybenzoyl chloride	*Diethyl fumarate
*Benzyl n-butyl ether	*Diisopropyl ether
*Benzyl ether	*1,1-Dimethoxyethane
*Benzyl ethyl ether	1,2-Dimethoxyethane
Benzyl methyl ether	*Dimethoxymethane
*Benzyl 1-naphthyl ether	*2,2-Dimethoxypropane
Bis(2-n-butoxyethyl) phthalate	*Dioxane
Bis(4-chlorobutyl) ether	*1,3 Dioxepane
1,2-Bis(2-chloroethoxy) ethane	2,4-Dinitrophenetole
Bis(2-chloroethyl) ether	*Di-n-propoxymethane
Bis(chloromethyl) ether	1,2-Epoxy-3-phenoxypropane
Bix(2-ethoxyethyl) ether	*1,2-Epoxy-3-isopropoxypropane
Bis(2-ethoxyethyl) phthalate	p-Ethoxyacetophenone
Bis(2-methoxyethyl) adipate	2-(2-Ethoxyethoxy) ethyl acetate
Bis(2-ethoxyethyl) adipate	2-Ethoxyethyl acetate
Bis(2-(2-methoxyethoxy) ethyl) ether	2-Ethoxyethyl o-benzoylbenzoate
Bis(2-methoxyethyl) carbonate	1-Ethoxynaphthalene
Bis(2-methoxyethyl) ether	o-Ethoxyphenyl isocyanate
Bis(2-methoxyethyl) phthalate	p-Ethoxyphenyl isocyanate
Bis(2-phenoxyethyl) ether	3-Ethoxypropionitrile
2-Bromoethyl ethyl ether	Ethyl ether
β-Bromophenetole	Ethyl β-ethoxypropionate
o-Bromophenetole	Ethyl vinyl ether
p-Bromophenetole	n-Hexyl ether
3-Bromopropyl phenyl ether	o-Lodophenetole
t-Butyl ethyl ether	p-Lodophenetole
t-Butyl methyl ether	*Isoamyl benzyl ether
n-Butyl phenyl ether	Isoamyl ether
n-Butyl vinyl ether	Isobutyl vinyl ether
*Chloroacetaldehyde diethylacetal	*Isophorone
*2-Chlorobutadiene	*β-Isopropoxypropionitrile
1-(2-chloroethoxy) 2-phenoxyethane	*Isopropyl ether
Chloromethyl methyl ether	Isopropyl 2,4,5-trichlorophenoxyacetate
β-Chlorophenetole	3-Methoxy-1-butyl acetate
o-Chlorophenetole	2-Methoxyethyl acetate
p-Chlorophenetole	β-Methoxypropionitrile
*Cyclohexene	Methyl p-n-amyloxybenzoate
*Cyclooctene	n-Methylphenetole
*Decalin	m-Nitrophenetole
*p-Dibenzyloxybenzene	Oxybis (2-ethyl acetate)
*1,2-Dibenzyloxyethane	Oxybis (2-ethyl benzoate)
p-Di-n-butoxybenzene	β-β-Oxydipropionitrile
1,2-Dichloroethyl ethyl ether	Phenoxyacetyl chloride
2,4-Dichlorophenetole	α-Phenoxypropionyl chloride

Table 12.1 (*Continued*)

Peroxide-Forming Chemicals	
m-Diethoxybenzene	*p*-Phenylphenetole
o-Diethoxybenzene	Phenyl *o*-propyl ether
n-propyl ether	Thiethylene glycol diacetate
n-Propyl isopropyl ether	Thiethylene glycol dipropionate
*Tetrahydrofuran	*1,3,3-Trimethoxypropene
*Tetralin	*Vinylidene chloride

*Denotes chemicals that form peroxide with ease.

Table 12.2. Incompatible Materials

DO NOT CONTACT:

Alkali metals, such as calcium, potassium, and sodium with water, carbon dioxide, carbon tetrachloride, and other chlorinated hydrocarbons.

Acetic acid with chromic acid, nitric acid, hydroxy-containing compounds, ethylene glycol, perchloric acid, peroxides and permanganates.

Acetone with concentrated sulfuric and nitric acid mixtures. Acetylene with copper (tubing), fluorine, bromine, chlorine, iodine, silver, mercury, or their compounds.

Ammonia, anhydrous with mercury, halogens, calcium hypochlorite, or hydrogen fluoride.

Ammonium nitrate with acids, metal powders, flammable fluids, chlorates, nitrates, sulfur, and finely divided organics or other combustibles.

Aniline with nitric acid, hydrogen peroxide, or other strong oxidizing agents.

Bromine with ammonia, acetylene, butadiene, butane, hydrogen, sodium carbide, turpentine, or finely divided metals.

Chlorates with ammonium salts, acids, metal powders, sulfur, carbon, finely divided organics or other combustibles.

Chromic acid with acetic acid, naphthalene, camphor, alcohol, glycerine, turpentine, and other flammable liquids.

Chlorine with ammonia, acetylene, butadiene, benzene, and other petroleum fractions, hydrogen, sodium carbides, turpentine, and finely divided powdered metals.

Cyanides with acids.

Hydrogen peroxide with copper, chromium, iron, most metals or their respective salts, flammable fluids and other combustible materials, aniline, and nitro-methane.

Hydrogen sulfide with nitric acid, oxidizing gases.

Hydrocarbons, generally, with fluorine, chlorine, bromine, chromic acid, or sodium peroxide.

Iodine with acetylene, fulminic acid, hydrogen.

Nitric acid with acetic, chromic and hydrocyanic acids, aniline, carbon, hydrogen sulfide, flammable fluids or gases, and substances that are readily nitrated.

Oxygen with oils, grease, hydrogen, flammable liquids, solids and gases.

Oxalic acid with silver or mercury.

Perchloric acid with acetic anhydride, bismuth and its alloys, alcohol, paper, wood and other organic materials.

Phosphorous pentoxide with water.

Potassium permanganate with glycerine, ethylene glycol, benzaldehyde, sulfuric acid.

Sodium peroxide with any oxidizable substances; for instance, methanol, glacial acetic acid, acetic anhydride, benzaldehyde, carbon disulfide, glycerine, ethylene glycol, ethyl acetate, furfural, etc.

Sulfuric acid with chlorate, perchlorate, permanganates and water.

6.4 Precautionary Measures Utilized by Well-Trained Quality Control Personnel

- Keep chemicals from the hands, face, and clothing.
- Keep hands and face clean. Wash thoroughly with soap and water whenever a chemical contacts the skin. Always wash face, hands, and arms after sampling and before leaving the laboratory.
- Do not smoke, drink, or eat in laboratories or chemical storage areas.
- Clearly label all containers. Do not use any substances in an unlabeled or improperly labeled container. Printed labels that have been partly obliterated or scratched over or crudely labeled by hand should be relabeled before using. Unlabeled containers should be disposed of promptly with care to avoid adverse reactions.
- Carefully read the label before removing a reagent from its container. Read it again as you promptly recap the container and return it to its proper location. Names of distinctly different substances are sometimes nearly alike and using the wrong substances can lead to accidents.
- To avoid violent reaction and splattering while diluting solutions, always pour concentrated solutions slowly into water or into less concentrated solutions while stirring.
- Become generally familiar with the biological and physical properties of materials, and practive good industrial hygiene principles to minimize exposure or safety risk.

GENERAL REFERENCES

Committee on Hazardous Substances in the Laboratory, Assembly of Mathematical and Physical Sciences, *Prudent Practices for Handling Hazardous Chemicals in Laboratories*, Washington, D.C.: National Academy Press, 1980.

CRC Handbook of Laboratory Safety, 2nd ed., Boca Raton, FL: CRC Press.

Juran, M. M., and F. M. Gryna, Jr., *Quality Planning and Analysis*, 2nd ed., New York: McGraw-Hill, 1980.

Juran, J. M. (ed)., *Quality Control Handbook*, 3rd ed., New York: McGraw-Hill, 1974.

Steere, N. V. (ed.), *Safety in the Chemical Laboratory*, Easton, PA: American Chemical Society, 1974.

CHAPTER THIRTEEN

Electroplating

Robert D. Soule

1   INTRODUCTION

Electroplating is the process by which a metal coating is applied through the action of an electric current. A metallic salt dissolved in an aqueous system dissociates into ionic species, that is, electrically charged chemical groups. If two electrodes are placed in this solution and an electrical potential difference is applied across them, the ions will move through the solution to the oppositely charged electrode. The positively charged ion, the cation, contains the metallic part of the salt. When this ion reaches the negatively charged cathode, it loses its charge and becomes a neutral metal atom that can deposit on the cathode. The operation of building up a layer of one metal on another by this process is called electroplating.

The metallic ions are removed from solution as they are neutralized and deposited at the cathode. Therefore, additional metal must be added to solution in order to make up for this loss and maintain a source of metallic ions. This can be accomplished by solubilization of the anode, if it is the same metal as that being deposited at the cathode, or by addition of metal salts to the solution.

The migration of ions in solution, in terms of chemical and physical principles, has been described in detail by Faraday in his well-known and precise laws of electrolysis. Although it is beyond the scope of this discussion to review these principles in detail, they state that the weight of metal deposited by electroplating processes is proportional to the quantity of electricity passed to the cathode, that is, the current density. Assuming the same efficiency for a given current density, the weight of metal deposited is proportional to its chemical equivalent. Since the rate of deposition will vary with current density, the higher the density, the faster the plating will occur. Thus, within limits, the thickness of deposit can be controlled by adjusting the current density and plating time. Another factor that is important is *throwing power,* a term referring to the ability of the system to deposit metal in remote areas and recesses on workpieces. Cyanide solutions have become common in use because they exhibit superior throwing power. Acid and alkaline solutions possess relatively good throwing power; chromium solutions are quite poor.

Almost any metal can be deposited on another. Copper and steel wire are often coated with tin. Automobile bumpers and other trim usually are nickel plated and then finished with a thin chromium plating. The complete process for the chromium plating of steel is depicted in Fig. 13.1; the pretreatment steps identified are discussed later in this chapter.

Although the electroplating process can be performed in a variety of ways, the basic operation consists of passing a direct electric current from an anode to the object to be plated through the electrolyte containing dissolved salts of the plating metal in solution. Figure 13.2 depicts a simple electroplating circuit. In addition to the metallic salt, other salts, acids, and alkaline materials usually are added to the solution to improve the electroplating action. Generally, the anode is a piece of pure metal of the type to be deposited, and the object to be plated serves as the cathode. Both anode and cathode are suspended in the plating solution in the tank by special racks. The electroplating tanks usually are lined with rubber, polyvinyl chloride, or a vinyl resin as protection against attack by the plating solution.

It is critical that the metal workpieces to be plated are in finished form. Electroplating is not a surface improvement process; a series of pretreatment steps always is required. Some of the more common ones are discussed in the following section. Several common electroplating processes are described in Section 3: cadmium, chromium, copper, gold, nickel, silver, tin, and zinc. Although essentially any metal can be electroplated, only a few are used on a production level. The potential health hazards associated with the various electroplating processes are discussed in Section 4; some unique hazards are mentioned in the section describing processes. Section 5 consists of a presentation of control measures necessary for safe operation of electro-

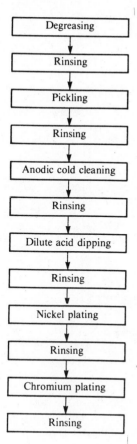

Figure 13.1. Flow diagram of chromium plating of steel.

plating processes; engineering controls, including local exhaust ventilation, mist reduction techniques, and isolation, as well as personal protection, are described.

2 BASE METAL PRETREATMENT

The metal workpieces that are to receive the plated metal must have been pretreated, sometime extensively. Physically, the workpieces must have been worked and formed, machined to desired configurations, and provided with appropriate finishing operations. Next, the pieces must be processed by one or more of several operations that are intended to remove surface contamination. These pretreatment steps are primarily chemical in nature and include solvent or vapor degreasing, alkali cleaning, acid cleaning, and pickling. Some additional physical work such as polishing and buffing might be necessary in some cases. Most of these unit operations are described elsewhere in this book. Although wide variations in methods have been used to prepare the base metals for subsequent plating operations, relatively simple cycles of operations are suitable for the vast majority of conditions encountered in the electroplating

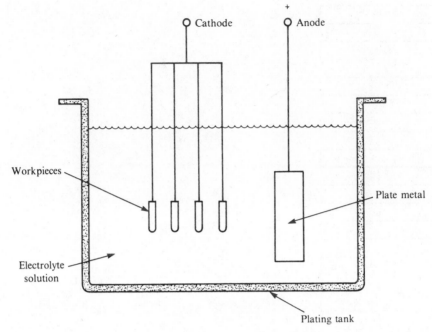

Figure 13.2. A basic electroplating circuit.

industry. These simplified, standardized steps are presented briefly in this section. Unless stated otherwise, the base metal preparation methods discussed in this section have been found to be appropriate for all forms of the indicated base metals.

2.1 Low-Carbon Steels

Depending on the nature and amount of soiling, low-carbon steel components usually are precleaned by vapor degreasing, soaking in emulsifiable solvent cleaners, or soaking in alkaline cleaning solutions. Spray cleaning in power washers may be used if there are appreciable amounts of soils or persistent drawing compounds or lubricants present. Precleaning is an important step in the overall process, and it is here that operating mistakes are made often. One common mistake is to overwork alkaline cleaning baths by using a single tank for removing the bulk of the soil and also for electrocleaning, discussed later. Another common mistake is to introduce parts into the electrocleaning solution before removing heavy rust deposits. Preferably, the rust should be removed by pickling in strong acids after the initial precleaning already mentioned. This limits the work to be done during the acid-dipping step that follows electroplating to simply neutralizing the alkali and removing any thin smut films that might be present.

Following precleaning, the low-carbon steels are subjected to electrocleaning in an alkaline solution. Anodic (i.e., reverse current) cleaning is preferable for the low-carbon steels. This step is relatively short in duration, typically 2 min or less. Fairly strong cleaners can be used at temperatures near the boiling point. If the metal parts are not cleaned sufficiently in this step, it is a clear indication that the precleaning was not thorough. Complete rinsing of the metal is essential following the electrocleaning step, preferably by using two, separate, water-rinse tanks.

Acid dipping is the final pretreatment step for low-carbon steels. This step, in which the metal is dipped into a hydrochloric acid bath for a few seconds, activates the surface and removes any thin smut films that might be present. Following the acid-dipping step, the metal is rinsed twice in cold water. Hot water is not used to avoid drying of the parts because of retained heat before they can be placed in the electro-plating tank. If it is not possible to take the parts directly from acid dipping to electroplating, they can be stored in a weak solution of either sodium cyanide or, if tin plating is to be done, sodium hydroxide.

2.2 High-Carbon Steels

High-carbon steels have a greater tendency to become embrittled in some of the pretreatment steps than do the low-carbon steels. These difficulties increase with increasing hardness of the steel, whether this is produced by the heat treating or cold working of the material. The hardened steels should be stress relieved prior to being prepared for electroplating. Heating for 30 min at about 400°F (200°C) usually is sufficient for this purpose. Case-hardened steels should be considered high-carbon steels since the low-carbon surface is transformed into a high-carbon material. Generally, steps in which hydrogen is generated, such as pickling and cathodic cleaning, should be limited to minimum duration.

Precleaning of high-carbon steel components is much more important than with other steels. Surface conditions are likely to vary over a wider range than with other metals. Thus, tumbling, abrasive blasting, wire brushing, and other pretreatments are used to obviate prolonged pickling treatments. Normally, vapor degreasing or solvent cleaning to remove any oil left on the parts should precede the pretreatment steps. If electrocleaning is used in the precleaning stage, it should be anodic (reverse current) to avoid hydrogen adsorption.

After precleaning, high-carbon steels should be cleaned anodically to avoid em-brittlement due to hydrogen. The cleaner should be used near boiling at moderate current densities and minimal cleaning times. The rinse following treatment should be done with heated water, but not so hot that metal surfaces are dried by the retained heat.

Acid dipping should be limited to the minimum needed to remove the final traces of oxides on the metal surfaces. Hydrochloric acid is preferred; minimal use of in-hibitors is recommended to avoid interference with adhesion in subsequent electro-plating steps. If a smut film results from the acid-dipping step, it can be removed by an anodic treatment in a cyanide bath. Caustic soda solutions or anodic electrocleaning also can be used if the amount of smut is small. In any event, thorough rinsing of the parts must follow this step.

Etching steps in the pretreatment process can assure good adhesion on high-carbon steels. An anodic treatment in sulfuric acid should be used with high current densities and essentially room temperature for short exposure times. Cold water should be used for rinsing following this step; addition of anhydrous sodium sulfate to the rinse bath is beneficial for many steels.

2.3 Stainless Steels

Precleaning of stainless steels presents complications due to the presence of a thin, transparent, and tenacious film on the metal. This film reforms quickly on exposure to air (or other oxidizing conditions) even after it has been removed. It is important to lay down an initial layer of plated metal before this film can reform. Thus, it is necessary to use an "activating" step immediately prior to the electroplating itself.

When the electroplating department receives stainless steel parts, they usually are in formed condition and, therefore, the heavy drawing and forming lubricants used to facilitate shaping must be removed completely. These precleaning steps may involve one or more operations such as vapor degreasing, power spray washing in alkaline baths, soak cleaning in alkaline solution, or solvent emulsion cleaning. After polishing and buffing, the removal of soil left from these operations can be accomplished by the normal techniques used with other metals.

Anodic cleaning is preferred for stainless steels, although it is a difficult operation. Where bright plating on stainless steel is desired, care must be taken to see that the alkalinity, current density, and temperature of the bath all are kept as low as possible, particularly when high chromium alloys are being treated. The degree of treatment should be only that needed to do a satisfactory job.

As already mentioned, stainless steel must be subjected to an activation step immediately prior to electroplating in order to remove the passive oxide film that is present. A variety of activation treatments is known, ranging from simple immersion processes to relatively sophisticated steps that simultaneously activate and electroplate the metal. Generally, the activation is accomplished with an acidic solution, most commonly sulfuric or hydrochloric acid.

Rinse waters used in conjunction with pretreatment of stainless steels should be maintained acidic. Normal carryover from the treatments in the sequence of pretreatment steps usually takes care of this problem. The rinsing steps should be conducted as quickly as possible to avoid the passivation of the surface.

2.4 Aluminum and Aluminum Alloys

Aluminum alloys usually preclude use of the normal preparatory treatments used for other metals because of the film of oxide that forms immediately on the surface of even freshly cleaned aluminum. The wide range of alloying elements added to aluminum, not all of which react similarly to a single treatment system, dictates some variations in treatment. The electroplater must know the type of alloy being treated to assure proper results.

The most common method for pretreatment of aluminum and its alloys is the immersion zincate treatment. With this process, the base metal is exposed to an alkaline system (sodium hydroxide and zinc oxide) with modifier additives in some cases, depending on the alloy composition. Generally, the metal is treated with the zincate solution twice; the first film being dissolved using nitric acid followed by rinsing and re-zincating. Although silver, brass, zinc, nickel, or chromium can be deposited directly over the zincate film, a Rochelle-salt-type cyanide copper solution ordinarily is used as a strike prior to the actual electroplating step.

Anodizing is used instead of zincating for some aluminum alloys; the deposits so produced tend to resist soldering temperatures. This process involves cleaning and deoxidizing as usual, immersion without current in the acid anodizing solution at room temperature, anodizing for a few minutes at low current density, rinsing, and treatment with an acid copper bath and copper plating. The metal then is ready for plating.

2.5 Zinc-Based Die Castings

The specific precleaning steps for zinc-based metal alloys depend on the type and character of the soiling to be removed, which can range from a fairly uniform, thin film of easily removed grease to the presence of hard, caked-on, partially carbonized buffing compounds packed into holes and recesses. Because these two extremes require different degrees of precleaning, it is not practical to outline an exact cleaning cycle. Precleaning operations that are used include vapor degreasing, emulsion cleaning, and cleaning in di-phase cleaners.

Electrocleaning incorporates both anodic and cathodic methods. The anodic process is preferred since any film produced during cleaning is removed more easily in the subsequent acid-dipping operation. The cleaning time should be kept to the minimum needed to produce the desired degree of cleanliness, and thorough rinsing should follow all alkaline cleaning operations.

Normally, acid dipping follows the electrocleaning step. The primary function of this step is to neutralize the alkali clinging to the metal from the preceding step. Thus, it is necessary to use only very mild acids such as dilute sulfuric, hydrochloric, hydrofluoric, fluoboric, or citric acids. Acid dipping should be followed by thorough rinsing.

2.6 Beryllium Copper

The pretreatment steps for alloys of beryllium copper typically consist of a fairly standardized sequence. The parts are immersed in sulfuric acid solutions at near boiling temperatures to loosen the dark scale that is present and then dipped in a strong nitric acid bath to remove the loosened scale. Next the parts are immersed in a solution of sulfuric acid, acetic acid, and hydrogen peroxide after which they are dipped into a chromic acid–sulfuric acid bath. Finally, the metal is immersed in concentrated phosphoric acid at a temperature slightly above room temperature. The sequence described provides a pickling and brightening of the metal surface. The metal usually is dipped into a sodium or potassium cyanide solution and copper strike prior to actual electroplating.

2.7 Other Base Metals

The most common base metals and alloy systems on which electroplating is performed have been mentioned in the preceding sections. Most other metals can be electroplated, and many are. The most significant among the remaining metals are lead and lead alloys, leaded brass and bronze, molybdenum, nickel, titanium, and tungsten. To a large extent, and with the exception of lead-bearing systems, the pretreatment of these metals resembles that for stainless steels. The typical treatment for lead-containing alloys involves the following sequence of steps: cathodic alkaline cleaning, thorough rinsing, anodic alkaline cleaning, thorough rinsing, dipping in fluoboric acid, water rinsing, and final electroplating.

3 COMMON ELECTROPLATING PROCESSES

The fundamental electroplating system consists of a tank that contains a metal salt of the metal to be applied, dissolved in an aqueous solution. Two electrodes powered by a low-voltage (4–12 V), direct-current power supply are immersed in the electrolyte. The anode is either an inert electrode or a bar of the metal to be deposited, and the cathode is the workpiece on which the metal is to be plated. When the power is applied, the metal ions deposit out of the bath on the cathode or the workpiece. The anode may be designed to replenish the metallic ion concentration in the bath. In addition to the salt containing the metallic ion, the electroplating bath may contain salts to adjust the electrical conductivity of the bath, additives that govern the type of plating deposit, and/or buffers for control of the pH.

Individual parts or racks of small parts are hung manually from the cathode bar. If many small pieces are to be plated, the parts usually are placed in a perforated plastic container in electrical contact with the cathode bar, and the container is immersed in the bath. The parts are tumbled during the electroplating process to achieve a uniform deposition of metal on the workpiece.

The purpose of this section is to describe briefly the basic processing conditions of eight of the most common electroplating processes and to discuss the peculiar health hazards associated with the operations. Finally, some of the other electroplating processes not discussed in detail are mentioned from the standpoint of their industrial hygiene problems.

3.1 Cadmium

Most cadmium plating is done using a cyanide solution, usually made up in a separate mixing tank and transferred to the plating tank when needed. For still plating, a cathode current density of 20–50 A/ft^2 (200–500 A/m^2) is used, depending on the extent to which the bath is agitated. The maximum anode current density usually is about 20 A/ft^2 (200 A/m^2). The temperature of the bath is maintained just above room temperature, about 80–85°F (27°C). The metal content of the plating bath must be held between the desired limits determined for the type of operation necessary, such as high throwing power, high efficiency, and speed of plating. Free cyanide content must be controlled within fairly narrow limits, more cyanide being required with each incremental increase in bath temperature. For maximum throwing power, there is a need to balance three operating parameters: cadmium metal content, total cyanide content, and current density; typical values for these, for maximum throwing power, are 2 oz/gal (14 g/L), 17 oz/gal (120 g/L), and 10 A/ft^2 (100 A/m^2), respectively.

For barrel plating, the operating voltage must be larger than for still plating, typically in the range of 6–12 V. All other operating conditions are the same as outlined for still plating. Depending on the concentration and temperature, the current density usually is about 6 A/ft^2 (60 A/m^2). Under these conditions, deposition of a sufficient amount of cadmium requires about 30 min.

Brighteners of a variety of types are used: gelatin, coumarin, furfural, dextrin, milk, sugar, aromatic aldehydes, and metallic types containing nickel, cobalt, or selenium compounds.

Some cadmium plating is done using fluoborate solutions rather than the cyanide systems previously described. The fluoborate bath has the advantage of plating at very high current densities, greater than 50 A/ft^2 (500 A/m^2). It is a very stable bath, has essentially 100% current efficiency, and results in virtually no hydrogen embrittlement. These solutions are used commonly where barrel plating is to be done since control of the processing conditions is straightforward. The major disadvantage of the fluoborate system is poor throwing power. The fluoborate bath contains typically cadmium fluoborate, ammonium fluoborate, and boric acid, in order of decreasing concentration in the bath. Cast aluminum anodes are recommended. For still plating, agitation of the bath during plating, either mechanically or with air, is desirable when high current densities are used. Cast or malleable iron parts can be plated with excellent results with this solution, provided there are no deep recesses on the parts.

The nature and extent of the contaminants released from the cadmium electroplating processes differ significantly between the cyanide and fluoborate systems. Because of the high efficiency of the cyanide-based system, there is very little gassing and therefore little contamination of the atmosphere above the plating tank. In fact, unless elevated temperatures and/or unusually high current densities are used, there frequently is minimal need for local exhaust ventilation at the plating tank itself; pretreatment tanks may require local control, however. In the case of fluoborate electroplating of cadmium, the contaminants released by the operation consist mainly of the fluoborate-based mist. In the typical cadmium electroplating operation, application of the control measures discussed elsewhere in this chapter can provide adequate control of workers' exposures.

3.2 Chromium

Perhaps the most versatile and popular of the electroplating processes is chromium plating because of the range of results and applications possible. Applications include preparation of black chromium surfaces, protective coating on various base metals, and decorative finishing of metal workpieces. Although the chemical solution used in chromium plating is very simple, the mechanics of the plating process are complex and beyond the scope of this presentation. Basically, the plating bath is chromium trioxide (chromic acid), water, and a catalyst without which chromium would not be deposited. The catalyst typically is a sulfate, usually provided in the form of sulfuric acid, or a fluoride, provided as fluosilicic acid. The ratio of chromic acid to catalyst usually is in the range of 100:1; lower ratios are preferred by some hard chromium plating operations, whereas the higher ratios are employed by decorative plating operations in order to achieve better coverage over some base metals. The total amount of chromic acid and catalyst in the plating solution is determined on the basis of desired rate of deposition, conductivity, and covering power.

Many electroplaters make use of proprietary baths, which tend to be self-regulating and simplify the analyses required to perform the plating. A typical conventional fluoride bath contains about 24 oz/gal (170 g/L) CrO_3, 0.2 oz/gal (1.4 g/L) sulfate, and 0.1 oz/gal (0.7 g/L) fluoride. The fluoride serves as a supplemental catalyst; sulfate must be present in the solution. The basic composition of this bath can be varied to meet particular plating requirements.

3.2.1 Black Chromium Plating. Black chromium plating refers to the deposition of a gray-black plate on a base metal, using a chromic acid-based solution. Much of the material treated with black chromium is intended for use in military application where nonreflective coatings are needed. The typical black chromium plating bath contains chromic acid, acetic acid, barium acetate, and water. Moderate current densities, commonly about 50 A/ft² (500 A/m²) and slightly elevated temperatures, approximately 110°F (43°C), are used. The throwing power of this system is very good, comparable to that of conventional chromium plating baths.

Many proprietary black chromium solutions are available, many of which are capable of doing a much better job than the acetic acid formula. Generally, the coverage and final color of the deposit are better, and process control is much easier. These packaged systems can be used over nickel-plated base metal to give excellent corrosion protection as well as to serve the many functions that a black coating performs.

3.2.2 Decorative Chromium Plating. Decorative chromium plating refers to the process of providing a nontarnishing film over an appropriate base metal. Most such plating is done with copper–nickel or nickel base, the nickel usually being of the bright variety. Chromium provides a good corrosion-resistant film directly over stainless steels, nickel–silver, or Monel. The electroplating solution used for decorative plating is that described for general chromium plating. The use of the fluoride-based baths, because of their self-activating features, simplifies the plating over nickel as compared to the conventional sulfate bath.

3.2.3 Hard Chromium Plating. Hard chromium is applied for functional reasons rather than for decorative purposes, as was the case with the previous processes. Chromium plate is hard, dense, has a coefficient of friction lower than any other metal, has excellent corrosion resistance to most materials, and exhibits good antigalling properties as long as it is not used against itself. These properties can be imparted to a metal piece, the base metal properties dictating the thickness of the chromium deposit needed to achieve the desired results. Despite the hardness of the chromium plate, it should not be used as a general substitute for the hardening of base metals; to do so requires proper engineering of the process.

3.2.4 Barrel Chromium Plating. Barrel-plating operations, in which small parts are coated in bulk, present unique plating problems when using chromium solutions. To achieve complete coverage in the chromic acid-based solutions, precise control is needed. Solutions used for barrel plating must have the ability to plate bright through numerous current interruptions. Thus, some type of fluoride-containing solution is normally used. The fluosilicate form is used almost universally, although other forms of fluoride can be used. A typical barrel chromium solution contains chromic acid, fluosilicate, and sulfate. The bath is operated at room temperature, voltage of 6–12 V, and current densities depending on the type of contact between metal pieces and the electrolytic circuit. A decorative deposit usually can be achieved in about 10 min under these conditions.

3.2.5 Health Hazards. Chromium electroplating solutions are notoriously poor in terms of the cathode current efficiency. Whereas efficiencies of most electroplating operations are 90% or better, efficiencies of about 20% are common with chromium baths. This means that a significant portion of the energy to the electrode is devoted to gassing with consequent release to the atmosphere, entraining chromic acid and other bath components in the resulting mist. The significance of contaminants released from these plating solutions is discussed elsewhere in this chapter. As a general rule of good practice, any tank that contains chromium plating baths should be provided with local exhaust ventilation capable of maintaining a capture velocity of about 150 fpm (46 mpm) at the most remote point of contaminant release from the bath. In addition to ventilation, the other elements of a total worker protection program detailed later in this chapter are normally necessary.

3.3 Copper

By proper selection of plating formulation, copper can be plated onto a variety of base metals, depending on the desired properties of the part. With acid copper plating, the copper usually is present as copper sulfate, and the solution is maintained acidic by sulfuric acid. Alkaline copper plating is done by using plating baths containing copper cyanide or copper pyrophosphate, the first being moderately alkaline and the latter only slightly alkaline.

3.3.1 Acid Copper Plating. Copper plating has been accomplished with a variety of acid-plating formulations; sulfate–oxalate–boric acid, sulfate, sulfate–oxalate, fluosilicate, and sulfamate formulations all have been used. The sulfate and fluoborate baths are used most extensively since they are simple, easy to operate, and straightforward to control; the following discussion is limited to these two systems.

The copper sulfate baths contain copper sulfate and sulfuric acid; the relative and absolute amounts of these ingredients vary depending on the application of the plating operation. Additives include thiourea, dextrin, hydrochloric acid, molasses, and wetting agents, all in very low concentrations. Temperatures range from room temperature to about 120°F (49°C), voltage ranges between 1 and 4 V, and current densities range from 20 to 200 A/ft² (200–2000 A/m²), the extent of agitation present in the tank determining the density required. The additives mentioned are used for such purposes as grain refinement, leveling, hardening, brightening the deposit, increasing limited current density, minimizing treeing, and minimizing the effects of impurities that may get into the solution. Arsenic and antimony can cause brittle and rough deposition of copper; gelatin tends to inhibit the codeposition of these metals with copper. Nickel and iron reduce the conductivity of the plating bath; silicates tend to precipitate onto the work.

The acid copper plating solutions are capable of releasing the copper salt and sulfuric acid mist into the atmosphere. Electrode efficiencies generally are quite high, often greater than 95%; thus, the gassing and consequent creation of acid mist is minimal. Copper plating in acidic solution is conducted with low current densities, and agitation of the bath normally is not needed. Under these conditions, the need for local exhaust ventilation may be minimal; however, when high current densities or agitation is used, significant contamination of the air above the tank can occur, and local exhaust ventilation must be provided.

The fluoborate plating baths contain copper fluoborate as the active component; the conventional solution used for still plating contains about 30 oz/gal (210 g/L). When barrel plating is done, the concentration is increased by half again, and, when electroforming or wire plating is done, the concentration is doubled. The fluoborate bath usually is operated slightly above room temperature, voltage usually is in the range of 3–12 V, depending on the application, and current density typically is about 125 A/ft² (1250 A/m²). Since electrode efficiencies are nearly 100%, maintenance of the bath is simple. The only major adjustment needed is occasional addition of fluoboric acid to maintain the pH of the bath in the desired range. The copper fluoborate bath is capable of operating over a wide range of concentration, temperature, current density, and solution pH, thus making it possible to modify the deposits to meet the conditions imposed by a particular job requirement. Usually, it is not necessary to use additives although they can be used to modify the deposits.

The high efficiency of the electrodes tends to minimize the release of contaminants from the plating baths, fluoborate salts and water being the primary components. The general concerns for worker health discussed for sulfate-based solutions apply here also.

3.3.2 Alkaline Copper Plating. When alkaline systems are used for copper plating, either cyanides or pyrophosphates usually serve as the source of the metal. The cyanide baths consist of three types: plain copper cyanide (low efficiency), medium Rochelle copper cyanide (medium efficiency), and high metal copper cyanide (high efficiency). The first two have excellent throwing power, good coverage, and can deposit adherently directly on steel and zinc workpieces. In fact, one of their major functions is to serve as a "strike" to form a good bonding medium for subsequent deposition of copper by high-efficiency, pyrophosphate, or acid plating processes.

The pyrophosphate bath has general characteristics similar to the high-efficiency cyanide bath, that is, electrode efficiency is essentially 100%, the plating rates are substantial, and thick deposits are produced easily. The throwing power is quite good but, as with the high-efficiency cyanide bath, a strike is needed and control is difficult. Adhesion produced by a pyrophosphate bath over a strike is not as good as with high-efficiency cyanide baths.

PLAIN CYANIDE BATHS. The plain cyanide baths, composed of copper cyanide, sodium cyanide, and sodium carbonate in aqueous solutions, are of limited use. They are very sensitive to changes when operated at low concentrations, low temperatures, and low efficiencies. Usually, it is best to increase the temperature to favor higher cathode efficiency. If problems of control continue, it is probably best to change to a Rochelle bath.

ROCHELLE BATHS. The limitations of the plain cyanide bath just discussed can be overcome by use of the Rochelle bath, typically containing copper cyanide, sodium cyanide, sodium carbonate, sodium hydroxide, and Rochelle salt in aqueous solution. This bath serves as either a plating bath or a strike. Anode corrosion and finer-grained deposits are promoted by the Rochelle salt, and there is a need for closer control of processing limitations. Elevated temperatures, about 150°F (60°C), are used; current densities of approximately 15 A/ft² (150 A/m²), are used; current densities of approximately 15 A/ft² (150 A/m²) at the anode and 30 A/ft² (300 A/m²) at the cathode are common. Continuous plating is done in Rochelle baths, where the anode and cathode efficiencies are closer than in the plain cyanide baths. Higher current densities, lower metal concentrations, and lower temperatures may be used to favor striking. Increasing the metal concentration and temperature tends to increase the overall efficiency of the plating process over that achievable with plain baths.

Close monitoring of the anodes should be provided to avoid polarization and loss of anode efficiency. Steel anodes, equal to about 5% of the total anode area, can be used to avoid excessive polarization of the copper anodes. The bath may be operated at temperatures up to 170°F (77°C) to increase plating rates and anode efficiency. However, the higher temperatures will increase decomposition of cyanide and formation of carbonate. Overall efficiencies usually can be improved with cathode agitation.

HIGH-EFFICIENCY BATHS. The high-efficiency baths in which the concentrations of metal and free cyanide are optimized overcome the anode problems by increasing the cathode efficiency to equal that at the anode, virtually 100%. This is done with high metal content, low free cyanide, high temperature, and agitation. The high-efficiency bath contains copper cyanide, sodium cyanide, and sodium hydroxide; no carbonate or Rochelle salt is added. Potassium salts can be substituted for sodium salts; in fact, potassium salts offer some advantages over sodium. Conductivity and cathode efficiency are improved by using potassium salts. Frequently, mixed potassium and sodium salts are used. The high-efficiency baths tend to be more susceptible to contamination by organic materials than are the lower concentration baths.

When a cyanide bath is used for strike or bright copper plating, both cyanide and alkaline mist are released from the bath, the tendency increasing as the concentrations of metals and cyanides are increased. The potential for release of significant concentration of these materials into the workroom atmosphere is great enough that installation of local exhaust ventilation should be required. The general engineering controls and administrative measures for worker protection apply here.

PYROPHOSPHATE BATHS. Copper pyrophosphate forms highly soluble and conductive complexes when dissolved in potassium pyrophosphate solution. These solutions, which are only slightly alkaline (pH = 8.4), can produce commercial plating with almost 100% efficiency and good throwing power. In use and application, the pyrophosphate baths are similar to the high-efficiency cyanide baths. The pyrophosphate bath must be preceded by a strike before the plating of steel or zinc. A cyanide strike or a pyrophosphate strike may be used. In most other aspects, pyrophosphate systems resemble the high-efficiency cyanide baths.

3.4 Gold

Gold plating solutions of many types have been used in the past, the most common one being the alkaline cyanide bath. For the most part, these systems were used to deposit decorative finishes on base metals; therefore, the gold was deposited in very thin layers, called "flash" coatings. With advances in the electronic and allied industries during the past two decades, there has developed a need for thicker coatings of gold. In addition, the coatings are required to have greater hardness, wear resistance, corrosion resistance, and lower porosity. The following discussions distinguish between plating systems for decorative and industrial purposes.

3.4.1 Decorative Gold Plating. In plating of jewelry and other decorative items, thin deposits, or flash, are applied usually to bright nickel-plated brass or other base metal parts to achieve a desired color and appearance. Gold flashing also is done over jewelry parts fabricated from rolled gold plate or gold-filled strip or sheet, consisting of a thin layer of 10- or 12-karat gold over a nickel–copper–zinc alloy. Gold is flashed over these parts to cover edges that do not contain gold, to prevent the low-karat gold from tarnishing, and to obtain desired color.

Many plating bath formulations are available for decorative gold-plating applications. Almost all flashing baths are based on alkaline cyanide solutions containing gold, free cyanide, and one or more of a variety of other metals. Proprietary solutions are available to produce essentially any desired end result. Because of the high cost of gold, it is important that the gold deposit be uniform over the surface of the plated part. Current densities should be as low as possible and proper anode area and spacing is important, particularly for rack plating. The speed of rotation of the barrel in barrel plating is important. The fundamental design of the barrel to accentuate good mixing of the parts, transfer of plating solution, and contact with the cathode is critical. Rack design is more important than with other metal plating. The solution selected for the plating of particular parts should be custom-fitted to the base metal and alloy used.

Virtually all of the gold electroplating processes are based on alkaline cyanide solutions. As discussed earlier, whenever cyanide baths are used, there is a potential for generation and release of highly toxic contaminants. Compounding the situation somewhat in the case of gold-plating processes is the elevated temperature normally associated with the plating operation, increasing the ease with which contaminants become airborne.

3.4.2 Industrial Gold Plating. Gold deposits are produced for industrial applications as either pure (24 karat) gold or gold alloys. These deposits may be hard or soft and may be of high purity or may contain traces of codeposited metals. The processes available for electroplating pure gold include hot cyanide, neutral, and acid baths. Plating temperatures typically are in the 120–150°F (50–65°C), range and current densities are low, rarely in excess of 5 A/ft^2 (50 A/m^2). The relative amounts of the bath components can be varied in order to produce the desired physical properties in the plated metal.

The principal advantage of the hot cyanide baths is economy since the potassium gold phosphate represents gold in its most economical form. The disadvantages of these solutions include poor bath stability, solutions that cannot be used with printed circuit boards, deposits that are subject to staining, and solutions that tend to become contaminated due to attack on copper alloy substrates. To overcome the disadvantages of the hot cyanide solutions, baths with lower pH values have been developed. Neutral solutions, based on potassium gold phosphate, sodium phosphate, potassium phosphate, and nickel as potassium nickel cyanide, have been used successfully. Generally, the acid baths have reduced cathode efficiencies, particularly if the baths become contaminated with base metals or organic impurities. Advantages of the acidic baths include brightness of deposits, relative ease of plating, no staining problems, and low porosity of the deposit.

Gold alloy plating processes have been developed to obtain gold-based deposits with specific characteristics. Generally, deposits of gold alloys are harder, brighter, and less ductile than pure gold deposits. Although the basic plating processes are similar to those described for pure gold plating, the systems are modified by addition of metallic salts needed to provide the metal to codeposit with gold.

Correct preparation of the base metals prior to gold plating is critical since the parts often are subjected to severe test conditions and the cost of rejects is high. All parts should receive an initial gold strike from an acid gold bath prior to gold plating. Nickel alloys require strong pickling with hydrochloric acid and, sometimes, cyanide activation prior to plating. Copper alloys often require acid pickling followed by conventional bright dipping or cyanide dipping prior to gold plating. Many techniques have been developed to assure proper treatment of base metals prior to gold plating.

For best results, the preparation process should be as simple as possible to remove all traces of soil and oxide from the part and to activate the metal surface so that it will readily receive gold with good adhesion.

3.5 Nickel

The conventional nickel-plating solution, referred to as Watts bath, consists of nickel sulfate, nickel chloride, and boric acid. In addition to these essential components, antipitting agents or wetting agents are used to improve surface uniformity of electrodeposited nickel and minimize surface pitting. Typical nickel-plating conditions are bath temperatures of about 140°F (60°C), current density of 50 A/ft² (500 A/m²), and pH of about 4. Agitation of the solution during electroplating is recommended.

The basic Watts bath can be modified to produce various desired properties of the plated piece. In an all-chloride bath, which contains nickel chloride and boric acid, corrosion-resistant materials must be used in components of the tanks, heaters, and pumps. An all-sulfate bath, incorporating nickel sulfate and boric acid, is used when insoluble anodes are required, as when plating internal surfaces of tubing or pipe. Under such conditions, chlorides are avoided to prevent evolution of chlorine gas and corrosion of the insoluble anode, usually lead. A sulfate-chloride bath, which contains equivalent amounts of sulfate and chloride ions, also is used. Corrosion-resistant equipment must be used because of the high chloride content. Bright nickel-plating baths, based on the Watts formulation and containing the same basic constituents, are operated and controlled in the same way as conventional Watts baths already discussed. Modern bright nickel-plating baths rely on a combination of a number of additives since one compound alone cannot provide the degree of brightness, leveling, and internal stress control required for practical nickel plating. Efforts to improve organic additives continue to be made in order to increase the ease and reliability of production-scale bright nickel plating.

In addition to the nickel sulfate-based solutions, nickel sulfamate baths are in extensive use. The primary constituents are nickel sulfamate, boric acid, and nickel chloride. Additives may include wetting agents, hardeners, and stress relievers. The formulations are of the Watts bath type, with nickel sulfamate replacing nickel sulfate. Operating conditions typically are bath temperatures of about 140°F (60°C), current densities in the range of 20–40 A/ft² (200–400 A/m²), depending on whether agitation is used as recommended.

Many special-purpose baths are used to produce nickel deposits with interesting and useful properties. These include formulations for barrel plating, baths for producing purely decorative, black nickel deposits, and baths for producing nickel–phosphorus alloys electrolytically.

From an industrial hygiene viewpoint, it can be generalized that all baths discussed are operated at moderate temperatures, low to moderate current densities, and have cathode efficiencies of approximately 95%. The contaminants likely to be released via the mist above the plating solutions include nickel sulfate, sulfamate, and borate, fluoborate, or hydrofluoric acid mist, depending on the exact bath composition being used. Usually, the gassing from the plating solutions using sulfate and/or chloride baths is low, and the need for local exhaust ventilation under such conditions may be minimal. However, good industrial hygiene practice would dictate that the overall control measures discussed in the following section should be considered until there is strong evidence that the potential health hazards are being controlled adequately.

3.6 Silver

In recent years, much attention has been given to improving the physical properties of electrodeposited silver to meet specific functional requirements, particularly in the electronics industries. Most of these improvements have been in bright silver-plating and proprietary processes for providing platings with improved characteristics; their use is increasing in both the technical and decorative applications.

Conventional silver-plating solutions contain potassium silver cyanide, free or uncombined potassium cyanide, and potassium carbonate as the major constituents. Most base metals will precipitate silver by immersion in the regular silver-plating baths, thus resulting in poorly adherent deposits. To avoid this problem, it is common to use strike solutions of very much lower silver content and high free cyanide content as an initial treatment. Strike solutions are operated at room temperature, voltages of 4–6 V, and current densities of about 20 A/ft^2 (200 A/m^2). Articles made of nickel–silver usually are given a strike in a nickel strike prior to silver striking. Stainless steel must be activated by application of this strike or by cathodic treatment in a solution of sulfuric acid.

The regular silver-plating solutions commonly used fall into two basic groups: conventional low-metal baths and high-metal, high-speed baths. Conventional baths contain silver cyanide, potassium cyanide, potassium carbonate, and various additives primarily for brightening. These baths are operated at room temperature and with agitation provided by cathode rod oscillation or by pumping the solution through a perforated pipe at the bottom of the tank. Sufficient silver anode area should be provided so that it is at least equal to the surface area being plated. Anode current density should not exceed about 12 A/ft^2 (120 A/m^2) to avoid polarization effects that can lead to undue decomposition of cyanide, carbonate buildup, and lowering of silver content in the bath. In the absence of brighteners, the deposit is mat. Brighteners, such as carbon disulfide and ammonium thiosulfate, should be added in carefully controlled amounts.

High-speed baths contain silver cyanide, potassium cyanide, potassium carbonate, and potassium hydroxide. These solutions are used almost exclusively for building up heavy deposits, such as those required in bearings and electroforming work. Brighteners such as ammonium thiosulfate and other proprietary materials are commonly used. These baths are operated at elevated temperatures, moderate-to-high current densities, and with substantial amounts of agitation. The potassium hydroxide aids in increasing the stability of the solution, improving the silver deposit, and in minimizing anode corrosion.

The bright silver-plating solutions differ from conventional baths in using a higher silver content and a significantly higher free cyanide content. Two additives usually are involved also, one a primary brightener and another serving to enhance the action of the brightener. These systems are proprietary in nature, but are readily available commercially.

Close monitoring of the silver and free cyanide content is necessary for proper operation of any silver-plating solution. This applies to both the strike solutions and, even more importantly, the bright silver solutions. Temperature control is important and frequently requires use of cooling water circulating through tubing in the bottom of the plating tanks. In potassium solutions, carbonates can begin to show adverse effects on the current density. Removal of excess carbonates can be accomplished by use of barium cyanide. Much of the trouble experienced in silver plating can be traced to poor housekeeping practices. Adequate attention must be paid to cleanliness of connections at anode and cathode bars. Good distribution and maintenance of sufficient anode area are important. Regular or continuous filtration of the solution to remove dirt and suspended matter frequently is necessary.

The silver-plating processes all are operated at nearly room temperature and low current densities and are characterized by cathode efficiencies of nearly 100%. Although cyanide baths are used, the combination of operating conditions just described results in very little gassing, and the release of contaminants from the plating baths is negligible. However, it is prudent to conduct a complete assessment of the potential health hazards in any plating operation.

3.7 Tin

Electroplating of tin usually involves the use of one of three common plating solutions: acid baths, alkaline (stannate) baths, or fluoborate baths. Each of these fundamental solutions is described in the following section.

3.7.1 Acid Baths. The composition of the acid bath used for electroplating of tin is based on the stannous sulfate electrolyte. The bath contains stannous sulfate and sulfuric acid, along with an agent for inhibiting the oxidation of the stannous ion in solution and organic agents necessary for producing smooth deposits. Examples of antioxidation additives are phenolsulfonic acid and cresolsulfonic acid; among materials used to produce smooth deposits are gelatin and β-naphthol. The plating operation is conducted at room temperature, using a voltage of less than 1 V, and cathode current densities ranging between 10 A/ft² (200 A/m²) for still baths and 100 A/ft² (1000 A/m²) for agitated baths.

The acid bath maintains its composition over long periods of time, except for the consumption of the additives. The only significant losses are from dragout and slow oxidation of the stannous sulfate. The anode efficiency normally is slightly greater than the cathode efficiency; this offsets the loss of stannous sulfate through dragout. Thus, in controlling the bath composition, the additive materials are the most important considerations.

3.7.2 Stannate Baths. The tin-plating solutions based on stannate are alkaline, using sodium and potassium stannates, and are used extensively in general job plating and manufacturing. Depending on the application intended, the plating solution will contain potassium stannate, or sodium stannate, potassium hydroxide, or sodium hydroxide, and many additives needed to produce specific desired properties in the plated coating. Conventional plating conditions include bath temperatures of the range 140–180°F (60–80°C), voltages in the range of 4–6 V, and current densities in the range of 20–100 A/ft² (200–1000A/m²). Bath temperature and current density do affect the character of the deposit; the effects become of some importance when very thick deposits are being produced. Although high temperatures favor efficiency at both anode and cathode, they also tend to make the deposits somewhat nodular.

3.7.3 Fluoborate Baths. Because of the high solubility of stannous fluoborate, electroplating baths of very high metal content can be used, a condition of great importance where extremely high plating rates are desired. The baths are capable of operation over a wide range of current densities at, or slightly above, room temperature. The throwing power of the solution is very good and, since the current efficiencies at both anode and cathode are essentially 100%, the bath is practically self-sustaining. Typically, the fluoborate baths contain stannous fluoborate, fluoboric acid, and additives such as peptone and β-naphthol. Voltage on the tank is low, rarely greater than 3 V, and current densities for practical operation range between 25 and 125 A/ft² (250–1250 A/m²). Higher current densities can be used if the bath is provided with mechanical agitation.

3.7.4 Health Hazards. Contaminants released into the air from tin-plating solutions depend on the plating solution used and the operating conditions employed. Among the general contaminants likely to be released are the tin salt mists and steam (from alkaline baths), fluoborate salts and fluoboric acid (from stannous fluoborate baths), and tin halides (from some acid baths). Because of the relatively high electrode efficiencies and the inherent low toxicity of inorganic tin compounds, the potential health hazards associated with tin-plating operations are minimal. However, the general strategy for control of worker exposures outlined in the following section of this chapter should be followed.

3.8 Zinc

Zinc plating is used extensively for application of protective coatings on strip steel, pipe, wire, wire mesh, steel rod, and many other forms. Most of the zinc-plating solutions are based on acid sulfate baths, although cyanide and fluoborate baths are used as well.

The acid sulfate baths tend to be less expensive to operate than other plating baths, but their application is somewhat limited due to their poor throwing power. Zinc deposits from acid baths are whiter than deposits from cyanide baths but are susceptible to metallic impurities in the electrolyte which, almost always, codeposit with zinc. This situation can be detrimental because of the possibilities of galvanic corrosion. On the other hand, deposits from acid baths are more adaptable to malleable iron castings and spring steel parts. In most formulations, the anode and cathode efficiencies are nearly 100%; efficiencies of alkaline baths are much lower.

The acid-plating baths contain zinc sulfate, other sulfates such as sodium, magnesium, or ammonium salts, and various additives used to achieve particular properties and brightening. The basic solution, using sodium or ammonium sulfate, produces a whiter deposit with smaller crystal structure. The addition of chlorides of zinc, sodium, and aluminum in place of sulfate salts makes the plating process more efficient but results in a more corrosive system. Zinc fluoborate, along with fluoborate and chloride salts of ammonia, is used for barrel plating of malleable iron parts. Regardless of the chemical composition of the acidic baths, the operating conditions are essentially the same. Bath temperatures are held at or slightly above room temperature, the pH of the solution is maintained between 3 and 4, and current densities are moderate, ranging from about 50 to 150 A/ft² (500–1500 A/m²).

The cyanide-based baths consist of solutions of zinc cyanide with certain amounts of sodium cyanide and sodium hydroxide. The exact amounts of the components vary. In general, the cyanide baths are characterized by much higher throwing power than is achievable with acid sulfate solutions. Although there is a wide range of adjustment possible among the zinc metal content, free cyanide content, and various additives, there usually must be a trade-off between overall plating efficiency and speed and throwing power or coverage from the bath. The cyanide bath is recommended for use with still plating operations, although it can be used for barrel plating.

Aside from the hazards associated with the cyanide baths, potential health hazards are minimal with zinc-plating operations. Airborne contaminants likely to be released from solution, although varying from one system to another, include fluoborate mists and steam from fluoborate baths and chloride and/or sulfate mists from acid baths. Unless unusually high current densities or bath temperatures are used in conjunction with acid-plating baths, the need for local exhaust ventilation may be minimal. On the other hand, when cyanide baths are used, the efficiency at the cathode drops substantially from that achievable with acid baths, resulting in greater amounts of gassing and misting from the solutions. Cyanide and alkaline mists are likely to be released from these latter solutions, and the control measures discussed in the final section of this chapter should be implemented.

4 POTENTIAL HEALTH HAZARDS IN ELECTROPLATING

The pretreatment and electroplating processes for the various base metals and electroplating metals contain many potential health hazards. The overall health hazards posed by a particular operation depend on these treatment solutions and the metallic composition of the workpieces. The hazards associated with the pretreatment processes described are, for the most part, covered elsewhere in this book in discussions of other specific unit operations such as pickling, degreasing, and alkaline cleaning. The primary focus of this section is on the health hazards associated with the electroplating operations themselves; only passing reference is made to pretreatment steps.

The primary source of air contamination in electroplating operations is the release of components of the electrolyte by virtue of the gassing that takes place in the bath. The health significance of the mist generated by the electroplating process depends on the contents of the bath. The following sections discuss the more significant specific contaminants associated with common electroplating operations.

4.1 Acid/Alkali Mists

Aqueous solutions of several inorganic acids and caustic materials are used extensively in electroplating, either as necessary pretreatment steps or in the actual electroplating step. Acids used routinely include boric, hydrochloric, hydrofluoric, and sulfuric acid; alkaline solutions are based primarily on either sodium or potassium hydroxide.

Detailed discussions of the health hazards associated with acid and alkaline mists appear in the section on metal-cleaning operations. The primary effect of these materials is irritation of the eyes, nose, and upper respiratory tract. Of greater significance, usually, is the fact that the mist containing the acid or alkali also contains other substances that, depending on the composition of the electrolyte, may be of greater hazard.

Sampling and analysis for acid mists can be accomplished by collecting the mist in distilled water with subsequent analysis using a specific-ion electrode. In some cases, it may be meaningful to collect samples of airborne mist in water and conduct titrametric analysis. Usually, however, the presence of several acidic or basic components makes this approach unworkable. Direct-reading indicator tubes and reagent kits are available for limited use in measuring levels of specific acid gases. In the majority of plant operations involving acids or alkaline materials, it probably is safe to say that other substances in the environment are of greater concern, and specific sampling for the acidic or alkaline materials is done only rarely. Control of these materials of greater concern provides inherent control of the acid or alkali mists as well.

4.2 Chromic Acid

Because of its relatively high toxicity and the popularity and extensive use of chromium plating for protective and decorative finishing of a variety of base metals, special mention is made in this section of the health hazards associated with chromic acid, CrO_3. In comparison to other plating processes, chromium-plating operations are very low in efficiency. Depending on specific operating conditions, as much as 90% of the total energy provided to the electrolytic bath may be directed to the dissociation of water at the electrodes. This results in severe gassing from the solution with consequent high potential for exposure of the operators to chromic acid mist. Although other constituents of the bath may be entrained along with the gas, chromic acid normally is the most significant in terms of potential health hazard. The compositions of the electroplating solutions used for chromium electroplating are discussed earlier in this chapter.

Chromium can exist in various oxidation states in the airborne contaminants released from the chromic acid-plating solutions. Of these, the hexavalent chromium is of greatest concern toxicologically. Sampling and analysis techniques for assessing quality of the workroom environment must ensure that hexavalent chromium is collected and determined correctly. Such methods include both wet collection, using either distilled water or alkaline solutions in impingers or bubblers, and dry collection in which the chromic acid mist is collected on filters. The use of the liquid scrubbing solutions is not practical for personal, breathing zone sampling in many cases; filtration offers the greatest collection efficiency and ease in the majority of cases. Several "wet-chemistry" methods are available for analysis of collected samples; however, they tend to be tedious. Atomic absorption spectrophotometry is not subject to the interferences from materials such as cyanides, organic matter, and reducing agents (iron, copper, and molybdenum) but determines the total chromium present in the sample. Distinction between hexavalent chromium and chromium in other oxidation states cannot be made from a sample applied directly to the spectrophotometer. The most reliable analytical method is one in which the collected sample is treated initially to separate the hexavalent chromium from chromium in other oxidation states and then presented to the atomic absorption unit.

Since the control measures applicable for chromic acid mists are common to most of the potential health hazards associated with electroplating operations, they are discussed generally in the following section of this chapter. The controls discussed must be implemented to a high degree of effectiveness for operations handling chromic acid.

4.3 Dusts

Dusts can be dispersed into the atmosphere of an electroplating shop in one of three basic ways: during the handling and use of powdered, granular, or flake materials used as additives or electrolyte compounds; as the result of evaporation of water from the mist released from the tanks; and due to the secondary dispersal or resuspension of settled dust. Much of the airborne dust can be identified qualitatively, and specific sampling and analytical methods should be used to document airborne concentrations and determine the potential health hazards. Permissible exposure limits have been developed for many of the materials likely to be present and should be used in interpreting the results of sampling. However, even for the most innocuous dusts, some minimum level of dust control should be considered. The philosophy of nuisance dust control expressed in the publications of the ACGIH's Threshold Limit Values Committee should be administered.

Sampling for airborne dusts can be done quite easily using common battery-powered pumps and filters of various available matrices, chosen to be compatible with the analytical requirements. For assessment of total dust concentrations, simple gravimetric analysis may be sufficient. In cases where the composition of the dust is known, specific analyses for the various components may be conducted. Many of the dusts present in the electroplating shop atmosphere can be analyzed by determining the amount of the metallic element present. Various wet-chemistry techniques also are available.

Control of the dust in the electroplating department depends on the nature of the processes involved and the manner in which the dust is released to the workroom air. Generally, a combination of local exhaust and general ventilation as the primary control features is used. Local exhaust ventilation is critically required at locations where solid materials are transferred and at the tanks where gassing and misting are particular problems. The secondary dispersal of settled dust into the environment can be minimized by use of effective local exhaust ventilation in the first place, and reliance on a program of general good housekeeping throughout the plant. These considerations are discussed in more detail in the following section of this chapter.

4.4   Hydrogen Cyanide

Electroplating operations make extensive use of cyanide solutions. These solutions are prepared and adjusted as necessary by addition of sodium or potassium cyanide salts. These salts are highly toxic, and contact by workers, particularly by way of ingestion during weighing and transfer operations, must be avoided. Of extreme concern is the potential for generation and release of hydrogen cyanide or cyanide salt mists from the electroplating process. The potential for formation of hydrogen cyanide exists whenever the cyanide compound comes into contact with acidic solutions or materials.

Common sources for inadvertent release of hydrogen cyanide from the electroplating plant exist. One example is mixing of acids and cyanides in storage, weighing, or transfer activities prior to addition to the electroplating solutions. Another is the potential release from rinse tanks used in the plating process, particularly if proper control of pH is not enforced. Thus, it is critical that storage of raw materials used to prepare electroplating solutions and treatments be done so as to avoid contact between cyanides and acids; invariably, this requires some type of isolated facility. Similarly, it is important that the rinse operations following treatment of metal pieces in cyanide solutions be operated so as to prevent the acidification of rinse water. Because of the possibility of inadvertent contamination of rinse tanks and the serious consequence of release of hydrogen cyanide, local exhaust ventilation should be installed at these operations.

Air sampling and analysis have been accomplished in a variety of ways, most of which present some difficulties to the investigator. Of available methods, the most useful involves collection of the sample by passing the air through an impinger or bubbler containing an appropriate alkaline solution. The resulting sample then can be analyzed by use of a specific-ion (cyanide) electrode. Although this is the most common technique used for cyanide sampling, it presents two significant problems. First, it is not practical to sample in a worker's breathing zone because of the liquid system used to collect the sample. Strategic positioning of the bubbler in the approximate breathing zone can overcome this problem. The second problem is interference by sulfide ion in the analytical procedure. Although sulfide might be collected along with the cyanide in sampling workroom air, it can be removed from the sample solution prior to analysis if its presence is suspected.

4.5   Metallic Salts

A significant portion of the mist generated above the various electroplating baths consists of metallic salts. These mists contain salts of the plating metal and the anionic constituent of the bath. Industrial hygiene assessment of plating operations should devote considerable attention to these metallic species.

Permissible exposure limits have been developed over the years for most metals and many metallic salts. With the exception of gold, all plating metals discussed in this chapter have had acceptable exposure limits recommended by at least one organization involved in establishing such limits. These limits vary greatly from one metal to another, and it is not possible to generalize about the level of control necessary.

Determination of airborne concentrations of metallic substances is a relatively straightforward process with modern sampling equipment and analytical support. Samples can be collected by drawing air through high-efficiency membrane filters and analyzed readily, with little preparation, by means of atomic absorption spectrophotometry.

Determination of the metallic salt in the total airborne mist is relatively insignificant in some cases. The presence of other substances in the mist frequently dictates the nature and extent of control measures needed. Thus, control of the most troublesome contaminant provides inherent control of the metallic compounds as well. Control measures common to electroplating operations are discussed in detail later in this chapter.

4.6   Skin Contact

Essentially all electroplating operations are conducted with solutions that are acidic, alkaline, or composed of moderately to highly toxic materials. Therefore, it is imperative that direct contact of the skin with these solutions be prevented. The results of exposure of the skin to plating solutions can range from a relatively mild direct irritation, such as that due to dilute alkaline solutions, to allergic dermatitis, after prolonged exposure to sensitizing materials like nickel, chromium, and their compounds, to a potentially cancerous lesion, reportedly possible following prolonged contact with chromic acid solutions.

Regardless of the effect, contact with plating solutions should be kept to an absolute minimum. Critical to the control of these chemicals in the workplace is a comprehensive program of issuing and requiring use of personal protective equipment in addition to the installation of effective engineering controls. Encouragement of workers to maintain a high level of personal hygiene is another important element of the overall program.

4.7 Other Potential Hazards

Additional hazards exist in most electroplating operations. One of these is the hydrogen and oxygen gassing that takes place at various rates because of dissociation of water occurring at the electrodes. This rate varies from one electrolytic system to another, but the potential for release of hydrogen and oxygen, and the consequent fire hazard, is present with essentially every plating process.

Although electroplating is an electrolysis process and one might anticipate electrical shock hazards, such is not the case. Very low voltages, typically in the range of 2–6 V, of direct current are used.

Normally, worker exposure to noise at plating operations is within acceptable limits. However, plating operations often are located within a large building adjacent to metalworking or metal-machining operations that are inherently noisy. Similarly, heat stress is a problem only in southern climates or in northern plants where the building encloses and houses other hot processes. Radiation and vibration are problems only rarely in electroplating operations.

The major hazards in the electroplating shops result from the handling of chemicals necessary for the process. Chief among these are the concentrated acids and alkaline materials when preparing the baths and the accidental mixing of acids with cyanides or sulfides during electroplating, bath preparation, and waste disposal with the formation of hydrogen cyanide or sulfide.

5 CONTROL OF HEALTH HAZARDS

The electroplating process is quite simple and has become standardized, varying essentially only with the plating solution and base metal to be plated. The health hazards of the materials released from the electroplating baths depend directly on the contents of the baths. However, the control strategy for electroplating operations tends to be the same from one shop to another, differing more in the degree to which control is necessary than in actual approach. Each of the components of a hazard control program is discussed in this section, presented in decreasing order of appropriateness. Table 13.1 summarizes the typical control elements required for the various electroplating systems discussed earlier in this chapter. For additional information, see Chapter 5.

5.1 Local Exhaust Ventilation

Clearly the most effective means of controlling airborne contaminants released from electroplating operations is local exhaust ventilation, positioned at or very near the tank containing the electroplating and auxiliary processes. The extent to which the ventilation must be employed will depend on the contents of the tank being ventilated, the base metal being electroplated, and the efficiency of the plating process. The latter consideration is important because it dictates the amount of gassing that will take place at the electrodes; the more efficient the plating process, the less gassing. For a

Table 13.1. Typical Controls Required for Various Electroplating Systems

System (Subsystem)	Local Ventilation	Mist Control	Personal Protection
Cadmium (cyanide)	Minimal	No	Yes
Cadmium (fluoborate)	Minimal	No	Yes
Chromium (black)	Maximal	Yes	Yes
Chromium (decorative)	Maximal	Yes	Yes
Chromium (hard)	Maximal	Yes	Yes
Chromium (barrel)	Maximal	Yes	Yes
Copper (acid)	Moderate	No	Yes
Copper (alkaline)	Moderate	No	Yes
Gold (decorative)	Moderate	No	Yes
Gold (industrial)	Moderate	No	Yes
Nickel	Minimal	No	Yes
Silver	Minimal	No	Yes
Tin (acid)	Minimal	No	Yes
Tin (stannate)	Minimal	No	Yes
Tin (fluoborate)	Minimal	No	Yes
Zinc (acid)	Minimal	No	Yes
Zinc (cyanide)	Moderate	Yes	Yes

discussion of the conditions associated with a specific electroplating system, refer to the discussions of the processes in Section 3.

The specific design of a ventilation system for an electroplating solution tank usually begins with reference to standardized practices that have developed over the years. The American Conference of Governmental Industrial Hygienists (ACGIH) through its Ventilation Committee, the American National Standards Institute (ANSI) through its Z9 Committee's efforts, the National Institute for Occupational Safety and Health (NIOSH), and, in fact, the Occupational Safety and Health Administration (OSHA) in its general industry standards, all prescribe specific local exhaust ventilation criteria for open-surface tanks. Briefly stated, these recommendations and, in the case of OSHA standards, requirements permit use of various hood configurations, such as enclosing hoods, lateral exhausts, and canopy hoods (under some conditions), provided the minimum capture or control velocity is established and maintained. The best arrangement typically consists of lateral slot-type exhaust systems positioned along a single edge or with two slots, one on each of the long sides of the tank, capable of maintaining slot velocities in the range of 1800–2000 fpm (550 mpm) and capture velocities ranging between 50 and 150 fpm (15–45 mpm), depending on contents of the tank and the amount of external air disturbance.

In addition to the tank contents, the manner in which the workers must interface with the process will influence the nature and extent of local exhaust ventilation. In small job-shop operations, the metal workpieces usually are transferred manually from tank to tank as needed for the plating operation. In production-scale operations, automatic transferring units, programmed to cycle the parts from tank to tank, are more likely to be used. In these latter situations, the worker's primary function is to load and unload the racks or baskets. The more automated the entire process becomes, the easier it is to incorporate enclosing hoods on the tanks, thus providing more effective control of the contaminants. Additionally, with the more automated systems, the worker typically is stationed at one loading position, is not required to monitor the activity in the tanks themselves and thus is not exposed to the same extent as the worker in a manually operated plating operation.

As mentioned earlier, the amount of local exhaust ventilation needed will depend on the bath contents and the amount of gassing that results. For example, gassing is almost negligible with nickel-plating solutions based on sulfates or chlorides. Therefore, the need for exhaust ventilation may be minimal on these tanks. At the other extreme, whenever a cyanide bath is used, particularly with alkaline systems, cyanide can be released easily, and local exhaust ventilation capable of providing a control velocity of 150 fpm (45 mpm) or greater must be used. In the final analysis, the proof of the effectiveness of the ventilation systems can be demonstrated by air sampling in the breathing zones of the operators. It is not uncommon to find situations where the ventilation system complies with recommended design, and yet there is evidence of overexposure of workers because of improper positioning of the worker in the air flow pattern created by the ventilation system.

As with all ventilation systems, periodic testing of the systems installed in electroplating shops should be performed. Qualitative tests, using smoke tubes or other means of determining air flow characteristics, and quantitative measurements, such as capture velocity, slot velocity, and hood static pressure, should be obtained on a regular basis. The effects of room currents and other sources of external air disturbance near the ventilation system should be considered in these periodic evaluations. In many cases, partitions can be installed to minimize the disruptive effects of cross drafts and thereby increase the effectiveness of the ventilation system. Finally, external features of the overall ventilation system should be evaluated; makeup air provisions, backflow dampers on any combustion devices, and air-cleaning components on the exhaust system should all be monitored on a regular basis.

5.2 Mist Reduction

Various attempts to minimize the amount of mist released from the electroplating tanks have been shown to be successful. Basically, the mist reduction approach involves either the addition of chemicals to the bath solution to reduce the surface tension of the plating solution or floating lightweight chips of plastic or other inert materials on the surface of the bath.

Many proprietary bath additives are now available to reduce the surface tension of the electrolyte, thereby reducing the misting. By reducing the surface tension of the plating solution, the surface-active substances help to retard mist formation and carryover by the hydrogen bubbles generated during plating. Fluorocarbon surface-active agents are available commercially for this application.

A layer of plastic chips, beads, or balls on the surface of the electrolyte bath provides a mechanism for trapping the mist released from the solution and causing it to drain back into the bath. To make this approach more effective, tanks using the layer of plastic chips should be provided with covers to reduce bath loss. Tests have shown the use of plastic chips or balls to reduce the concentrations of airborne contaminants to very low levels and, at the same time, result in substantial conservation of the bath solution.

Although these mist suppression techniques are helpful, they must not be considerd the primary control measure. Local exhaust ventilation normally is required to ensure healthful work conditions, even where plastic chips or other mist suppression methods are used.

5.3 Isolation of Reactive Chemicals

Special mention is made here, as part of the overall control plan, of the importance of physically isolating cyanide salts and acidic materials because of the critical consequences of inadvertent mixing of these materials. Although this is a critical feature in the design of storage areas, it applies as well to the need for thorough rinsing of workpieces after treatment with the cyanide solution to prevent the cyanide from coming into contact with any acidic medium.

5.4 Personal Protective Equipment

The issuance and informed use of personal protective equipment by electroplating operators is an important component of an overall health protection program. Generally, the minimum protective equipment includes gloves, apron, and boots made of rubber or other impervious materials, and chemical workers goggles. Aprons that come below the top of the boots should be worn. A complete change of clothing at the workplace should be available for all workers. If plating solutions are splashed onto the work clothing, the clothes should be removed immediately, the skin washed with copious amounts of water, and the worker should change into clean clothing. Showers and eyewash stations, serviced with tempered water, should be located strategically throughout the work area. Because of the wide range of chemicals handled, often in an open fashion, there is a hazard of skin contact by these chemicals. Skin problems can occur if good housekeeping and personal cleanliness are not maintained at a high level.

Various protective ointments, or skin barrier creams, can be helpful if used properly. For workers who operate chromium plating systems, for instance, an ointment consisting of lanolin and soft paraffinic wax sometimes is used to minimize the effect of chromium on the skin, including the nasal passages and other moist surfaces. Although these creams can be useful, they must be used in conjunction with good personal hygiene practices, for which there is no substitute.

As part of the personal protection package, personnel with proper medical training and knowledge should be involved in the health surveillance program providing periodic examination of the skin and respiratory system. Preemployment medical examinations of the hands, arms, and nose sometimes can identify persons who are likely to be sensitive to chemical action of the skin. If subsequent periodic examinations show that a worker is being affected by something in the environment, the worker should be removed from contact with that environment until the offending agent has been identified and controlled or other adequate protection provided. Where effects of skin contact by plating baths is particularly critical, such as with chromium plating baths, all cuts and abrasions on the hands and arms should be appropriately protected against contact.

GENERAL REFERENCES

Many excellent references are available for addressing the general industrial hygiene concerns (toxicology, air sampling and analysis, protective clothing, and many others) associated with the various electroplating processes discussed in this chapter. The following list consists of recent publications in this field, with particular emphasis on engineering control of electroplating processes.

Development of Design Criteria for Exhaust Systems for Open Surface Tanks, NIOSH Publication 75–108, 1975.

Emergency Eyewash and Shower Equipment, American National Standards Institute, ANSI Z358.1-1981.

Fundamentals Governing the Design and Operation of Local Exhaust Systems, American National Standards Institute, ANSI Z9.2-1979.

Good Work Practices for Electroplaters, NIOSH Publication 77–201, 1979.

Health and Safety Guide for Electroplating Shops, NIOSH Publication 75–145, 1975.

Huebener, D. J., and R. T. Hughes, Development of Push-Pull Ventilation: Plating, *Am. Ind. Hyg. Assoc. J.* **46,** (5), 1985.

Open Surface Tank Design Data, Industrial Ventilation: A Manual of Recommended Practice, 19th ed., ACGIH, Ventilation Committee, 1986.

The Recirculation of Exhaust Air . . . Symposium Proceedings, NIOSH Publication 78–141, 1978.

Validation of a Recommended Approach to Recirculation of Industrial Exhaust Air—Vol. 1 (Chrome Plating), NIOSH Publication 79–143B, 1979.

CHAPTER FOURTEEN

Fluidized-Bed Drying

Paul F. Woolrich

1 INTRODUCTION

Fluidization is the suspension and agitation of a bed of solid particles by a rising stream of gas. Because every particle is surrounded by gas, there is rapid heat transfer between the gas and the particles (material) and a mass transfer of water from the material to the gas.

As the gas velocity through the bed is increased, the pressure drop through the bed increases, and the bed expands. When the pressure drop equals or slightly exceeds the weight of the bed, the entire bed is suspended or is in the fluid state. Superficial space velocities from 0.5 to 10 or more fps (0.15–3 mps) are commonly used.

The fluid-bed process has other useful applications as well as drying of materials. Although this chapter deals only with fluidized-bed drying, other applications exist:

- Agglomeration: It is possible to agglomerate materials ranging from fine powder to coarse grain into granules. A suitable liquid is sprayed onto the fluidized material causing the particles to stick together. The granules thus formed are then fixed by drying.
- Coating: Individual particles can be coated to change the product properties or for protection. The coating substance is dissolved in a suitable liquid and is sprayed onto the fluidized bed. In the aerospace industry, fluidized-bed coating

has been utilized to fuse silicon to alloys that would otherwise oxidize at temperatures of 3000°F (1650°C). This chapter deals only with fluidized beds for drying purposes.

- Chemical reaction: Fluidized beds are widely used for gas-phase/solid-phase reactions.

2   OPERATION

2.1   Pressure Operation

The schematic shown in Fig. 14.1 depicts a fluidized-bed dryer that utilizes indirect steam heat for *pressure* operation. Air flows from the fan or compressor (a), through the heater (b), into the plenum chamber (c), up through the fluidized bed (d), and exhausts through the dust collector (e). Solids flow into the fluidization chamber via the feed connection and out a discharge nozzle and from the dust collector. A distribution plate at the bottom of the fluidized bed is specially designed for the specific operation.

Typical applications for pressure operation fluidized-bed dryers include chemicals, polymers, fertilizers, and other granular or powder materials. Various methods of feeding, discharging, and collecting the dust are used, depending on the particular material handled.

2.2   Suction Operation

The schematic shown in Fig. 14.2 depicts a fluidized-bed dryer that utilizes indirect steam heat for *suction* operation. A suction fan draws the air from outside into the machine. This air first passes a prefilter unit (HEPA filter system) where foreign particles are removed. While passing through the cooler, the air temperature is reduced. The mist filter demists the air and retains water droplets. Subsequently, the air heater will heat the air flow.

The mixing flap enables the drying temperature to be accurately regulated. The

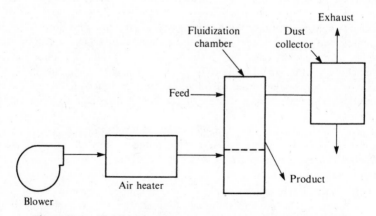

Figure 14.1.   Fluidized-bed dryer components—pressure operations.

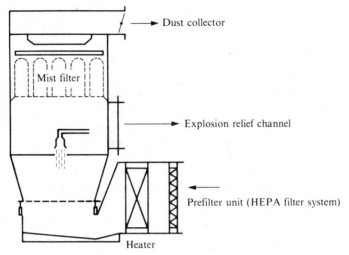

Figure 14.2. Fluidized-bed dryer components—suction operations.

warm and cold air streams are mixed proportionally in the subsequent air mixer to achieve an even temperature across the whole air duct.

This conditioned air then passes from the bottom upward through the moist material in the product container removing the moisture. There is a distribution plate in the floor of the product container above which a stainless steel wire mesh is fitted. The filter installed above the product container prevents fine particles of dust from escaping. The dust particles are automatically returned to the fluid-bed zone throughout the whole working process by a time-controlled shaking device.

3 APPLICATIONS

There are a large variety of products that use fluidized-bed dryers, including (a) pharmaceutical products and products requiring high purity; (b) foods (sugar, cocoa, coffee, tea, soups, starch); (c) microorganisms (yeast, soil bacteria); (d) organic chemicals (detergents, enzymes, intermediate products); (e) inorganic chemicals (Kaolin, gypsum, salts, metal oxides); (f) agrochemicals (fungicides, herbicides, insecticides, fertilizers); (g) plastic materials (polyolifines, PVC, accelerators, stabilizers); (h) dye stuffs (organic and inorganic pigments); and (i) natural products (flours, coal powder).

4 ADVANTAGES

The following are the principle advantages of fluidized-bed dryers:

- Specific product treatment can be optimized to save costs of investment, building, and installation.

- Good manufacturing practices of cleanability and avoidance of cross contamination can be fulfilled.
- Low maintenance is required.
- Organic solvent vapors, gases, and dust-laden emissions are easily contained using appropriate equipment.
- Optimum product quality is obtained because of short handling times and thermal load.
- Particle size classification is possible.
- Automatic control and supervision of the process are routine.

In the pharmaceutical industry, the product coming from a blender can be directly fed into a product container and hydraulically lifted into the dryer. After drying, the product can be transported to a blender and then to the tablet or other dosage form equipment.

5 LIMITATIONS

One disadvantage of fluidized-bed dryers is the fundamental requirement that the product can be fluidized. This means free flowing and, for good fluidization, a top particle size of approximately $\frac{3}{8}$ in. (1 cm). When extremely low gas velocities are required, the size of the fluidizer chamber may become exorbitant.

6 INDUSTRIAL HYGIENE ASPECTS

In industries where fluidized beds are utilized, the industrial hygienist can play a vital role. Industrial hygiene problems arise primarily from the handling of materials and nonenclosed operations that can create potential health, environmental, and safety risks.

The industrial hygienist should work closely with research and development, manufacturing, and other groups until potential health and safety risks are eliminated or effectively reduced. Emphasis is placed on removing the risk by engineering means through, for example, the use of an enclosure with local exhaust ventilation, thermally insulated and heated transfer lines to prevent plugging, safety valves and switches, absorption columns, air cleaners, and alarms. The workroom atmosphere should be frequently monitored for toxic materials and physical risks.

Standing operating procedures should detail:

- An emergency procedure,
- First aid instruction,
- Protective equipment requirements,
- Safety precautions,
- Equipment checking sequence list,
- A procedure for starting up, making a run, handling the materials charged and removed from the dryer, and shutting down.

Those operations creating or suspected of creating health risks are reviewed by the industrial hygienist, and studies are conducted when necessary. Where health risks are apparent, specific directives are prepared, and personnel are advised when areas are sufficiently free from significant risk to permit reentry. Employees are always kept informed regarding the risk potential of all materials.

Fluidized-bed drying involving toxic chemicals that are released to the atmosphere also requires the attention of the industrial hygienist.

6.1 Special Considerations

6.1.1 Static Electricity. Electrostatic buildup is of concern when it occurs in an environment containing flammable vapors or explosive dusts. Preventing the generation of excessive amounts of static electricity is a difficult task when one considers that every movement of one object against another, such as clothes against skin, air against plastic, soles of shoes against the floor, and liquids dropping through the air, is an efficient generator of static electricity.

Since the fluid-bed process is based on the principle that superficial space velocities suspend the bed, the effect is that of liquids falling through the air (movement of objects against each other) and the resultant generation of static electricity.

Reasonable controls can minimize the generation of static electricity and control it to within acceptable limits, including:

- *Bonding and grounding.* All objects in a system or group should be bonded together and the entire system grounded to earth. This allows the static electrical potential to easily equalize itself not only to each object but to earth. Assuming the worker is also bonded, excessive static electricity potential difference generation is unlikely.

- *Adding moisture to the air to increase humidity.* The insulating value of room atmosphere can be reduced by the introduction of moisture into that atmosphere. The atmosphere then becomes more conductive, allowing a slow and safe bleed-off of some of the static electric potential without generating an ignition spark.

- *Equipping personnel with conductive safety shoes and the facility with conductive floors.* One without the other does not effectively bleed off static electrical charges.

- *Avoiding the use of clothing made from material with a high dielectric characteristic.* Cotton is the most satisfactory for outer wear. Since the static electrical phenomenon occurs on the surface of objects, the material used in underclothing is usually not critical.

6.1.2 Preventing and Controlling Explosions

The combination of explosive dusts and static electricity has caused concern with fluidized-bed dryers. This has resulted in optimizing the safety measures to be taken in their design and operation. The measures to be taken are:

- Motors and electrical switching in explosion-proof construction.
- Fluid-bed machines designed for a pressure shock resistance.

- Relief openings in accordance with standards.
- Safety clamps that prevent the separation of the product container and filter housing in case of an explosion.
- Operation with inert gas.

The explosion relief channel is constructed to withstand the pressure shock resulting from an explosion. The pressure wave and flame following an explosion can escape through pressure relief flaps through the channel into the open air. The diameter of this safety device has been precisely designed, and the exit opening should never be made narrower. The channel should be no longer than 18–20 ft (6 m), lead directly into the open, and contain no bends.

GENERAL REFERENCES

Fluid Bed Diffusion Drier, Manesty Machines Limited, Speke, Liverpool 24, England.

Perry, R. H., and C. H. Chilton, *Chemical Engineer's Handbook,* 5th ed., New York: McGraw-Hill, 1973.

Technical Air Installations and Apparatus, Aromatic Ltd. Farnsburgerstrasse 6, Muttenz-Basle, Switzerland.

CHAPTER FIFTEEN

Industrial Centrifuging: Separators, Clarifiers, and Filters

Paul F. Woolrich

1 INTRODUCTION

A centrifuge is a machine having a hollow rotor into which a slurry or a mixture of liquids may be fed in such a fashion that the mixtures will be separated into component parts by centrifugal force. In industry, centrifuges are used as classifiers, clarifiers, concentrators, and liquid–liquid phase separators; to remove solid particles that are suspended in a liquid; and to deliquify solid matter. Centrifuges that filter, that is, cause the liquid to flow through a bed of solids held on a screen, are commonly called *centrifugals* or *centrifugal filters*. They are also known as wringers, extractors, or dryers.

2 GENERAL CLASSIFICATION

There are hundreds of applications in the chemical, petroleum, food, and other industries for centrifuges. Therefore, to meet specific purposes, centrifuges are classified

327

according to solid or perforated wall; rotational axis; overdriven or underdriven; continuous, cyclic, or batch operation; and shape or configuration of the rotor.

2.1 Imperforate (Solid) Bowl Centrifugal Separators

In the solid wall or bowl (imperforate bowl) centrifuge, the bowl or basket has no drain holes, and solids are separated from liquids by sedimentation (inhanced by inertial forces). Liquids are separated according to their specific gravity. Solid-wall centrifuges are used as classifiers, clarifiers, and liquid–liquid phase separators. This type can be used to throw out sediment from sludges that are impossible to filter. In a sedimentation centrifuge, the particle size and its distribution (in most applications) is such that the separation follows Stokes's law. They include batch and continuous machines:

Continuous	Batch or Intermittent
Screw conveyor type	Upright disk tube supercentrifuge
Disk bowl separator (DeLaval)	
Disk bowl recycle clarifier, nozzle discharge	

2.1.1 Continuous Operations

SCREW CONVEYOR TYPE. This is a continuous decanter centrifuge with a helical conveyor in a large horizontal bowl. Since the bowl is rotating rapidly, the helical conveyor must also do so at the same or slightly less or more than the rotation speed of the solid-wall bowl.

DISK BOWL SEPARATOR. This is a DeLaval-type apparatus with a stack of cones inside a solid-wall bowl. The cones spin with the bowl and speed up clarification by, in effect, shortening the distance that the slurry must travel to strike the wall. Cream separators work on this principle. Sludge is collected on the inside wall, or periphery, of the bowl.

2.1.2 Batch Operations or Intermittent Operations

DISK TUBE SUPERCENTRIFUGE. Mechanically, this centrifuge is known as a super-centrifugal. The fluid mixture is fed continuously into the bottom end of the rotor [an upright tube about 4 in. (10 cm) in diameter] spinning at 15,000 rpm. Centrifugal force caused by the spinning tube makes the heavier liquid go toward the wall while the lighter liquid stays in the middle of the tube, issuing at separate ports at the end of the rotor. This centrifuge is useful for separating emulsions or liquid and solid sludges that would otherwise be impossible to filter. Fine slurries are separated in the same way. When enough cake has accumulated, the tube must be taken down and cleaned.

2.2 Perforated Bowl Centrifugal Filters

These are centrifugal filters in which the solid phase is supported on a permeable surface or membrane through which the liquid phase is free to pass. In principle, they are all the same: in each type, a cake of granular solids is deposited on a filter medium held in a rotating basket, washed, and spun "dry." They differ in whether the feed is batch, intermittent, or continuous; whether they are overdriven (suspended) or underdriven; and in the way the solids are removed from the basket. See Fig. 15.1.

2.2.1 Continuous or Cyclic (Horizontal Axis). In the *push-type* continuous centrifuge, the slurry is fed in continuously. A reciprocating pusher, the same diameter as the basket, moves in and out and pushes the cake a couple of inches toward the edge of the basket. When the pusher returns, the basket is refilled with slurry.

The *scrapper-type* semicontinuous centrifuge is a continuous solid discharge centrifuge that has a mechanically driven plow or scrapper that moves up into the cake and scrapes it off.

2.2.2 Batch Operations (Vertical Axis). The vertical basket centrifuge is available in a variety of configurations, all rotating on the vertical axis. The principle variants are the location of the drive with respect to the basket: underdriven with the drive below the basket and overdriven (suspended), wherein the prime mover is offset from the axis of rotation with drive through belts, or suspended, with the drive mounted on an upper cross member from which the basket is suspended and driven by a relatively long spindle. For mechanical reasons, the suspended drive permits operations at higher rotative speeds. Both configurations are made with solid-bottom baskets, as is the case with solid-bowl sedimentation-type centrifuges. The solid bottom requires the basket to be brought to a rest and its solids content manually unloaded. The open bottom permits unloading with a mechanical knife or plow, while the basket is revolving at low speed (10–50 rpm).

The solid-bottom batch basket centrifuge is used for small-scale operations: when it is desired to preserve the integrity of a basket-size batch of solids, when the dewatered solids cannot tolerate mechanical handling, or when the traces of solids remaining in a more automated centrifuge would be subject to decomposition or spoilage. It can be either under- or overdriven.

3 INDUSTRIAL HYGIENE AND THE CENTRIFUGE

Since a variety of potential health-risk chemicals are processed in centrifuges, there are potential industrial hygiene problems to be reckoned with, particularly as regards the batch basket centrifuge (Fig. 15.2).

- The operator may be exposed to high concentrations of vapors during the charging, washing, and spinning cycles when volatile liquids are processed.
- A problem may exist when the centrifuge is unloaded because of the residual vapor or gas in the filtered material.
- Operators may also be exposed to vapor during transfer of product to containers.
- The operator may be exposed to spray, mist, or wet solids during the filtration and wash cycles. Skin contact or inhalation problems of industrial hygiene significance may result from such exposure.
- Skin contact with filtered solids during centrifuge unloading and the transfer to containers may constitute a health risk.

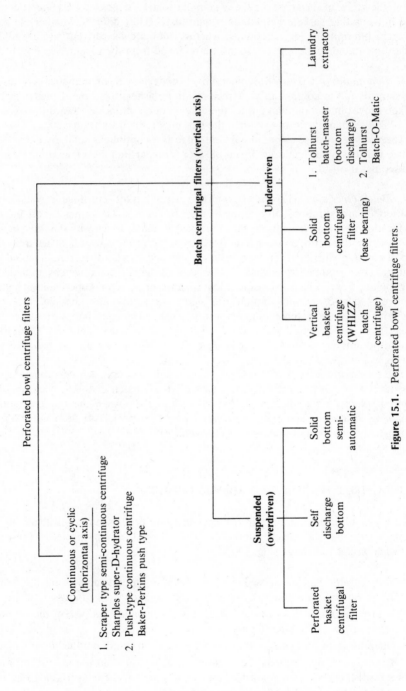

Figure 15.1. Perforated bowl centrifuge filters.

Typical draw-through ventilation system for the control of gas and vapor liberated during centrifuge operation. A tight cover is required because the centrifuge acts as a fan and gas may escape if the cover is loose.

Control velocity during discharging: 100 fpm
Duct velocity: 2000 fpm

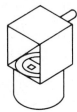

Hood enclosure for control of gas or vapor liberated during centrifuge operation.

Control velocity: 100 fpm
Duct velocity: 2000 fpm

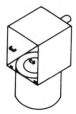

Hood enclosure for control of materials that present a health hazard through skin contact as well as inhalation.

Control velocity: 100 fpm
Duct velocity: 2000 fpm

Combination slot type and draw through ventilation system to control gas and vapor liberated during centrifuge operation. Slot hood controls gas and vapor during centrifuging. It can be moved aside and draw-through ventilation used during plowing.

Control velocity: Draw through, 100 fpm; Slot, 1000 cfm
Duct velocity: 2000 cfm

Hood enclosure for control of gas and vapor liberated during centrifuge discharge.

Control velocity: 100 fpm
Duct velocity: 2000 fpm

Figure 15.2.   Ventilation of vertical axis basket centrifuge filters.

3.1   Industrial Hygiene Control

- Whenever possible, it is desirable that a tight-fitting cover be used, as the centrifuge acts as a fan and will force air out along the edge.
- When the filtered material or the residual liquid in the filter cake may constitute a health risk by way of skin contact, the operator should be furnished with rubber gloves, apron, and other protective equipment that the operation requires.

- In operations where absorption through the skin, as well as by inhalation present a problem, a complete enclosure of the centrifuge may be necessary to minimize exposure.
- Control of residual gas or vapor may be accomplished by the use of a draw-through ventilation system. This is especially desirable if the centrifuge is large and if it is necessary for the operator's head to be near the opening.
- If the transfer of product is made to a drum by scoop or shovel, a simple slot hood will provide excellent control of gas or vapor evolved during transfer of wet solids to the drum.
- If discharge is done mechanically through a bottom opening into drums or tote boxes, an enclosure of the container with mechanical exhaust will provide the necessary control.
- Figure 15.3 is an engineering sketch of an overhead automtic centrifuge providing essentially an enclosed process during the entire cycle from charging the centrifuge to dumping in the tote bin.
- The Occupational Safety and Health Administration has no regulations that pertain specifically to centrifuges. Section 1910.212(a)(1) requires rotating parts to be guarded. Section 1910.212(b) requires the centrifuge to be secured (not necessarily bolted) to prevent uncontrolled movement. Section 1910.219 requires the mechanical power transmission apparatus to be guarded. The electrical requirements are located in Subpart S, as amended.

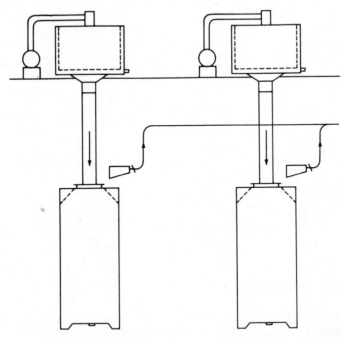

Figure 15.3. Engineering sketch of an overhead automatic centrifuge providing essentially an enclosed process during the entire cycle from charging the centrifuge to dumping in the tote bin.

For additional information, see Chapter 5.

GENERAL REFERENCE

Industrial Ventilation, a Manual of Practice, 19th ed., American Conference of Governmental
   Industrial Hygienists, Committee on Industrial Ventilation, Lansing, MI: Edwards Bros.,
   1986.

CHAPTER SIXTEEN

Liquid Filtration and Clarification

Paul F. Woolrich

1 INTRODUCTION

There are five commonly used physical methods of separating liquid from liquid or removing solids from liquid: gravity settling, filtration, centrifugation, distillation, and crystallization. Any one of these may prove to be the most economical solution to a particular separation problem.

Filtration is one of the most important unit operations in industry. Since filtration affects both chemical yield and purity of chemical products, approximately 40% of every capital dollar spent in the chemical industry is for filtration equipment.

1.1 Definition

A filter is a device that allows solutions (slurry) to drain, be drawn (vacuum), or forced (pressure) through a medium to remove solid particles (cake) from the solution. The filtrate is commonly called *mother liquor*. Either the cake or filtrate may be the desired product, and sometimes both are valuable.

Cake filtration is useful for collection of a product; clarification relates to quality or purity of filtrate through removal of fine particles from the fluid.

1.2 The Media

The filtration media utilized in industry include:

- Woven fabrics of cotton, wool, glass, synthetics (like Dacron, Nylon, Orlon, Teflon), and metal wire screens.
- Porous solids, sintered glass or metal, sheet metal sponge, carborundum, alundum, and ceramics.
- Nonwoven matted materials like paper and cotton, synthetic, or woolen felts.
- Granular materials such as beds of sand, coal, or gravel.
- Precoats of filter aids such as diatomaceous earth.
- The cake itself, which is a most effective filter medium.

The medium is backed by a structural support that provides grooves, holes, or channels to allow the filtrate to run off. Coarse granular crystals can form a thick porous cake that generally presents no problem in filtering, and little filtration area is required.

As particles get finer, they tend to penetrate the pores of the filter medium, pack tightly together, and have slow filtration. Very fine, sticky, or slimy particles can form an impassable cake layer; a higher filter force only packs the layer more tightly and makes filtration worse. Pretreatment can be used to aggregate, coagulate, or precipitate these particles or to condition the slurry for better filtration.

Filter aids form a porous cake with channels through which liquids can flow. Certain filter aids are used to precoat the filter or are added to the slurry to make a looser and more porous cake. Particles are kept from sticking to each other and from flogging the filter.

1.3 Filtration Rate

Filtration rate is the number of gallons that can be filtered through one square foot of filtration surface per hour. Filters are generally sized according to the number of square feet of filter area. Some filters present a continually renewable filter surface through self-cleaning or utilization of a mechanical scraping device.

2 CAKE FILTERS

Cake filters operate under the impetus of a hydrostatic head (gravity), pressure imposed by a pump, or a vacuum. They function in batch, intermittent, or continuous operation modes.

2.1 Gravity Batch Filters

Gravity batch filters include (a) pot filters (Nutsche), (b) boat or box filter (Nutsche), and (c) gravity sand filter. Gravity filters consist of a tank equipped with a false bottom (perforated or porous) that may support a filter medium or may itself act as the septum.

In a gravity Nutsche, the slurry is filtered under its own hydrostatic head. The filtrate collects in a sump beneath the filter or it is sewered directly. Gravity filters are used only for freely filtering materials and in those cases where the highest rates are not required.

The advantages of gravity filters are:

- Their extreme simplicity.
- Their dependency on simple accessories.
- Low first cost.
- Construction of almost any material.
- Quick settlement of large particles to provide a low resistance precoat for the fine particles.
- Effective and efficient washing of the cake.

The disadvantages of gravity filters are:

- Low rate of filtration.
- Excessive floor area occupied per unit of filtration area.
- High labor intensity.
- Difficult housekeeping problems.
- Poor containment exposing operators to solvent and product vapors.

2.2 Pressure Batch Filters

Pressure batch filters include (a) pressure Nutsche, (b) pressure sand filter, and (c) plate and frame filter press.

Nutsches may be enclosed and converted to pressure or vacuum Nutsches. The closure makes it even more awkward and undesirable to operate (high labor intensity); however, operator exposure to product or solvent vapors is lessened.

The plate and frame filter press is the simplest of all pressure filters and is still one of the most widely used. The plate and frame press is an assembly of alternate solid plates, the faces of which are studded, grooved, or perforated to permit drainage, and hollow frames in which the cake collects during filtration. A filter medium, usually a fabric, covers both faces of each plate. The plates are hung in a vertical position on a pair of parallel support bars. During filtration, the plates are compressed to a watertight closure between two end half-plates. One of the end half-plates is fixed; the other is moveble by capstan screw, ratchet and tommy bar, gear and pinion, or hydraulic ram.

The advantages of the filter press are less floor area occupied per unit of filtration area and few moving parts to get out of order.

The disadvantages of the filter press are it requires labor to open, clean, and reset, which is conducive to operator exposure to product and solvent vapors, and it is not suitable for cases requiring frequent cleaning.

2.3 Intermittent Pressure Leaf Filters

Intermitttent pressure leaf filters include the Kelly, Niagra, Sweetland, Vallez, and Sparkler filters. The advantages of pressure leaf filters are (a) considerable flexibility, (b) low labor intensity, and (c) basic simplicity.

The disadvantages of pressure leaf filters are that it (a) requires intelligent and watchful supervision, (b) is unable to form as dry a cake as the filter press, (c) has a tendency to form misshapen cakes unless the leaves rotate, and there is a (d) pressure limitation to 75 psi (3900 mm Hg) or less.

2.4 Continuous Vacuum Filters

Continuous vacuum filters are widely used for liquid–solids separation in a number of process industries such as chemical, petrochemical, coal, fertilizer, food, and wastewater treatment centers.

Continuous vacuum filters fall into three classes: (a) drums, (b) disks, and (c) horizontal filters (table, pan, and belt types). Although there are differences in design and application of continuous vacuum filters, they all have the following features in common: (a) a filtering surface that moves from a point of slurry application where a cake is deposited under the impetus of a vacuum to a point of solids removal where the cake is discharged by mechanical or pneumatic methods and thence back to the point of slurry application, and (b) a valve that regulates the pressure below the surface at various stages of travel.

Continuous vacuum filters are by far the predominant choice for filtration installations, and they handle more tonnage of solids than all other kinds of filters combined. The advantages are (a) low labor intensity, (b) efficient adjuncts to continuous processes, that (c) the filtering surface can be open to the atmosphere providing accessability with minimal operator exposure, and (d) low maintenance costs.

The disadvantages are (a) a vacuum system must be maintained; it (b) cannot be used with filtrates that are volatile whether because of low vapor pressure or high operating temperature; it (c) cannot handle compressible solids; and (d) inflexibility does not allow good performance if the feed stream changes with respect to rate, consistency, or character of solids.

3 CLARIFYING FILTERS

Clarifying filters are used to separate liquid mixtures that contain only small quantities of solids. Compared with cake filters, clarifying filters are of minor importance to pure chemical process work. Their greatest use is in the field of beverage and water polishing, pharmaceutical filtration, fuel and lubricating oil clarification, electroplating-solution conditioning, and dry cleaning solvent recovery.

Most cake filters can be operated as clarifiers, but not necessarily with efficiency. Clarifying filters have been developed that can be used for no other purpose than clarifying or straining.

Clarifying filters are classified as (a) disk and plate presses, (b) precoat pressure filters, (c) cartridge clarifiers, and (d) miscellaneous types.

In-line filters/clarifiers are called *ultrafilters* and are usually leaf-type (Adams and Silas), disk, or plate types. Operators of in-line clarifiers are alert to resistance pressure in the cartridge housing and change the medium or cartridge frequently. Since the filters are changed by hand, there is a potential risk associated with contact and inhalation of toxic materials.

Safe filtration requires grounding and inerting to protect against static-electric buildup and flammable solvent ignition. Operators must also be alert to valve misdirection of the mother liquor to the clarifier.

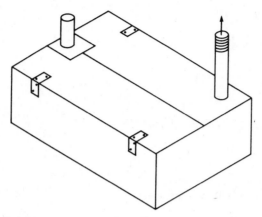

Figure 16.1. Ventilation of a filter box (Nutsche filter).

4 INDUSTRIAL HYGIENE CONTROL

4.1 The Filter Box (Nutsche Filter)

The gravity, pressure, or vacuum Nutsche filter is a circular or rectangular tank from which the filter cake is removed by means of shovels or scoops. Solvent vapors liberated from the filter box may constitute a health problem during filtration and/or cake removal. If the health problems exist only during filtration, the method of control is to provide a removable cover for the filter box with a minimum opening for piping and observation. Figure 16.1 shows a typical draw-through ventilation scheme. The volume of air required is the sum of that required to maintain a velocity of 100 fpm (30 mpm) across openings and the volume of vapor evolved per minute.

If the health problem exists during the removal of the filter cake, a hooded enclosure with one side open may be used; however, laterial exhaust is best suited for ventilating under these conditions since it is sometimes necessary for the employee to enter the filter box to remove filter cake (See Fig. 16.2.)

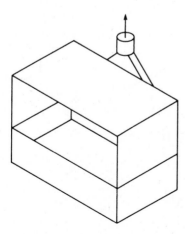

Figure 16.2. Hood enclosure for control of vapors during filtration and unloading of a Nutsche filter.

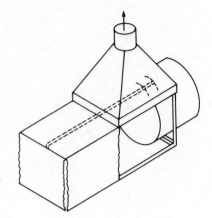

Figure 16.3. A canopy hood for control of vapor liberated during cleaning of horizontal leaf pressure filters.

4.2 Enclosed Filters

Enclosed filters are vapor tight and thus do not constitute a problem during filtration and washing. A health problem may exist during cleaning and setting-up operations if residual liquids or solids are toxic. This potential hazard can be controlled by purging the filter using water or low-toxicity solvent.

An alternate method of control is through utilization of local exhaust ventilation. Considered here are both horizontal and vertical leaf or plate filters. The covers of horizontal enclosed filters or leaf presses should be replaced after the filter core has been removed for cleaning to reduce evaporation from this source. A canopy hood for control of vapor liberated during cleaning of horizontal leaf pressure filters is shown in Fig. 16.3.

The cleaning is accomplished through openings on both sides of the leaf press. Cleaned leaves or plates are stored in the curtain-covered portion of the frame. The use of bottom panels would further reduce the maximum hood opening.

Figure 16.4 shows a hood for control of the vapors liberated during the cleaning of leaves or plates from a vertical leaf pressure filter. In this arrangement, the leaf

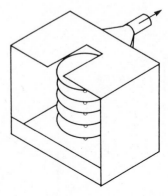

Figure 16.4. Hood for control of the vapors liberated during the cleaning of leaves or plates from a vertical leaf pressure filter.

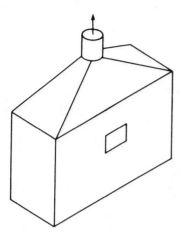

Figure 16.5. Ventilation of plate and frame press when only filtration is a health risk.

assembly is removed from the filter housing and placed in the hood. A control velocity of 100 fpm (30 mpm) should be maintained across the hood opening. A transport velocity of 2000 fpm (600 mpm) should be maintained in the exhaust duct to the prime mover. The tray is used to contain the "cleanings" and "washings." A hoist cable may be attached to the hood for ease in moving to various locations.

4.3 Plate and Frame Press

Solvent vapors liberated from the press may constitute a health problem during filtration, cleaning, and setting-up operations. The canopy hood is not suitable when a health problem exists during press cleaning and setting-up operations, as the updraft ventilation will pull the toxic substances through the operator's breathing zone. The canopy hood can be used if a health risk exists only during the actual filtration, yet,

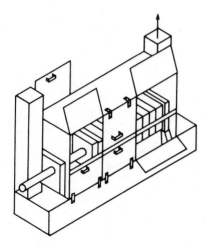

Figure 16.6. Ventilation of plate and frame filter press with operator outside enclosure.

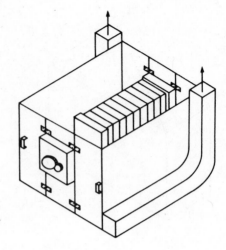

Figure 16.7 Open-top downdraft enclosure for plate and frame press.

because of its inefficiency and requirement of large air volumes for control, it is recommended only when other methods are not feasible.

Figure 16.5 is a filter press enclosure for use when a health risk exists only during filtration. Drop or removable sides and canvas or plastic curtains can be utilized. The press is completely enclosed except for observation openings. The sides or curtains are removed during cleaning or setting up. Control velocities of 100 fpm (30 mpm) across the maximum working opening should be maintained. Transport velocities of 2000 fpm (600 mpm) should be maintained in the duct to fan.

For operations where a health risk prevails during cleaning and setting-up operations as well as during filtration, a press enclosure and downdraft ventilation should be used. Three ventilation hood enclosure designs may be used:

- Enclosure with operator inside.
- Enclosure with operator outside.
- Partial enclosure with enclosure top open.

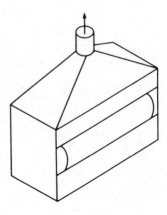

Figure 16.8. Ventilation of a rotary drum filter.

Only on small presses where cleaning may be accomplished without bending over the press can an enclosure and updraft ventilation (Fig. 16.5) be used. Figure 16.6 shows ventilation for a plate and frame press enclosure when the operator is outside. Downdraft ventilation is utilized. Hinged doors provide access for press cleaning and setting up. The control velocity required across the maximum working opening is 100 fpm (30 mpm). The transport velocity in the duct to fan is 2000 fpm (600 mpm).

Figure 16.7 shows an open-top downdraft enclosure for the control of vapors during filter press operation. The operator cleans and sets up the press inside the enclosure. Control air is drawn through the top at a minimum control velocity of 100 fpm (30 mpm). Transport velocity in the duct to fan is 2000 fpm (600 mpm).

4.4 Rotary Drum Vacuum Filters

Figure 16.8 shows a typical hood enclosure for the control of vapors during the operation of rotary drum vacuum filters. The hood opening should be provided adequate working and observation space. The control velocity across the hood opening should be 100 fpm (30 mpm). Where large volumes of steam or solvent vapors are evolved, an additional exhaust volume to compensate for this added volume should be exhausted. The duct-to-fan velocity is 2000 fpm (600 mpm).

4.5 Cartridge (Clarifying) Filters

Since cartridge filters are vapor tight, they do not constitute a health problem during filtration and washing. However, a health risk may exist during cleaning and setting-up operations. As with other types of enclosed filters, control of potential health risks can be effected by purging the filter with water or a solvent of low toxicity before cleaning.

Figure 16.9 shows hood and ventilation requirements for control of gasses and vapors during cleaning and cartridge replacement in these filters. A control velocity of 100 fpm (30 mpm) should be maintained across the working opening of the hood. A transport velocity of 2000 fpm (600 mpm) duct-to-fan velocity is maintained. For additional information, see Chapter 5.

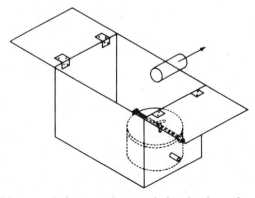

Figure 16.9. Hood for control of gases and vapors during cleaning and cartridge replacement.

GENERAL REFERENCES

Kent, J. A. (ed.), *Riegel's Handbook of Industrial Chemistry*, 7th ed., New York: Van Nostrand Reinhard, 1974.

Kirk–Othmer, (eds.), *Encyclopedia of Chemical Technology,* New York: Wiley: Interscience, 1964.

Perry, R. H., and C. H. Chilton, *Chemical Engineer's Handbook*, 5th ed., New York: McGraw-Hill, 1973.

CHAPTER SEVENTEEN

METAL CLEANING

Thomas J. Walker

1 DESCRIPTION OF PROCESS: HOW CONTAMINATION IS REMOVED

Cleaning of metals usually is the first of many important steps in preparing a final product. Cleaning in some manner almost always follows the metal forming process (rolling, extrusion, and forging). Cleaning is necessary so that the product can receive the next treatment, such as painting or plating. The main reason why flaws occur in plated or painted metals is because the metal was improperly cleaned. Therefore, in order to minimize flaws and produce quality product, metal cleaning is extremely

345

important. In addition, the cost of metal cleaning is extremely expensive; if it is performed incorrectly at this stage of manufacture, the financial loss of scrapping the metal is significant. Therefore, it cannot be overstressed that the cleaning process contributes significantly to the quality and cost of the product. As a result of these costs, management is extremely reluctant and resistant to make changes in processes.

The industrial hygiene, safety, and environmental considerations for metal cleaning are related to the toxicology and hazards of the chemicals used in processing the metal rather than to the toxicology and hazards of the fundamental metal itself. The chemicals requiring these considerations include acids and bases, salts, halogenated hydrocarbons, aliphatic and aromatic hydrocarbons, and process oils and lubricants. The toxicology and hazards of these chemicals, as well as their cleansing abilities should be studied and understood thoroughly in order to provide the best solution to the problem.

There are four types of soil or contamination that the manufacturer wishes to remove: grease soil, loosely adherent soil, adherent soil, and moisture. These forms of contamination, and the types of problems they produce for the user, are analogous to problems caused by grease, dirt, and moisture that we may have experienced at some time when we tried to write on a common blackboard.

Grease soil includes normal lubricating oils, oil used during machining, grease used in protecting components during storage, and natural oils from fingerprint contamination. In our blackboard analogy, fingerprints on the blackboard provide sufficient grease to cause chalk to skip.

Loose soil includes dust, fine turning, and grinding materials produced during machining. In our blackboard analogy, chalk itself can be considered as the loosely adherent soil.

Adherent soil includes common scale and rust, but also includes burnt-on oil or paint. Regarding the blackboard, any old painting on the board is comparable to adherent soil that causes the chalk to skip.

Moisture is ever present from the air from water-soluble lubricants and washing. In our blackboard example, writing over freshly washed board causes a distorted chalk mark.

Table 17.1 provides a summary of different processes used in metal cleaning and the kind of metal on which they are most often used. Acid treatments have only minor use for aluminum because the adherent oxide is not massive, as is the case of iron oxide (rust), and is easily removed during alkali treatment. Each of the treatments is used primarily to clean, and secondarily to provide an appropriate finish to the metal.

1.1 Abrasive Cleaning

1.1.1 Abrasive Blast Cleaning. One of the simplest and most effective techniques used in the cleaning of metal is abrasive blast cleaning. It is effective (1) in removing adherent soil such as scale and undesirable paint, as well as some of the grease soil; (2) in preparing the surface for further surface treatment; and (3) in improving the surface of the metal by removal of small burrs and minor imperfections. Simple materials such as washed silica, silica flour (amorphous), silicon carbide, iron grit, steel shot, crushed walnuts and plumstone, aluminum oxide, garnet, and glass beads have been commonly used. The grit is propelled by making use of an impeller, compressed air, or water under pressure. The type of abrasive used depends on the desired surface. The manufacturer will be extremely reluctant in the simple substitution of

Table 17.1. Cleaning Treatments

Process	Category	Metal
Abrasive blasting	Abrasive	All
Acid descaling	Acid/alkaline	Steel
Alkaline descaling	Acid/alkaline	Aluminum/steel
Aquablast	Abrasive	All
Barrelling	Abrasive	All
Degreasing	Solvent	All
Descaling	All	All
Dry blasting	Abrasive	All
Drying	Other	All
Electrocleaning	Other	All
Emulsion cleaning	Solvent	All
Flame cleaning	Other	Steel
Flame descaling	Other	Steel
Flame scaling	Other	Steel
Fluxing	Solvent	All
Glass bead blasting	Abrasive	All
Grift blasting	Abrasive	All
Liquid honing	Abrasive	All
Needle descaling	Abrasive	Steel
Nitralizing	Other	Steel
Pickling	Acid	Copper/steel
Plumstone blasting	Abrasive	Aluminum/steel
Polishing	Abrasive	All
Refrigerated vapor degreasing	Solvent	All
Sand blasting	Abrasive	All
Sawdust drying	Other	All
Scouring	Abrasive	All
Scratch brushing	Abrasive	All
Shot blasting	Abrasive	All
Sodium hydroxide	Alkali	All
Solvent cleaning	Solvent	All
Solvent degreasing	Solvent	All
Ultrasonic cleaning	Ultrasonic	All
Vapor degreasing	Solvent	All
Wire brushing	Abrasive	All

one grit for another without considerable research and development. If the size and shape of the grit is important to quality of the product, screens or other particle separators, including cyclones and baghouses, are used.

Aquablast, or wet blasting, employs the use of water (another solvent may be used in special applications). Besides reducing the dust problem, water blasting provides higher velocities than with dry blasting and can provide a superior surface. This is sometimes known as liquid honing.

1.1.2 Barrel Finishing. In this low-cost process, components, which are frequently small stampings, castings, or machine parts, are tumbled in contact with chips of inert material such as walnut shells, hardwood pegs or sawdust, fine abrasive compounds such as sand and pumice, and liquids, usually water, with detergent or wetting agents added, and sometimes paraffin. The purpose is to deburr, burnish, or form a smooth surface. Some metals, such as aluminum, may undergo an additional step, such as bright dipping and anodizing.

The process initially consisted of wooden barrels, similar to beer barrels, rotating at an angle of approximately 45° in which components were tumbled with water, paraffin, or no lubricant at all. The process was time consuming, up to 5 days, and the product quality was uncontrollable. Modern barrels are hexagonal and operate horizontally with or without abrasive compounds. The use of abrasives to produce a high-quality finished product is relatively new technology. In this process, deburring is confined to the period when the components are sliding or falling down the slope against the direction of the barrel movement. A pH of about 8 is usually required for aluminum. Other metals have other requirements.

Vibrating barrels are the most recent innovation in the art of barrelling. They have the appearance of a bathtub, and are spring-mounted with a motor-driven eccentric that imparts a vibratory motion designed to give a corkscrew motion to the parts that are in constant vibrating contact with chips and abrasive materials. Deburring takes 1–3 hr for most components, compared to 4–8 hr for horizontal barrels, and 4–5 days for the original barrels. Modern chips are preformed ceramics in a large variety of shapes and sizes.

1.1.3 Polishing and Buffing. Polishing is a manual or automated mechanical method of improving the surface finish of materials, and uses two distinct methods in succession to achieve the desired effect. First, cutting materials, such as abrasives, are used to cut the surface being polished with parallel scores. Successive scorings are made with finer grit until a reasonably polished surface is prepared. Abrasive polishing is generally required for a higher-standard finish and will almost invariably be the technique used where there is a rough surface. Second, the surface is smoothed or smeared using special hardened-steel tools that do not cut into the surface of the metal but produce a highly polished surface by bending the peaks or high ridges over and filling in the valleys. This is sometimes referred to as burnishing.

Buffing is a specific type of polishing using a high-speed disc made from layers of cloth, leather, or plastic impregnated with an abrasive in liquid slurry or solid form to press against the object to be buffed.

1.1.4 Wire Brushing. Wire brushing is used to descale as well as produce satin finishes using revolving hand-held brushes with wire bristles usually made from hard-drawn stainless steel, although mild or low-alloy steel may also be used. Wire brushes are often cleaned of grease and oxides by placing a pumice stone in contact with the rotating brush.

1.1.5 Electrolytic Polishing. This nonabrasive form of polishing that produces a bright surface with a highly reflective finish, usually for decorative purposes, is generally associated with some form of metal finishing such as anodizing, plating, or lacquering. By the judicious choice of electrolytes and current, a smoothing action is produced by removing the peaks of the metal. In electrolysis, more metal is removed from peaks than from valleys. Solutions are usually acids, such as phosphoric, sulfuric, and occasionally, chromic acids.

1.1.6 Scouring. This labor-intensive process, sometimes used in plating shops, could be used prior to painting or just to achieve an attractive surface finish. An abrasive material, such as sand or pumice, is used on a moistened cloth and the surface is abraded or scoured to remove adherent dirt and oxide. The material must be completely removed from the surface if the metal is to receive an additional treatment. This method is seldom used today because of cost.

1.2 Inorganic Chemical Cleaning

1.2.1 Acid/Alkali Acid and alkali treatment of metal removes oxidation or corrosion that rapidly forms on metal surfaces. Oxidation occurs rapidly at elevated temperatures but also at room temperature. Chemicals, primarily acids, bases, and salts, oxidize the surface of metals extensively. Electrolysis of metals also causes oxidation. Cathodic protection using sacrificial anodes uses this latter effect to prevent oxidation of one metal by allowing another to oxidize rapidly. Removal of oxidation using acids or alkalies is really a very complex metallurgical process.

1.2.2 Acid Descaling. An alternative name for pickling, this process uses acid to dissolve oxide and scale.

1.2.3 Aldip. This is a process for coating ferrous metal with aluminum and aluminum oxide for corrosion and heat resistance. Parts are first cleaned of grease and dirt in an alkaline rinse followed by pickling in acid, rinsing, and furnace drying. They are then placed in a preheating salt bath at about 750°C for 4–5 min, dipped into a molten aluminum bath, covered with flux for about 1 min, and returned to the preliminary salt bath, slowly removed, and air cooled. An aluminum and oxide surface remains. Excess metal or roughness is removed by shot blasting or similar abrasive action.

1.2.4 Alkaline Descaling and Etching. Alkaline descalers are used for all metals and contain sodium hydroxide solution, sometimes with additives. Temperatures and concentrations depend on the metal and condition; temperatures typically are 50–200°C and concentrations can be as high as 50%. Alkali attack on steel occurs within specific parameters; its attack on aluminum is so rapid and continuous over such a wide range that inhibitors or chelating agents are frequently added to decrease the chemical activity. Attack on magnesium, copper, and zinc have similar but much slower activity. Alkaline descaling is often preferred to acid because it does not cause hydrogen embrittlement. Springs, which have high tensile strength, would become brittle and later fail if descaled in acid.

1.2.5 Anodizing. Anodizing is the process of forming a film of surface oxide electrolytically. Used primarily for aluminum and its alloys, but also for magnesium and titanium, its purpose is to improve resistance to corrosion, prepare the surface for an additional coating such as paint, increase the surface hardness, and aid in detection of surface cracks. The reaction occurs very rapidly at first until a surface coat of aluminum oxide is formed; then the electrolysis slows and finally ceases. There are many different anodizing processes that impart different surface characteristics, including color. Most of them are variations of those that follow.

Sulfuric acid anodizing, at 10% solution, typically uses 10–20 V, maintaining a temperature less than 30°C. Since sulfuric acid dissolves the oxide, a lower temperature must be maintained. The lower the temperature the blacker the color becomes. This is sometimes referred to as the black oxide process.

Chromic acid anodizing, at 3.5% solution, typically uses about 10 V at the beginning and increases to about 60 V at a temperature of 20°C or higher (temperature is not critical). Historically, chromic acid, which has color, has the additional advantage of acting as a dye, penetrating surface imperfections such as pores or cracks and making them visible. Recent advances in the use of synthetic dyes have lowered the use of chromic acid.

Oxalic acid has been used, but because of the danger of explosive by-products, the method is not popular.

1.2.6 Etching. Etching is a special-purpose technique used to prepare surfaces for electroplating, remove metal from printed circuits, highlight surface defects, and show metallurgical structure.

1.2.7 Emulsion Cleaning. Emulsions of water and special chemicals and wetting agents can also be used to clean metal parts. Each component of the emulsion has the capability to clean independent of the other. These are highly technical products that are usually proprietary.

1.2.8 Fluxes. Fluxing is used to remove the oxidations so that it does not interfere with the end product. Fluxing is seldom used alone, but in combination with other processes. Fluxes are electrolytes, usually metal chlorides, fluorides, and sulfates that dissolve the oxide surface of the metal component.

1.2.9 Immersion Coating. Following metal cleaning, the metal can be immersed into another metal, such as occurs when steel is immersed into molten zinc (galvanized steel), and when metal components are immersed into paint.

1.2.10 Phosphating. Phosphating is commonly used on steel and is the usual pre-treatment for painting. The process is also used for aluminum and zinc. In phosphating, the metal surface converts to the metal phosphate, which readily accepts painting if applied within 4–6 hr; the time is critical and should not exceed 24 or more hours. The temperature is usually between 60–90°C and the concentrations usually less than 10% phosphoric acid. Immersion time is usually less than 5 min, although for some applications, 30 min may be the required time. Most phosphatizing applications are continuous, rather than batch processes, because time considerations are frequently critical. Chromate sealers, or other metal treatment processes, may be applied following phosphating.

1.2.11 Pickling. Pickling is a chemical treatment to remove oxide or scale from the surface of any metal. The term describes the use of hydrochloric or sulfuric acid on iron or steel during hot-forming of products; less commonly, it is used to describe etching prior to electroplating.

Aluminum seldom requires pickling because the oxide film is thin and adherent. With aluminum, caustic soda is the most common chemical used for descaling or oxide removal. Hydrofluoric acid and sodium fluoride, chromic acid, and sulfuric acid are frequently used as conversion coatings.

Titanium is pickled with hot hydrofluoric acid with various additives. Titanium must be descaled to remove the oxide and the surface films of nitrogen, hydrogen, and oxygen, which produce metal fatigue.

Sodium hydride is used for pickling, but the hazards involved from molten sodium and hydrogen are great, and the process has high operating costs and large capital investment.

1.3 Organic Solvent Cleaning and Vapor Degreasing

Solvents are effective in the removal of oils and greases from the surface of metals by wiping, dipping, and vapor degreasing. The principal solvents used are halogenated to prevent fires: trichloroethylene, perchloroethylene, methyl chloroform, and chlorofluocarbons. Flammable substances such as Stoddards solvent are seldom used.

1.3.1 Vapor Degreasing. Vapor degreasers are the most commonly used method of cleaning with solvent. In a vapor degreaser, solvent is heated to produce a vapor in which the components are placed. When the oil-laden solvent vapor condenses, the solvent is distilled from the oil and the process repeats. Refrigerated coils are sometimes used in the degreaser to assist in condensing the solvent vapor into a trough that transports the materials to the still where distillation occurs. These degreasers are so effective in removal of the solvent at and near these coils, that refrigerated coil degreasers can be operated without a cover. This type of solvent degreaser is a form of continuous distillation.

Solvent degreasers usually use trichloroethylene, although methyl chloroform, chlorofluorocarbons, and similar halogenated hydrocarbons can be used. Solvents cannot be substituted at will in vapor degreasers because the distillation temperature is set for a particular solvent, as well as for trapping the vapor condensates.

1.3.2 Dipping. The process of dipping parts into solvent in a small tank is frequently used to remove large amounts of dirt and oil. The tanks are usually covered when not in use. The contaminants in the solvent are distributed evenly over the surface of the components, which are subsequently cleaned using different methods.

1.3.3 Wipe Cleaning. In wipe cleaning, a worker will moisten a cloth in solvent and wipe or rub the surface of the component while using fresh cloth to adsorb the dirt and not contaminate other areas. This procedure is seldom used where the surface requires a high and consistent quality of cleanliness.

1.4 Ultrasonic Cleaning

Ultrasonic cleaning makes use of expensive energy-generating equipment to remove dirt adhering tightly to the surface of simple shapes in water or solvent at a frequency of 16 kHz or higher. Ultrasonic cleaning rapidly loosens and removes surface contamination such as dust, but it is not effective for continuous films. The cyclical nature of the energy causes a rising and falling pushing pressure followed by a pulling pressure to the surface, dislodging the dirt. This process works effectively on simple surfaces. The direction or attitude of the piece to the wave generator is also important. Ultrasound is produced in pencil-thin beams that are extremely efficient if the component or surface is in the line of the beam of energy. Outside of this narrow beam, the degree of cleaning will be low or nil. This narrow beam provides a big advantage when the location and attitude of components can be maintained in a constant direction. For this reason ultrasonic cleaning is extremely effective in the removal of surface oil and dirt from aluminum welding wire in a solvent degreaser.

1.5 Other Cleaning

There are a few metal cleaning treatments that cannot be classified in the preceding format. Short decriptions of these are provided here.

1.5.1 Drying. Drying of metal can be a very costly activity. The purpose of drying is to remove moisture. Although heating in a drying oven would appear to be easily accomplished, moisture, which is detrimental to metals, must be removed from the drying ovens. Also, water could leave ugly waterspots behind. Sawdust is sometimes used to remove the largest part of the moisture. If oil is present, the metal must either be dried at a low temperature or dried in an inert gas oven so the oil does not oxidize and leave behind an ugly spot.

1.5.2 Electrocleaning. Electrocleaning is an efficient cleaning process in which an electric current is passed between the components being cleaned and an inert electrode. The components can be placed in the anode or cathode position, depending on what action is required. In the anodic position, oxygen will be produced at the surface of the component and released, providing scrubbing action. When the solution is alkaline, the oil and grease are saponified and pass into solution. When the component is cathodic, hydrogen is given off, providing a more active scrubbing action than anodic oxygen. Anodic cleaning usually results in contaminants in the solution migrating to the anode and forming a badly contaminated surface. Therefore, in practice, electrocleaning always consists of cathodic cleaning followed by anodic.

1.5.3 Flame Cleaning and Descaling. This process is limited to use with steel and iron, and uses oxyacetylene and other gas burners designed to produce a high-velocity, oxidizing flame with a large area. It is used to burn off old paint, rust, and similar materials on plate such as ships, and I-beams such as bridges. In industry, the process is usually highly automated and used in combination with wire brushing.

1.5.4 Nitralizing. Sheet steel is immersed into a bath of sodium nitrate at about 500°C to produce an active surface for vitreous enamelling.

2 Industrial Hygiene, Environmental, and Safety Controls

The business of metal cleaning usually requires many skilled workers to operate expensive pieces of equipment and many unskilled workers to provide labor-intensive manual material handling and housekeeping. There are numerous operational and housekeeping activities performed in which the worker can be injured or be exposed sufficiently to develop an occupational disease.

In the metal cleaning industries exposure of workers to physical and chemical hazards results in acute and chronic diseases from a variety of chemicals including solvents, acids, bases, fluxes, and fumes; from a variety of physical stresses including noise and heat, and a variety of safety problems involving manual material handling. The chemicals used in metal cleaning usually do an excellent job quickly without adverse effect on the product. However, because they have adverse effects on people and the environment, they usually have stringent controls.

In order for workers to perform duties safely, they need specific training in safe work practices regarding to movement of equipment, supplies, and products to reduce potential for physical injury to themselves and others. They must receive hazard communication relative to the use of chemicals, use of personal protective equipment including respirators, gloves, aprons, eye protection, and hearing protection. And they must know how to operate equipment effectively so that exposures are minimized.

2.1 Industrial Hygiene: Work Practices, Ventilation, and Engineering Controls

2.1.1 Work Practices
EXPOSURE CONSIDERATIONS: SOLVENTS, ACIDS, AND BASES. Although ventilation is extremely important in maintaining low exposures, a maximum effort must be made to assure proper work practices. Workers must be conscious of skin exposures and must continuously guard against skin contact to acids, bases, and solvents. They must never allow the use of solvents for purposes of removing debris from skin, no matter how small. They must be trained in what to do in case of spills. Training in all aspects of manual handling, as it applies to these jobs, as well as in understanding of material safety data sheets, cannot be overstressed. Finally, work must be performed according to the procedure. Shortcuts eventually will result in accidents and illnesses.

2.1.2 Ventilation and Engineering Controls.
The ACGIH *Ventilation Manual* provides several approaches to ventilation engineering controls for dip tanks and vapor degreasers. When designing new systems, care must be taken to approach the problem with a systems viewpoint, including optimum sizes of loads to be placed into tanks, amount of ventilated space needed for drainage, clearances, dripages, local exhaust considerations including cross drafts, and manual material handling. Far too frequently the industrial hygienist is called in to resolve an exceedingly difficult problem caused by a very bad design. Fans must be chosen with consideration of noise exposure.

2.2 Environmental Considerations

Nearly all of the chemicals used in metal cleaning have environmental impact. The acids, bases, heavy metals, solvents, and halogenated hydrocarbons, such as trichlorethylene, typically have environmental limits for soil and groundwater. Many problems have been caused by workers abusing the environment or not taking adequate safeguards to protect the environment. Workers must be made aware of these regulations and hazards to the environment. Procedures must be developed so that workers properly dispose of these materials in hazardous waste containers rather than into the wastewater or directly on the ground, even in remote areas.

2.3 Safety

Safety training must include recognition and control of those activities in which injuries can occur. These include driver training, watchfulness regarding mobile equipment, manual material handling, use of personal protective equipment, and operation of equipment to minimize hazardous exposures to noise and chemicals, burns, lacerations, muscular skeletal strain and sprain, and slips and falls.

GENERAL REFERENCES

Encyclopedia of Chemical Technology, 3rd ed., New York: Wiley, 1978.

Harris, J. C., *Metal Cleaning Bibliographical Abstracts*, Philadelphia: ASTM, 1953.

Wolf, K., *Chlorinated Solvents: The Regulatory Dilemma*, Santa Monica: Rand, 1986.

CHAPTER EIGHTEEN

Metal Working and Forming

Robert D. Soule

1   INTRODUCTION

This chapter covers a broad range of industrial processes characterized as unit operations associated with conversion of products of the primary metals industry (ingots, blooms, billets, and slabs) into metal components that are either products themselves or raw materials for metal fabrication industries. Coverage in this chapter is limited to those operations involved in forming or working metal after leaving the molten state. Specifically excluded, therefore, are smelting, refining, and foundry operations, as well as processes secondary to metal forming such as cleaning, cutting, machining, electroplating, riveting, soldering, welding, and various finishing operations.

Dozens of industrial processes in which metal is formed or worked are described in the literature. Differences among processes can be very subtle in some cases, amounting to not much more than relatively minor changes in processing conditions. Particularly frustrating to the lay reader is the widespread confusion in the popular literature and, to lesser extent, even in the trade literature. For example, "stamping" can refer, correctly so, to such diverse unit operations as extrusion and drawing, operations inherently different in that one places the metal under compressive force and the other employs a tension or stretching action.

Recognizing the complexity of the metalworking industries, therefore, and yet wanting to present descriptions and discussions of representative processes, this chapter is organized on the basis of the process temperature necessary to perform the work. Although there remains some difference of opinion as to specifics, there is

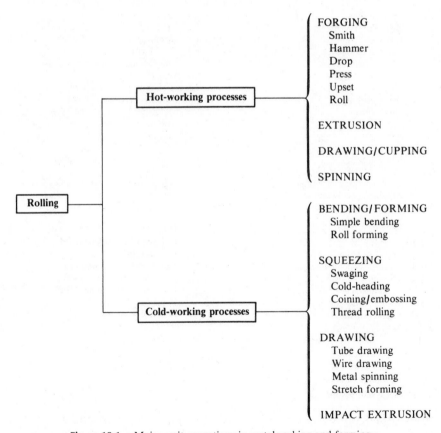

Figure 18.1.   Major unit operations in metalworking and forming.

general reference to, and distinction made between, hot working and cold working of metals. This basic subdivision is used to identify and organize material presented in this chapter.

Some differences between hot and cold working of metals are not easy to define. However, the distinction is not completely arbitrary either. All metals and alloy systems are crystalline in nature and are composed of irregularly shaped bodies, called "grains." These grains, easily seen under a microscope, are composed of atoms of the metal in an orderly arrangement, or lattice. The orientation of the atoms in a given grain is uniform but can be substantially different than in adjacent grains. All metal systems are characterized by a range of recrystallization, defined by temperature, within which the grain structure of the metal can be refined since the metal is in a plastic state. For our purposes, when work is performed on a metal at temperatures above the lower limit of the recrystallization range, it will be referred to as "hot work." Work done on a metal at temperatures below the recrystallization range is "cold work." The effects on metal structures due to hot working and cold working are discussed in more detail in the respective portions of this chapter.

The specific unit operations covered in this chapter are identified and categorized in Fig. 18.1. Processes discussed here are not intended to be representative of the total metalworking industry. It is believed that those presented here are commonplace in application across a broad range of end-use and metal systems. Where a particular operation is applicable for a limited range of metals or alloys, it is so identified in the text.

Within this chapter, discussions are limited to the metalworking or forming operations themselves and are not extended to support operations such as material transfer, warehousing, and shipping/receiving. Finally, since the impact of the various unit operations on the quality of the workroom environment is similar from one process to another, a separate section of this chapter is devoted to discussion of potential health hazards on a generic basis. Where a particular unit operation presents a unique hazard, mention of it is made during description of the process.

2 HOT WORKING OF METAL

As discussed earlier in this chapter, the hot working of metal is accomplished at temperatures above the recrystallization range of the particular alloy being processed. During all hot-working unit operations, the metal is in a plastic or nearly fluid state. As a result, the following advantages of hot working of metal over cold working usually are present in the unit operation:

- The metal can be formed or shaped readily by application of suitable pressures.
- Because of the high pressures involved, any porosity that might have been present in the original ingot is minimized or eliminated entirely.
- Since hot working is done in the recrystallization range, the original coarse grain structure is refined resulting in improved strength properties.
- Any minor impurities that might exist in forms of inclusions are broken up and distributed homogeneously throughout the metal.
- Since the metal is in a plastic state, the energy needed to shape it is far less than that required for similar cold working of the metal.

Of course, hot-working unit operations present some disadvantages that must be considered. There is very rapid oxidation and scaling of the surface of the metal because of the high temperatures involved, generally resulting in poor surface quality. Therefore, unit operations involving hot working of metal are not used where there are relatively close tolerances on the finished piece. The processing equipment and the maintenance costs are very high but are somewhat offset by the lower operating costs, relative to cold-working operations. Overall, the hot-working operations are comparable economically to those conducted at lower temperature.

As indicated in Fig. 18.1, the unit operations associated with the hot working of metal include rolling, forging, extrusion, drawing, spinning, and several somewhat specialized although common methods. Each of these is discussed in detail in the following sections.

2.1 Rolling

The unit operation that serves as a transition between the primary metals industry and the metalworking and forming industries is rolling. The metal ingots that represent the product of the basic metal industry are converted into the feedstock materials for the metalworking industry, usually in two steps. The ingots are rolled into intermediate shapes (blooms, billets, or slabs) that, in turn, are formed into plates, sheets, bar stock, particular structural shapes, or foils. The process of converting the ingots into appropriate feedstock configurations is referred to as rolling.

After pouring the metal into molds and allowing sufficient solidification to enable handling of the ingots, they are placed in holding furnaces, usually gas fired, where they are kept until a uniform working temperature is attained; for steel alloys this is about 2200°F (1200°C). Ingots then are taken to the intermediate rolling mill where blooms, billets, or slabs are formed. A bloom is square or circular in cross section with a minimum size of 6 in. (15 cm). A billet is smaller than a bloom, usually a minimum dimension of 1.5 in. (4 cm) and also is square or circular in cross section. A slab, which may be rolled directly from an ingot or from a bloom, is rectangular in cross section with the width being at least three times the thickness. Slabs usually are at least 1.5 in. (4 cm) thick and 10–15 in. (25–38 cm) wide. A progression of this intermediate rolling operations is used to produce sheets, bar stock, foils, or other desired forms.

Most primary rolling is done in a two-high reversing mill in which the piece passes between two rolls that are then stopped and reversed in direction. This back-and-forth movement is continued until the desired reduction in cross section of the piece is accomplished. The metal is turned one-quarter at frequent intervals in order to keep the cross section uniform and to maximize the refinement of the metal throughout. A minimum of about 26 passes usually is required to reduce a large ingot into a bloom. The two-high rolling mill is quite versatile since it can be adjusted easily to accommodate a wide range of piece size and rate of reduction. Its primary limitations are the length of stock that can be processed and the inertial forces that must be overcome each time the direction of movement is reversed. The latter limitation can be eliminated by using a three-high mill in which the middle roll serves as the bottom roll for movement of the piece in one direction and the top roll for movement in the reverse direction. Obviously, an elevating mechanism is needed with the three-high mill to raise the piece to the upper level; the piece falls by gravity to permit positioning in the lower two rolls. Although there is some minor difficulty because of incorrect speed for all passes, the three-high mill is less expensive to construct and install and has a higher output than the simple reversing mill.

Arrangements of rolls in mills other than the simple reversing mill and the three-high mill usually take advantage of four or more rolls with the extra rolls serving as backup for the two that are doing the actual rolling. For example, a four-high mill typically consists of four rolls aligned vertically with the two inner rolls serving to work the metal piece and the outer two providing backup or support for the rolling process. In addition, many special rolling mills take previously rolled stock and fabricate it into finished products such as rails, concrete reinforcement rods, and bars, by means of appropriately shaped grooves on both the upper and lower rolls of the mill. Although these mills frequently are referred to by the product being run (e.g., bar mill, rail mill), they are similar in appearance and operation to the mills used for rolling blooms and billets.

Billets can be rolled to size in the large mills used for blooms; however, this usually is uneconomical, and they more often are rolled from blooms in a continuous billet mill consisting of eight or more rolling mills in a straight line. The piece makes only one pass through each mill and emerges from the last with a final billet size that is appropriate for raw material for many final shapes, such as bars and tube stock. Typically, a 4-in. (10 cm) high by 4-in. billet can be reduced to $\frac{3}{4}$ in. (1.8 cm) round bar stock in approximately 10 passes through the finishing mills.

A major positive effect of the rolling operations is the grain refinement brought about by the recrystallization as the coarse, random structure of the grain is broken up and elongated by the action of the rolling mill. Because of the high temperature, recrystallization starts immediately and small grains begin to form and grow until recrystallization is complete. If further work is not carried out on the piece, growth of these grains continues until the lower temperature of the recrystallization range is reached. The wedging action on the piece is overcome by frictional forces acting on the points of contact between roll and piece. The metal emerges from the rolls traveling at a higher speed than the entering piece. Most deformation takes place in thickness, although there is some slight increase in width of the piece. Temperature control and uniformity is critical in all rolling operations since temperature directly affects metal flow and plasticity.

2.2 Forging

As a unit operation, forging can be defined as the process of plastically deforming metal or alloys to a desired shape by means of compressive force. This force can be exerted by a hammer, a press, rolls, or an upsetting machine. The process of forging clearly is among the oldest of the metalworking unit operations. Primitive man, when hammering metal tools and weapons into crude shapes using heat from an open fire, was forging.

Forging is being treated in this chapter as a hot-working operation, although it is recognized that some metals can be forged at room temperature. Forging at elevated temperatures improves the plasticity of the metal and reduces the needed forces. Each specific metal composition has its own plastic range, fairly wide for some alloys but quite limited for others. Alloys that do not possess a plastic range are not considered to be forgeable. As a point of reference, several metal and alloy systems are listed in Table 18.1 along with the range of temperatures suitable for forging operations.

Forged parts are characterized by a fibrous structure, generally attributable to the

Table 18.1. Temperatures for Hot Working of Common
Metals and Alloys

Metal/Alloy	Temperature Range	
	(°F)	(°C)
Copper and copper alloys	700–850	370–450
Low-carbon steel and alloys	2000–2400	1100–1300
Aluminum	700–850	370–850
Magnesium	600–700	315–370
Titanium and alloys	1350–1950	730–1100

effects of hot working. When metal is forged, both strength and ductility increase significantly in the direction of metal flow, resulting in a characteristic flow pattern that is unbroken and follows the contour of the piece. Thus the main objective of forging design is to control the lines of metal grain flow so as to put the greatest strength and resistance to fracture where it is needed, maximizing the strength-to-weight ratio.

Forging can be accomplished with either open or closed dies. Open-die forging applies local compressive forces progressively on different parts of the metal stock. Simple forged shapes can be obtained by either hammering or pressing the piece held between two flat surfaces between dies of relatively simple configuration. Parts produced by a closed-die forging operation are formed by applying force to the entire surface, causing the metal to flow into a die cavity that has been cut to a specific desired shape. More complex and precise parts can be produced with closed dies at production rates greater than with open dies.

The necessary compressive force can be applied by a number of methods; the two main methods are hammering and pressing. Although both methods are used together in some equipment, hammering causes the metal to change shape by repeated blows of a freely falling ram, and pressing changes the shape by the squeezing action of a slowly applied force. The basic forging processes consist of the following: smith forging, hammer forging, drop forging, press forging, upset forging, and roll forging. These and a few specialized techniques are discussed in more detail in the following sections. Modern forging strategies often use combinations of the unit operations described here. For example, a component part might be preformed to a rough approximation of the part geometry using a conventional hammer forging process and completed to its final shape on a forging press. It is not unusual for a modern forging shop to involve processing of a piece through three or more different types of forging equipment.

2.2.1 Smith Forging. Smith forging is the most basic of the conventional forging processes. It consists of forming the shape of a heated metal part by applying blows with a hand-held hammer, the piece being supported on a flat die. The desired shape is maintained by the forger; the desired length and cross section are adjusted manually by positioning the part on the flat surface. The village blacksmith represents the classic application of smith forging.

The quality of the forging is dependent totally on the skill of the smith. The nature of the process is such that close accuracy is not obtained and the shapes must be relatively simple. Accordingly, smith forging is not considered to be a production process normally. However, it seems reasonable to assume that this old, proven method of metalworking will continue to be used in its present form even though much more sophisticated forging methods are available.

2.2.2 Hammer Forging. Hammer forging differs from smith forging in that the heated metal part is shaped by impact of a mechanical hammer, operated either by steam or air. As a result, larger, heavier work is possible since much higher pressures can be exerted on the piece. Hammer-forged parts generally have a more uniform structure because of the consistency of the blows from the hammer. Considerable skill on the part of the operator is still necessary because, with the exception of the blow itself, requirements for processing the metal part are the same. The piece must be positioned properly and special tools, dies, and punches, used for producing holes, notchings, and other features of the part, must be used correctly. Mechanical manipulators are used to hold and move work that is too heavy to be positioned by hand. With these limitations, therefore, parts produced by hammer forging are limited to relatively simple shapes and in small quantities. With the possible exception of applications for preparation of special preforms, hammer forging is not generally considered a production method.

2.2.3 Drop Forging. Drop forging and the rest of the methods discussed in this chapter use closed-impression dies. The dies are made in sets or halves; one-half of the die is attached to the hammer and the other is positioned on the stationary anvil. The forging is produced by impact or pressure that forces the hot and pliable metal to conform to the shape of the dies. Repeated blows on the metal piece result in drastic flow of the metal. To ensure the proper flow of the metal during intermittent blows, the process is divided into a series of steps; parts made by drop forging cannot be formed by a single blow of the hammer. Each successive blow of the hammer changes the shape and configuration of the piece slightly, thus controlling the flow of metal until the final desired shape is obtained. The number of steps (blows) required is determined by the size and shape of the piece, the forging qualities of the metal, and the tolerances required. For large or complicated pieces, preliminary shaping operations using more than a single set of dies may be required. Depending on the size and complexity, parts may be drop forged at rates between 10 and 500 pieces per hour. Tolerances of 0.15–0.75 mm for relatively small pieces (less than 2.5 kg) and 0.75–2.30 mm for pieces weighing approximately 50 kg are easily obtained.

There are two principal types of drop forges, differing in the mechanism by which the force is delivered: steam hammer and gravity-drop hammer. In the former, the ram and hammer (upper, movable die half) are lifted by steam and the force of the blow is controlled by throttling the steam. Steam hammers range in capacity from 500 to 50,000 lb (200–20,000 kg) pressure and can deliver over 300 impacts per minute. For a given weight ram, a steam hammer can develop twice the energy at the die as can be achieved from a gravity-drop hammer.

With the gravity-type hammer, the impact pressure is developed by the free-falling ram and die as it strikes the lower fixed die half. Steam or air pressure is used to raise the ram to the desired height, and the pressure is released, allowing the ram to free-fall against the anvil. The capacities of the gravity-type machines are similar to those of the steam hammer type (i.e., 500–10,000 lb, or 20–4500 kg). The force of the impact delivered by the gravity units is dependent on the weight of the hammer. Gravity hammers are used extensively in industries that manufacture hand tools, cutlery, automotive and aircraft parts, and cutting surfaces such as scissors, chisels, mower blades, and similar implements.

By necessity, a slight excess of metal is processed; this results in each forging having a thin projection of the excess metal, called flash, extending around it at the parting line between the two die halves. Most forgings can be trimmed of the excess metal while cold, provided care is taken not to distort the part. The forged piece usually is held uniformly by the die in the ram and is pushed through a trimming die. Some industrial processes combine this trimming step with simple punching operations performed with the single movement of the ram.

Following the trimming operation, all forgings must be cleaned to remove the scale that develops on the surface of the piece. Normally, this is done by pickling in acid, shot peening, or tumbling, unit operations that are discussed in more detail in other sections of this book.

Between hammer blows, lubricants usually are applied to the die face and the exposed workpiece positioned in the bottom die. The purpose of the lubricant, applied by swab or spray, is to minimize wear of the die; however, it also serves to prevent stocking or fusing of the parts of the die, improve flow of the metal, and act as a parting agent. Prior to application of the lubricant, some operations utilize compressed air to blow-off or clean the die surfaces. The combination of lubricant application and blow-off frequently results in potential contamination of the workroom atmosphere. There usually is some overspray, if lubricant is applied as a spray. In addition, when oil-based lubricants hit the hot surfaces of the forging equipment, a portion is burned off, generating oil mist, particulate, and vapors associated with the more volatile components of the oil. The composition of these lubricating oils and their impact on the quality of the workroom are discussed in Section 4.4. Only under unusually severe conditions would the concentrations of any of the major constituents of these lubricating oils reach significant levels in the workroom. Most forging shops utilize local exhaust ventilation or internal air circulation, via pedestal fans and extensive general ventilation.

2.2.4 Press Forging. The forging methods described to this point all take advantage of impact blows from a hammer of some design to deform a metal piece. With press forging, no impact blows are struck; instead a squeezing action is used to force the metal to flow throughout the cavities of the die. The presses operate in a vertical orientation and are either mechanically or hydraulically operated. Forces of as much as 50,000 tons can be exerted. Pressures necessary to form steel, for example, at its forging temperature typically are in the range of 3,000–27,000 psi, based on the cross-sectional area of the forging across the surface of the die at the parting line.

With the press-forging operation, the primary function of the operator is to load and unload the press; the operator has very little influence over the quality of the forged parts. Most press-forged parts are preformed initially on other machines such as forging rolls, upsetters, benders, or other equipment. This enables the part to be finished on the forging press with a minimum of excess metal, resulting in somewhat more accurate forgings as a rule. The structural quality of press-forging products generally is considered to be at least as good as, often better than, that produced by drop forging. Press forging is faster than drop forging and the operating costs are lower. Press forging is best suited for production of pieces that are symmetrical in shape, with smooth surfaces and relatively close tolerances. Irregular and complicated shapes usually can be produced more economically by drop forging.

Of particular significance is the lower noise level associated with press forging relative to drop forging. Much of the energy associated with the impact of the hammer in drop forging is absorbed by the machine itself and its foundation from which it can be reverberated as noise and vibration. With press forging, a much greater proportion of the total energy put into the machine is transmitted directly to the metal part.

2.2.5 Upset Forging. Upset forging refers to the process by which feedstock, usually cylindrical, is held firmly in place by a section of dies and pressure is applied to the heated end resulting in it being upset or formed to the shape of the die. Thus, metal deformation takes place at one end of the feedstock, the formed piece is processed further if necessary, the finally shaped piece is removed, and the feedstock bar or rod is advanced to allow upsetting of the next piece.

The upsetting operation consists of placing heated feedstock into a stationary grip die, one-half of which is movable to allow advance of the feedstock. The heading die, with attached punch, moves forward and presses against the piece, displacing the stock until it conforms to the shape of the die impression. The heading die then releases the piece and returns to its original position. Depending on the configuration of the finished piece, the die cavity may be entirely contoured in the grip die, the punch, or both. Although a punching or shearing operation might be necessary, trimming usually is not required.

The punch on the heading die can be used to progressively pierce the part so as to displace the internal portion of the metal. This procedure is used to produce parts such as casings for artillery shells, engine cylinder housings, and even large cylinder barrels. Parts produced by the upset forging process range in size from small, fractional-gram units to massive pieces weighing several hundred kilograms. The dies are not limited to use in upsetting equipment and can be used for piercing, punching, trimming, or extrusion operations.

2.2.6 Roll Forging. Roll forging is used primarily for reducing the cross section of relatively short lengths of bar stock; typical products of roll forging are rifle barrels, levers, axles, leaf springs, and many simple preformed shapes for further processing. The function of the roll-forging machines is to reduce or taper short lengths of stock. The machine itself is similar in configuration and operation to the rolling mills discussed in Section 2.1, except that the rolls on the machine are not completely circular; usually at least one-quarter of the roll is removed to permit the stock to enter between rolls. The circular portion of the rolls is grooved according to the shaping that is desired. When the rolls are in the open position, the operator places a heated bar between them, using tongs. As the rolls rotate, the bar is gripped by the grooves on the circular portion and is pushed toward the operator. When the rolls rotate again to the open position, the operator either pushes the bar back into the machine or places it in the next of a series of grooves for subsequent forming. By rotating the bar one-quarter turn with each pass, the cross section is kept uniform and there is no flash. This process is repeated progressively, sending the piece between grooves of decreasing cross section, until the desired size and shape is obtained. As was the case with initial rolling operations, an improved fiber structure is developed in parts made by roll forging because of the progressive squeezing and hot-working action of the roll dies on the metal piece. Both straight and tapered work can be roll forged; surfaces that are smooth and free of scale pockets can be obtained.

2.3 Extrusion

The principle of extrusion has been used in a wide range of industrial processes for manufacture of such items as brick, hollow tile, even macaroni. Metals that can be hot worked can be extruded to uniform cross-sectional shapes by the aid of pressure. Some metals, primarily lead, tin, and aluminum, can be extruded cold (a process discussed later in this chapter), whereas others require application of heat to render them plastic before extrusion. Although the specific extrusion processes vary somewhat depending on the metal and application, the basic operation consists of forcing metal, confined in a pressure chamber, out through specially formed dies. Metal extrusion products include rods, tubes, molding trim, structural shapes, brass cartridges, and metal-clad cables.

Most extrusion presses are horizontal in configuration and are operated hydraulically. Depending on temperature and material, operating speeds vary from a few feet per minute up to 1000 fpm (300 mpm). The advantages of extrusion include the ability to produce a variety of shapes of high strength, good accuracy and surface finish at relatively high production speeds, and with relatively low die cost. More deformation of shape change can be achieved by this process than by any other hot-working operation discussed in this chapter. Almost unlimited lengths of continuous cross section can be produced easily; this, together with low die costs, may justify selection of extrusion processes. In general, extrusion processes are about three times slower than roll forming processes.

Extrusion processes can be divided into direct and indirect methods. In the direct extrusion process, a heated billet is placed into the die chamber and a dummy block is placed against the billet at the ram end of the extruder. The metal then is forced by the ram through the die opening at the opposite end of the billet. The metal is forced through the die until only a small amount is left; this is sawed off next to the die and the butt end is removed. Indirect extrusion is similar to direct extrusion except that the extruded material is forced through a die in the ram stem. Thus the extruded metal leaves the extruder in a direction opposing the motion of the ram; with direct extrusion, the extruded metal leaves the extruder moving in the same direction as the ram. Less force is required with indirect extrusion since there is no frictional force between the billet and the billet chamber. However, since the extruded metal moves through a hollow ram stem, the ram itself is weaker than in the case of direct extrusion. It also is more difficult to support the extruded material as it leaves the die since it is moving through the ram stem.

Tubes, when made by hot working of metal, usually are produced by a direct extrusion method, using a mandrel to shape and support the inside of the tube. The heated billet is placed inside the pressure chamber, and the die containing the mandrel is pushed through the die and around the mandrel. The entire operation must be rapid; speeds of 10 fps (3 mps) have been used.

2.4 Drawing/Cupping

Drawing refers to a process for making tubes or closed-end cylinders (cups) by literally punching the center from a metal blank piece. For products that cannot be made by conventional seamless rolling mill equipment, this process often is used. The piece is heated to forging temperature, and, with a piercing punch operated in a vertical press, the metal is forced into a closed-end hollow forging. The piece is reheated and placed in a heated drawbench consisting of several dies, of successively decreasing diameter, mounted in a single frame. The punch, usually operated hydraulically, forces the heated cylinder through the full length of the drawbench. For long, thin-walled tubes or cylinders, repeated heating and drawing operations may be necessary. Deep drawing refers to a more severe action of the punch, resulting in the draw (or length of displacement of the metal) being greater than the diameter or cross section of the piece.

If the final product is to be a tube, or other open-ended component, the closed end is cut off, and the remaining shell is sent through finishing and sizing rolls of proper configuration. To produce closed-end cylinders, such as those used for storage and transfer of compressed gases, the open end is swaged to form a neck or is reduced by other processes such as hot spinning.

2.5 Hot Spinning

Metal spinning as a hot-working unit operation is used to form heavy circular plates, or dishes, by use of a rotating form or to neck down or close the ends of tubing. In any case, a form of lathe is used to rotate the part rapidly, with the shaping being done with a blunt pressure tool or roller held in contact with the rotating part. This causes the metal to flow and conform to the mandrel of desired shape. Once the operation is begun, there is generation of considerable heat due to friction; this aids in maintaining the metal in a plastic state. For tubular products, hot spinning can be used to reduce the diameter of the tube, form the end to some desired contour, or close the end completely.

2.6 Other Hot-Working Operations

Seamless tubing can be produced by a hot-work process in which the billet or metal is forced over an appropriate mandrel held between two conically shaped rolls turning in the same direction. The solid billet first is center punched, brought to forging temperature in a furnace, pierced, and fed through a series of processing rolls. An alternate squeezing and bulging action opens the billet in its center, the size and shape of which is controlled by the mandrel. As the relatively thick-walled tube emerges from the piercing mill, it passes through a set of mills that converts it into a longer tube with specified wall dimensions. Still at working temperature, the tube passes through a reeling machine that straightens and final-sizes it, as well as gives it a smooth surface.

Another special feature becoming common in hot working of metals is the use of heated dies to obtain thinner sections in forgings. With use of the proper lubricant, closer tolerances can be obtained, surface oxidation is minimized, the work remains pliable for a longer period of time, and the production rate can be increased. However, heating of the dies decreases their useful life. There is also additional operating cost associated with heating of the dies. This process is justifiable only when very thin-section components are desired.

High-energy rate forming (HERF) usually is considered a cold-work unit operation and is discussed in more detail in Section 3.4.5. However, some high-velocity presses are driven by various high-energy mechanisms such as explosive discharge or capacitor discharge, most parts being formed in a single blow. Since this operation is very fast, thin sections can be forged before the heat is lost from the metal piece. Again, because the impact load is so great and the temperature increase at the die is rapid and extreme, die life is very short. Therefore, this process finds practical application only in the forging of high-temperature difficult-to-form alloys.

Finally, metal that is difficult to forge, such as titanium, can be formed in a press surrounded by inert gas. This operation is sometimes referred to as environmental hot forming. It virtually eliminates oxidation and scaling and tends to maximize life of the dies. For very large forgings, the inert gas is introduced and maintained only in the forming area of the press. Small presses can be enclosed completely in a cabinet, and the entire cabinet purged with the inert gas.

3 COLD WORKING OF METAL

Cold working of metal refers to the rolling, extrusion, or drawing of a metal at temperatures below the recrystallization range for the metal. Most metals are cold worked at room temperature, with elevation of temperature resulting from the action on the metal. In contrast to the effect of hot working to refine the grain structure in the metal, cold working tends to distort the grain and does virtually nothing to its size. On the other hand, cold work generally results in increased strength, machinability, dimensional accuracy, and surface finish of the metal. In addition, since oxidation is much less of a problem with cold working, much thinner sheets and foils can be processed than by comparable hot-working operations.

Many of the processes and much of the equipment used for cold working of metal are similar to those used for hot-working operations; however, the forces required and the mechanisms for dissipating the heat usually are quite different. Generally, much greater pressures are needed for cold working than for hot working. The metal, being in a more rigid state, is not deformed permanently unless and until stresses exceeding the elastic limit are reached. Since, by definition, there can be no recrystallization of grains in cold working, there is no recovery from the grain distortion or fragmentation. Thus, as grain distortion continues, greater resistance to this action is built up internally, resulting in the increased strength and hardness of the metal. The piece is said to be strain hardened by this action; this is the only means of changing hardness and strength properties of some metals that do not respond to conventional heat treatment.

The amount of cold work a metal can withstand is dependent on its ductility; the greater the ductility, the more cold working it can receive. Generally, pure metals can withstand greater deformation than alloys since alloying increases the tendency toward rapidity of strain hardening. Also, large-grain metals are more ductile than small ones and are therefore more desirable from a cold-working standpoint.

When metal is deformed by cold work, severe residual stresses are set up in the metal. These stresses are not desirable and, to remove them, the metal must be reheated. Below the recrystallization range, this heating renders the stresses ineffective without significant change in physical properties or grain structure of the metal. If heating is taken into the recrystallization range, however, the positive effects of cold working are eliminated and metal reverts to its original condition. In some special cases, it might be desirable to retain some residual stress. For example, the fatigue life of small parts sometimes can be improved by shot peening (Section 3.4.4), which causes the surface of the metal to be under compression while the material beneath is in tension.

Cold working of materials frequently is done as a finishing step following hot rolling. Hot-rolled strips and sheets are relatively soft, have surface imperfections and usually lack dimensional accuracy and other specific physical properties. Cold-rolling operations reduce size slightly, permitting accurate dimensional control. In addition, no surface oxidation results, a very smooth surface is obtained, and strength and hardness properties are maximized. In summary, the following effects generally result from cold-working operations.

- Stresses are set up in the metal; these remain unless removed by subsequent heat treatment.
- Distortion of the metal grain structure results.
- As the ductility of the metal is decreased, strength and hardness are increased.
- For some metals, the lower temperature of the recrystallization range is increased.
- The quality of the surface finish is improved.
- Very close dimensional tolerance can be achieved.

The cold working of metal, as defined here, is essentially the same as the generic process commonly referred to as presswork. A more general and more vague term, stamping, is used almost interchangeably with presswork. In any event, presswork refers to the wide variety of processes by which workpieces are formed, usually from metal sheets as feedstock. The overall process of presswork normally is understood to include four major subdivisions: cutting/shearing, bending/forming, extrusion/squeezing, and drawing. Since the first of these processes results in no change in the grain structure of the metal piece, it is not covered in this chapter but is discussed as a metal machining unit operation in Chapter 19. The remainder of this section consists of discussions of the various unit operations that comprise the bending, squeezing, and drawing processes, all of which are considered cold-working operations.

3.1 Bending and Forming

Although the terms *bending* and *forming* often are used interchangeably, there are distinct differences between these metalworking processes. Bending consists of uniformally straining flat sheets or strips of metal around a linear axis. As a result, metal on the outside of the bend is stressed in tension beyond the elastic limit and metal on the inside of the bend is compressed. The metal tends to become slightly thinner at the bend because of the tension of the metal fibers as they are stretched plastically around the outside of the bend. Forming, on the other hand, involves virtually no metal flow. As a result, the thickness of the metal and the area of the original blank are not changed substantially. Precise right-angle bends are difficult to form on high-volume production, particularly when harder materials and/or heavier gauge metal are used. Because of the high possibility of cracking, parts to be made with right-angle bends should be laid out so that the bends are formed across the strip. The largest radius of curvature practicable should be used; as a general rule, the inside radius of the bend should be at least equal to the stock thickness. The shape that is to be formed on the sheet or strip depends on the shape of the punch and die used to produce the piece. When the desired configuration is complex, the shape may be developed over a progressive sequence of operations.

3.1.1 Simple Bending/Forming. Simple bending or forming operations usually are performed on crank, eccentric, or cam-operated presses, or by press brakes. The presswork equipment used for bending and forming is the same as that used for cutting purposes. The only difference is that the configuration and travel of the punch is limited so that fracture of the metal pierce does not occur. A tremendous variety of punches and dies is used in bending/forming presses.

Press-brake equipment is a special type of press arrangement for forming straight and relatively narrow workpieces, the lengths of which usually exceed 2 ft (0.6 m). Press-brake forming operations also are used for limited production quantities when close tolerances are not needed. This versatile and economical process may be used for stock ranging from thin foil sheets up to plates over 1 in. thick. Press brakes are available in sizes sufficient to accommodate stock over 20 ft (60 m) in length. The action on the ram (punch) is provided either by mechanical or hydraulic means, with capacities of up to one stroke per second.

As indicated earlier, the action of a bending or forming machine is essentially the same as a slow-speed punch press. In both cases, suitable dies and punch configurations are used to form the desired shape in the workpiece. With the arrangement of punch and die, the presses and press brakes can be used to blank, pierce, shear, lance, straighten, emboss, corrugate, or flange the metal workpiece.

3.1.2 Roll Forming. The other major category of bending and forming processes is roll forming, which, in turn, is subdivided into two important production methods: contour roll forming and three-roll forming.

CONTOUR ROLL FORMING. Contour roll forming is a high-production process in which flat sheets or strips of metal are fed continuously through a line of contoured rolls producing forms of a wide variety of uniform cross-section shapes. The feedstock may be precut lengths of sheet or may be fed from a coil, the latter being more economical. When the sheet is fed from a coil, an automatic cutoff device is necessary to obtain the desired lengths of the roll-formed pieces. As the metal passes through the machine, each pair of forming rolls, usually referred to as a station, produces a partial change in the cross section. As the piece moves from station to station, the flat stock is transformed into the desired final shape. Of course, the number of stations that are required on the line depends on the complexity of the desired shape, the material being processed, metal thickness and hardness, and tolerances on the piece. Some simple shapes can be produced easily on a two- or three-roll machine, whereas over 40 stations have been used to produce fairly intricate pieces.

Contour roll forming has been applied successfully in the manufacture of a wide range of products. Shapes ideal for this process are those that are symmetrical about a vertical center line, since they require minimum straightening. Products commonly produced by contour roll forming include door and window trim, gutters, downspouts, siding, structural shapes, channeling, appliance framing, automotive frame and body parts, and tubing.

Although pieces could be made by other processes, such as press-brake methods, these usually are not competitive unless the desired piece can be made in a single stroke of the press. For sections requiring two or more forming strokes, contour roll forming generally is the most economical method. Although tooling costs are much higher with contour roll forming than with conventional presswork, roll forming usually can be justified on the basis of increased production rates.

THREE-ROLL FORMING. As implied by the name, this process consists of forming various cylindrical shapes by passing sheet metal blanks through a set of three rolls. Two of the rolls are held in fixed position; the third is adjustable. As a metal plate or sheet passes through the rolls, its final radius of curvature is determined by the position of the adjustable roll. The closer the adjustable roll is to the two fixed rolls, the smaller will be the diameter of the rolled piece. Some roll forming machines provide drive to all three rolls; these tend to hold the work very firmly in place, and parts with greater dimensional stability can be produced.

With use of appropriate dies on the top roll of the three-roll forming machines, special irregular shapes can be formed. For example, corrugated round cylinders, flattened or elliptical cylinders, even truncated conical shapes are relatively simple to produce.

Most metals are work hardened by cold, roll-forming operations. In severe cases, such as in forming small-diameter cylinders, intermediate annealing usually is necessary. Although cold forming always is preferred over hot forming for economic reasons, a modified cold-forming process known as "warm forming" may be used for severe forming requirements. Warm forming usually is done at a temperature equivalent to the tempering temperature of the metal and is done where hot forming would alter the mechanical properties of the metal.

For the roll-forming processes, the skill of the operator and the condition of processing equipment play key roles in maintaining the desired roundness of the workpiece. To impart additional strength and stiffness to the completed part, features such as beads or flanges are incorporated where possible into the workpiece design. Common examples of this approach are oil drums and other similar containers.

SEAMING. In the manufacture of metal drums, pails, cans, and various other products made of light-gauge metal, various types of seams are used as locking mechanisms to join the ends of the workpiece together. The seaming machines are, in fact, bending machines, and are used to produce the seam once the basic shape of the container has been formed. Many different types of seams are possible. The most common one probably is the simple lock seam, which is, in effect, the interlocking of U bends formed at the end of each piece being joined.

Seaming machines typically are made to receive the container, form the joints on the mating edges of the part, and close the seal either by means of a powerpress or manually. These simple seams usually are used for longitudinal seams. Seams on the bottom, and sometimes the top, of the container can be made either flat or recessed and usually are actually double seams. Double-seaming machines may be hand operated, semiautomatic, or fully automatic. Here, semiautomatic implies that the operator must load and unload the machine but the seaming itself is done automatically.

3.2 Metal-Squeezing Operations

In the case of bending and forming operations just discussed, the metal piece is subjected to very little distortion at the grain level, except when severe bending is done. In contrast, metal-squeezing operations result in substantial flow of metal to produce a desired shape or configuration of the piece. The four major subdivisions of squeezing processes that are common enough to be included in this section are swaging, cold heading, coining/embossing, and thread rolling.

3.2.1 Swaging. Swaging is a general term that may be appropriately applied to a number of metal forming operations in which the cross section of a piece is changed. On a production basis, swaging operations usually are performed on rotary swaging machines. These machines progressively reduce the cross-sectional shape of bar, rods, tubes, or wires by impacting blows from pairs of opposing dies. The blows displace metal and form the blank to the shape of the dies.

Most rotary swagers have either two or four dies. The sequence of operations begins when the metal piece is introduced into the die opening. A revolving spindle throws the rotating dies outward by centrifugal force against a series of rollers held in a cage that surrounds the spindle. As the die supports, or backers, strike the rollers, they rebound to force the dies inward and strike against the workpiece. After impact, the dies open and release the workpiece when the backers lose contact with the rollers. The cycle then is repeated. The time interval between opening and closing of the forming die halves depends on the speed of the spindle rotation, distance between adjacent rolls, and the radius of the backer. A part undergoing swaging may be subjected to as many as 5000 impacts per minute. Mechanical pencils, metal furniture legs and umbrella poles and struts are examples of parts made by this process.

Swaging machines are somewhat difficult to keep clean and present several disadvantages over other competitive means of tube or bar stock size reduction. Tooling must be changed in order to produce stepwise reduction. Excessive surface contamination, such as oxides or scale, must be removed from the piece to prevent loading the die and interfering with moving parts of the equipment. Swaging must be done on a part before plating or painting because of the significant changes made on the physical properties of the swaged end. The process usually is not economical for short-run productions. It is very difficult to form steep tapers or shoulders on parts. In fact, when multiple, or stepwise, reductions are made on a piece, die marks often are noticeable and have to be removed by appropriate machining processes. From an environmental standpoint, one major objection to the swaging machines is the extremely high noise output, a condition difficult to control by engineering means, thus resulting in most operators having to rely on personal hearing protection in some form. For these and other limitations, swaging should be selected in a production scheme only if it is the most practical method of producing the necessary shape; competitive methods such as press forming, spinning, or machining, may be better suited. In fact, it is common to combine swaging with some other process, such as machining, to achieve increased production, decrease tooling costs, and/or obtain improved tolerances or surface finish.

3.2.2 Cold Heading. Cold heading, also known as cold upsetting and axial-flow forming, has become a very important high-volume method of metalworking. It is one of the fastest and most economical methods for producing assembly components such as nails, rivets, and bolts. In a sense, cold heading is a form of swaging in that the work on the piece results in a change in the shape and configuration of the end of the metal piece. Cold-heading machines usually are distinguished on the basis of the means by which the feedstock is positioned in and held by the die and are referred to as either "solid-die" or "open-die" units. In the case of the solid-die machines, the die is simply a hardened cylinder with an axial hole through its center. The hole is slightly larger than the outside diameter of the feedstock. On the automatic machines, feed parts are moved to a cutoff station where slugs of proper length are formed and forced into the die by a punch. An ejector pin, inside the die, stops the slug, permitting the correct amount of the blank to project out the end of the die. One or more blows by the punch against the protruding blank then reshape, or head, the metal, causing it to flow plastically into the die impression. The die impression can be in the stationary die, in the punch entirely, or partially in both the die and the punch. In any event, when the heading operation is complete, the ejector pin advances to knock the finished part from the die cavity.

Open-die machines use dies composed of two halves with matching grooves machined to the desired configuration. The machine is loaded by feeding stock, usually wire or small-diameter rod, into the die halves and against a stop. The die halves then close to grip the part and shear it to correct length, after which the part is headed to final shape by one or more blows of a punch. After the operation is complete, the die opens and the incoming feedstock serves to eject the finished part. Fully automatic machines of this type can perform at a rate of about 500 parts per minute.

Cold heading lends itself readily to combinations with other metalworking processes. An important category of special cold-heading machines includes the so-called bolt-makers in which cold heading, trimming, pointing, and threading operations all are performed on a single machine. Completely finished hexagonal head and socket-head cap screws are now made routinely on these bolt-makers. These units can accept feedstock over an inch (2.5 cm) in diameter and produce in excess of 300 parts per minute.

Almost all metals can be cold headed. The low- and medium-carbon steels are probably the most commonly used materials for this process, although some grades of alloy steels and even stainless steels are used for some products. The more common of the nonferrous materials used in cold-heading applications are brass, copper, bronze, nickel alloys, and aluminum alloys.

3.2.3 Coining and Embossing. Coining and embossing are somewhat similar methods belonging generically to the squeezing category of cold working of metals. The principle of coining is precisely that used for production of coinage. A blank is placed upon or in a stationary die that confines the metal and restricts its flow in a lateral direction. A punch then impacts the blank, producing relatively shallow configurations on the surfaces of the piece. The configurations on opposite sides of the piece normally are different, such as with common coinage, and there is no matching of the punch and die outlines. Special high-pressure presses are required for this type of operation; as a result its use is limited to fairly soft alloys.

Embossing actually is more of a drawing or stretching operation in that there is relatively little squeezing of the metal and practically no change in the thickness of the metal. The punch is relieved so that it touches only the part of the blank being embossed. Of course, the punch and die conform to the same configuration, and the result of the embossing operation is a design that is raised from the parent metal. (Debossing refers to the case where the design is pressed into the parent metal and is used where a raised pattern might create production or other problems.) The major application of embossing operations is making license plates, nameplates, medallions, and decorative work on thin sheet metal or foil.

3.2.4 Thread Rolling. Threads can be rolled into any material that is sufficiently plastic to withstand the forces of cold working. Materials can be characterized as to their "rollability," their behavior during the rolling process, which is dictated by such basic physical properties as hardness, ductility, internal friction under deformation, and yield point.

In thread rolling, the metal feedstock is a cylindrical blank that is cold forged under

considerable pressure by the rolling action of the blank between either rotating cylindrical dies or reciprocating flat dies. The surface of the dies has the reverse form
of the thread that is rolled. Rolling under pressure results in a plastic flow of the metal
in such a way that the die penetrates to form the root of the thread and the displaced
metal flows upward to form the crest. This approach requires slightly less material
than would be needed for operations in which the thread is cut into the blank.

When the piece is rolled between two flat dies, both dies are provided with parallel
grooves cut to the size and shape of the desired thread. One of the dies is held stationary
while the other reciprocates and rolls the blank between the dies. The other thread
rolling method uses either two or three grooved roller dies. In the two-die machine,
the blank is placed on the work stop between two parallel, cylindrical rotating dies,
and the right-hand die is fed into the blank until the correct size is reached, after
which it is returned to its starting position. The three-die machine uses cylindrical
rotating dies mounted on parallel shafts driven synchronously at the desired speed.
They advance radially into the blank by cam action, dwell for the necessary time, and
then withdraw.

3.3   Drawing

The third category of cold working of metals discussed in this chapter is referred to
as drawing, a process that can be considered a stretching of the metal, resulting in a
reduction in the dimension of the piece in one direction and corresponding increase
in another direction. For example, in the case of tube finishing, the tube is drawn,
resulting in a reduction in the cross-sectional area or diameter but an increase in the
length of the piece. Thus, this basic process is significantly different than either bending/
forming operations or squeezing processes discussed previously. Although there are
many variations of the basic principle of drawing present in industrial operations, only
four unit operations within the category of drawing process are presented here: tube
drawing, wire drawing, metal spinning, and stretch forming.

3.3.1 Tube Drawing. Tubular products normally require dimensional accuracy, smooth
surfaces, and superior physical properties. The finishing of tubing products commonly
is done either by cold drawing or by use of a tube reducer, which also is a drawing
process. Tubing usually is first, or rough, formed by hot-working processes such as
hot rolling (Section 2.1). The tubing then is treated by pickling and cleaning to remove
scale. Just prior to the drawing operation, a lubricant is applied to prevent galling,
reduce friction, and increase the smoothness of the surface. The actual tube drawing
is accomplished on a drawbench. To do this, one end of the tube is reduced in diameter
by an operation such as swaging in order to permit the tube to enter the die. This
reduced end is gripped in tongs fastened to the chain or other conveyance mechanism
on the drawbench. In this operation, the tube is drawn through a die smaller than the
outside diameter of the tube. The inside surface and diameter are controlled by a
fixed mandrel over which the tube is drawn if the nature of the inside diameter is not
important, the mandrel may be omitted. The typical medium-to-large industrial drawbench may have a total length of up to 100 ft (30 m) and require drawing power of
up to 300,000 lb (135,000 kg).

The process of drawing a tube obviously is very severe, requiring the metal to be
stressed above its elastic limit to permit plastic flow through the die. Maximum reduction for one pass through the drawbench is about 40%. This operation increases
the hardness of the tube so much that, if several stepwise drawings are desired, the

tubing must be annealed after each pass. This process can be used to produce tubing having diameters much smaller and walls much thinner than can be obtained by hot rolling. For instance, hypodermic tubing, having outside diameters of less than 0.005 in. (0.013 cm) can be produced readily by this process.

The tube reducer differs from the drawbench in that it has semicircular dies with tapered grooves through which the feedstock tubing is alternately advanced and rotated. The dies rock back and forth as the tubing moves through them. A tapered mandrel inside the tubing regulates the size to which the tube will be reduced. The tube reducer can make the same reduction in one pass that might take four or more passes through a conventional drawbench. The primary advantage of the tube reducer over the drawbench is the much greater lengths of tubing that can be processed. Tubing finished by either method has all of the advantages normally associated with cold-working metals and, additionally, can be made in longer lengths and with thinner walls than is possible by any hot-working operation.

Although this basic process is used for production of tubing (i.e., open-end cylindrical pieces), a modification of the drawing process permits the forming of small, closed-end units such as metal cans and containers. The actual process is a hybrid of drawing and stretch forming, discussed in Section 3.3.4. Basically, this process involves positioning a flat feedstock over a die and having a punch press into the metal. As a result, parts of the metal are stretched and others are compressed. As with any drawing process, the metal is hardened by this action, and it sometimes is necessary to include annealing operations between stages of forming. This is particularly true for deep-drawing operations where the depth, or length, of the draw is greater than the diameter of the piece.

3.3.2 Wire Drawing. Wire is finished by a process similar to tube drawing but obviously without concern for the internal configuration of the piece since it is solid. Wire is produced in its finished form by colddrawing a wire rod through one or more dies, each being slightly smaller in cross section than the preceding one. The feedstock to the wire-drawing operation typically is rolled from a single billet, cleaned in appropriate acid baths to remove scale and rust, and is coated with a material to prevent oxidation, neutralize any remaining acid, and act as lubricant or a coating to which a later applied lubricant can cling.

Both single-draw and continuous-drawing processes are in use. In the case of single-draw operations, a coil of rough-finished wire is placed on a feed reel, and the end of the wire is pointed so that it can enter the die. The end is grasped in the same way as tubing was on the conventional drawbench, and the wire is forced through the die to such a length that permits winding the end of the wire around a drawing block/takeup reel. From this point on, the rotation of the draw block pulls the wire through the die and forms it into a coil. These operations are repeated with smaller dies and blocks until the wire is drawn to its desired size.

In continuous drawing, the wire is fed through several dies and draw blocks arranged in series. This permits drawing the maximum amount in one pass before annealing is necessary. The number of dies in the series will depend on the kind of metal or alloy being processed and may involve as many as 12 successive drafts. The dies usually are made of tungsten carbide, although diamond dies can be used for drawing very small diameter wire.

3.3.3 Metal Spinning. Simply stated, metal spinning is the process of shaping thin sections of metal by pressing them against a form while it is rotating. The nature of the process limits its application to symmetrical articles of circular cross section. This type of work is accomplished on a lathe, similar to an ordinary wood lathe except that a mechanism for holding the metal against the form replaces the usual tailstock. The forms used to shape the piece commonly are turned from wood and attached to the face plate of the lathe. However, for high-volume, long production runs, it is recommended that smooth steel chucks be used. Steel chucks are less likely to develop interior imperfections and, in fact, can be more economical than wood if surface finish is a consideration.

Practically all parts formed by the metal-spinning process are made with the aid of blunt hand tools that are used to press the metal against the form. A hand or tool rest is positioned in front of the workpiece to provide support for the hand tools; additionally, a support for a trimmer cutter or forming roll is placed at the rear. Parts may be formed either from flat metal blanks or from feedstock blanks that have been partially drawn in a press. The latter is used as a finishing operation for many deep-drawn pieces. Most spinning work is done on the outside diameter although inside work is possible.

Bulging work, such as on metal pitchers, vases, and similar pieces, is done by means of a small roller, supported from the tool rest, operating on the inside and pressing the metal out against a form roller. In such an operation, the part must first be drawn and possibly given a bulging operation prior to spinning since spinning cannot be done near the bottom of the piece.

Tool friction is minimized in this operation by use of lubricants such as soap, beeswax, white lead, and linseed oil. Since metal spinning is another cold-working operation, there is a limit to the amount of drawing of the metal possible before it becomes necessary to anneal the piece. Spinning lends itself to relatively short-run production jobs although it has some application in quantity production work. Advantages of metal spinning over conventional presswork include lower tooling costs, the ability to bring a new product to production much sooner, and the minimal capital costs in comparison to what a press capable of doing the job would cost. On the other hand, labor costs are much higher for spinning than for conventional presswork, and production rates are probably much less. This process is used frequently in the manufacture of bells on musical instruments, light fixtures, kitchenware, reflectors, funnels, and large chemical processing vessels.

When thick metal plates are processed by metal spinning, power-driven rollers must be used in place of the conventional hand tools. When powered rollers are used, the process is referred to as shear spinning. The metal usually is formed from flat plate that initially is held securely against the mandrel by holders of appropriate configuration. Roll formers, the powered counterparts of the hand tools, force the plate to conform to the mandrel, maintaining a uniform wall thickness from the starting point until completion. Reduction in wall thickness of up to 80% is possible by this process although, in most cases, the reductions are much smaller.

In conventional metal spinning, the wall thickness remains about the same throughout the operation. The hand tools are used merely to bend or flare the metal into a new contour; there is no, or very little, plastic flow of the metal or reduction in wall thickness. In shear spinning, however, the metal is reduced uniformly in thickness over the mandrel by a combination of the rolling action and extrusion of metal in front of the rollers.

Most metal can be formed by shear spinning. Although heat is sometimes applied throughout the cycle to facilitate the flow of metal, it is not required for most steel alloys and other nonferrous metals. Some advantages of shear spinning over other methods of forming partial spheres or conical shapes include increased strength of the part, material savings, reduction in manufacturing costs and a good smooth finish to the surface.

3.3.4 Stretch Forming. One operation that is difficult to accomplish by most metal-working methods is the forming of large sheets of thin metal involving nonsymmetrical shapes or double-curved bends. For this unique type of forming operation, stretch forming can be used effectively. In stretch-forming machines, a single die is mounted on a ram that is positioned between two slides that grip the metal part. The die moves in a vertical direction, and the slides move horizontally outward from the ram center. Jaws grip the metal blank and stretch the part at the same time as the ram forces the die downward and into the metal. Large forces of 100 tons or more must be provided for the ram and slides. In the process of stretching, the sheet is stressed above its elastic limit while conforming to the shape of the die. Some thinning of the sheet occurs as well. The overall action is such that there is little springback of the metal once it is formed. The process is adaptable to both short-run jobs and long-run production schemes.

Dies of inexpensive materials such as wood, plastic, or steel can be used effectively. Large double-curvature parts, difficult to form by other methods, can be made easily with this process. Many alloys, difficult to process by conventional methods, can be shaped by stretch forming. There is very little localized stressing of the metal, and the problem of unequal metal thinning is minimized. On the other hand, scrap loss is quite high since material must be left at the ends and sides of the piece to permit the jaws to grip the piece, and this must be trimmed and finished.

Equipment combining stretch and draw forming has been developed and used to produce many special pieces. There is some evidence that most metals become unusually ductile after being stretched a few percent; as a result, they can be formed with one-third the force required normally.

Despite the relatively high scrap loss, stretch forming and stretch-draw forming appear to be gaining in application. These processes are used for such purposes as short-term production of aircraft parts of aluminum and in the automotive industry to make steel roof panels, hood covers, rear-deck lids, and door posts. Titanium and stainless steel sheets can be formed easily in this manner.

3.4 Other Cold-Working Methods

The preceding unit operations comprise the most common process by which metal and alloys are formed and worked at or near room temperatures. There literally are dozens of other methods, many being relatively minor modifications of those already discussed, in use in manufacturing processes. Several additional methods are described briefly in the following section, either because they appear to be gaining in popularity or they hold promise of becoming used more routinely.

3.4.1 Hobbing. Hobbing is a cold-working process similar in many respects to coining and embossing processes. However, the purpose of the hobbing operation primarily is to produce mold cavities for use in the plastics and die-casting industries. The process consists of forcing a hardened steel form, the "hob," into a softer metal thus producing the cavity of the mold. The hob is machined to the exact form of the piece to be molded. It then must be heat-treated in order to obtain the necessary hardness and strength required to withstand the tremendous pressures involved. Pressing the hob into the blank requires much care and skill. Several successive pressings and annealings are necessary before the typical hobbing job is complete. During the hobbing operation, flow of metal in the blank is restrained from any appreciable movement in a lateral direction by means of a heavy retainer ring placed around it. The actual pressing is done usually in hydraulic units having capacities from a few hundred to several thousand tons.

The primary advantage of hobbing is that many, identical mold cavities can be produced economically. The surfaces of the cavities have a highly polished finish, and no finishing or machining work is needed other than to remove surplus metal from the top and sides of the piece.

3.4.2 Cold Extrusion. The more ductile metals and alloys can be extruded under cold-working conditions, although the process usually involves high-energy impact on the part. The latter process, cold-impact extrusion, is discussed in Section 3.4.3. For simple cold extrusion, techniques and equipment similar to those discussed for conventional hot-extrusion operations would be used. A major difference would be the need for intermediate annealing of the cold extruded part because of the extensive work hardening of the metal that results.

Application of simple cold extrusion usually is limited to such tasks as reducing the diameter of elongated metal parts, such as axles or motor shafts. For such operations, the billet feedstock, with previous upsetting completed if there is to be special configuration at the ends of the shaft, is loaded into the extruder, after which the process is essentially automatic. The basic process is one of reducing the diameter of the piece by forcing the shaft through a ring-type die, slightly smaller in diameter than the piece. Several stages, or passage through a series of such ring-dies, are needed to achieve the desired final diameter of the part.

Parts completed by this process typically have an improved surface finish that extends the fatigue life of the material. Other advantages of this process include the work-hardening of the surface that improves the physical properties of the material and the relatively simple finishing operations required for the final product since only minimal amounts of stock have to be removed.

3.4.3 Cold-Impact Extrusion. Cold-impact extrusion is used as a high-volume production process for the manufacture of an impressive variety of commercial products. The basic process involves the shaping of a metal blank in a die by the forceful blow of a punch. Parts produced by this process tend to be characterized by fairly long, tubular (i.e., constant cross section) regions. In cold-impact extrusion, the shape of the metal blank, or slug, is rearranged by way of movement as a viscous fluid when it is struck by the punch. The shape of the final part results from the plastic deformation of the slug under the tremendous pressures of the compressive forces, often as high as 300,000 psi.

Parts can be produced by cold-impact extrusion by one of three methods that differ in the nature and direction of plastic deformation relative to the position and action of the punch. These methods are referred to as backward, forward, and combination (backward/forward) extrusion. The specific method of extruding chosen for a particular application is selected on the basis of the desired shape of the part, although other factors such as the material, accuracy requirements, cost, and quantity desired also are considered.

In backward extrusion, parts are extruded from a solid slug in a closed-bottom die so that a significant portion of the slug flows backward over the descending impacting punch in a direction opposing that of the punch. As the punch is withdrawn, the completed extrusion is removed by a stripping device positioned above the die. This method is used frequently to make cup-shaped parts such as cans and shells. Projections or recesses can be formed readily inside on the base of the part by proper design of the die. The bore size of the die determines the outside diameter of the finished part, and the clearance between the punch and die determines the wall thickness. Collapsible tubes used for toothpaste, lotions, creams, and paint pigments are made routinely by this process with production rates typically of one part per second. In some applications, the sequence of operations can be extended by including additional operations such as piercing, coining, or cold heading. For simple parts, the only operation required after extruding the desired shape is trimming excess stock from the piece.

In forward extrusion, as the punch moves into the die, the impact forces the slug to flow out through an opening in the end of the die. Slugs may be short cylinders, small disks, washers, short lengths of tubing, or cups. The clearance between the punch and die walls is too small to permit any backward flow of the metal. Thus, the metal flows along the path of least resistance, which is in the same direction as the motion of the punch. Forward extrusion can be used to form hollow or solid, round or nonround, straight, twisted, or ribbed parts. This wall tubing with one or both ends open (e.g., copper radiator tubing) can be extruded easily.

With combination extrusion, parts are produced by the simultaneous action of both forward and backward extrusion during a single impacting stroke of the press. Instead of flowing in one direction only through a single opening, the confined metal is made to flow in two directions, one in the direction of punch travel and one opposing it. Solid or tubular shapes or combinations of the two can be produced with this method. After extrusion, the punch is withdrawn and the completed part is ejected from the die.

A wide variety of commercial products is produced by cold-impact extrusion operations. Metals and alloys used extensively for these products include aluminum, brass, copper, lead, tin, zinc, magnesium, titanium, low-carbon and low-alloy steels, and stainless steel. Materials specified for impact extrusion should have the lowest yield strength and the optimum extrusion ratio (length-to-diameter) that are compatible with the requirements for the finished part. Among the products made by the impact-extrusion method are food and beverage cans, cosmetic cases, thermos cases, cartridge casings, fire extinguisher shells, shielding cans for electronic equipment, and seamless tubing.

3.4.4 Shot Peening. This method of cold working improves the fatigue resistance of metal by setting up compressive stresses in the surface of the part. This is accomplished by blasting or hurling small shot at very high velocities against the surface being peened. As the shot strikes, small indentations are produced causing a slight, but finite, plastic flow of the surface metal to a depth of a few thousandths of an inch. This stretching of the outer fibers is resisted by those beneath the surface, the result being a return to the original length. The net effect is creation of an outer layer of metal having compressive stress while layers underneath are under tension. The surface is hardened slightly and strengthened by the cold-working operation. Since fatigue failures result from tension stresses, having the surface in compression greatly offsets any tendency toward such a failure.

Shot peening is done either by air blast or by mechanical units that utilize centrifugal force for hurling steel shot upon the metal work at high velocities. These machines are similar to those used for cleaning forgings or castings. The surface roughness or finish can be varied according to the size of the shot used. Stress concentration, due to the roughened surface, are offset because the indentations are close together and no sharp notches exist at the bottom of the pits. Of course, excessive peening is not desirable since it may cause weakening of the metal.

This process adds increased resistance to fatigue in working parts and can be used either on parts of irregular shape or on localized areas of larger parts that may be subjected to concentrations of stresses. Surface hardness and strength also are increased and, in some cases, the process is used to produce a suitable commercial surface finish. However, it is not effective for parts subjected to reversing stresses, nor is it effective on heavy sections of metal.

3.4.5 High-Energy Rate Forming. High-energy rate forming (HERF) processes are those by which parts are formed at a rapid rate under extremely high pressures. Perhaps these processes are described more accurately as high-velocity forming since it is the metal deformation velocities to which reference is made. If high velocity can be imparted to the workpiece, the size of equipment necessary to form large parts can be reduced substantially and certain materials that do not lend themselves to conventional forming methods can be processed.

Simply stated, these processes involve some mechanism for setting up high-velocity motion in the vicinity of the die. Assuming sufficient pressures are involved, the energy can be used to deform a metal blank positioned between the energy source and the die. Die costs are low, good tolerances can be maintained, and the production costs are minimal. Although much of the process development has been focused on forming relatively thin metal parts, applications of HERF methods are possible in a wide range of processes, including compacting metal powders, forging, cold welding, extruding and cutting.

As mentioned, it is necessary for the metal deformation velocities to be generated at rates orders of magnitude above conventional processes in order for the HERF methods to offer significant advantage. Whereas the deformation velocities for actions of standard hydraulic presses, brake presses, and drop hammers rarely exceed 1 fps (0.3 mps), HERF methods produce deformation velocities of more than 100 fps (30 mps). Three types of HERF processes are explosive forming, electrohydraulic forming, and magnetic forming.

EXPLOSIVE FORMING. Explosive forming has been proven to be an excellent method of utilizing energy at a high rate since the gas pressure and rate of detonation can be controlled. The process involves positioning an explosive charge in air or suitable liquid near a metal bank held in position over a suitable die. As the charge is detonated, the force presses the metal into the shape of the die. Both low and high explosives can be used in various applications. With low-explosives cartridge systems, the expanding gas is confined and pressures can build up to 100,000 psi. With high explosives, which detonate with much higher velocities, pressures of up to 20 times that of low explosives can be attained.

Explosive charges, whether exploded in air or a liquid, set up intense shock waves in the medium between the charge and the workpiece. These waves decrease in intensity as they spread over a large area. Springback problems exist but can be minimized in explosive forming by the use of sheet explosives positioned close to the workpiece, high clamping forces on the holddown areas, and the absence of lubricants. Suitably high gas pressures can be obtained by use of powedred explosives, expansion of liquified gases, explosion of hydrogen–oxygen mixtures, spark discharges, or the sudden release of compressed gases.

ELECTROHYDRAULIC FORMING. Electrohydraulic forming, sometimes referred to as electrospark forming, is a process whereby electrical energy is converted directly into work. High pressure is obtained from a spark gap instead of an explosive charge. In practice, a bank of capacitors first is charged to a high voltage and then is discharged across a gap between two electrodes in a suitable, essentially nonconductive, liquid medium. This generates a shock wave that travels radially from the arc at high velocity supplying the necessary force to form the workpiece to desired shape, dictated by the die used. This process is relatively safe to operate, has low die and equipment costs, and allows close control of the energy rates available.

MAGNETIC FORMING. Magnetic forming is the final example of the direct conversion of electrical energy into useful work. A charging voltage is supplied by a high-voltage source into a bank of capacitors connected in parallel. The amount of energy stored can be varied either by adding capacitors to the bank or by increasing the voltage, the latter being limited by the insulating ability of the dielectric material on the coils. The charging operation is very rapid, and, when complete, a high-voltage switch triggers the stored electrical energy through the coils establishing a rapid high-intensity magnetic field. This field induces a current into the conductive workpiece placed in a die or near the coil that, in turn, produces a force on the piece. This force causes permanent deformation of the metal when it exeeds the elastic limit of the material.

4 POTENTIAL HEALTH HAZARDS

The potential health hazards associated with a particular plant operation depend on the specific materials being processed as well as the manner in which they are handled and used. The processes described earlier in this chapter all are directed toward the same basic end result, that is, conversion of metal or alloy in some bulk form into a usable finished form. It is expected that the potential health hazards associated with one of these operations will be similar, if not identical, to another for the processing of the same material. Therefore, the potential health hazards that are common to all or most operations are reviewed in this section.

For each of the health hazards discussed, the primary sources of release within the unit operation, general strategies for evaluation of the magnitude of the problem, and general reference to the nature and extent of control measures typically required all are presented. Specific hazards presented include metal fumes and dust, nuisance dusts, products of combustion, die lubricants, hydraulic fluids, oil decomposition products, acids and alkalis, dermatitis-causing chemicals, noise, heat stress, radiation, and vibration. There will not be exposure to all of these stresses in all metalworking plants; however, a modern fully integrated plant will incorporate processes that will be characterized by all of these. Unique problems, beyond those covered in this section, are mentioned in the descriptions of unit operations.

4.1 Metal Fumes and Dust

Metal fumes and dusts released from the materials being processed obviously represent major contributors to the airborne material present in metalworking plants. The toxicological properties of most of the common metals have been studied in some detail, and occupational exposure limits supported by these studies have been published. The range of metals used today is extensive, from the relatively benign iron to highly toxic beryllium alloys. Qualitatively, the concerns of the industrial hygienist responsible for metalworking plants will be the same as those in the primary industry, that is, foundries and other casting operations. Quantitatively, however, the concern will be much lower since, by definition, the metal is not handled in the molten state in the metalworking plant. Consequently, the vapor pressure, volatility, and overall ability of the metal to become airborne in the first place is much less than in that operation where the same metal or alloy is handled in molten form.

The potential hazard associated with metal fumes and dusts is related directly to the composition of the metal system being processed. The extent of the problem will depend on the metal composition, the vapor pressure of the metal at processing temperatures, metal flow characteristics, and the oxidizing potential of the metal. In plants where only conventional steels are processed, the concentrations of iron in the air usually are insignificant, and dilution ventilation can be relied upon to control the environment. Where stainless steels, or other alloys incorporating the more toxic metals such as chromium and nickel, are processed, extensive air sampling should be done to document exposure potential. More elaborate engineering controls usually are required in these cases; greater enclosure of the unit operation with local exhaust ventilation and capture velocities in the range of 100–200 fpm (30–60 mpm) are standard approaches. The degree of enclosure and ventilation should be extended even further when very toxic metals (i.e., beryllium, mercury, and uranium) are subjected to metalworking processes.

Although many metals are incorporated into various alloys, the same basic approach can be used to determine the airborne concentrations of most of them. Samples of airborne dust can be collected by standard filtration techniques and analyzed subsequently by means of atomic absorption spectrophotometry. Some special analyses (e.g., total sulfur or distinguishing between hexavalent and trivalent chromium) require methodology other than atomic absorption. When sampling is done for multiple contaminants and analyses are mutually incompatible, it is better to do replicate, parallel sampling than to attempt to split the single sample.

4.2 Nuisance Dusts

A significant portion of the total airborne particulate present in metalworking facilities is nondescript. It represents material released from the initial processing of the metal, for example, scale sluffed off in roughing mills, as well as dusts that have been resuspended as a result of the movement of workers, vehicles, and machines. The dispersal of this dust, even if it is inert, should be minimized, since it poses additional inhalation exposures for the workers. Some of the materials present in the dust can be defined qualitatively, and, to the extent that specific exposure limits have been developed for them, they should be dealt with as specific contaminants and are covered elsewhere in this section.

Occupational exposure limits have not been developed for all of the constituents likely to be present; therefore, the industrial hygienist must use other guidelines. As a rule, however, the limits for nuisance dusts can be used as a minimum standard in such situations.

Conventional sampling techniques, incorporating battery-powered sampling units and filters of appropriate characteristics, can be used. Collected samples can be analyzed, usually simply by gravimetric methods, either for total particulate or for respirable fraction if size-selective sampling is done.

4.3 Products of Combustion

Generation of heat is required for the metalworking in most of the unit operations described in this chapter, and burning of some type of fuel, usually gas or oil, is necessary. Holding or stock furnaces are used to maintain rough stock (blooms, billets, and slabs) at proper temperature for hot-working; annealing furnaces are required for intermediate treatment of metal being subjected to extended cold-working; burners frequently are positioned at the dies of larger hot-working equipment to provide supplemental heating; and gas-fired heaters are used commonly to provide heating of the local work stations. All of these represent potential sources of release of products of incomplete combustion of the fuel—carbon monoxide, aldehydes, nitrogen oxides, sulfur oxides, and larger hydrocarbon molecules including polynuclear aromatics— depending on the fuel being used. When natural gas is used, the combustion products tend to be limited to carbon monoxide and other low-molecular-weight carbon–oxygen compounds. When kerosene or heavier oils are used, however, the other combustion products mentioned are possible. For example, sulfur dioxide would be a reasonable contaminant with fuel oils containing 1 or 2% sulfur.

The primary sources of combustion products in most plants are atmospheric furnaces, using ambient air as the source of oxygen. Some special operations may require the use of a controlled-atmosphere furnace in which special systems are used to produce oxidizing, reducing, or inert atmospheres within the furnace. The potential health hazards associated with these special applications are discussed elsewhere in this book.

Frequently, the combustion products are released directly into the workroom air, even from furnaces. In most situations, the large volume of the workplace and the thermal draft created by the heating process result in these contaminants being diluted and carried upward, where they are discharged to the outdoors by large exhaust fans located in the roof. However, the concentrations of combustion products should be determined as part of the total industrial hygiene effort, particularly of other sources, such as gas- or propane-fired industrial trucks, also are present. Several direct-reading instruments are available for determination of concentrations of the gaseous products of combustion. Significant findings with these instruments should be followed by integrated sampling efforts to document time-weighted average exposure conditions.

4.4. Die Lubricants

The various materials applied to dies in order to facilitate the processing of the metal represent major contributions to the contamination of the atmosphere in metalworking shops. Although the severity of the problem is somewhat greater in hot-working operations than in cold-working shops, it is present in all metalworking processes. The lubricants are applied to facilitate the action on the metal and to maximize the life of the dies; therefore, the lubricant must be such that it remains on the die through the working of the metal.

The first lubricants used consisted of graphite blended into animal fat and oil.

Various additives have been, and are being, used in attempts to improve the process. These additives include coke, clays, sawdust, talc, asbestos, and compounds of sodium, tin, aluminum, antimony, bismuth, and arsenic. Die lubricants in common use today include materials such as aluminum stearate, molybdenum disulfide, and lead naphthenate. The vehicle, or carrier, for the lubricants may be water, light petroleum distillates, low-viscosity mineral oil, heavier fuel oil, black oil, or heavy residuum oil, depending on the type of metalworking being done.

In assessing the industrial hygiene aspects of exposure to the die lubricants, the chemical composition of the lubricant must be ascertained first. It might be desirable to use a specific material in the formulation as an index of exposure to the total system. Often, the carrier for the system, particularly if it is a medium-weight oil, is analyzed by collection on filters and analysis by X-ray fluorescence or other suitable technique.

The die lubricant usually must be applied between impacts of the die and metal piece. This application, depending on the size and complexity of the process, can be done by manually swabbing the die or automatically by a mechanical swabbing or spray action. In any case, there is almost always some overspray and the concentration of oil mist in the vicinity of the operation can be significant. It is not uncommon to observe a blue haze in the atmosphere above metalworking operations, particularly where hot working of the metal is being done.

Properly designed, ventilated enclosures around the dies or, on smaller equipment, around the entire machine, can be used effectively to control the airborne lubricant materials. In some cases, and perhaps more commonly, there is reliance on the pattern and quantity of dilution air exhausted through the work space. In large-volume, open-bay shops, it is common to use pedestal fans or other air movers to direct the oil mist out of the worker's breathing zone (and provide some cooling). Contaminants that are not subsequently captured by local exhaust systems are removed eventually by the roof exhausters.

To some extent, the oils and other materials used as hydraulic fluids can contribute to the total contamination in the vicinity of the hydraulically operated equipment. However, these tend to be controlled to much greater degrees and should represent minor elements of the total oil mist problem.

4.5 Oil Decomposition Products

A portion of the oil-based lubricant that comes into contact with the hot die or heated workpiece can be volatilized or burned off. This will result in not only oil mist, but highly carbonaceous, that is, sooty, particulates and a range of vapors that would be characteristic of the incomplete combustion of the oil involved. The oils used in die lubricants have flash points ranging from 300 to 600°F (150 to 320°C), depending on the type of metalworking being done. In the hot working of ferrous metal, for example, the workpiece may be heated to 2000°F (1100°C) or more, and ignition of the oil is straightforward.

The products of combustion of these oils are many and varied; attempts to systematically investigate the characteristics of oil decomposition are incomplete and inconclusive at this time. Depending on the specific lubricant and its application, air sampling for decomposition products can be done using a variety of conventional techniques. Because of the variety of degradation products likely to be present, it usually is necessary to sample with a number of collection devices in series, for example, filter followed by an impinger followed by a freeze-out trap, or by means of replicate, or parallel, sampling in order to permit the necessary workup for analytical procedures. Of particular and recent concern has been the nature and extent of polynuclear aromatic hydrocarbon compounds in the decomposition products. Various techniques have been suggested for quantifying exposures to such materials. On one end of the spectrum is specific analysis of individual compounds likely to be present; this is a tedious and very expensive approach. At the other end of the spectrum is identification and use of an indicator; benzene-soluble fraction and specific determination of a key compound such as benzo(a)pyrene have been used for this purpose.

Whenever hot working of metals is performed at temperatures high enough to result in ignition of the die lubricant oil, the resulting emission of decomposition products should be controlled by local exhaust ventilation at the point of release. Because of the increasing concern for air contamination from the oil-based die lubricants, water-based compounds, introduced into the industry over 30 years ago are continuing to gain widespread acceptance and application where their use is compatible with production requirements.

4.6 Acids and Alkalis

As indicated earlier, many of the unit operations discussed in this chapter require a careful treatment of the metal surface prior to processing. Often, this treatment involves cleaning, pickling, or other surface action resulting from exposing the metal to various acids and/or alkaline solutions. This metal treatment is a separate unit operation and is treated elsewhere in this book. The potential health hazard of exposure to acids or alkalis is mentioned here only for sake of completeness; the reader is directed to Chapter 17, which deals with metal cleaning, for detailed discussion of this potential hazard. Exposures of metal workers to acids and alkalis are relatively simple to control by proper application of fundamentals of enclosure and ventilation. It is unusual for exposures to approach the permissible exposure limits for these materials.

4.7 Dermatitis-Causing Chemicals

Another problem associated with the metalworking industries is dermatitis, particularly that resulting from contact of the skin with various cutting fluids. Although some of the oils used in the metalworking and forming industries are very similar to those used in metal-machining operations, the dermatitis problem is more characteristic of the latter. Therefore, the detailed discussion of the dermatitis hazard is presented in Chapter 19 discussing metal-machining unit operations.

4.8 Noise

The most prevalent potential health hazard in metalworking operations is exposure to excessive noise. Forging operations, for example, were among the first to be recognized as high-risk processes from a noise-induced hearing loss perspective. As a result, a wealth of information has been developed about the characteristics of noise generated by the unit operations and the impact on the workers' ability to hear.

Depending on the nature and complexity of the plant operation, high noise levels across the entire spectrum of frequencies to which the ear responds can be produced. The furnaces and large hydraulic equipment normally are associated with low-frequency noise. The extensive and pervasive metal-on-metal contact and resulting noise tends to be in the mid-frequency range. Noise associated with operation of steam-operated or pneumatic equipment usually is characterized by major contribution in the high frequencies.

In addition to the broad range of frequencies produced in the operations, the nature of exposures differs from one operation to another. Noise produced by furnaces, motors, and conveyor systems, for instance, is of a continuous nature. Much of the equipment used to form or work the metal, on the other hand, produces distinct impacts with intervals between impacts ranging from fractions of a second to minutes. Complicating the picture somewhat are those conditions in "job shops" where the work being done, and therefore the noise produced, varies from day to day depending on customer orders. It becomes impossible for such operations to define a typical exposure pattern because no two days may be exactly alike.

Regardless of the specific production setting, most metalworking shops have implemented comprehensive hearing conservation programs consisting of noise monitoring, audiometric testing of workers, and noise abatement efforts. Documentation of worker exposure to noise can be done using standardized techniques with either appropriate sound level meters or audiodosimeters. To extend the noise monitoring into application for engineering controls normally requires use of additional equipment such as octave band analyzers and real-time noise analyzers.

From an engineering control standpoint, the high-frequency noises associated with release of air or steam are the simplest to attenuate. For these processes, there are many commercially available muffler configurations that can be adapted to almost any unit operation. In some cases, it is advantageous to duct or pipe the air or steam away from the operator's position so that it is released in an unoccupied area. Commercially available mufflers can be installed on equipment where air is used to clean the dies and/or eject finished parts.

The noise associated with hydraulic systems is somewhat more difficult to reduce significantly. However, since this noise is characterized more by the lower frequencies, it contributes much less to the overall noise exposure of the workers, due to the discrimination of the human hearing mechanism toward low-frequency noise. Some reduction in noise levels from hydraulic equipment is possible by application of vibration dampening, enclosure, and decoupling of the equipment from the basic support structures.

The most pervasive contributor to the overall noise level in most metalworking shops is the metal-on-metal noise. This results primarily from the necessary contact between the workpiece and the dies. Although it is possible in some cases to partially enclose some unit operations with high-density barriers or curtains, effective reduction of noise produced by the unit operation itself has so far eluded the engineer. Accordingly, it is possible, and necessary, to rely on personal protective equipment for effective control of worker exposure to noise in these shops.

For additional information, see Chapter 8.

4.9 Heat Stress

Heat is a necessary input to most metalworking operations (even cold-working processes where intermediate annealing is needed) and, therefore, heat stress in workers must be considered. In high-volume, hot-working operations, such as a large forging shop, heat stress can be a significant problem. Of course, the geographical location of the shop will influence this problem as well. The heat loads imposed on workers are associated not only with convection and radiation from the processing equipment but with their metabolic loads, particularly if extensive manual handling of materials is required.

Various heat stress indices have been developed to permit evaluation of environmental conditions and correlate them to predicted levels of stress in workers. Generation of the raw data needed for calculation of these indices is simple; interpretation of resultant indicators in terms of severity of heat strain is straightforward.

Where excessive heat stress exists, protection of the workers is accomplished by engineering controls, work practice controls, or a combination of both. Only under unusually severe and infrequent conditions would personal protective equipment be recommended. For the most part, engineering controls consist of strategic installation of either reflective or absorptive shielding on or around the processing equipment since the radiant heat load tends to dominate. If convective heat loading is significant, appreciable benefit can be achieved with either properly designed general ventilation or a system of spot cooling at specific work stations.

4.10 Radiation

Limited potential exists for worker exposure to either ionizing or nonionizing radiation in metalworking operations. These hazards are more significant in other related unit operations and, therefore, are covered elsewhere in this book. They are mentioned here for completeness.

Radiographic techniques are used in some metalworking operations as a component of the quality assurance or inspection effort. In such cases, where radioactive materials such as ^{60}Co and ^{137}Cs are used to observe the internal structure of finished metal parts, the work normally is done in a regulated area, isolated from the primary production area. Workers in this operation have been thoroughly instructed in proper use of the equipment and receive personal monitoring of exposure by means of film badges or other dosimeter. Security measures associated with handling and use of the radioactive materials are strict and enforced effectively in most instances. As a result, there is minimal exposure of workers to ionizing radiation.

Handling of "red-hot" metal provides an opportunity for exposure of workers in some of the hot-working operations to nonionizing radiation in the form of infrared radiation. Although the effect of exposure of the skin is not of great concern, prolonged exposure of the eye to infrared energy can result in cataract formation, depending on the intensity of the radiation, duration of exposure, and health status of the worker. Although of much greater concern in the glass industry, exposure to infrared radiation remains a potential health hazard in the metalworking industries as well.

4.11 Vibration

Virtually all metalworking operations result in exposure of workers to vibration. Of the stresses discussed here, probably less is known about the long-term effects of exposure to vibration, particularly in conjunction with other chemical and physical stresses, than other hazards. Although exposure limits have been proposed by some organizations, none are in common use. Within the metalworking industries, several activities are associated with exposure to whole-body vibration: overhead crane operators, industrial truck operators, furnace operators, rolling mill operators, press operators, and users of hand tools. As interest in, and concern for, this hazard continues, evaluations of exposure of the metalworking industries are likely to be performed routinely.

5 CONTROL OF HEALTH HAZARDS

The metalworking and forming processes described and discussed in this chapter are extremely varied in terms of their principles of operation and the consequent release of, and potential exposure to, airborne contaminants. The specific nature of the contaminants depends directly on the metal or alloy system being processed and, to significant degree, the substances (such as cutting oils) used to facilitate the process. Thus, as general strategy, the control of workroom contaminants associated with metalworking and forming processes is a combination of containment or isolation of the process and ventilation of the workplace, locally, generally, or both. These concepts are discussed in the following section. Table 18.2 is an attempt to generalize and summarize the feasible combination of controls that pertains to metalworking and forming processes as a function of the metal or other significant contaminant potentially released.

5.1 Ventilation

Ventilation in this application refers to an overall engineering concept for providing control of a workplace environment by strategic use of air flow. This air flow can be used to remove a contaminant near its source of release into the environment or to dilute the concentration of a given contaminant to acceptable levels. Thus, ventilation

Table 18.2. Engineering Controls Typically Required for Metalworking and Forming Processes

Forming/Working Contaminant	Process Isolation	Local Ventilation	General Ventilation
Aluminum	No	No	Yes
Beryllium	Yes	Yes	Yes
Cadmium	No	Yes	Yes
Chromium	No	Yes	Yes
Cobalt	No	Yes	Yes
Copper	No	Yes	Yes
Iron	No	No	Yes
Lead	Yes	Yes	Yes
Mercury	Yes	Yes	Yes
Nickel	No	Yes	Yes
Nuisance dust	No	No	Yes
Oil mist	Yes	Yes	Yes
Steel	No	No	Yes
Uranium	Yes	Yes	Yes
Zinc	No	No	Yes

is by far, or should be, the most important engineering control principle used for contaminant control in the workplace. Detailed discussions of the principles of ventilation design are beyond the intent of this chapter; many excellent references are available for the reader who is unfamiliar with these principles. In this section, application of ventilation to metalworking and forming processes is addressed.

5.1.1 Local Exhaust Ventilation. As indicated in Table 18.2, local exhaust ventilation should be used whenever there is significant opportunity for release of toxic contaminants into the work environment. In addition to the consideration of metal or alloy system being formed or worked, the most significant consideration is processing temperature. As a rule, the greater the processing temperature, that is, for those hot processes, the greater the need for ventilation control at the source. In most situations, however, the critical component in the process is the metalworking fluid or oil. Control of workroom concentrations of oil mist to acceptable levels almost assuredly provides inherent control of other potential contaminants. Therefore, the following discussion centers on control of metalworking fluids.

Metalworking oil can be applied to the point of contact manually, by an air-carried mist, or by continuous flooding. Manual application is the simplest and least costly. Misting, using a high-velocity stream of air to form and carry the mist, is somewhat inefficient, with much of the oil evaporating on contact with the hot surface of the machine. The most common method of metalworking fluid application is "flooding" in which a low-pressure pump delivers the fluid to a nozzle situated over the metalworking zone. In all cases, the fluid is collected in a catch pan, returned to a pump sump, and recirculated.

Oil mist from the fluid can be generated in three different ways, depending on the specific metalworking process being considered: mechanically (by centripetal force of a rapidly spinning workpiece), thermally (by vaporization and subsequent condensation of the fluid), and intentionally (by application of the fluid as an atomized mist).

As already stated, the most effective method of controlling mist from these operations is a combination of enclosure and local exhaust ventilation. Specific detailed descriptions of application of this approach are contained in the General References section. In most cases, the exhaust air is passed through an electrostatic precipitator and returned to the workroom, although filters and centrifugal mist collectors also are used. Typically, these units incorporate individual air cleaeners mounted directly on a given metalworking machine. Caution should be used with this approach since large numbers of air-cleaning units distributed throughout a large work space can present substantial maintenance problems, become very expensive, and interfere with the overall general ventilation pattern. Use of the individual units should be limited to relatively small facilities with few metalworking machines. In the larger, integrated facilities, it is best to incorporate a plenum-type system, using centrally located air-cleaning provisions and branch ducts to discrete points of contaminant release.

Exhaust hoods should be designed so that the machine can be serviced easily. In fact, the structural sides of the hoods can serve as splash guards, preventing any oil thrown from the operation from getting into the atmosphere. Typically, exhaust volumes of 400–600 cfm (11–17 m³pm) are necessary to assure control velocities of at least 100 fpm (30 mpm) at the point of operation. With careful design of the hood configuration and its orientation with respect to the application of the fluid, oil vapor and mist can be reasonably prevented from dispersing into the air. The reader is encouraged to read the various design references included at the end of this chapter for a more thorough understanding of exact criteria for local exhaust ventilation. Also see Chapters 7 and 10.

5.1.2 General (Dilution) Ventilation. As used in this chapter, *general ventilation* refers to the practice of supplying and exhausting relatively large volumes of air throughout a work space. In the metalworking industries, it frequently is done to achieve comfortable work conditions (temperature and humidity control) as well as to dilute concentrations of airborne contaminants to acceptable limits. Properly used, it cannot only be effective but, in some situations, is an important component in the overall engineering control strategy.

When a large number of metalworking machines, representing individual sources of contaminant release, are distributed in a workroom, general ventilation can be used effectively. If the total amount of metalworking fluid mist generated is small, a general ventilation system, by itself, can provide effective control if the workers are far enough away from the mist sources during their routine duties. Of course, since general ventilation refers to dilution of contaminated air, the rate of oil mist generation must be known to determine the required volume of dilution air. Unfortunately, this information usually is not readily available nor can it be estimated easily; thus, in practice, the best data come from operations comparable to one of interest rather than application of rules of thumb. Simply stated, the role of dilution ventilation should be to provide a minimal volumetric flow of air through the workplace to adequately dilute fugitive releases from processes for which local exhaust ventilation systems are not completely effective by themselves.

In connection with dilution ventilation systems, and to some extent with large local exhaust systems, the recent increasing tendency toward "air conditioning" of metalworking facilities has been accompanied by greater recirculation of air. Thus, there is a greater need to incorporate effective control and cleaning of air contaminated by metalworking and forming processes. Also see Chapter 7.

5.2 Isolation/Containment/Enclosure

Specific mention is made here of "isolation" as an additional engineering concept important in the overall strategy for control of airborne contaminants in metalworking operations. In the simplest of situations, the fundamental of isolation is used in large-volume, spacious shops where physical distance between metalworking equipment from workers can be utilized.

Most often, however, the application of the principle of isolation is done to maximize benefits of other engineering control concepts, notably local exhaust ventilation. For example, it is common for metalworking unit operations in a large, integrated manufacturing facility to be physically separated from the main production areas. In doing so, the volume of air that must be controlled by ventilation is minimized and the effectiveness of the latter effort is maximized.

In narrowing the concept of isolation to individual units, one more often refers to enclosure of the equipment. This is consistent with the principles of design of local exhaust ventilation systems, that is, maximizing enclosure of the source and ventilating what cannot be enclosed. Again, the reader is referred to more detailed coverage of design principles in the General References section.

5.3 Personal Protection

The primary concern from the standpoint of personal protective equipment is the potential for dermatitis and other skin-related diseases potentially resulting from contact with metalworking fluids. However, such risk can be minimized by proper work practices and use of personal protective equipment. Generally, cleansing of soiled skin at regular intervals, followed by thorough drying and application of emollient cream, is a hygienic procedure appropriate for all situations in which there is potential exposure to metalworking fluids. Use of petroleum solvents and abrasive cleaners as skin cleansers should be avoided.

Oil-resistant gloves provide protection both against direct contact with metalworking fluids and nicks and scratches from metal handling. Protective creams or ointments are of limited value only. Water-soluble creams are needed to protect against the oil; no available cream has proven effective against a typically diversified work environment where the hands are exposed to a combination of oils, solvents, and water-based fluids.

GENERAL REFERENCES

General industrial hygiene aspects of many occupational settings are discussed in a wide range of excellent references today. The following list of references is comprised of recent publications that address directly the specific industrial hygiene concerns outlined in this chapter, with particular emphasis on engineering controls in the metalworking and forming industries.

Assessment of Selected Control Technology, NIOSH Publication 79–125, 1975.

Control of Exposure to Metalworking Fluids, NIOSH Publication 78–165, 1978.

Fundamentals Governing the Design and Operation of Local Exhaust Systems, American National Standards Institute, ANSI Z9.2-1979.

Health and Safety Guide for Metal Stamping Operations, NIOSH Publication 75–174, 1975.

Occupational Health Control Technology for the Primary Aluminum Industry, NIOSH Publication 83–115, 1983.

Recommended Industrial Ventilation Guidelines, NIOSH Publication 76–162, 1976.

Working with Cutting Fluids, NIOSH Publication 74–124, 1974.

Wu, Weh S., Douglas K. Arai, Mark Z. Nazar, and David K. Leong, Determination of Nitrites in Metal Cutting Fluids by Ion Chromatography, *Am. Ind. Hyg. Assoc. J.*, **43** (12), 1982.

Metal Machining

Robert D. Soule

1   INTRODUCTION

Machining of metal surfaces is done routinely for several reasons: to produce geometrically true surfaces, to correct minor surface imperfections, to improve dimensional accuracy, or to provide close fit of surfaces that are intended to be in intimate contact. Simply stated, machining is a "chip-removal" process, that is, small incremental pieces of the surface are removed under controlled conditions until the desired dimension or finish has been obtained. This chapter is limited to processes based on this chip-removal principle; other chapters in this book address other techniques such as ultrasonic, electrical discharge, laser, electrochemical, electric arc, and plasma arc methods of surface finishing.

Review of the various metal machining operations shows many similarities among processes. Equipment used for machining includes various power-driven machines used to "cut" metal. These machines operate on either a reciprocating or rotating basis; that is, either the tool or the metal workpiece reciprocates or rotates. For example, with a planing machine, the workpiece reciprocates past the tool, which is held in a stationary position. In other machines, such as shapers, the work is stationary and the cutting tool moves back and forth. Similarly, with rotating machines the workpiece is rotated and the tool is stationary in the case of lathes, whereas the workpiece is held in place and the tool rotates with common drill presses.

Although all of the machining processes remove some metal from the workpiece, the amount varies considerably from one process to another. Grinding, for instance, results in a moderate amount of metal being removed from the part as well as providing a relatively good finish. In other processes, such as honing and lapping, the intent is to remove small scratches with little change in the dimensions of the piece. Regardless of the specific application of the machine, there are basic elements common to all of the cutting equipment.

Each machine is characterized by a frame or structural support configuration, the arrangement of which tends to be indicative of the type of action provided by the machine. There must be an appropriate drive mechanism to provide the reciprocating or rotary motion needed for the cutting action; the drive may be electrical, hydraulic, mechanical, or pneumatic. There must be a device for holding the workpiece securely in position for the machining; mandrels, chucks, vises, and collets are examples. Most machines have some separate mechanism for handling the workpiece prior to and following the machining work. Finally, various control mechanisms are possible: entirely manual, mechanical, hydraulic, cam action, timing cycle, or more sophisticated numerical control.

In contrast with the metalworking and forming processes, operators of machining equipment necessarily are in close proximity to the point of direct contact between the metal and the cutting surface. The operator of a simple drill press, for instance, standing directly in front of the work being done is more likely to be affected by, or at least exposed to, any stress released from or created by the operation than the operator of a process that can be controlled from some remote location. Since machining usually is done for purposes of refining the dimensional character of the workpiece, it is understandable that the operator must be in direct line of sight of the work. This condition results in a more fundamental reliance on localized control measures in the machining operations, as compared to metal forming/working where broader controls, such as general ventilation, often are used effectively.

Although metal machining is a process of simply removing small amounts of metal from the surface of a workpiece, there are literally dozens of operations that have developed to accomplish this. Those most common include cutting, abrasive grinding, drilling/boring, planing, shaping, slotting, broaching, honing, lapping, mass finishing, and superfinishing. Each of these common processes is described in the following section. Since the potential health hazards associated with these processes are similar, varying only in degree, the next section is a review of industrial hygiene concerns generally applicable to machining processes. The final section of this chapter is a discussion of control strategies, with particular emphasis on engineering controls, appropriate for metal machining operations.

2 PREPARATION OF WORK FOR MACHINING

Metal machining is a step in the total manufacturing process for products containing metal components. Many unit operations necessarily precede machining when this total process is considered. The appropriate metal or alloy system must be created, the rough forms of the metal components must be cast or forged, the rough forms must be formed or worked into the final desired configurations, and particular desired properties must be imparted to the metal base or its surface by metal treatment, cleaning, or surface coating. For full understanding of the industrial hygiene aspects of these preceding unit operations, the reader is referred to appropriate chapters in this book.

3 METAL MACHINING PROCESSES

A tremendous array of processes is utilized in industry to place the finishing touches on metal work. This section briefly describes the more common of them in terms of principles of operation and application. The potential health hazards associated with these processes consist of the same basic spectrum of stresses and are discussed collectively later in this chapter. Similarly, the control of health hazards in metal machining operations is addressed somewhat generically at the conclusion of this chapter.

3.1 Cutting

Cutting, or shearing, is a process in which a metal piece, usually in plate, strip, or sheet form is subjected to shear stress beyond its ultimate strength by positioning it between two sharp edges. Cutting operations normally are associated with preliminary phases of metal machining work and therefore are discussed in this chapter.

As a press operation, cutting can be performed in a variety of related but distinctly different ways. The basic action of the press is the same in all cases. As the punch descends onto the metal, the pressure causes a plastic deformation to take place at the point, or along the line, of contact between the metal and the cutting surface. As this deformation continues, fracturing of the metal begins. When the ultimate strength of the material is reached, the fracture progresses and, if the clearance between the punch and die is correct, the fractures meet at the center of the sheet, and the material is parted. The clearance between punch and die is a critical feature in design of cutting machines. If improper clearance is used, fractures must progress across the entire sheet thickness, rather than meeting at the center; this results in much greater power requirements for the cutting work. Similarly, flat punches and dies require a maximum of power. To reduce the shear force, the faces of the punch and die should be made at an angle so that the cutting action is progressive. This results in distribution of the shearing action over a greater length of the stroke and significantly reduces the power required.

Metal cutting action can be classified on the basis of the purpose of the cut and can be subdivided into three categories: blanking, piercing, and edge improvement. Blanking operations include such actions as cutoff, parting, shearing, lancing, notching, and nibbling. Piercing or punching operations include slitting, perforating, and extruding. Trimming, shaving, and fine-edge blanking are examples of edge improvement operations utilizing a cutting action. Each of these categories of unit operation is discussed in the following sections.

3.1.1 Blanking. Almost all metal machining and many metalworking operations begin with the blanking of a piece of material from some form of flat stock. In general, the punch penetrates the material a distance equivalent to less than one-half the stock thickness, the penetration required being a function of the metal brittleness, and the part simply fractures and breaks off. As the punch continues its stroke, it pushes the part into and below the face of the die. The sheared edges of the blanked part are characterized by ridges of torn metal, referred to as die break. In a single stroke of a blanking press, a complete outline of an individual flat workpiece is produced. The blanking process can be applied to a variety of purposes discussed in the following sections.

CUTOFF. When individual flat metal blanks are produced by cutting along a line that extends across the entire width of the stock, the blanking operation is referred to as cutoff. Special cutoff punch-and-die sets are designed for this particular application. Generally, the dies used are relatively simple and among the least expensive to produce and maintain.

The cutoff line may extend across the stock in a straight line or along a broken or curved path, depending on the desired shape of the workpiece. The relationship between the punching action and feeding of the metal stock is designed so that one or more identical blanks are produced with each stroke of the press. By careful design of the die layout, it is sometimes possible to virtually eliminate scrap, even with fairly intricate shapes. Cutoff operations typically use single-action presses capable of operating at about one stroke per second.

PARTING This operation consists of die cutting the desired flat blank shapes by generating scrap in the form of metal that lies between adjacent blanks. Each stroke of the press produces one complete part but, unlike cutoff operations, sacrifices some stock. Parting is a means of producing blanks with outlines too complex to permit compact nesting of the blanks as can be done with regular cutoff methods. Such design is justified when it is necessary to space blanks at intervals along a strip of material to avoid a tendency of stock distortion, to obtain greater accuracy of the outer contour, or to allow sufficient room for bulky tool movement. Generally, parting tools are the simplest of the stamping tools, with each stroke of the press producing one complete part.

SHEARING. This process involves the cutting of flat metal forms from sheet and plate by the action of two blades working in opposite directions. Thus shearing differs from cutoff or parting methods where punch-and-die arrangements were used. Shearing can be subdivided broadly based on the type of blade or cutter used; either straight-blade or rotary-blade arrangements are possible.

Straight-blade shearing is used primarily to square or otherwise cut flat stock to the shape and size of the desired blank. The operation consists of holding the stock rigidly in place and severing it by the action of an upper blade that moves down past a lower, stationary blade. The degree of penetration of the upper blade into the stock is determined by the ductility and thickness of the metal. As is the case with conventional blanking, the sheared edge is relatively smooth where the blade penetrates; the texture along the torn portion is considerably rougher. As a rule, it is not possible to cut multiple layers of stock with a single cut since each layer prevents the necessary breakthrough of the preceding workpiece. Straight-blade shearing is used frequently to prepare the blanks prior to production-scale press operation. It is a relatively simple operation and can produce stock within reasonably close tolerances. Squaring shears are available commercially with mechanical, hydraulic, or pneumatic mechanisms.

During rotary shearing, flat stock is cut by means of two revolving circular cutters, tapered to provide the cutting edge. This process is used most often to cut circular or curved shapes, although it is possible to make straight-line cuts. With these machines, only the upper cutter pinches the workpiece and causes it to rotate between the two cutters. Usually, special holding fixtures are provided to rotate the workpiece so that the desired shape is obtained; a straight-edge fixture is used for straight-line cutting. Commercially available equipment can process stock up to 1 in. (2.5 cm) thick and can cut circular blanks between about 6 in. (15 cm) and 10 ft (3 m), the larger-diameter blanks requiring special attachments to the machines.

LANCING. Lancing refers to a cutting operation in which a single line is made through the thickness of the stock but only partway across the metal. In the processes discussed to this point, there has been separation of blanks from the sheet stock or a continuous cut across the entire width of the stock. Since there is no metal removed from the stock, there is no scrap associated with lancing operation. Lancing often is used as a means of forming small tabs or protrusions that may later be used for spring retainers, for assembly purposes, or for louvers on certain parts. The small knockouts common on electrical circuit-breaker panels, junction, and outlet boxes are examples of application of the lancing process.

In special cases, lancing can be used to partially cut the outline of a flat part that will be separated from the work metal in a progressive die at a final operation by blanking. In other cases, lancing may be specified for some parts in order to offset the possibility of fracture in operations where metal must be drawn or formed into a restricted area.

NOTCHING. Notching is a presswork operation by which metal pieces are removed from the edge of a piece of flat stock. Depending on the configuration of the punch used, the notches may have almost any desired shape. Metal removed by notching almost always is scrap.

Notching sometimes is performed on edge configurations of sheet metal blanks that might be difficult to cut by other means. Another application of notching is the pretreatment of flat pieces prior to forming a concave radius on the part. This helps to prevent the formation of wrinkles in the metal, but also weakens the part at the point of the curve. Production-scale notching presses are high-speed, short punch stroke machines.

NIBBLING. Nibbling is an economical and versatile method of cutting out relatively small numbers of flat parts that can range from simple to very complex contours. The process sometimes is substituted for blanking when the cost of special tooling would be difficult to justify. Most nibbling machines use standard round, triangular, or rectangular punches rigidly clamped to a tool adapter. The small punch moves in and out of its mating die at a rapid rate, as high as 900 strokes per minute. The sheet metal may be positioned manually or clamped and guided by mechanical or electro-hydraulic means. The shape of the desired cutout is cut progressively by the action of continuous, overlapping punch strokes that slice the moving metal along the desired contour. There is virtually no scrap.

3.1.2 Piercing or Punching. Piercing or punching, is a cutting operation by which various-shaped holes are sheared in blanks. Piercing is similar to conventional blanking in reverse; the work metal that surrounds the punch is the workpiece, and the slug that is cut out is scrap. Usually, piercing is the fastest method of making holes in blanks, sheets, or strips of metal. Single-operation dies are used when piercing is the only operation required or where compound dies cannot be used because of limitations imposed by holes that are too close to the edge of the stock. Compound dies are particularly appropriate when close accuracy of hole positions is important. Piercing can be done for slitting, perforating, and extruding.

SLITTING. Strips or coiled stock can be cut lengthwise by passing the stock through spaced, circular blades in a process called slitting. Continuous strips of stock in various widths may be obtained by adjusting the spacing of the various slitters to the desired settings. The slitting line, or manufacturing sequence for dividing a coil into a number of narrower coils, consists of an uncoiler for holding the original coiled stock, one or more slitters, and a recoiler that has spacers corresponding to the various widths of the slit stock. Stock may be driven or pulled through the slitting line. Usually, the thicker stock is pulled through the slitter and thin sheet or coiled stock is pushed or driven through to avoid the tendency of tearing the metal.

Coil stock can be slit practically in widths as narrow as 1 in. (2.5 cm) or less; if slit to widths of less than $\frac{1}{2}$ in. (1.2 cm), the strip is referred to as flat wire. Burrs usually are formed when slitting coiled stock and must be removed prior to further machining or finishing.

PERFORATING. Perforation is used to describe the process of creating multiple holes in flat work material. The holes may be of almost any conceivable shape. Perforation implies that the holes are close and regularly spaced. The tooling, normally consisting of an arrangement of multiple punch and dies, is essentially the same as for conventional piercing operations.

EXTRUDING. Another piercing operation performed on flat stock is the special process of extruding, so called because there is formation of a flange on a flat part by drawing the stock out of a previously made hole. In general, extruding involves rather severe stretching of the metal. Extruded holes often are made on sheet metal parts to provide added thread length for holes tapped for assembly purposes, to increase a bearing surface, or to produce a recess into which the head of a flat-head machine screw or rivet can be fit.

Flanged holes can be produced by forcing a punch of the desired hole diameter through a small prepierced hole. In some cases, a shouldered or pointed punch that both pierces the hole and flanges it can be used. As a general rule, a flange with a depth of up to one-half the hole diameter can be produced without splitting the metal. Greater depths have been achieved by initially forming an indentation, or small cup, punching a small hole in the bottom, and extruding the flange.

3.1.3 Edge Improvement. The cutting operations discussed so far have been directed toward producing desired cuts in the interior of a workpiece. Edge improvement processes discussed in this section are used to cut along the outside edges of the work in order to create a smoother, more well-defined surface.

TRIMMING. This process consists of die cutting unwanted, excess material from the periphery of a previously formed workpiece. Usually, a trimming is combined with one or more other pressworking operations, such as notching, in a compound die. It is often possible to design special-purpose dies that accurately control the relationship between the trimmed outline of the workpiece and some other feature of the workpiece, such as a hole or slot. Deformed or uneven metal on the edges of blanked workpieces and the flash that results from casting and forging operations usually are removed by trimming.

SHAVING. Shaving improves the quality and accuracy of blanked parts by removing a thin strip of metal from along the periphery of the piece since edges usually are unsquare, rough, and uneven when initial working is done. In most shaving operations, the material removed is very thin, a few thousandths of an inch, and the scrap resembles the chips that are produced in conventional machining processes.

On heavy stock, the edges of the piece may have a burnished area on both the punch and die sides, with an irregular recess of torn material in between. Two or even three shaves may be required to improve the edge straightness and finish in such cases. The design and production of blanked pieces take into account the fact that a small amount of extra stock is needed to permit the subsequent edge improvement operations. Thus, it is helpful to the metal machining process operators to have those surfaces designated for which shaving will be necessary. Shaving can be done as a separate operation or incorporated into one station of a progressive die. In any case, the blank must be located precisely over the die because of the very small amount of metal that is actually removed.

FINE-EDGE BLANKING. Fine-edge blanking, a relatively new process, is a variation of conventional blanking in which it is practical to produce a smooth edge on workpieces during a single press stroke. It involves use of a special impingement ring that is forced against the stock prior to blanking to lock it tightly against the die. The metal ring is positioned very closely to the outline of the part being punched, and the stock is forced to flow toward the punch that, in effect, extrudes the part out of the strip without any fracture. Thus, no die break appears on the sheared edge of the piece. Relative to conventional blanking, the die clearance necessarily is very small and the speed of the punch much slower.

Both ferrous and nonferrous parts can be worked by this process with edges comparable in quality to those produced by shaving or finish machining. Clean and accurate holes of various shapes can be pierced by this method, thereby eliminating the need for broaching or milling operations. The small burrs formed by fine-edge blanking can be removed by simple finishing methods. Significant cost savings, resulting from elimination of secondary finishing operations, are possible on high-volume parts such as levers, cams, and gears.

3.2 Sawing

Another means by which preliminary work can be done on metal being machined is by sawing operations. The cutting surface, and its relationship to the piece being sawed, can be used to categorize metal saws into three configurations: reciprocating, circular, or continuous (band) machines.

3.2.1 Reciprocating Units. Reciprocating saws, which may vary in design from light-duty, crank-driven saws to heavy-duty, hydraulically driven units, have been in common use for some time because of the simplicity of design and low operating cost. These machines can be designed for manual, semiautomatic, or fully automatic operation and vary in the manner in which the saw is fed into the metal workpiece and the type of drive used.

Generally, the saw can be fed into the metal either by positive- or uniform-pressure methods. With positive feed, there is an exact depth of cut for each stroke of the saw; the pressure on the blade, therefore, will vary directly with the degree of contact between the blade and the workpiece. In cutting round pieces, the pressure is light at the start and finish of sawing and is maximum at the center. With uniform-pressure feed, the pressure is constant at all times, regardless of the contact between saw blade and metal. Here the depth of cut varies inversely with the degree of contact between blade and workpiece. Many machines have incorporated both systems into their design with automatic control. In any case, the pressure is released on the return stroke in order to eliminate wear on the saw blade.

The simplest type of feed is gravity in which the blade is forced into the metal by virtue of the weight of the saw and frame. Uniform pressure is exerted on the work during the stroke, but some control is provided over the depth of feed for a given stroke of the blade. Additional cutting pressure can be provided by clamping weights on the frame or spring loading the frame. Positive screw feed, with overload features, can provide a definite depth of cut for each cutting stroke, whereas hydraulic feeds afford excellent control of the cutting pressure. The simplest drive for the saw frame uses a crank rotating at constant speed. The cutting action takes place only half of the time since the duration of the return stroke is the same as the cutting stroke. A link mechanism can be used to provide a quick-return action, thus improving this design. With automatic feed systems, the usual cycle of operations consists of the following: the piece is moved forward through an open vise, the vise is clamped, the cut is made by the saw, the saw blade is raised to its idle position, the vise is opened, and the cycle is repeated until the entire length of stock is processed.

It is important that a proper lubricant be used for all power saw cutting, not only to lubricate the tool but to flush away the small chips that might otherwise accumulate between the saw teeth. In most sawing operations, the heat generated is minimal. Therefore, the lubricant or cutting oil should be chosen from the standpoint of lubrication qualities, rather than cooling effects. The potential health hazards associated with the cutting oils are described later in this chapter.

3.2.2 Circular Unit. Cutting machines using circular saws commonly are referred to as "cold sawing" machines. The saws are quite large in diameter and operate at low speeds of rotation. Solid blades used in circular sawing machines are limited to about 16 in. (40 cm) in diameter because of manufacturing cost and the fact that broken or worn teeth cannot be replaced. Most large-diameter cutters have either replaceable inserted teeth or segmental blades in which the segments are grooved to fit over a tongue on the disk and are riveted in place. Both inserted teeth and segmental-type blades are economical and have the additional advantage that worn teeth can be replaced. Cutting speeds for circular blades range from as low as 25 fpm (8 mpm) for ferrous metals to as high as 4000 fpm (1200 mpm) for some nonferrous metals. The life of the saw is extended when the peripheral speed can be reduced. Again, use of an appropriate lubricating fluid is recommended for all circular sawing.

Another type of circular unit takes advantage of a steel disk operating at high peripheral speeds; these are used as a means of rapidly cutting through structural steel members. When the disk is rotated at rim speeds of about 20,000 fpm (6000 mpm), the heat of friction quickly melts a path through the part being cut. The disks are ground slightly recessed, or hollow, in order to provide clearance on the side of the disk when cutting through large sections. Water cooling is recommended when these units are used. These friction cutting units are not satisfactory for cutting nonferrous metals since these tend to adhere to the disk and not break away readily as a result of the disk action. The hardness of ferrous metal is not a limiting factor with these friction cutters. The cutting ability depends more on the structure of the metal and its melting characteristics than on metal hardness. In cutting, the tensile strength of the steel is lowered rapidly as the temperature increases due to friction. The metal finally is weakened to the extent that the friction pulls it away from the colder metal. A 24-in. (60-cm) I-beam can be cut through in about 30 sec with this process.

A third circulating cutter replaces the steel friction disk with a wheel having a cutting surface of an abrasive material. The cutting action with these units depends entirely on the abrasive material in the wheel and is not influenced by any softening of the metal that takes place. However, the efficiency of the cutting action, in terms of energy consumption, is increased with increased wheel speed because of the softening of the metal that takes place. These units can be operated for either wet or dry cutting.

3.2.3 Continuous Units. The reciprocating and circular units discussed previously are designed for making straight-line cuts primarily and are used for cutoff operations in most cases. Saws that incorporate a continuous blade, or band, can be used for cutoff operations but, in addition, can cut irregular curves in metal. This widens the field of application for the band saw since it is capable of doing a variety of work that otherwise could be done only with other machining tools. Intricate contour sawing of dies, jigs, cams, templates, and other parts that previously had to be made by hand or with other machining tools at greater expense now is done routinely with band-sawing machines that resemble those used for wood, with obvious differences in the type of saw and cutting speed. Most machines are designed with the saw operating in a vertical position and the work supported on a horizontal table, usually with a tilting mechanism to allow the cutting of angles. Another type is similar to the ordinary hacksaw machine; the work is held in a vise and the band saw is passed through the metal, with cuts of up to 10 ft. (3 m) being possible. As with virtually all sawing machines, there are provisions for automatic application of cutting oil to the point of contact with the blade. The oil may be applied either as a continuous stream or as a mist.

3.3 Abrasive Grinding

A process used for the rough machining of many metal parts and the preliminary finishing of most is abrasive grinding. It consists of removing metal chips from a workpiece by bringing it into contact with a matrix of appropriate abrasive grains bonded to a moving wheel, disk, or belt. Each of the individual, irregularly shaped grains acts as a cutting tool. Application of the grinding process frequently is subdivided into "rough grinding" and "precision grinding."

Rough grinding is used commonly for removing excess material from castings, forgings, and welded pieces, or as a means of removing burrs, fins, sharp corners, or other unwanted projections on a piece of metal. Small metal parts usually are handheld and moved into contact with the moving abrasive grinding tool over the workpiece surface.

A potential health hazard unique to the rough grinding operations and not common to the other processes discussed in this chapter is exposure to silica. It is not uncommon for the dust removed from rough castings to contain appreciable amounts of silica as a result of the partial solubilization of silica from the casting molds into the outer portions of the metal. Silica contents of 10% of the airborne dust released by the grinding process can be expected. In addition to the other hazards discussed earlier in this chapter, rough grinders or castings cleaners should be monitored for exposure to silica-containing dust. Conventional sampling methods involve the collection of respirable dust by passing the airborne material through a cyclone, analysis of the collected respirable dust using gravimetric tehcniques, and determination of the silica content using X-ray diffraction equipment. The current standard for occupational exposure to silica utilizes a formula that takes into account the silica content of the dust in the air.

The major application of abrasive grinding as a unit operation in metal machining is in precision grinding. This is a principle production method for cutting materials that are too hard to cut or machine by other conventional tools. It is used extensively to produce surfaces on parts to tolerance and to provide finishes more exacting than can be achieved by other methods.

The cutting action is provided by the abrasive material contacting the metal. No single abrasive is best suited for all grinding applications. An important element in the design of an abrasive grinding process is the correct matching of abrasive material with the material being machined. Abrasive substances are necessarily both hard and rough. Efficient cutting can take place only when the abrasive is hard enough to penetrate and scratch the workpiece. Maximum cutting is obtained when the dulled abrasive grains are fractured continually, thereby exposing a succession of fresh cutting edges. Both natural and synthetic materials are available for use as abrasives. The natural substances, such as sandstone, quartz, emery, corundum, garnet and diamonds, are used either for rough grinding or for relatively specialized purposes. Very few production line grinding operations use natural abrasives; therefore, the precision grinding operations on which this chapter focuses tend to utilize synthetic materials.

The most common synthetic abrasives used in machining, in decreasing order or prevalence, are silicon carbide, aluminum oxide, synthetic diamonds, and boron nitride. Silicon carbide is very sharp, extremely hard, but brittle. Because of these properties, silicon carbide is particularly well suited for grinding on soft, nonferrous metals such as brass, copper, bronze, magnesium, aluminum, or cast iron. Aluminum oxide is softer but tougher than silicon carbide. As a result, it is used commonly for grinding on high- or medium-carbon steels, nonferrous cast alloys, and annealed malleable and ductile iron. On a volumetric basis, aluminum oxide abrasive is the most used material. Diamonds are the hardest and most expensive of the abrasive materials. Unlike the other abrasives, natural diamonds do not fracture readily, and their application in industry has thus been limited in scope because of the tendency of dulling and glazing of the cutting faces. Synthetic diamonds, actually more costly to produce than the natural material, are processed to provide better breakdown characteristics, thereby giving a cooler and more efficient cutting action. Present diamond systems with special bonding features are proving to be economical replacements for silicon carbide and aluminum oxide abrasives for the grinding of stainless steels, cast iron, and tool-and-die steels. Boron nitride and other similar materials are being used increasingly on workpieces of tool-and-die steels and many hard, high-alloy steels. Significant advantages reported are superior surface integrity, ability to maintain close tolerances, and increased life of the abrasive.

Beyond the abrasive material itself, the effectiveness of the grinding operation is determined by the size of the abrasive grains and the materials used to bond the grains to the support surface. Bonding systems generally consist of one of the following six types: vitrified, silicate, resinoid, metal, rubber, and shellac. Vitrified bonding is the most common. The abrasive grains are baked with a claylike material and shaped by hydraulic pressure in molds. They are strong, porous, and rigid and are not affected by water, oils, or acids. Silicate bonding consists of a mixture of the abrasive material with sodium silicate, compressed into a mold and baked. These systems wear faster than vitrified systems because the grains are released more easily. Silicate-bonded abrasives are used to sharpen cutting tools and in applications where heat must be minimized. Resinoid systems consist of a mixture of abrasive and synthetic resins. These systems can be made very strong and tough and are in common use for high-speed grinding applications. Metal bonding is used for diamond wheels and for electrolytic grinding. A variety of rubber-bonded systems is available with superior properties that result in their use for plastics, glass, and procelain, as well as practically all metals, particularly for cutoff operations and for rough grinding on castings and forgings. Shellac-bonded wheels are used to produce superior finishes on many materials because of the tough and relatively elastic nature of the system.

Manufacturers of grinding wheels and other abrasive systems have adopted a standard system for identifying and marking their products. These data, marked on each grinding wheel, identify the type of abrasive, the grain size, the bonding type and grade, the concentration of abrasive (a hardness rating), and any particular modifications and manufacturer's record.

Grinding fluids are used to cool the workpiece and maintain the face of the wheel, disk, or belt. Heat generated by the grinding action is dependent on the sharpness of the grains; gradual wear, without fracture of the grains, is undesirable because the face becomes dull and glazed and heat builds up in the wheel. The grinding fluids carry the chips away and improve the surface finish of the workpiece. In conventional dry grinding, the fluid is usually the surrounding air. Oxygen in the air oxidizes the workpiece surface; this is sufficient to prevent welding back of the chips onto the ground surface. Water-based fluids, consisting of soluble oil emulsions or synthetic compounds, are effective in limiting wheel wear and heat generation but are somewhat messy and constitute a fire hazard. The grinding fluids generally are applied in copious amounts directly to the grinding area.

There are three basic categories of precision grinding machines for production applications: surface grinders, external cylindrical grinders, and internal cylindrical grinders. Surface grinding refers specifically to production of flat surfaces. Surface grinders can be classified according to the orientation of the wheel spindle as either horizontal or vertical. Both types are available with a variety of contours of the grinding face, and each can be provided with either a rotary or reciprocating table. External cylindrical grinders can be subdivided into centertype and centerless machines. Most of these machines are used for grinding relatively small parts with shapes that are adaptable for holding in a chuck or collet. Chucking grinders are particularly useful for parts with shapes that cannot be readily held between centers.

Internal cylindrical grinders are designed to produce internally ground surfaces or holes on metal parts. There are three types of internal grinding machines: one that holds the workpiece in a rotating chuck, one in which the workpiece is rotated by the outside diameter between rolls, and the planetary type, for heavy-duty work, in which the workpiece is held stationary.

From an industrial hygiene standpoint, studies have indicated clearly that most of the airborne particulate associated with grinding operations is released from the workpiece being ground. Relatively little of the airborne dust originates from the abrasive material matrix. From time to time, however, grinding wheels may load or plug and the wheel must be dressed, with either a steel roller or diamond tool. During this process, significant amounts of the wheel materials are removed, some of which may become airborne. The overall hazard potential from grinding operations depends on the specific operation, the metal of the workpiece and its surface coating, and the abrasive system in use. Ventilation requirements for grinding operations have received extensive investigation. Subjectively, there appears to be a difference in need for local exhaust ventilation from one type of metal to another: Aluminum parts appear to release relatively little dust, titanium seems to be quite dusty, and most steel alloys are intermediate. Although the specific circumstances will dictate the actual need for engineering controls, it is reasonable to assume that local exhaust ventilation will be needed under the following circumstances:

- Work on metals or alloys containing metals of high toxicity; beryllium, chromium, cobalt, lead, nickel, and vanadium are examples.
- Rough grinding on castings produced by sand molding; the sand fused into the castings can be released during such action resulting in potential exposure to silica.
- Grinding on metal that is coated with toxic materials, ferrous metals coated with lead- or chromate-based paints are cases of interest.

Extensive guidelines for effective control of grinding dusts have been published by the American National Standards Institute, the American Conference of Governmental Industrial Hygienists, and the National Institute for Occupational Safety and Health. Specific standards for grinding operations exist in the general industry regulations promulgated by the Occupational Safety and Health Administration (OSHA).

Evaluations of exposure of grinding machine operators to airborne dusts can be performed easily using the personal monitoring techniques and analytical procedures discussed in Section 4.

3.4   Drilling and Boring

Drilling, as a unit operation, is a means of producing round holes in metal workpieces. Many metal products are assembled with bolts, screws, and rivets placed in holes drilled into or through various parts. Perhaps the simplest of the machining tools is the drill press, a machine that produces the hole in an object by forcing a rotating drill through it. The machine consists of a base, column, stationary table, spindle, and power head. The workpiece is clamped securely to the table, and a manually operated feed lever forces the rotating tool into the work. Industry makes considerable use of the drill press; drills are set up frequently in gangs so a sequence of operations can be performed without changing tools. Special drilling jigs and fixtures can be mounted on the table to hold and handle mass-produced items. Heavy upright drilling machines, similar to the drill press, can be equipped with turret attachments; a sequence of drilling operations can be carried out by simply rotating the turret hand-wheel. When the tool is in position, it is automatically connected to the spindle. Of course, production-scale drilling operations normally are conducted with powered tables. The workpieces are clamped against stops and then carried by the table to various positions for tool operation. In these cases, the tools are mounted in a turretlike holder, called a positioner, and the tool to be used for a certain operation is rotated into position and automatically connected to the spindle. Speeds and feeds of the drills and other tools, as well as the movement of the table, all can be controlled automatically.

Whereas drilling generally refers to the process of producing the initial hole in the workpiece, boring is a process of enlarging a hole that has already been drilled. In essence, it is an operation of truing a hole that has been drilled previously, usually with a lathe-type tool rather than a drill press. When this operation is conducted on a drill press, a special boring tool and holder is necessary.

Other operations in which the basic principle of drilling is used include counter-boring, spot facing, countersinking, and reaming. Counterboring is the process of enlarging one end of a drilled hole. The enlarged hole, usually concentric with the original one, is flat on the bottom. The tool is equipped with a pilot pin that fits into the drilled hole to center the cutting edges. Counterboring is used principally to set bolt heads and nuts below the surface. Spot facing refers to the finishing off of a small surface area around a drilled hole. This is a customary practice on rough surfaces to provide smooth seats for bolt heads. Countersinking is the process of beveling the top of a drilled hole in order to accommodate the conical seat of a flat-head screw. Finally, reaming is the enlarging of a machined hole to proper size with a smooth finish. A reamer is a more accurate tool than the others mentioned in this paragraph and is not designed to remove much metal. Although all of the operations just mentioned can be done on a drill press, other machine tools are equally well adapted to perform them.

To obtain the best performance and longest life for cutting edges on the drills, cutting fluids of various types are used. The cutting fluid improves the cutting action between the drill and the work, facilitates removal of chips, and cools the work and the drill. In production-scale drilling, the matter of cooling is critical. To ensure maximum life of the tool, the cutting oil should be selected that dissipates the heat at essentially the same rate as it is generated. Mineral oil–lard oil mixtures are used commonly with aluminum, brass, copper, and magnesium. Water-soluble oils are used extensively with bronze, copper, malleable iron, and both soft and tool-strength steels.

3.5 Milling

Milling refers to a machining unit operation in which metal is removed from a work-piece when the work is fed against a rotating cutter. The milling cutter has a series of cutting edges on its circumference, each of which acts as an individual cutter in the cycle of rotation. The work is held on a table that controls the feed against the cutter. In most machines, the table can move in three ways longitudinally, crosswise, and vertically; some tables may also possess a rotational movement for greater versatility.

The milling machine is probably the most versatile of all machine tools. Flat or formed surfaces can be machined with excellent finish and accuracy. Angles, slots, gear teeth, and recess cuts can be made. Drills, reamers, and boring tools can be held in the socket by removing the cutter and arbor. Since all table movements have micrometer adjustments, holes and other cuts can be spaced accurately. Most operations that can be performed on shapers, drill presses, gear-cutting machines, and broaching machines can be done on milling machines. In fact, they produce better finishes and holes to accurate limits with greater ease than do shaping machines discussed later in this chapter.

The cutters used with the milling machines generally are classified according to their shape or the type of work they are intended to perform. A plain milling cutter is a disk-shaped cutter having teeth only on the circumference. The teeth may be either straight or helical. Wide helical cutters, for heavy-duty work, may have notches in the teeth to break up the chips and facilitate their removal. Side milling cutters are similar to plain cutters except that they have teeth on the side. When two side cutters operate together, each cutter is plain on one side and has teeth on the other. Side milling cutters may have straight, helical, or staggered teeth. A metal slitting cutter resembles a plain cutter except that it is made very thin, usually $\frac{3}{16}$ in. (0.5 cm) or less. Plain cutters of this type are relieved by grinding the sides to afford clearance for the cutter. Angle milling cutters are either single- or double-angle cutters. The single-angle cutter has one conical surface whereas the double-angle cutter has cutting teeth on two conical surfaces. Angle cutters are used for cutting rachet wheels, dovetails, flutes on milling cutters, and reamers. Form-milling cutters have teeth arranged with special spaces and configurations: convex, concave, fluted, and corner rounded. End-mill cutters have an integral shaft for driving and have teeth on both periphery and the end. End mills are used for surfacing projections, squaring ends, cutting slots, and in recess work such as die making. T-slot cutters resemble small plain cutters that have an integral straight or tapered shaft for driving. They are used for milling T-slots. Finally, inserted tooth cutters are cutters that have teeth made of expensive material inserted into less expensive steel. As cutters increase in size, is it economical to use these cutters since teeth can be replaced when worn out or broken.

Again, the overall potential health hazard associated with milling operations is dependent on the metal being milled, the processing conditions, the nature and extent of controls, and the cutting oil used. Cutting oils commonly used for milling processes include the mineral-oil- and fatty-oil-based products for the soft steels and copper alloys, chlorinated paraffinic fluids for stainless steels, and the "extreme pressure" compounds for cast iron and titanium alloys. The potential health hazards of these materials are discussed in Section 4.

3.6 Lathing, Planing, Shaping, and Slotting

This group of machines includes a number of processes in which material is removed from the workpiece by moving the part, in either a reciprocating or rotary motion, against a single-point cutter. Parts to be machined can be held between centers, attached to a faceplate, supported in a chuck, or held in a draw-in-chuck or collet. Some machines, such as the lathe, are particularly adapted to cylindrical work, whereas others, such as planing machines, are best suited for providing a plain, flat surface. Each of these machine types is described briefly in the following subsections.

3.6.1 Turning Machines (Lathes). Because of the many variables in the size, design, method of drive, and purpose of the various turning machines, it is difficult to classify them. Most are named according to some outstanding design characteristic. Thus, there are speed lathes, engine lathes, turret lathes, and other special-purpose units.

Speed lathes, the simplest of the turning machines, consist of a bed, a head stock, a tail stock, and an adjustable slide for supporting the tool. Usually, it is driven by a variable-speed motor built into the head stock. Because hand tools are used and the cuts are small, the lathe is driven at high speed, the work being held between centers or attached to a faceplate on the head stock. The speed lathe is used principally for turning wood, centering metal cylinders prior to further work on an engine lathe, and metal spinning.

The engine lathe derives its name from early models that obtained power from engines. It differs from the speed lathe in that it has additional features for controlling the spindle speed and supporting and controlling the feed of the cutting tool.

Bench lathes refer to small lathes that are mounted on a work bench. In design, they have the same features as speed or engine lathes and differ from them only in size and mounting. They are limited to use with small workpieces, parts having a maximum swing of about 10 in. at the faceplate.

Turret lathes are the most common in use for production-scale work. Whereas the lathe types mentioned earlier require a highly skilled operator and take appreciable time to produce pieces, the turret lathes has operator skill "built into it," making it possible for inexperienced operators to reproduce identical parts. The principal characteristic of this group is that tools for consecutive operations can be set up for use in a desired sequence. Although it requires considerable skill to set and adjust the tools, little skill is required to operate them; many parts can be produced before adjustments are necessary.

3.6.2 Planers. A planing machine is designed to remove metal by moving the workpiece in a straight line against a single-edge tool. Similar to the work done on a shaper (Section 3.6.3) a planer is adapted to much larger work. The cuts, which are mainly plane surfaces, can be horizontal, vertical, or at an angle. In addition to machining large work, the planer is used frequently to machine multiple small parts held in line on a platen. Planers are no longer important for production work, their work more commonly being done now by milling, broaching, or abrasive machining units. Drive mechanisms for planing machines include gear drive, hydraulic drive, screw drive, belt drive, and crank drive. The first two of these are the most common. Planers are classified into four types, based on general construction of the machines: double housing, open side, pit type, and edge or plate.

Double-housing planers consist of a long, heavy base on which the table or platen reciprocates. The upright housing, near the center on the side of the base, supports the crossrail on which the tools are fed across the work. These machines can be fed manually or by power in either a vertical or crosswise direction.

Open-side planers have the housing on one side only. The open side permits the machining of wide workpieces. Most planers have one flat and one double-V way, which allow for unequal bed and platen extensions.

The pit-type planer is a relatively massive machine and differs from an ordinary planer in that the bed is stationary and the tool is moved over the work. Workpieces 15 ft (4.5 m) in width and over 30 ft (9 m) in length can be planed in typical pit-type machines.

The plate or edge planer is a special type of planer devised for machining the edge of heavy steel plates for pressure vessels and armor plate. The plate is clamped to a bed and the carriage supporting the cutting tool is moved back and forth along the edge. A large screw drive is used for moving the carriage. Most edge planers use milling cutters instead of conventional planer tools for greater speed and accuracy.

3.6.3 Shapers and Slotters. A shaper is a machine with a reciprocating cutting tool, of the lathe type, that takes a straight-line cut. By moving the work across the path of this tool, a plane surface is generated, regardless of the shape of the tool. Perfection is not dependent on the accuracy of the tool as it is with most other machining processes. Shapers can be classified according to their general operational design: horizontal, vertical, and special purpose (for cutting gears).

The horizontal-type shaper is used commonly for production and general-purpose work. It consists of a base and frame that support a horizontal ram. The ram, which carries the cutting tool, is given a reciprocating motion equal to the length of the stroke desired. A quick-return mechanism is incorporated into the ram so that the return stroke of the shaper is faster than the cutting stroke, thus reducing the idle time of the machine to a minimum. The tool head at the end of the ram, which can be swiveled through an angle, is provided with means for feeding the tool into the work. The cutting tool pivots upward on the return stroke so as not to dig into the workpiece.

Vertical shapers are also referred to as slotters and are used principally for internal cutting and planning at angles. They also are used for operations that require vertical cuts because of the position in which the work must be held. The shaper ram operates vertically and has the usual quick-return feature found on the horizontal units. Work to be machined is supported on a round table having a rotary feed in addition to the usual table movements. The circular table feed permits the machining of curved surfaces, a process that is particularly desirable for many irregular parts that cannot be turned on a lathe. Plane surfaces are cut by using either of the table cross feeds. A special type of vertical shaper, known as a keyseater, is used to cut keyways in gears, pulleys, and cams.

Although both planers and shapers are adapted to the machining of flat surfaces, there is not much overlapping in their fields of application. When the two machines are compared, the following differences can be seen. The planer is especially adapted to large work; the shaper can do only small work. On the planer, the work is moved against a stationary tool; on the shaper, the tool moves across the work, which is stationary. On the planer, the tool is fed into the work; on the shaper, the work is usually fed across the tool. The drive on the planer table usually is by gear or hydraulic means; the shaper normally utilizes a quick-return mechanism. Most planers differ from shapers in that they approach more consistent velocity in their cutting speeds.

3.7 Broaching

Broaching is the operation of removing metal by means of an elongated tool that has a number of teeth of successively increasing size that cut in a fixed path. A part is completed in one stroke of the machine, the last teeth on the cutting tool conforming to the desired shape of the finished surface.

In most machines, the broach or cutting element is moved past the workpiece, although equally effective results can be obtained if the tool is held stationary and the workpiece is moved.

The machine itself may be hand, electromechanically, or hydraulically operated. Both the cutting tool and the workpiece are held in rigid fixtures, the principal function of the machine being to provide sufficient speed to permit the cut to be made. The feed, and therefore the material removal rate, is regulated by the broach, the hardened steel bar with a series of cutting teeth. Each tooth is made progressively higher than the preceding one, to remove successively larger amounts of material. Thus, the cut deepens as the operation progresses. Cutting fluid is applied generously along the cut.

The process of surface finishing by broaching is roughly the same as planing except that in broaching a multitoothed cutting tool is used. In either case, the cutting tool may be either passed across a fixed workpiece or the tool may be held stationary and the work moved in a continuous stroke. In some machines, the tool is pushed along the surface being broached, while in others, the broach is pulled through or over the workpiece.

Generally, broaching tools are specially designed and made for specific use and for specific machining operations. However, certain types of broaches can be identified, based on the type of work they do. Internal broaching tools are used to enlarge and finish a wide variety of holes previously rough formed by casting, forging, punching, drilling, boring, or other method. Most internal broaching is done with pull broaches since they can take longer cuts and consequently can remove more stock than push broaches. Special internal broaches, called burnishers, are used on parts when surface finish and accuracy are critical.

Special sizing broaches are pulled or pushed through semifinished holes to remove the remaining amount of stock faster and more efficiently than is possible with a conventional tool. Burnishing generally is restricted to soft, ductile metals. Most internal broaching is done with a straight-through action of the cutting tool; however, helical splines, rifling in gun barrels, internal gear teeth, and other similar features can be produced by rotating the broach as it is pulled through the workpiece.

External broaching tools may be of one-piece construction or may consist of a series of segmented sections, assembled end to end on a moving slide. Sectional construction results in a broach that is easier and cheaper to construct and sharpen. Broken sections can be replaced readily without discarding other sections. The broach itself may be made with either straight or angular teeth.

The machines that house the broaching operation can be oriented either horizontally or vertically. Horizontal machines are general-purpose machines and may be used for high or low production quantities. The majority of horizontal machines are used for internal broaching operations. These machines may be as long as 50 ft (15 m), although there is convenient access to any part of the machine. Vertical machines are available as push-down, pull-down, or pull-up versions. They are adaptable to both external and internal broaching tasks.

Broaching has been adopted for a variety of mass-production operations because of the many advantages offered. Both rough and finishing cuts are completed with one pass of the tool. Production rate is high; actual cutting time is a matter of seconds. Rapid loading/unloading keeps production time at a maximum. The process is useful for both internal and external finishing. Any form that can be broached can be machined. Tolerances can be maintained; finishes comparable to those achieved by milling processes can be obtained. In fact, burnishing shells incorporated on the broach can improve the surface finish.

The limitations of broaching as a production-scale operation include the following. Tool cost is high, particularly for large or irregularly shaped tools, thus making short-run jobs undesirable. Parts to be broached must be supported rigidly and be able to withstand the broaching forces set up; there can be no obstruction on the surface to be broached since a straight-line cut is made.

3.8   Honing

Honing is a unit operation in metal machining that utilizes various abrasive materials bonded to stones or sticks and mounted on a metal mandrel. The cutting action is very similar to that associated with conventional abrasive grinding. However, material is removed at much slower speeds. Heat and pressure are minimized, resulting in excellent control of the process. All honing gives a smooth finish with a characteristic crosshatch appearance. The depth of these hone marks can be controlled by variation in pressure, cutting speed, and type of abrasive material used.

The most common application of the honing process is removal of stock from internal and external surfaces on cylindrical parts. The abrasive materials used include aluminum oxide, silicon carbide or oxide, and in some special cases, diamond grains; bonding systems typically include vitrified or resinoid materials. As is the case with conventional grinding processes, each abrasive grain protruding through the bond contacts the part being honed and acts as a tiny cutter to remove small chips of the metal. This tends to be a self-sharpening process since the grains break out of the bond when they become dull, allowing other sharp, fresh surfaces to contact the workpiece. The size and shape of the workpiece determines the number of stones and their dimensions, to be spaced at regular intervals around the periphery of the mandrel. Unlike grinding operations, in which there is a line of contact between abrasive and the workpiece, a relatively large portion of the honing stones is in contact with the metal surface. A drive shaft provides a combined reciprocating and rotating action to the stone, with contact between stone and metal being controlled by positive pressure. Although the amount of metal removed by each cutting edge is small, the collective action of the many surfaces, working together, provides a relatively fast and accurate means of stock removal.

Internal honing is the most common application of this process. The cutting tool aligns itself in the bore while rotating and reciprocating within the workpiece, held securely in place. The rotational and reciprocating actions can be controlled independently, actually being set at odd ratios to one another. In this way, the individual grains of abrasive material are able to act on the workpiece in a continuously changing pattern. Honing is used extensively to correct a variety of inaccuracies remaining from previous operations producing holes. High spots, chatter marks, taper, out-of-roundness, and deviations in axial straightness can all be corrected easily by means of honing action on the internal surfaces. Of course, this action can be used only to improve the size and shape of the feature since the honing tool follows the neutral axis on the workpiece initially established by the preceding operation. It cannot correct errors of hole location or alignment, nor can it improve concentricity with other diameters.

As mentioned earlier, honing produces a characteristic crosshatch finish. This normally has a functional use in parts made for use as a load-bearing surface. Each of the minute scratches in the crosshatching serves as a reservoir for lubricants, thus minimizing the friction and heat generated and maximizing the life of the part.

Most honing is done with cast iron and steel materials, although parts composed of titanium, copper, bronze, carbides, and many nonmetallic materials can be honed.

3.9 Lapping

Lapping is a relatively fine finishing operation, basically of the grinding type. In the basic procedure, workpieces are rubbed on a perfectly flat surface that has been loaded with a fine abrasive material. Typical lapping machines use a special rotating table on which conditioning rings ride; these conditioning rings continually grind the table surface flat and also contain holders for the workpieces. A heavy disk is placed on top of the parts to press them against the table surface. In operation, the rotating table and workpieces are flooded with an abrasive suspended in a liquid. All lapping is done at very low speeds. As such, it is not considered to be a stock removal process but rather a final finishing step. Although used most often for microfinishing of flat or cylindrical surfaces, it is adaptable to spherical or specially formed surfaces.

Hand lapping of individual pieces can be done on flat or cylindrical surfaces, although this work requires highly skilled operators to produce consistent and accurate results. Hand lapping is used commonly for finishing carbide die parts and the exceptionally hard materials. For flat work, a serrated lapping block or plate charged with a fine-grain, loose abrasive material with a suitable oil- or water-based vehicle is used. Silicon carbide, aluminum oxide, or boron carbide are excellent abrasives for metal lapping work. Most laps are made of soft cast iron or other material softer than the workpiece. The process of hand lapping consists of manually rubbing the workpiece with a constantly changing motion over the accurately finished surface of the lapping plate. The rate of stock removal and the ultimate quality of the work depend primarily on the pressure exerted by the operator.

Mechanical lapping machines are characterized by relatively light work pressures, moderate speeds, and generation of little or no heat during the process. Single-lap machines are used to process single or multiple parts when only one surface needs lapping. Lapping can be performed either with cast-iron laps using loose abrasives or with bonded abrasive wheel laps. The work may be held manually against the lap or work holders may be used to loosely hold the work and guide it into the proper motion.

Match-piece lapping is a special application of this unit operation by which pairs of mating parts may be equalized by rubbing one against the other. Loose abrasive compound, mixed with a lubricant, is applied to the contact surfaces of parts, such as gears, plungers, cylinders, valves and valve seats, pin and hole sets, and other mating pieces. Unwanted small irregularities in surface geometry caused by heat treatment, tool chatter, cutter makers, or machining can be eliminated, thus increasing the service life of moving parts. Matched-piece lapping is used commonly by manufacturers as a method of forming tight, leakproof seals on engine heads and blocks or on pump components. Mating parts processed in this manner are stocked in pairs, for obvious reasons.

Flat surfaces on some individual parts are finished by manipulating them by hand using fine-grit abrasive paper. A rotating lapping plate may be used or the operation may simply consist of manually rubbing the workpiece against a sheet of abrasive paper placed on a flat surface. The main purpose is to produce a bright reflective surface generally on softer materials. No significant amount of metal stock is removed.

Essentially all materials can be lapped. Common metals and alloys for which lapping is done as a finishing step include steels, stainless steels, cast iron, brass, bronze, aluminum, magnesium, and some specialty alloys.

3.10   Mass Finishing

Mass finishing refers to unit operations involving the final finishing of large quantities of relatively small parts. These operations have resulted in making hand-finished operations, such as deburring, descaling, and general surface refinement, obsolete. The mass-finishing operations use a mixture of abrasive grain media, principally aluminum oxide or silicon carbide, together with special compounds and water. The size of the various media is determined by the nature of the work to be performed, the workpiece materials, and the type of machine to be used. The media may be randomly shaped abrasive grains or nuggets, grains on carriers such as steel, steel balls, or slugs, granite, gravel, sand, or even sawdust. Compounds added to the system keep the media functioning properly, modify the cutting properties, and prevent rusting, pitting, and other forms of corrosion on the workpiece. Among the most common mass-finishing processes are barrel finishing or tumbling, vibratory finishing, and spindle finishing.

3.10.1 Barrel Finishing. Barrel finishing sometimes is considered to be a refinement of the more basic tumbling process. In practice, however, there is very little difference. In a typical barrel finishing machine, the unit is loaded with parts to be worked, abrasive media, and other necessary compounds to maximize the action of the media. Loading normally means filling the unit to about half, or slightly more, of its rated capacity. As the machine, essentially a horizontal barrel, is rotated, parts and abrasive media are carried up the side until gravity causes them to slide down, during which the abrasive work is performed. This type of finishing is very slow; during the time that a part of the load is being carried up to the start of the slide, no work is being done. The barrel cannot be filled since there would then be no sliding action and therefore no work accomplished. Similarly, there is a maximum speed of the barrel for any given machine and parts/media mixture. If the barrel rotates too fast, parts begin to cascade, thereby not only avoiding the sliding action but resulting in nicking and marring of the pieces. Optimum speed is just below the point at which cascading begins. Another disadvantage of this operation is the fact that work cannot be examined in progress; the machine must be stopped and parts removed for inspection. Once the time cycle has been determined for a particular application, however, this limitation is not of practical significance.

The barrel finishing process abrades edges and exposes surfaces much more than recesses and inside openings. Large corner radii can be formed on parts easily, and large burrs can be removed without roll-over common to some methods. Barrel finished pieces have a distinct finish pattern caused by the sliding and rubbing action of the media. Since the barrel is closed, it is possible to build up thick suds consisting of the mixture and water, facilitating very high finishes.

Barrel finishers are simple machines, are easy to repair, and require very little maintenance. They are, however, batchtype units and, therefore, are not compatible with in-line processing. On the other hand, the size of the units, relative to the size of the parts being processed, makes these machines very productive. Productivity can be increased by use of compartmented barrels, a series of smaller barrels using a common pair of rolling rails.

3.10.2 Vibratory Finishing. In vibratory finishing, the parts, abrasive material, additive compounds, and water are charged into a container, or tub, that is vibrated.

The vibrating action causes motion of the entire contents, usually in an elliptical path, the contact between abrasive and workpiece resulting in the removal of burrs, rounds, corners, and other unwanted projections. The action itself is faster and much more aggressive than barrel finishing; accordingly, the cycle times are much shorter.

In contrast to the barrel machining, which produces a relatively long scratch pattern, the vibrator produces a short, choppy pattern. This is characteristic on the finished pieces. The work can be examined at any time during the cycle, but only by stopping the unit and removing the parts from the tub. An advantage of the vibrator over the barrel is a much lower tendency of parts to become entangled. Barrel action results in a random sliding motion of the mass of parts and media; in the vibrator, parts and media retain their distances, moving in a continuous and unbroken path.

Although not a very complicated piece of equipment, the vibratory finisher is subjected to much more wear than the barrel finisher. However, it lends itself more readily to in-line processing applications since it is easier to load, cycle, and unload than corresponding barrel equipment. Most applications of the vibratory units are for relatively small parts. Generally, anything that can be loaded into the machine can be finished.

3.10.3 Spindle Finishing. The third type of mass finishing unit in common use is the spindle finisher. The work is chucked over one or more spindles and lowered into a tub containing the abrasive grain. The spindles, with the attached parts, are rotated slowly to expose all surfaces to a high-velocity stream of the abrasive media. During processing, the tub spins rapidly in a direction opposing the rotation of the spindle. Control of the finishing process is achieved through variation of several elements of the cycle: tub rotational speed, spindle rotational speed, depth of submersion of the parts, angle of the spindle, duration of the cycle, and type and size of the abrasive medium. These elements are all interrelated. For example, if the tub is rotated too fast, the grain will have a peening action rather than a cutting action and, among other things, will tend to roll over any burrs rather than cut them and round the edge.

To be effective, the part must be covered by the abrasive at all times during the cycle. The abrasive material is thrown up in the shape of a wave by the motion of the spindle, the depression following the wave being important. The higher the tub rotational speed, the deeper the depression, and the deeper the part must be submerged to assure contact with the abrasive.

The abrasive material in most common use in spindle finishing is aluminum oxide grain, wetted down with water and detergent periodically charged into the tub. The water flows onto the top of the tub in a steady stream and out the bottom in a similar manner, its only purpose being to keep the abrasive grain wet. The detergent softens the water for more effective flushing of the sludge and spent abrasive material.

This process is practical for use on parts, such as fragile components, that are difficult to finish by other methods and where close control of deburring and edge breaking is critical. The configuration of the part is not important as long as it can be chucked.

3.11 Superfinishing

All of the machining and grinding operations described in this chapter leave the metal surface coated with fragmented, noncrystalline, or smeal metal that, although easily removed by sliding contact, can result in excessive wear, increased clearances, noisy operation, and difficulties in maintaining adequate lubrication of the part in service. Superfinishing is a process that removes this undesirable fragmentation metal, leaving a base of solid crystalline metal. In principle, it is similar to honing processes in that an abrasive stone is used. It differs, however, in the type of motions given to the stone. Superfinishing is, in a sense, the ultimate finishing process; it is not used for dimensional improvement. This process can be superimposed on virtually all other commercial finishing operations.

When two mated parts are in service, any high spot or fragmented metal particle extending from one contact surface will tend to wear down rapidly and result in a loose fit. Such minute surface projections also are objectionable because they tend to penetrate the thin film lubricant used to separate the moving parts in service and thus contribute to excessive wear. The major purpose of superfinishing is to produce a surface on a workpiece that is capable of sustaining an even distribution of load by improving geometrical accuracy. As a result, the life of parts superfinished to maximum smoothness is extended considerably.

During superfinishing, a flood of low-viscosity oil, often a lightweight mineral oil, is applied in the work area between the abrasive stones and the workpiece. The lubricant–coolant carries away the abraded particles and keeps the work at a uniformly cool temperature. On most machines, there is a large area of abrasive stone in contact with the work. On cylindrical surfaces, for example, the width of the abrasive stone may be two-thirds of the diameter of the part. Only a short time is required to complete surface improvements since most of the finishing involves removal of minute projections. Processing time frequently is measured in seconds, although a few minutes may be required to develop sufficient cutting action to obtain an improved surface of extremely close tolerances.

The abrasive stones are self-dressing as is common with abrasive finishing operations. Aluminum oxide is the most common abrasive medium for superfinishing steels and silicon carbide abrasives generally are used for cast iron and most nonferrous metals. Vitrified bonded stones are used most often in preference to the slower cutting action associated with shellac or resin-bonded systems.

Either flat or spherical operations can be performed on the same type of superfinishing machine, using vertically opposed upper and lower spindles. The upper spindle has a spring- or hydraulically loaded quill on which a cup-shaped stone is mounted. The lower spindle contains a circular table that carries the workpiece. As the lower spindle and the work are revolved, the end face of the cup-shaped stone in the rotating upper spindle is brought into contact with the work. Some degree of offset is given the cup stone with relation to the work so that the path of any one grit is rarely repeated. The outer portion of the cup wheel overhangs and cuts free of the workpiece surface, resulting in the self-dressing action for the wheel face. When the ends of the two spindles are exactly parallel, the combined effort of the rotation of both stone and work results in a very true, flat surface. Spherical shapes can be generated by adjusting the upper spindle to some angle with the lower. Variables that determine the results obtainable by superfinishing include the grit size and grade of abrasive stone, surface speed of the workpiece, pressure on the stone, viscosity of the stone lubricant, and the reciprocation speed. Typical applications of superfinishing operations include calender rolls in paper mills, computer memory drums, sewing machine parts, and many automotive parts, for example, cylinder bearings, pistons, clutch plates, and tappet bodies.

4 POTENTIAL HEALTH HAZARDS

With few exceptions, the metal machining operations discussed in the preceding section of this chapter are characterized by the same set of potential health hazards for a given metal or alloy system. The magnitude of the potential for exposure of machine operators will vary from one process to another, however, depending on the manner in which the operation is conducted and the nature and extent of control measures instituted at the site of release of contaminants. Where appropriate in the discussion of specific unit operations, reference was made to health hazards peculiar to those processes. In this section, for each of the common health hazards, the primary sources of release within the unit operation, basic approach for evaluation of the magnitude of the stress, and general discussion of the need for and nature of control measures— all are presented.

4.1 Metal Dusts

The principle action of the various machining operations described in this chapter is to shear or abrade metal from the workpiece, forming thin running coils that usually break into small chips. Although to a lesser extent than with the metalworking and forming operations, extremes of temperature and pressure occur at the point of contact between the cutting tool and the metal. The airborne particles generated by these processes depend on the type of metal being machined and the cutting tool used, the dust-forming tendencies of the metal, the machining technique, and the nature and method of application of lubricants or coolants.

The type of metal being machined is of primary interest. Such metals range from the mild steels, characterized by a very low potential health hazard, to the highly toxic beryllium alloys, representative of metal work for which the ultimate industrial hygiene controls are warranted. Within this spectrum of concern are the high-temperature and stainless steels, incorporating metals recognized as being moderately toxic, such as chromium, nickel, and cobalt.

In most cases, conventional machining operations using cutting tools or abrasive media do not lead to significant concentrations of airborne metals, provided reasonable control measures are utilized.

Another source of airborne metal dust in machining operations is the release of metallic components from the cutting surfaces. A variety of specialty alloys has been developed for use in the manufacture and fabrication of the cutting tools. High-carbon steel with alloying metals (such as vanadium, chromium, and manganese), so-called high-speed steels containing manganese and tungsten, special cobalt steels, cast alloys of tungsten, chromium, and cobalt, and tungsten carbide are among the materials used for manufacture of these cutting surfaces. With conventional machining, there is very little loss of material from the cutting tool itself; the contribution of dust from the cutting tools to the total dust loading is minimal. On the other hand, when the cutting tools are subjected to grinding or dressing operations in preparation for cutting applications, significant exposure to these toxic dusts could result. Appropriate controls, including use of local exhaust ventilation, should be used during these operations.

Evaluation of exposure to airborne metallic substances is rather straightforward. Small, battery-powered sampling pumps can be used to collect the airborne dust by drawing the air through appropriate filter media. Analysis for specific metals can be performed easily using atomic absorption spectrophotometry techniques.

4.2 Silica

Mention is made of the potential health hazard associated with exposure to silica because of the potential for release of silica from metal castings and forgings during some unit operations, particularly rough grinding on fresh, unworked metal parts. Silica, fused into the outer layers of the metal part as a result of solubilization into the metal during solidification, can be released easily by grinding action on the periphery of the part. The silica content of airborne dust generated by such processes can be as much as 10% and more, depending on the specific circumstances of the pouring process and the molding sand systems used.

In addition to the evaluation of exposure to metallic species, then, it is necessary to determine if, and to what extent, the dust contains crystalline silica. Gravimetric analysis of the samples collected with a cyclone preselector to trap the respirable portion of the dust can determine the amount of dust present. Subsequent analysis of these samples for quartz will indicate the silica content. Current standards for exposure to silica dusts are based on a formula that is a function of the silica content of the dust.

Because of the significance of chronic exposure to silica-containing dusts, operations that have been shown to release appreciable quantities of silica should be conducted under carefully controlled conditions. Maximum enclosure of the process and installation of local exhaust ventilation should be done for these unit operations.

4.3 Cutting Fluids

Cutting fluids are important in metal machining operations. They are used to cool and lubricate the point of contact between the cutting tool and the metal workpiece, as well as to flush away the chips removed by the machining. The composition of these fluids has varied significantly over the years. Presently, there are three types of cutting fluids: straight mineral-oil-based systems, soluble oils and oil emulsions, and various other synthetic compositions. Two major potential health hazards are associated with use of cutting fluids: dermatitis due to extensive skin contact and inhalation of oil mists. Additional problems are associated with some special formulations of cutting fluid. A general discussion of the typical compositions, potential health hazards, special-handling requirements, and general guidelines for control of cutting fluids, is presented in this section.

4.3.1 Composition. Perhaps the most common group of cutting fluids is that based on mineral oil. These fluids typically are composed of a base containing at least 60% mineral oil. A variety of polar additives is available, depending on the intended use of the fluid. Animal and vegetable oils, fats, and waxes are added to wet and penetrate the interface between the cutting tool and the workpiece. Lubricants added to the base oil include esters, fatty oils, fatty acids, and complex alcohols. For application to machining operations where high pressures are anticipated, special lubricants are added. These include sulfurized mineral oil or sulfurized fats, chlorinated waxes, chlorinated long-chain esters, combination chlorinated and sulfonated fatty oils, and organic and metallic phosphates. In addition, cutting fluids usually are provided with some type of germicide or bactericide to suppress growth of bacteria in the oils; phenolic and quaternary ammonium compounds are used commonly for this purpose.

Some mineral-oil-based cutting fluids incorporate other additives to provide desired performance under special conditions of use. To provide antioxidant and/or anticorrosion properties, additives such as alkyl zinc, aryl dithiophosphate, alkylated phenols, phenylenediamine, and nitrites of chromium and sodium are sometimes added. Detergents and dispersants are typically methyl alkyl sulfonates or alkyl phenolates.

The mineral-oil-based cutting fluids are used in a wide range of application. They are used extensively for turning, such as lathing, milling, drilling, and reaming operations on magnesium, aluminum, copper, a wide range of steels, and titanium alloys. Drilling and finishing grinding of magnesium alloys also utilize mineral oil systems.

The soluble oils or, more correctly, oil emulsions represent the second most common type of cutting fluid in use. These liquids, which tend to be milky in appearance, contain mineral oil in the base concentrate, perhaps to the extent of 80% or more. However, in use these concentrates are diluted with water in ratios of as much as 1:50. This dilution, plus addition of emulsifying agents, produces a soluble oil system consisting of the oil suspended in a water emulsion. Emulsifiers include petroleum sulfonates, amine-based soaps, rosen soaps, and naphthenic acids. Polar additives and lubricants also may be included in the formulation, depending on the intended use of the fluid. Similarly, corrosion inhibitors such as hydroxylamine and germicides such as the phenolic compounds usually are added.

The soluble oils are used across the full range of machining operations discussed in this chapter: grinding, milling, turning, drilling, broaching and reaming. They can be used effectively with magnesium, aluminum, copper, steels, stainless steels, nickel alloys, cast iron, and titanium alloys.

The synthetic cutting fluids are clear, transparent liquids, the concentrates containing 50–80% water. The concentrates are diluted further with water in use; dilution ratios of as much as 1:200 are used. Whereas a truly synthetic oil contains no oil, some synthetic cutting fluid concentrates contain mineral oil in the range of 5–20%. The remainder of the synthetic oil concentrate is composed of surfactants, lubricants, corrosion inhibitors, and germicides or bactericides.

The synthetic oils have found more limited, but increasing, applications within the metal machining unit operations. Grinding operations on most metals can utilize the synthetic oils; use in other operations such as milling and lathing tends to be limited to the softer metals.

4.3.2 Dermatitis. The most significant potential health hazard posed by the cutting fluids appears to be that associated with extensive skin contact by operators of machining equipment. Hundreds of thousands of cases of dermatitis occur each year as a result of contact with cutting fluids, according to NIOSH estimates. In addition to dermatitis, contact with some of these oils can result in acnelike lesions, infection of sweat pores, and folliculitis. In some instances, there appears to be a sensitization or allergic reaction developed in particularly sensitive workers. In general, the primary dermatitis problem is associated with cutting fluids in which the major oil constituent is relatively low boiling. The folliculitis and acne problems tend to result from contact with the heavier, higher-boiling oils.

Of substantial concern in recent years has been the indication that much of the adverse response of the skin due to contact with various cutting fluids may be precursor stages of skin cancers. There appears to be somewhat higher incidence of skin-related concerns among machinists and other workers who routinely contact the cutting oils.

Whether the cause-and-effect relationship is established or not, the potential for skin diseases certainly exists, and due concern should be shown for control of the exposures of workers to the cutting fluids.

4.3.3 Inhalation of Oil Mists. The application of the cutting fluid to hot, rotating, or reciprocating parts can generate an oil mist. The odor characteristic of machine shops is due to the pervasive oil mist in the air. Compounding the problem has been the introduction of cutting oil application systems in which the fluid is sprayed onto the work in mist form. Although there have been conflicting opinions reported in the literature regarding the significance of oil mist levels typically present in the machine shop, a strong case can be built for implementation of a rigorous control program.

4.3.4 Other Potential Hazards. Some investigators have suggested that much of the skin problem associated with cutting oils is due to bacterial contamination of the fluids because they are used for extended periods of time in recirculation systems. Although there has been no identification of bacteria in these cutting fluids that are harmful to humans, this potential should be recognized, along with other hazards.

It would be expected that cutting fluids may become contaminated with trace quantities of the base metal on which the machining work is being done as the fluid is used. Some of these metals may dissolve into the cutting fluid coming into contact with the skin of workers as they encounter the cutting oil. It is conceivable that such contact with metals from alloys containing chromium, nickel, and cobalt—all known sensitizers—contributes to the overall dermatitis problem.

4.3.5 Control Measures. There is a need for relatively high degrees of control of the various cutting fluids used in metal machining operations. Of particularly great importance is the need for a high level of personal cleanliness in these operations. Workers should be encouraged to avoid all unnecessary contact with the cutting fluids, wear clean, dry clothing and remove any obvious skin contamination immediately.

From an engineering standpoint, the cutting fluids should be used at the lowest temperature compatible with proper working of the machines. Local exhaust ventilation can be installed on many of the high-temperature or high-energy machining equipment. It is the unusual case where concentrations of oil mist in the workroom atmosphere cannot be controlled to acceptable limits by means of appropriate enclosure of the process and local exhaust ventilation. It frequently is necessary to use a general ventilation system to provide total control of the environment, diluting to acceptable concentrations any oil mist that is not controlled effectively at the point of generation.

4.4   Other Surface Treatment Chemicals

As a metal component is transformed from raw form, such as an ingot or rough casting, to final form, ready for assembly and final finishing, there may be a need to treat the surface to assure proper effect of the unit operation. Other surface cleaning or preparation operations, using solutions and solvents (other than acids or alkaline compounds), are used depending on the characteristics desired in the workpiece. Those metal treatment processes are considered separate unit operations and are discussed in other chapters. They are mentioned here only for sake of completeness. Generally, they do not constitute a significant health hazard in the metal machining processes because of the limited extent of involvement and the adequacy of controls normally in place.

4.5 Noise

The most significant of the physical stresses present in the typical machining operation is noise. All of these unit operations involve the direct, moving contact between the metal workpiece and either a metal cutting tool or an abrasive material surface. This metal-on-metal or other abrasive action varies in intensity from one process to another but is characteristic of all. In addition, noise generated by the power sources for the equipment—compressed air, pneumatic mechanisms, and hydraulic systems—contributes to the overall noise level in the machining areas. The consequent existence of a potential hearing loss problem must be recognized and dealt with accordingly.

The noise associated with compressed air and pneumatic systems is characterized by predominance in the high frequencies. For the most part, these noise sources can be provided with appropriately designed muffler systems. Modern units can incorporate noise control features that minimize the exposure of the machine operators by directing the points of air release out of the operator's position and using mufflers to dissipate the noise. Air that is used to clean, or blowoff, the working surface of a metal part can be directed through commercially available "silencers" that dissipate some of the noise and yet retain the effect of the blast of air on the workpiece.

The motor used on hydraulic systems present in some metal machining equipment represents another source of noise in the process. The movement of the hydraulic fluid through the system contributes some, although relatively little, additional noise. For the most part, noise from the hydraulic systems is of low-frequency origin; this poses a much lower degree of concern because of relative insensitivity of the human ear toward low-frequency noise.

The most significant of the noise sources in metal machining operations is that due to the more vigorous metalworking actions, typified by rough grinding of metal castings or forgings. The grinders are available in a wide range of sizes, from small portable units to large semiautomated machines. The noise level produced by these machines will vary with the type and size of grinder as well as the material being machined. As a general rule, the harder metals and alloys will produce higher noise levels than the softer, more ductile metals, other things being equal.

Noise produced by metal machining equipment, in general, and grinders, in particular, can be minimized by a continuing, effective maintenance program. Of particular importance, in all cases, is the sharpness of the cutting tool or the integrity of the abrasive material matrix. Many machining operations can be provided with partial enclosures, lined with sound-absorbing material. Of course, on production line equipment, there must be continuous passage of material through the machine, and this material transfer requirement limits the degree to which enclosure can be used as a control strategy. However, noise barriers can be arranged on most equipment in such a way that the exposure of the operator is reduced.

Much of the noise present in the metal machining operations is similar in nature to that found in metalworking and forming processes. The reader is directed to that chapter for additional discussion of the noise problem in metal processing operations.

Documentation of worker exposure to noise can be done using standardized techniques and equipment. In most instances, noise surveys conducted with sound-level meters can be used as the basis for determining the time-weighted average exposure of the workers. In some situations, where the exposure of an individual might be extremely variable or borderline, in terms of compliance, it is advantageous to use noise dosimeters to document actual exposure. Surveys conducted for purposes of designing appropriate engineering controls will invariably require the use of more sophisticated equipment such as octave-band analyzers or real-time analyzers.

It would not be unusual to find evidence of overexposure to noise in some operations in a metal machining facility. Therefore, in most plants, it will be necessary to implement a complete hearing conservation program consisting of noise monitoring, audiometric testing of exposed workers, and a noise abatement program.

For additional information, see Chapter 8.

4.6 Vibration

Another potential health hazard present in many metal machining operations is vibration. All of the equipment used for the machining of metal, described in preceding sections of this chapter, operates on either a reciprocating or rotary action. Depending on the support structure for the equipment, the vibrations resulting from the action of the machinery can be propagated through the support and walking/working surfaces of the employees, ultimately to workers' positions. Exposure of the workers to whole-body vibrations can result.

Of perhaps greater concern are the operators of portable grinders and other metal machining equipment that is handheld. These workers are subjected to segmental vibration applied locally to the hands and arms. Whereas the whole-body vibration resulting from the machining equipment is relatively low in intensity, the segmental vibration effects on operators of vibrating handheld tools can be significant. Extensive use of this equipment, particularly in cold environments, can result in Raynaud's phenomenon, or "white fingers." This condition is characterized by numbness and blanching of the fingers, often with lessened muscular control and reduction of sensitivity to heat, cold, and pain. The significance of chronic exposure to vibration currently is receiving much attention by investigators.

5 CONTROL OF HEALTH HAZARDS

In general, the engineering control strategy for metal machining operations consists of a combination of enclosure of the process to the feasible maximum and optimal orientation and use of local exhaust ventilation. Although this might sound like the same approach used for metalworking and forming processes, the need for localized control is much greater with machining processes because of the proximity of the machine operator to the points of release of contaminant. The reader should review the general discussion of controls for the metal forming processes, presented in Chapter 18, in addition to those described below for machining operations. Because with machining processes an operator is near the point of contaminant release, much greater use of enclosure of the process is made and much less reliance on general, or dilution, ventilation is acceptable.

5.1 Enclosure

As mentioned earlier, a general need exists for at least partial enclosure of most machining operations because of the necessary presence of the machine operator near the point of release of contaminants. Contaminants comprise a spectrum of possibilities: dusts and fumes of the metal being machined, vapors and mists of oils and other fluids used for cooling and lubricating purposes, and, in some instances, gases such as ozone and nitrogen dioxide generated by interaction of high-temperature machining processes with air.

Enclosures for metal machining processes can take either of two basic forms: fixed, rigid-walled structures or movable, flexible "shrouds" at specific points of contaminant release. As a general rule, those machining processes that hold the workpiece in place and function by movement of the cutting surface lend themselves more readily to enclosure by flexible curtains or equivalent. This allows the point of operation to move as needed to produce the surface action on the metal. On the other hand, the processes that involve movement of the workpiece into or against the cutting tool are more easily provided with fixed-position, rigid enclosure. Thus, processing equipment such as drill presses, boring machines, grinders, and hobbing equipment can be fitted easily with rigid ductwork systems, whereas it is more difficult to install such systems on lathes, milling machines, and similar equipment.

Frequently, partial enclosures consisting of drapes, curtains, or similar flexible materials are installed as "splash guards" in attempts to minimize release of coolants or cutting fluids from the operations. Although intended to serve primarily as cutting fluid containment, these provisions establish a basis for incorporation of local exhaust ventilation systems. Thus, the installation of splash guards invites extension of the control concept to ventilation since they undoubtedly are positioned at points of mist generation and release, the point of operation of the machine.

5.2 Ventilation

With few exceptions, metal machining operations should be enclosed and provided with exhaust ventilation to maintain air concentrations of contaminants below permissible exposure limits. The hood serving as partial enclosure and entry into the exhaust system is critical in its design, configuration, and orientation relative to the points of release of contaminants. Depending on the nature of the contaminant and condition of release, one of three basic types of hood is most appropriate: receiving hood, enclosing hood, or low-volume, high-velocity (LVHV) hood.

Receiving hoods are essentially "open" hoods; they must employ sufficient capture velocity to reach beyond their physical boundaries, be positioned strategically to take advantage of the "throw" of contaminant (such as on grinding machines), or both. The hood opening (face) can be any shape needed to accommodate the particular process. The minimum exhaust volume through the hood can be calculated easily using the standard design approach outlined in several of the general references listed at the end of this chapter; capture velocity is the most significant parameter to be considered. In cases where an elongated hood configuration is best, slotted hoods are most effective. Obviously, the greater the extent to which the contaminant release point can be enclosed partially by the hood, the more effective will be the control.

The LVHV hood is a special type of receiving hood, designed to produce very high capture velocities at points of contaminant release. LVHV hoods are used to achieve greater effectiveness in contaminant control or to control contaminants released with high initial velocities. LVHV hoods generally are designed to be small in size and are located in close proximity to the contaminant release point. Hoods of this type require relatively small exhaust volumes to achieve contaminant control. To be effective, these hoods must produce capture velocities exceeding the contaminant release velocity.

Table 19.1. Typical Local Exhaust Ventilation Requirements for
Metal Machining Operations

Hood Type	Example Contaminants Released from Processes	Minimum Velocity	
		(fpm)	(mpm)
Receiving	Aluminum, iron and iron oxide, nuisance dust, steel, and zinc	100	30
Low volume, high velocity	Cadmium, chromium, cobalt, copper, nickel, nitrogen dioxide, oil mist, and ozone	200	60
Enclosure	Beryllium, lead, mercury, and uranium	300	90

When LVHV hoods are used, the minimum exhaust flow requirements are realized. Proof of overall effectiveness has been demonstrated by personal breathing zone monitoring on operators equipped with LVHV systems. Positioning of the hood, that is, the actual location of the pickup point of the exhaust system, can be established either by installing a suspended hood at the point of operation of the machine or by mounting the LVHV hood on the cutting tool itself.

The ultimate hood arrangement is one in which the point of operation and contaminant release is virtually completely enclosed and the enclosure ventilated, the basis of the "enclosure" hood. These units are designed to prevent the projection of contaminants outward through any openings in the hood.

Table 19.1 summarizes the preferable hood type and minimum exhaust ventilation rates as a function of the most significant potential contaminant associated with the machining process being considered.

5.3 Personal Protection

The most significant aspect of personal protection associated with metal machining operations is prevention of contact with the various cutting fluids. Although there is relatively low hazard associated with inhalation of vapors or mists of the cutting fluids, skin contact with the fluids is one of the most common causes of industrial dermatitis. Causative factors include (1) synthetic cutting fluids that are potent defatting agents, as a rule, (2) soluble oil emulsions that provide a breeding ground for bacteria (bactericides are added primarily to prevent decomposition and odor formation, not prevention of skin infections), and (3) fluid additives that can cause either primary irritation or hypersensitive dermatitis.

For a discussion of the personal protection elements of the total hazard control program, the reader is referred to the chapter on metalworking and forming since the potential problems and appropriate corrective action are similar.

GENERAL REFERENCES

Control of Exposure to Metalworking Fluids, NIOSH Publication 78–165, 1978.

Fundamentals Governing the Design and Operation of Local Exhaust Systems, American National Standards Institute, ANSI Z9.2-1979.

Health and Safety Guide for Tool, Die and Precision Machining Industry, NIOSH Publication 77–198, 1977.

Industrial Ventilation: A Manual of Recommended Practice, 18th ed., Am. Conf. of Govt. Ind. Hyg., Lansing, MI, 1984.

Recirculation of Exhaust Air, NIOSH Publication 76–186, 1976.

Recommended Industrial Ventilation Guidelines, NIOSH Publication 76–162, 1976.

Validation of a Recommended Approach to Recirculation of Industrial Air (Spring Grinding), NIOSH Publication 79–143A, 1979.

Validation of a Recommended Approach to Recirculation of Industrial Air (Metal Grinding), NIOSH Publication 79–143B, 1979.

Ventilation Control of Grinding, Polishing, and Buffing Operations, American National Standards Institute, ANSI Z43.1-1966.

Ventilation Requirements for Grinding, Buffing, and Polishing Operations, NIOSH Publication 75–107, 1975.

Working with Cutting Fluids, NIOSH Publication 74–124, 1974.

Weh S. Wu, Douglas K. Arai, Mark Z. Nazar and David K. Leong, Determination of Nitrites in Metal Cutting Fluids in Ion Chromatography, *Am. Ind. Hyg. Assoc. J.* **43** (12), 1982.

METALLIZING

Robert D. Soule

1 INTRODUCTION

Metallizing is the process by which molten metal is sprayed onto a surface to form a coating. A wide variety of metallic elements, compounds, and mixtures can be deposited on metal, glass, ceramics, cloth, and other materials to build up, fill voids, or change surface characteristics of the substrate. In welding and brazing, the substrate temperature must be sufficiently high to hold the filler material at or above its melting point for a finite time. In flame-spraying processes, molten particles are cooled upon, or immediately after, application to the substrate; hence, it remains relatively cool. This process creates new material structures in the deposit bonded to the substrate. Machining and grinding techniques differ from those applied to the substrate because the deposit is more porous and brittle and tends to layer. The increased porosity is fortuitous for bearing surfaces since the pores act as oil reservoirs for lubrication.

The broad scope of metallizing can be indicated by examination of just a few of the common applications. For example, metallizng is used to:

- Build up worn bearings.
- Fill voids in castings.
- Improve refractory properties for missile heat resistance.

- Coat ceramic insulators for electrical/thermal conductivity.
- Coat iron and steel with aluminum for corrosion resistance.
- Coat jet engine combustion chambers for heat resistance.
- Coat cloth with aluminum or zinc for electrolytic plates.

2 METALLIZING PROCESSES

Metallizing processes, differing in the method of heat generation and the form of coating material, are subdivided into three classifications: wire process, flame-spray, and plasma flame-spray. All processes require clean surfaces, completely free of oil, for intimate bonding of the coating. The substrate surface usually must be roughened and may need grooving or threading for thick deposits (greater than about 10 mils). Undercutting may be required as for shafts whose bearings surfaces are to be built-up. A separate bonding deposit, usually molybdenum or nickel aluminide, may be applied by one of the three processes. The deposit may require sealing, machining, grinding, or no further finishing.

Specially designed spray guns may be hand-held or automated, while material to be coated is usually fixed or turned on a lathe. Distance from the gun to the work varies from about 3 to 10 in. (8–25 cm).

2.1 Wire Process

The first application of metallizing involved a continuous roll of zinc metal wire fed through an oxygen–acetylene flame with an airstream propelling molten metal onto a surface. This process is still in wide usage as an economical means of coating nails, automotive exhaust valves, voids in castings, ceramics, and many other components. Most metals that can be shaped into a uniform wire can be applied by this process. Although oxygen–acetylene is the usual fuel, oxygen–propane can be used.

2.2 Flame-Spray Process

The flame-spray process supplies coating material in a powder form rather than continuous wire. Its development permitted the use of many compounds and mixtures that were not previously feasible. Application blossomed from primarily zinc coating to both zinc and aluminum for corrosion resistance and steel and bronze for machine elements. Heat is provided by oxygen-enriched acetylene or propane or electric arc.

2.3 Plasma Flame-Spray Process

The plasma flame-spray process was developed primarily to permit metal spraying of materials with melting points exceeding 5000°F (2760°C). A plasma state exists when vapors of a material are raised to a higher energy level than the ordinary gaseous state. This occurs when the diatomic molecules of elements such as nitrogen or hydrogen are disassociated with subsequent separation of some of the electrically charged particles of the atom. Monatomic gases such as argon and helium can also be raised to the plasma state by the application of electrical energy. The plasma torch using electrical energy produces an arc in a chamber pressurized by a carrier gas. The gas is disassociated and ionized, emerging as a plasma flame. Nitrogen or nitrogen with about 10% hydrogen is the usual carrier gas because of its low cost and high energy content in the plasma state. About 75% of its energy is released as the atoms return to a neutral state and reunite into the diatomic molecule with a temperature decrease from about 18,000°F (9980°C) to 8000°F (4430°C). Argon, argon–hydrogen, helium, or argon–helium are other acceptable plasma gases. The inert carrier gas has considerable shielding effect from atmospheric oxygen. With a hydrogen mixture, the hydrogen burns to further reduce the oxygen content. Even more stringent nonoxidizing atmospheres can be effected by the introduction of inert gas through jet rings mounted on the spray guns. Development of the plasma flame-spray process opened a new dimension in spray coating for elements like tungsten (melting point of approximately 6170°F, 3410°C), high-melting-point alloys, and other metal mixtures. It also brought greater latitude in traverse rates during deposition and smoother surfaces on the finished substrate.

3 POTENTIAL HEALTH HAZARDS

It is emphasized here that many of the pretreatment steps needed to prepare a metal substrate for metallizing pose some threat of exposure to airborne contaminants. Cleanliness, especially freedom from oil, is critical for bonding integrity and roughening of the surface to be coated improves the quality of the coating. Abrasive blasting with aluminum oxide, sand, or chilled cast-iron shot is common practice to clean and produce roughened surfaces. This operation produces high atmospheric concentrations of the abrasive particulates and any contaminant carried on the base material. Noise levels are potentially hearing hazardous, frequently exceeding 110 dBA. Abrasive blasting is preceded frequently by solvent cleaning for removal of heavy oil or grease contamination. The health hazards associated with these pretreatment processes are discussed elsewhere in this book.

3.1 Metal Fumes and Dusts

Application of the molten sprayed material releases overspray of condensed particulates that vary from low-toxicity metal fumes of zinc and aluminum to those that are highly toxic, such as lead or lead alloys, chromium, and cadmium. In addition, finely divided particles of any combustible material pose fire and explosion threats; magnesium and aluminum are exceptionally high-risk elements. This hazard is discussed later in this section.

Metal fume fever is a condition that commonly arises in workers exposed to fumes of zinc in galvanizing or tinning processes, in brass working, and in additional work on galvanized metal. It may also arise from exposure to other metals such as copper, manganese, and iron. Cases of zinc ague have been reported in cases of men engaged in spraying zinc. Lead spraying is obviously a process in which a serious health risk could arise if adequate precautions were not provided and followed.

The specific adverse health effects associated with metal fumes and dusts are discussed in more detail elsewhere in this book in sections pertaining to welding and brazing operations.

3.2 Reactive Gases and Vapors

Because of the high-energy systems commonly associated wtih metallizing processes, it is prudent to anticipate creation of ozone, nitrogen dioxide, and possibly other reactive gases as a result of reactions with air. The health hazards associated with these gases are discussed in Chapter 27.

As mentioned earlier, metallizing normally requires a substantial amount of pretreatment of the substrate, much of which involves cleaning with volatile solvents. Again, the health hazards attributable to these materials are discussed in great detail elsewhere in this book. See Chapter 17.

Exposure assessment of organic solvent vapor by charcoal adsorption, laboratory desorption, and gas chromatography, and of noise is similar to that discussed in other chapters of this book. Oxides of nitrogen can be collected and analyzed in the field by portable infrared spectrophotometry. Instrumentation is available for field evaluation of ozone; however, control of ozone may be presumed where control of oxides of nitrogen is verified.

3.3 Noise

Sound levels generated by conventional metallizing operations are substantial, frequently in the range of 110–115 dBA. Depending on the physical configuration of the area used for the metallizing process, only limited opportunity exists for engineering control of the noise. Improvements in nozzle configurations and air flow patterns have resulted in reduction of noise levels at the spraying point. However, beyond these modifications, very little can be accomplished from an engineering standpoint. Thus, there is necessary reliance on an effective hearing conservation program in most industrial situations where metallizing processes are operated routinely. The reader is referred to discussions of the noise problems associated with other metalworking processes for a more deailed treatment of the noise exposure problem (Chapter 8).

3.4 Fire and Explosion

Of much greater concern than the health hazards associated with metallizing processes is the potential for fire and explosion hazards in most settings. Such potential exists as a result of several independent but related sources. First, there is the fuel used to sustain the flame; acetylene and propane are the most common systems although others are possible. Second, is the potential for improper mixing of gases within the spray equipment; actual design of spray "pistols" has been the suspected cause of many gas explosions. Finally, combustible and/or explosive concentrations of metal dusts can be generated either as a result of powders used in the process or, more commonly, by the condensation of metal fume and accumulation of settled dust in the vicinity of the spraying operation; explosions have occurred when this dust has been disturbed, resuspended, and ignited.

3.5 Other Hazards

A variety of additional potential hazards is worthy of mention. Ultraviolet (UV) and infrared (IR) radiations are produced in plasma processes and in the electric arc flame process. The eye-damaging potential of these nonionizing radiations is well documented.

Burns from droplets of molten metal are continuing sources of potential hazard to operators of metallizing equipment. Persons working with metal spray guns should be supplied with adequate hand and arm protection, as well as eye and face protection.

4 CONTROL OF HEALTH HAZARDS

Engineering control by enclosure to the maximum extent feasible, combined with local exhaust ventilation, is basic for precleaning, final grinding, particulates, and gases released in spraying. Controls for precleaning are similar to those discussed elsewhere in this book. Metal spray application is best controlled by complete enclosure in a sound-attenuated booth or room. Exhaust ventilation is designed to remove gases and particulates and to handle the heat load. Provision for the introduction of makeup air is essential, and collection and salvage of particulates should be incorporated. With complete automation, full enclosure, and effective exhaust ventilation, workers may be protected completely with no requirments for personal protective equipment. For manual or partially automated processes, operators must utilize eye protection for UV and particulates, hearing protection, and protective clothing; respiratory protection may be required if exhaust ventilation is inadequate. Evaluation of fire and explosion potential should culminate in the formulation of policies and procedures for effective control. Emphasis should be placed on controls for acetylene, oxygen, aluminum, and magnesium.

When articles to be sprayed are too large to be accommodated in a conventional spray booth, the operation sometimes is performed in open air. When weather protection is provided, the housing in which the spraying is done should be open sided, designed to provide as much free access to open air as possible. Particular care should be given to cleaning of floors, walls, fixtures, and other surfaces, in order to prevent the dust from accumulating.

Metal spraying operations should not be permitted in or near rooms containing flammable materials; all fire and explosion hazards must be eliminated, as discussed earlier. Wood floors, if present, should be kept clean; perferably, they should be covered with metal or other noncombustible material to prevent hot metal or slag from falling thorugh cracks in the floor or into machine-tool pits.

ACKNOWLEDGMENTS

This chapter is a revision of the one prepared by Roland E. Byrd, deceased, for Vol. 2, *Industrial Hygiene Aspects of Plant Operations,* in recognition of his contribution to the profession of industrial hygiene.

GENERAL REFERENCES

Assessment of Selected Control Technology Techniques, NIOSH Publication 79–125, 1979.

Engineering Control of Welding Fumes, NIOSH Publication 75–115, 1975.

Health and Safety Guide for Metal Coating and Allied Services, NIOSH Publication 77–187, 1977.

Metallic Coating processes other than Plating, in *Metals Handbook: Heat Treating, Cleaning and Finishing,* American Society for Metals, New York, 1974.

CHAPTER TWENTY-ONE

Mixing and Blending

Paul F. Woolrich

1 PASTE AND VISCOUS MATERIAL MIXERS

1.1 Batch Mixers

1.1.1 Change-Can Mixers

PONY MIXER. Separate cans allow the batch to be carefully measured or weighed before being brought to the mixer. The mixer may also be used to transport the finished batch to the next operation or to storage. The identity of each batch is preserved and weight checks easily made. The pony mixer consists of a drum (can) that revolves in one direction and an agitator that revolves in the other. The agitator has four vertical blades and a scraper blade. The pony mixer is made in 40- to 125-gal (150–475 L) sizes. It is used for thin pastes.

PLANETARY MIXER. The can remains stationary while the agitator simultaneously rotates and moves in a circular path within the can.

1.1.2 Stationary-Tank Mixers

BANBURY MIXER. The Banbury mixer is a high-intensity mixer suitable for heavy-duty applications. This mixer is a completely enclosed kneader chamber with a bottom ridge in which two spiral-shaped rotors revolve in opposite directions toward each other at slightly different speeds.

PAN MULLER MIXERS. Muller mixers consist of a pan inside of which one or two Muller wheels of rollers revolve about a vertical shaft. There are rotating pan and stationary pan types. Pan Muller mixers can be used if the paste is not too fluid or not sticky. The main application of Muller mixers is in the foundry industry, in mixing small amounts of moisture and binder materials with sand particles for both core and molding sand.

NAUTA MIXER. The Nauta mixer utilizes an orbiting action of a helical screw rotating on its own axis to carry material upward, while revolving about the center line of a cone-shaped shell near the wall for top-to-bottom circulation.

2   MIXING/BLENDING OF DRY POWDERS AND GRANULES

2.1   Batch Mixers

2.1.1 Tumbler Mixers (Tumbling Barrel). The simplest tumbler mixer is the drum roller or tumbling barrel. The main drawback to such devices for blending solids is there lack of end-to-end flow. The drum roller is not recommended where the quality of mixing is critical.

2.1.2 Double-Cone Mixer. The double-cone mixer consists of two truncated cones joined to a cylndrical section, driven in rotation about an axis symmetrically located in a horizontal plane through the straight-sided portion (Fig. 21.1).

2.1.3 Twin-Shell Tumbler (P-K Blender). Two cylinders are cut and joined to form a V shape. The axis of rotation is horizontal, allowing charging at the top of each arm of the V and discharge from the bottom of the V. Mixing is accomplished by rotation of the V around the horizontal axis (Fig. 21.2).

2.1.4 Ribbon Blender. Ribbon blenders are used for mixing a wide range of products that may be solid, liquid, or slurrylike. They are best suited, however, for mixing dry powders with or without small amounts of moisture. They are used for mixing resins, plastics, pigments, dyestuffs, fertilizers, drugs, and food products. They can be used for crystallizing, drying, and treatment during agitation with liquids or gases to produce various chemical reactions. The ribbon blender is a stationary U-shaped tank, 2 to 3 times as long as it is wide. Running longitudinally through the mixer is a shaft on which are mounted arms that support a combination of slender spiral ribbons or helical screws.

2.1.5 Mushroom Mixer. The mushroom mixer is most often used as a tablet-coating pan in the pharmaceutical industry. Used as mixer or blender, it is simply a tumbler mixer. The rotating bed of a Ready-Mix concrete delivery truck is a mushroom mixer.

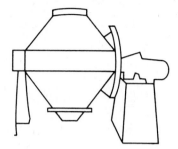

Figure 21.1. Double cone mixer.

2.1.6 Tote Blender. Figure 21.3 illustrates a device called a tote blender that makes it possible to blend directly in tote bins or containers. The bins can be filled directly from filters, centrifuges, pulverizers, and the like, placed on scales, and weighed, removed, or tumbled to blend the contents, then taken to the point of use. The bins can be transported and placed in the frame of the tumbler by means of a lift truck.

A hydraulic mechanism lowers the frame to clamp the bin securely before the tumbler is turned to a horizontal position, with the axis of rotation along diagonal corners. When the mixing is complete, the device is again tipped into an upright position, the clamps removed, and the bin lifted out with a lift truck. Following blending, the bin may also be used for drum or barrel loading while still in the blender frame.

3 INDUSTRIAL HYGIENE CONTROL

The principal industrial hygiene problem associated with mixers and blenders is dust control durng charging of the blender through the charging opening and filling drums directly from the blender. The exhaust system for blender charging and drum filling may be sized for control of one operation because only one operation can be performed at one time. A similar system can be applied for charging and emptying tumbling barrels, cone mixers, twin shell-tumbling blenders, ribbon blenders, and the Nauta mixer.

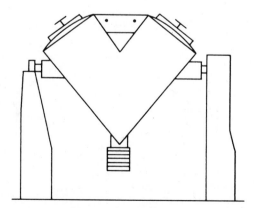

Figure 21.2. P-K blender.

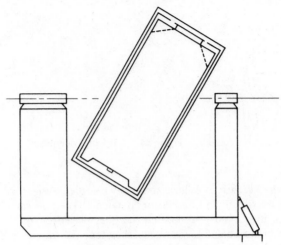

Figure 21.3. Tote blender.

3.1 Drum blenders

Drum blenders of various sizes are utilized to batch blend chemicals. For optimum efficiency it is desirable to use one hood to accommodate the various size drum blenders during filling and one hood for unloading operations. Portable-type hoods should be designed for each operation since blenders are large and heavy, necessitating utilization of chain or electric hoists for loading and unloading operations. The portable hood for charging drum blenders is designed with annular grooves on the underside to fit various diameter drums. The material to be blended is dumped through the hood opening hopper. The hood should be fabricated of aluminum for ease of handling (see Fig. 21.4).

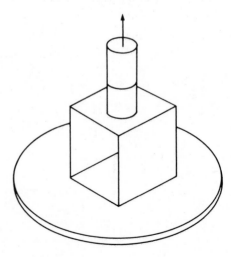

Figure 21.4. Portable hood for loading drum blender.

Figure 21.5 illustrates a portable drum blender unloading hood for control of dust during transfer of material from the blender to shipping container.

The hood has annular grooves to accommodate various blender sizes, and it covers only half of the drum opening. Air volume exhausted can be regulated by a damper in the sheet metal duct. A control velocity of 125 fpm (38 mpm) across the maximum opening should adequately control the dust. Unloading is accomplished by elevating the blender with a chain hoist, and the material is transfrred by hoe to a shipping container.

3.2 Ribbon Blenders

In the ribbon blender, the solid material is added through a charging opening in the blender cover, and exposure to toxic dusts may result. In this case, a hooded enclosure should be utilized.

When the material is added through a screw conveyor to a closed ribbon blender, a vent is required to remove the displaced air. This vent should be tied in with the exhaust system to produce a slight negative pressure. This will produce a slight indraft on the blender that will leak into the blender around the cover gasket should it become worn. Figure 21.6 shows a schematic drawing for dust control of a ribbon blender. It is also applicable to cone blenders and tumbling barrels.

As with other batch blenders, the ribbon blender drum dumping hood and drum loading hood will be used intermittently, so the system volume requirements may be determined from the maximum working opening on one hood. A control velocity of 125–200 fpm (38–60 mpm) should be maintained across working openings.

3.3 Mushroom Mixer

Slot-type ventilation for a mushroom mixer is shown in Fig. 21.7. Alternatives to the slot hood are extension of the hood to completely cover the hood manhole, using a hinged door for access to the mixer, or complete enclosure of the mixer.

3.4 Banbury Mixer

Figure 21.8 illustrates the ventilation of a Banbury mixer's charging gate and enclosed kneader chamber.

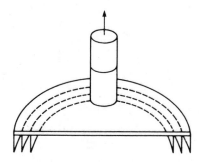

Figure 21.5. Portable blender unloading hood.

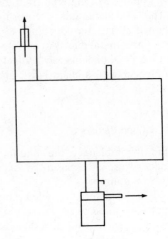

Figure 21.6. Dust control for ribbon blender, cone blender, and tumbling barrels.

3.5 Tumblers and Mullers

In the foundry industry, tumblers and Mullers are utilized to prepare mold sand and cores. Since high-crystalline silica is often used, dust control is essential. Figure 21.9 illustrates the ventilation of a hollow trunnion tumbler. Figure 21.10 illustrates the enclosing hoods for the foundry Muller. For additional information, see Chapter 5.

4 INDUSTRIAL HYGIENE PRACTICE FOR BLENDING VERY TOXIC OR POTENT MATERIALS

For blender operations involving very toxic or potent materials (pharmaceuticals, pesticides), dust concentrations of 0.02 $\mu g/m^3$ represents the minimum level theore-

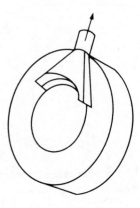

Figure 21.7. Ventilation for the mushroom mixer.

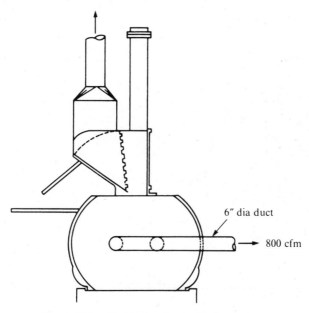

Figure 21.8. Ventilation for the Banbury mixer.

tically achievable. Employee exposures within guidelines of 0.1–0.3 $\mu g/m^3$ (required for some potent materials) can be maintained only through engineering control, enforcement of a very rigorous program of personal protective devices, and both biological and environmental sampling.

One of the main problems associated with preventing exposure is strict compliance by the individual worker with the established precautionary measures. The following procedures, if observed, will minmize the possibility of physiologic reactions:

Protection Equipment

> Rubber gloves
> Dust mask or air-supplied hood

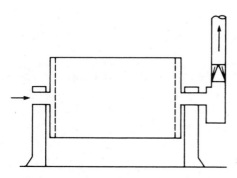

Figure 21.9. Ventilation of the hollow trunnion tumbler.

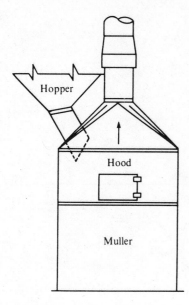

Figure 21.10   Ventilation of the foundry Muller.

Head cover

Long sleeved lab coat buttoned to the neck

Discard protective equipment at the end of the day or immediately after completing a job. Gloves should be rinsed, then washed and dried or permanently discarded.

Fresh protective equipment should be used each day (mask insert, gloves, head covering, coat); no "carryovers" should be permitted.

PROCESSING. All work should be conducted in a hood or in the vicinity of an operating exhaust duct.

Continuous surveillance, both environmental and medical, is essential for workers blending highly toxic and potent material. This is needed for assurance in preventing inadvertent exposures resulting from the inefficient operation of engineering control equipment or inadequate personal protective devices, and to detect any signs of a change in the workers' health profile at an early date.

GENERAL REFERENCE

Industrial Ventilation, a Manual of Practice, 19th ed., American Conference of Governmental Industrial Hygienists, Committee on Industrial Ventilation. Lansing, MI. Edwards Bros. 1986.

CHAPTER TWENTY-TWO

Nondestructive Inspection

John B. Feldman

1 INTRODUCTION

Nondestructive inspection (NDI) involves methods of examining materials or manufactured parts to determine their fitness for certain purposes, without destroying their usefulness. NDI methods are used to detect flaws such as cracks, porosities, discontinuancies, and the like that would affect the performance of the parts or materials in service. In highly stressed applications, such as aircraft and nuclear reactor components, where the reliability of the materials are paramount, NDI methods have become an indispensable tool. Common NDI methods include visual, liquid penetrant, magnetic particle, ultrasonic, and radiographic inspection. These methods are used throughout the aircraft, transportation, utility, and electronic industries to better assure the quality reliability of their products.

437

Visual inspection was the earliest NDI method used. It utilizes no chemical or physical agents other than cleaning materials and a suitable light source. But visual inspection is inherent with limitations. The defect must appear on the surface of the material being inspected and be of sufficient size to identify. Where large areas must be inspected under high magnification, the method is inefficient and subject to operator fatigue and decreased reliability. The methods described in this chapter were developed to better identify and define surface defects, as well as subsurface and internal defects hidden from the human eye. Since these methods often rely on indirect evidence (such as X-ray photographs), they must be applied and interpreted by trained and experienced technicians. The methods themselves must also be constantly audited and inspected to assure their reliability and consistency. Literature published by The American Society for Metals, American Welding Society, American Society of Non-Destructive Testing, and American Society for Testing and Materials is replete with NDI methodology, techniques, and standards.

2 PRECLEANING

2.1 Process Descriptions

Parts or materials for most NDI methods must be precleaned to remove foreign materials that may prevent detection of defects or confuse or alter defect indications. This topic is reviewed in general to avoid duplication in each procedural discussion. Typical cleaning procedures include:

- Alkaline—cleaning in tanks utilizing strong alkaline water solutions containing detergents, usually heated.
- Water—washing machines or spray wands using hot water. Detergents may also be used.
- Steam—large parts may be cleaned by pressurized application of steam containing alkaline detergents.
- Vapor degreasing—vapor phase degreasing using chlorinated or Freon solvents for removal of oils and grease. Preliminary removal of inorganic soil may be required using other methods.
- Ultrasonic—when cracks are filled with hard contaminants such as metallic oxides, carbon, or varnish, ultrasonic cleaning may be used to break up or remove the materials.
- Solvent—solvent cleaners may be selected for dipping or wiping parts in the absence of other procedures.
- Mechanical—abrasive blasting may be used to clean metals if the metal surface is not opened by the process and surface defects are not sealed or contaminated with abrasive particles.
- Paint removal—paint is removed from the metal surface by appropriate paint strippers. Vapor phase degreasing is usually performed following paint removal.
- Chemical—chemical etching with mineral acids may be used on some metal parts to superficially remove the surface layer.

Drying processes, with or without heat, follow cleaning and rinsing to assure the evaporation of water or solvents from cracks or other defects.

2.2 Industrial Hygiene Aspects

Hot alkaline cleaners in open surface tanks generally do not present inhalation hazards, unless they are agitated or maintained close to their boiling point where aerosols or mists are released. Under such conditions, caustic-containing aerosols or mists can cause irritation to the upper respiratory tract of employees working in the immediate vicinity. The primary hazards of such cleaners, though, involve skin irritation from contact and severe eye injury in the event of splashing from open surface tanks.

Common solvents for vapor degreasing include 1,1,1-trichloroethane (methyl chloroform), perchloroethylene, trichloroethylene, methylene chloride, and Freons. All have excellent grease and oil removal capability and, for the most part, are free of fire potential. Overexposure to the vapors of these solvents can result in drowsiness, central nervous system depression, and irritation to the eyes and throat. Prolonged or repeated skin contact dissolves natural oils of the skin resulting in dermatitis. The Freons and 1,1,1-trichloroethane are considered the least toxic of these solvents. The other solvents mentioned have been the subject of recent controversy concerning their carcinogenic potential in animal studies. There is no evidence to date, though, indicating they are human carcinogens. High-temperature sources of heat such as open flames, electric arcs, space heaters, and ultraviolet radiation should be avoided around degreasing operations where halogenated vapors are present. The formation of toxic vapor decomposition products such as hydrogen chloride, phosgene, and chlorine is possible due to interaction with these sources.

Ultrasonic cleaning solutions include 1,1,1-trichloroethane, Freons, and caustic detergents in water. The selection depends on the type of soil present and the material to be cleaned. The chemical hazards present relate to type of solution used and are similar to those discussed for degreasing and alkaline cleaning. Although most ultrasmic cleaners operate at frequencies above the audible range (20,000 Hz), lower octaves can develop that may require some degree of noise control.

Paint removers typically contain mixtures of strong alkalies, amines, methylene chloride, acetates, methyl ethyl ketone, phenols, or cresols. These compounds are usually toxic by inhalation and skin adsorption. They are corrosive to the skin and can produce serious eye injury on contact. In addition, many of those mixtures are flammable.

2.3 Controls

The preparation (dissolution) of strong alkaline cleaning solutions can result in exothermic reactions. Proper personal protective equipment (gloves, goggles, face shield, and possibly respirator) should be used when preparing or recharging solutions. Generally, ventilation is not required on alkaline open surface tanks unless they are agitated or heated to the extent that would cause the evolution of aerosols or mists. Under such conditions, local exhaust ventilation is required. Guidelines for the design criteria are available in the industrial hygiene literature.

Degreasing is best performed in a commercially available vapor phase degreasing tank with freeboard chilling that prevents the release of solvent vapors by condensation. Thermostats are located near the tank bottom and in the freeboard area. The former indicates the degree of accumulation of contaminants in the solvent by the increased boiling point and the need for tank cleaning. The latter is set to shut off the heating units when temperature in the freeboard zone rises to within about 5°C of the solvent boiling point, thus preventing escape of vapors due to inadequate cooling.

Manufacturers supply detailed procedures for operation and maintenance of vapor phase degreasers. Proper operation, maintenance, and a location free of drafts are essential to minimizing loss of solvent, thus reducing worker exposure. Operators must also be trained in proper work procedures (part orientation, soak times, and handling methods) to minimize solvent dragout and vapor blanket disruption.

There is no clear-cut industry practice concerning the need for local exhaust ventilation on vapor degreasers. Where degreasers are located in open spaces, maintained and operated properly, and there is minimal solution drag out, local ventilation is usually not required. Where ventilation cannot be avoided, it should be carefully designed to pull vapors away from the operator, but not increase solvent loss by disturbing the vapor blanket.

There has been a trend among air pollution regulatory agencies in recent years to reduce the use of certain halogenated degreasing solvents in an effort to curb ozone depletion in the upper atmosphere. In many precleaning operations, aqueous detergents can be substituted for haloginated solvents, particularly with metal parts. Though there may be some reluctance on the part of operating management to make such changes, aqueous cleaners generally are less hazardous than organic solvents and often less expensive.

Personal protective equipment must be employed to prevent eye and skin contact when using paint strippers. Chemical safety goggles and full-face shields are recommended for eye protection. Neoprene or other impervious gloves and aprons are appropriate for hand, forearm, and body protection. Deluge-type safety showers and eyewash stations in both the precleaning and inspection area should be readily available, clearly marked, and maintained in operable condition.

Noise exposures from ultrasonic cleaners can be controlled through equipment isolation or enclosure, vibration mounting, administrative measures, or use of personal protective equipment (ear plugs, muffs, etc.). The audible noise is usually of a high-frequency, pure tone nature, which can be quite irritating to personnel.

Abrasive blasting is usually carried out in an enclosed booth provided with exhaust ventilation and a dust collector. The operator is outside the booth with hands and arms inserted into the booth through heavy gloves and armlets. No further personal protective equipment is needed, with the possible exception of hearing protection, as long as the contaminant booth is properly sealed and maintained.

3 PENETRANT INSPECTION

3.1 Process Description

Liquid penetrant inspection (PT) is a fast and reliable NDI method for detecting surface flaws in many nonporous materials such as metal alloys, ceramics, and some plastics. Defects detectable by this method include overt and fatigue cracks, porosity, seams, lack of bond, intergranular corrosion, and stress defects.

The basic principle of PT is to increase the visible contrast between the discontinuity and its background by the following methods:

- Treating the whole object with an appropriate searching liquid of high mobility and penetrating power, which enters the surface opening of the discontinuity. The search liquid or penetrant usually contains a visible or fluorescent dye.
- Cleaning the surface of excess search liquid.

- Encouraging the liquid to emerge from the discontinuity to reveal the flaw, through use of a developer.
- Inspection under daylight conditions for visible dye penetrants, or under black light (near ultraviolet) for fluorescent penetrants. Figure 22.1 demonstrates a typical PT flow diagram.

Penetrant selection is based on factors such as penetrability, sensitivity, visibility, type defect anticipated, configuration of part, surface conditions, number of parts to

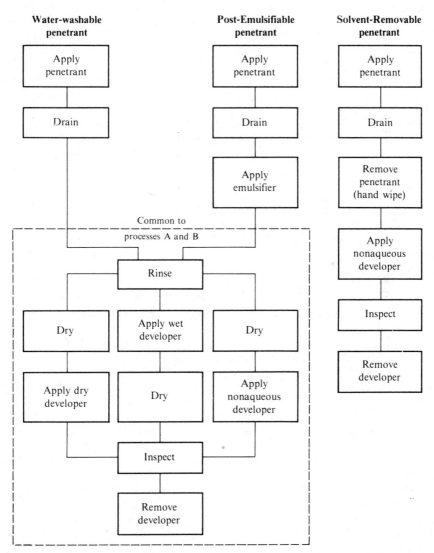

Figure 22.1. Liquid penetrant inspection flow diagram. (Reproduced with permission by courtesy of the American Welding society.)

be tested, effect of the penetrant chemicals on materials, and the like. Penetrant sensitivity is defined as the ability of the penetrant, along with compatible family items in its group to find defects of the type sought. Penetrants are applied by dipping, flow coating, or spraying.

Emulsifiers are liquid additives that, when applied to postemulsifiable penetrant on the surface of the part by dipping, spraying, or flow-on combine with the excess surface penetrant to render it water removable. Emulsifiers have very low penetrating properties, an essential quality to avoid the emulsifier's removal of penetrant from discontinuities. Their color contrasts with the postemulsifiable penetrant used to assure that all of the emulsifier has been removed during rinsing. The selection of emulsifiers is preclassified by group for the compatible penetrant. Dwell time—the time allowed for the emulsifier to mix with the surface penetrant in order to make it water removable—is critical. Optimum time may be determined by pretesting the part; rinsing follows immediately.

Developers consist of three types: dry, aqueous wet, and nonaqueous wet. They are used following removal of excess surface penetrant to provide a contrasting background to flaw indications and to absorb or weep out the penetrant at the defect, making it more visible.

Developers and penetrants are removed from parts upon conclusion of inspection using aqueous and organic solvent cleaning methods described in the preceding section.

3.2   Industrial Hygiene Aspects

Penetrants vary depending on their application of use and source. They usually contain a refined petroleum distillate base with additives including pine oils, glycols, phthalates, phosphate esters, and fluorescent dyes. Penetrants have been used for many years without any known major health problems. Spray application without adequate ventilation may be irritating to the mucous membranes. Prolonged or repeated skin contact can produce dermatitis. Most penetrants are combustible, but of relatively high flash point.

Emulsifiers also vary in composition depending on type of penetrant used and source vendor. They range from aqueous detergents to high-molecular-weight glycols and glycol ethers, diols, alkylamines, and sodium nitrate. Generally, most of these constituents are of low vapor pressure or are present in low concentrations that do not present serious health hazards. Spray application usually requires some form of local ventilation and repeated skin contact can result in dermatitis. Most emulsifiers do not contain amines and nitrates in combination, thus precluding concern over formation of nitrosamines.

Dry developers are usually fine powders consisting of such materials as amorphous silica, talc, alumina, magnesium oxide, iron oxide, and corn starch. Most are classified as nuisance dusts. The presence of asbestos fibers in talc developers raised concern in the early 1970s, but this problem has been eliminated in recent years through better control of talc sources.

Wet developers typically contain a solvent vehicle (chlorinated hydrocarbon, alcohol) with a powder suspension. Their hazards will depend on the type of solvent used and method of application. Generally these solvents present inhalation hazards as well as dermatitis concerns due to defatting of natural oils of the skin from repeated contact.

The inspection of fluorescent dyes requires ultraviolet light in the μV-A black light range. The light is usually produced by mercury vapor or phosphor-coated lamps, which are either hand held or fixtured in an inspection booth. The lamps are equipped with Kopp filters that restrict the passage of ultraviolet to the 330–390 nm range. The American Conference of Governmental Industrial Hygienists (ACGIH) and the National Institute of Occupational Safety and Health (NIOSH) recommend exposure to ultraviolet light in the μV-A range be limited to 1000 μW/cm^2, for periods over 1000 sec. Generally, intensities in this range can be expected in some inspection applications. There is no common evidence of skin or eye problems experienced by operators who routinely use such lamps. Some of these lamps (nonfluorescent types) may present thermal burn hazards to operators in darkened booths, when their housings become hot.

3.3 Controls

The type and degree of controls required in various PT processes depend on the constituency of the products used and the methods of application. Generally, dip and flow coat application of penetrants and emulsifiers do not require local ventilation. Wet developers with volatile solvent vehicles, require some type of local ventilation. Spray application of penetrants, emulsifiers, and developers are normally conducted under local ventilation. Dry developers must be applied (sprayed, dusted) within a ventilated or enclosed hood for housekeeping as well as industrial hygiene purposes. Eye protection should be used through the PT processes to protect against accidental splashing during liquid applications. Emergency eyewash stations should be readily accessible. Employees who have continual hand contact with penetrants, emulsifiers, and developers should wear impervious gloves to protect against skin irritation. Guidance on the selection of gloves can be found in material safety data sheets or in glove manufacturers' literature.

In many facilities, fluorescent black light inspection and developing are conducted in enclosed booths, with curtains to restrict outside light. Small quantities of solvent-based developers and cleaners are locally sprayed or brushed on parts. The confined nature of these booths usually requires local ventilation be provided. Smoking should be prohibited in the booths where solvents are used.

The ultraviolet black light used in fluorescent penetrant inspection is not harmful during routine operations as long as the filter is securely in place. Any cracked or broken filters should be replaced immediately to prevent exposure to higher intensity ultraviolet light.

Glycol ethers and phalates have received attention over the past several years as a result of animal tests indicating potential reproductive and carcinogenic effects, respectively. For the most part, these compounds when used in penetrants and emulsifiers are in low concentration and are of the high-molecular-weight variety. They present minimal exposure hazards by inhalation due to their low vapor pressure. Skin exposures can be avoided through the use of gloves and aprons.

4 MAGNETIC PARTICLE INSPECTION

4.1 Process Description

Magnetic particle inspection (MT) is a widely used NDI method for locating discontinuities in the ferromagnetic metals: iron, cobalt, nickel, and most steels, except those of high carbon content (austenitic steels). It is not applicable to aluminum, magnesium, or other nonmagnetic materials. The principle of the MT method is that a magnetic field can be created in ferromagnetic metals. This field will be interrupted by any discontinuity or defect in the material. The distorted patterns of the field become visible at the defect when finely divided magnetic particles are applied to the surface of the material. The field may be established by a permanent bar or horseshoe magnet or by passing an electric current directly through or in a coil around the part to be magnetized. The electric current, usually high-amperage low-voltage, may be alternating, direct, or rectified. This technique's primary application is detection of defects at or near the surface. Although subsurface discontinuities may be detected, ultrasonic and radiographic methods are more reliable. Optimum detection occurs when the magnetic field is perpendicular to the discontinuity. This condition exists when current flow is parallel to the defect since the magnetic field is always at right angles to the flow of current. A typical MT process flow chart is shown in Fig. 22.2.

Parts to be examined by the MT method must be free of dirt, grease, oil, rust, loose scale, paint, lacquer, or other coatings that may act as an insulator and prevent effective magnetizing. Vapor phase degreasing is the preferred cleaning procedure. Preliminary chemical, manual, or abrasive cleaning may be required to remove coatings, rust, scale, or heavy contamination.

The method of magnetization is selected by consideration of five interrelated factors:

- Elemental metal or alloy and its shape and condition.
- Type of magnetizing current.
- Direction of magnetic field.
- Sequence of operations.
- Value of flux density (strength of the magnetic field).

Magnetic particles are applied using either dry or wet methods. Dry particles—colored red, yellow, black, gray, or fluorescent for optimum contrast with the metal—are applied while the part is magnetized by a gentle air current using a gun or dusting bulb to float them onto the surface. The weak magnetic fields at the discontinuities attract the particles into the distorted patterns. Excess powder is gently blown off. Dry particles are more sensitive to weak fields than those in liquid suspension and are used normally with portable equipment. Wet particles, suspended in water or oil, are available in black, red, or fluorescent colors. The parts are coated with the suspension, then magnetized. The particles then form patterns of the magnetic lines of force and discontinuities at the defects are identified. The oil is usually a refined petroleum distillate in the mineral spirits fraction range with a minimum Tag, closed-cup flash point of 57°C.

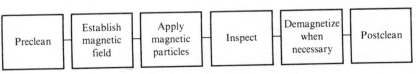

Figure 22.2.   Flow diagram for magnetic particle inspection.

Visual examination by trained inspectors follows immediately after magnetization is completed and the current secured. Good lighting improves the effectiveness of detection and evaluation of flaws. For fluorescent particles, a mercury vapor lamp with a Kopp filter is used to transmit viewing light peaking at 365 nm, between visible light and ultraviolet. Records of discontinuities may be documented by sketching or photographing, or they may be preserved by fixing with clear lacquer coating or transparent tape that is then transferred onto white paper. Wet particles must be dried before fixing or transfer.

Demagnetization is required where the part's residual magnetic field may:

- Affect the instrument of which it is a component.
- Interfere with removal of chips during subsequent machining.
- Interfere with subsequent MT where a reverse field is to be used.
- Seriously interfere with postcleaning of magnetic particles.
- Retain or attract magnetic particles on moving parts, when in use, resulting in excessive wear or friction.

Demagnetization may be accomplished by placing the part in a coil of three or four loops, applying alternating current of a higher value than the magnetizing current, then going through the stepdown demagnetizing cycle until the current is zero. Direct current may be used to demagnetize with a provision for reversing the current and using a stepdown cycle to attain a negligible current flow.

Most parts must be cleaned immediately after inspection to remove particulates and oils when used. The cleaning agent is water or the same type oil used as the vehicle for wet processing. Rust preventatives are usually applied after postcleaning, although some parts may be processed immediately without final cleaning.

4.2 Industrial Hygiene Aspects

The hazards associated with magnetic particle inspection involve contact with the oils and particles used, and potential exposures to ultraviolet light and magnetic fields. Generally, these are not considered serious hazards under normal process conditions.

Most of the oils used in wet particle suspensions are refined distillates and not particularly toxic. Repeated or prolonged skin contact can cause dermatitis. The particles are usually of the nuisance dust variety (iron oxide). Inspection of fluorescent particle-treated parts requires an ultraviolet black light in the μV-A spectrum. This range is considered the least harmful of the ultraviolet and is generally not a hazard at the intensities used. Magnetic fields of high flux density are generated by this process that may affect the operation of heart pacemakers and can magnetize watches. Although there has been controversy in the past several years over potential health effects from magnetic fields, there is no substantive evidence to date indicating exposures from such operations to be harmful.

4.3 Controls

Personal protective equipment (neoprene gloves, aprons, and goggles) should be used to minimize skin and eye contact when working with wet oil–particle suspensions. Local ventilation is usually not required with wet particle methods, but some degree of local ventilation may be required when using dry powder applications. Black light ultraviolet controls, similar to those discussed in Section 3.3, should be employed. Where energized circuits are used to drive the electromagnet and inspection lamp are in close proximity to the wet suspensions, proper design and maintenance of electrical equipment are critical. Ground fault circuits interruptors should be provided to minimize shock hazards.

5 ULTRASONIC INSPECTION

5.1 Process Description

In ultrasonic inspection (UT), high-frequency mechanical waves, above the range of human hearing, are electrically generated by a transducer and introduced through a coupling medium to the test part. The waves travel in a straight line until reflected by an interruption or defect in material continuity. The reflected wave propagates back to the transducer and a reverse electrical impulse is generated. It is electrically processed and displayed on a cathode ray tube (CRT), indicating the presence, magnitude, and location of the defect. Figure 22.3 depicts a flow diagram for ultrasonic inspection.

The transducer element, a piezoelectric element or crystal, expands and contracts as the applied high-frequency electrical current alternates, thus converting these pulses into vibrations or mechanical waves. Frequencies of 1–25 MHz are common for this widely used NDI method. Effective transmission of the sound waves requires a coupling medium of low accoustical impedance such as oil or glycerine at the interface between the transducer and test part. The transducer and part to be tested may also be immersed in a water tank for coupling purposes. The transmitting transducer or a separate receiver may be used to pick up reflected waves, convert them to electrical pulses, and transmit them to a signal processing unit for analysis and visual display on a CRT. Ultrasonic inspection is a sensitive procedure for the detection of extremely small discontinuities and flaws located deep within metal parts. Continuous inspection of nuclear power components relies primarily on this method.

5.2 Industrial Hygiene Aspects and Controls

Ultrasonic inspection is considered relatively free of industrial hygiene hazards. While there is little information in the literature on hazards associated with industrial use

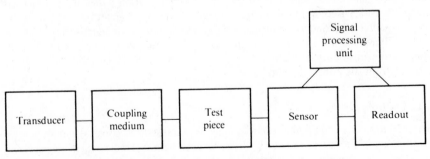

Figure 22.3. Flow diagram of ultrasonic inspection system.

of ultrasound, it has been a recent topic of concern in medical diagnostic applications. Here, ultrasound is used to differentiate soft tissue structure, in a similar manner to industrial inspection. It is also used therapeutically for deep heat and massage. Its high-frequency mechanical vibrations dissipate into thermal energy in body tissue via molecular friction. The biological effects of ultrasound appear to be energy related. Industrial inspection applications typical are at low power levels (10 mW/cm²). This is about one-thousandth the level used for therapeutic purposes, and one-tenth the level used in medical diagnostics. Studies of hazards related to medical diagnostic usage of ultrasound have indicated predominantly negative findings. There is no evidence to date to indicate that industrial ultrasonic inspection techniques present health hazards to operators.

Where water baths are used for transducer part coupling, additives are often utilized to better wet parts and prevent corrosion. The additives include polyglycols (wetting agents) and sodium nitrate (corrosion inhibitor). While such compounds present no particular health hazards at the low concentrations used, some employees have experienced mild dermatitis from repeated skin contact with the bath. This can be readily controlled with appropriate protective clothing, barrier, or moisturing creams.

6 RADIOGRAPHIC INSPECTION

6.1 Process Description

Radiographic inspection (RT) is an NDI method in which a radiation source is directed through a test piece that absorbs a portion of the energy.

The penetrating energy is quantified by a detection device that records images on photographic film, a fluorescent screen, or electronic instrumentation. Figure 22.4 depicts a flow diagram for radiographic inspection.

Radiation sources include:

- X rays produced by the flow of electrons from a filament (anode) heated by an electric current to the target (cathode), usually tungsten, all housed within an evacuated tube.
- Gamma rays that are constantly emitted from some radioactive elements, primarily cobalt 60 (^{60}Co) and iridium 192 (^{192}Ir), although cesium 137 (^{137}Cs) and radium 226 (^{226}Ra) have been used.
- Neutrons, particulate matter rather than electromagnetic radiation, are produced by accelerators, nuclear reactors, and subcritical assemblies and are constantly emitted by some elements. These are extremely high atomic number elements such as californium 252 (^{252}Cf) and some intermediate radioactive isotopes, such as antimony 124 (^{124}Sb).

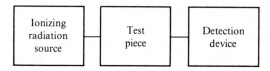

Figure 22.4. Flow diagram for X- and a-ray inspection.

X rays and gamma rays, both forms of ionizing radiation, are absorbed in materials in direct proportion to both density and thickness. Film is placed underneath or behind the test piece to record its image. If a defect is present, the radiation will be absorbed to a lesser degree in the area of the defect. This results in a difference in optical density on the processed film, thus identifying the location and extent of the defect. RT has limited sensitivity for small flaws, but along with ultrasonic testing, is widely used for the detection of internal flaws. Rdiographic inspection is commonly used for pressure vessels, piping, and other components of nuclear and conventional power plants, for welds, castings, electronic circuit boards, aircraft parts, and the like.

Portable X-ray units operate at 40–300 KVP and usually 4–5 mA. Exposure time varies with the density and thickness of the test part, and is measured in minutes as compared to fractions of a second for medical and dental X-ray procedures. Fixed X-ray units may range up to about 3000 KVP. Linear accelerators and betatrons permit much higher energy radiography. Penetrating capability of X rays is primarily a function of applied voltage to the tube anode. Gamma-emitting rdioactive isotopes continuously emit discrete energy radiation, specific for each isotope.

Neutrons are attenuated by interaction with the nuclei rather than electrons in the orbital shells of atoms, thus their application is to elements or materials in which the elements have effective capture capability for thermal neutrons. Since film is relatively insensitive to neutrons, the image is enhanced by:

- A conversion screen, such as gadolinium foil, that absorbs neutrons, becomes radioactive, and emits low-energy gamma rays to the film emulsion.
- A fluorescent screen that contains an isotope such as lithium 6 that emits an alpha particle when it absorbs a neutron. The alpha particle causes the phosphor to fluoresce, and the resultant light exposes the film to form an image. Figure 22.5 depicts a flow diagram for thermal neutron radiation.

Neutron radiography is used to inspect nuclear fuel elements, explosives, and plastics and has been used experimentally to detect the presence of water in the fiberglass honeycombing of helicopter blades. Development of the potential of this method is in its infancy.

6.2 Industrial Hygiene Aspects

The primary effect of ionizing radiation on absorbed material is the removal of an electron from the outer shell of the atomic structure. In living tissue, this interferes with the complex biochemical mechanisms at the molecular level. The clinical effects of acute and chronic exposure to ionizing radiation are well known and documented from studies of medical personnel, patients receiving X-ray and radioactive isotope

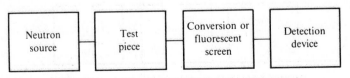

Figure 22.5. Flow diagram for neutron radiography.

treatment for malignancies, survivors of nuclear explosions, and extensive animal studies. The potential for serious exposure is an inherent possibility for all types of radiographic inspection.

Current Occupational Safety and Health Administration (OSHA) and Nuclear Regulatory Commission (NRC) standards limit whole-body exposure to X rays and gamma rays to 1.25 rems per calendar quarter (13 weeks). A rem is a unit of biological absorbed dose equal to 100 ergs of energy per gram of tissue. The standards include a number of variations and limitations based on such factors as the portions of the body exposed, age of employee, length of exposure period, and so forth. Expectant mothers should be limited to less than 0.5 rems during their pregnancy.

Film processing, a routine step in the RT method, may be automated or manual. Vapors from acetic acid, used in the processing, may cause irritation to the eyes, nasal membranes, and upper respiratory system. Direct contact with liquid developers, short stop, and hardeners may cause skin irritation or dermatitis.

6.3   Controls

Radiation exposures should be kept as low as economically and practically feasible, irrespective of allowable regulatory standards. For radiographic inspection operations, this can be accomplished through appropriate facility design, shielding, employee training, management overview, and strict adherence to applicable regulatory requirements.

The NRC has responsibility for regulation of man-made radioactive isotopes in the United States. The NRC does not have jurisdiction over naturally occurring radioactive elements or machine-made (X ray) radiation. This is regulated by OSHA. NRC and OSHA have entered into agreements with some state agencies to carry out regulatory activities. Agreement states must have equivalent or more stringent requirements. Man-made radioactive isotopes may be obtained only after licensing by NRC. The license specifies the isotopes by identity, maximum activity, usage, and controls. OSHA and state requirements for NDI application of X-ray radiation and naturally occurring radioactive isotopes are likely to be comparable. The intensity of radioactive isotopes for NDI application is certified at a specific date, and a decay curve accompanies it. These sources are shipped in depleted uranium or lead containers in compliance with Department of Transportation (DOT) requirements. When sources have decayed below a useful intensity, they are returned to the supplier in the same container, again complying with DOT regulations.

Control requirements for radioactive isotopes and ionizing radiation-producing equipment are contained in NRC Standards 10 CFR Parts 20 and 30, and OSHA Standards 29 CFR Part 1910.96. Most NRC licenses contain specific compliance conditions relating to the use of the by-product material by the licensee. The basic requirements for proper use of by-product material follow. These requirements, though, will vary depending on the specific type, quantity, and manner in which the material is used. The requirements are typically less rigorous for machine-made (X ray) sources, which vary in size from small cabinet units to high-power room sized facilities.

- Approved license for isotope or X-ray unit.
- Secure storage facility.
- Exposure room for fixed installation.
  - Adequate shielding for room.

- Audible and visual signals during exposure.
- Interlocks for exposure room doors.
- Wipe test at intervals specified for the isotope.
- Written detailed procedures for all exposures.
- Monitoring during all exposures.
  - Area monitoring with instrumentation.
  - Personnel monitoring by photodosimetry, thermoluminescent dosimeter (TLD), or personnel dosimeters.
  - At least annual notification to each person of exposure during the previous year and cumulative exposure at termination.
- Exposure areas clearly identified and posted.
- Limited access permitted.
- Qualified users only.
  - One responsible person identified by name for each usage.
  - Training given and documented to assure that users are knowledgeable of potential health hazards, detailed procedures, and control requirements.
- Appointment and functioning of a radiation safety officer; specific identification by name, training, and experience is required for NRC license.
- Medical examination of users prior to exposure and at termination of exposure.
- Complete recordkeeping of all pertinent information.

Local management in combination with the designated radiation safety officer have specific responsibility for assuring scrupulous conformance with all the NRC or agreement state requirements, safe work procedures, and proper recordkeeping.

Adequate darkroom ventilation should be provided for manual film processing to prevent eye and nasal membrane irritation. Eye and skin protection is needed during manual processing and mixing and replenishment of chemical solutions. An emergency eye wash in the darkroom is recommended.

7 LASER INSPECTION

7.1 Process Description

Laser inspection of moving sheets of material and of critical components produced in large numbers has accelerated the rapid growth of this versatile NDI technique. The simple application of helium–neon optical lasers to detect surface flaws and measure thickness is especially suitable for rapidly moving sheets of material. The beam continuously scans the width of the sheet from a head that compensates to maintain the appropriate angle of incidence. The reflected beam may be recorded on a photographic plate or by a standard television type of video system or an electronic image processing system. Figure 22.6 depicts the flow diagram for laser scanning. Processing systems may be automated for rejection or other appropriate response when unacceptable irregularities exceed established parameters. For thickness determination, the receiving system is located to detect and evaluate the transmitted light fraction through nonpigmented plastics and other transparent or translucent material.

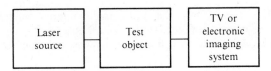

Figure 22.6. Flow diagram for laser scanning.

Optical and acoustical holography have become firmly established NDI techniques within the past decade. Holography is a two-stage process used to reconstruct a three-dimensional image of the subject test piece. First a hologram, the recording of both the amplitude and phase of any type of coherent wave motion emanating from the object, is encoded in a suitable medium. From the hologram, the wave motion is reconstructed by a coherent beam and processed into a true image, which serves as a quality assurance standard for the evaluation of deviations in similar objects. The laser beam is split with one portion, the reference beam, directed to a mirror then reflected through a spatial filter to a second mirror and reflected to a photographic plate. The second portion, the object beam, is directed through another spatial filter, strikes the object and is reflected to the same photographic plate that records the hologram. Figure 22.7 depicts the flow diagram for optical holography. For reconstruction, the hologram recorded on the plate is lighted by the reference beam and the diffracted beams produce the real and virtual images of the object. Either continuous wave (CW) or pulsed lasers may be the optical source for the holograph; however, only a CW laser is employed for the reconstruction. The most common holography source is the helium–neon laser with a wave length of 633 nm, orange-red range. Argon and ruby lasers are others in frequent use.

Acoustic, thermal, increased or reduced pressure, or mechanical stress applied to

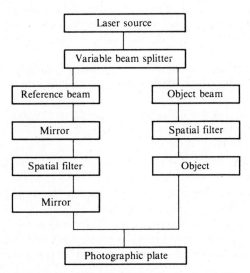

Figure 22.7. Flow diagram for optical holograhy.

an object produces changes in its shape that modify interferrometric fringe patterns that may be demonstrated and recorded by optical holography.

Examples of procedures for which these techniques are effective follow:

- Detection of unbonded areas in laminates.
- Detection of stress or corrosion cracks in metal components.
- Qualitative evaluation of hollow turbine blades; simultaneous recording of both sides of turbine blades is possible with a dual system.
- Evaluation of welding integrity.

7.2 Industrial Hygiene Aspects

Laser beams can cause skin, corneal, or retinal burns from acute exposure and corneal or lenticular opacities (cataracts) or injury to the retina from chronic exposure. None of the three most commonly used NDI laser systems emits radiation in the ultraviolet range; however, helium–cadmium lasers do have optical holography application and have an emission at 325 nm (ultraviolet). Electrical and fire hazards may exist with extremely high power laser systems, some of which also utilize cryogenic coolants. These coolants can cause severe burns, fire, or explosions and exposure to an atmosphere of reduced oxygen content caused by vaporization of the coolant, usually nitrogen.

7.3 Controls

The American National Standard Institute (ANSI) develops and publishes concensus guidelines whose scope covers protection against hazards for the maker, user, and the general public. These standards are not legally binding; however, they are the product of those substantially concerned with its scope and provisions.

ANSI Z136.1 classifies lasers and laser systems according to their relative hazards and specifies appropriate controls for each classification. This classification scheme is based on the ability of the primary beam or reflected beam to cause biological damage to the eye or skin. ANSI Z136.1 summarizes laser classification as follows:

> A Class 1 laser is considered to be incapable of producing damaging radiation levels, and is, therefore, exempt from control measures or other forms of surveillance.
>
> A Class 2 laser (or low-power system) may be viewed directly only under carefully controlled exposure conditions and must have a cautionary label affixed to the external surfaces of the device.
>
> A Class 3 laser (or medium-power system) requires control measures to prevent viewing of the direct beam.
>
> A Class 4 laser (or high-power system) requires the use of controls that prevent exposure of the eye and skin to the direct and diffusely reflected beam.
>
> A Class 2, Class 3, or Class 4 laser or laser system contained in a protective housing and operated in a lower classification (Class 1, Class 2, Class 3) shall require specific control measures to maintain the lower classification.

Laser or laser systems certified for a specific class by a manufacturer in accordance with the Federal Laser Product Performance Standard may be considered as fulfilling all classification requirements of this standard. Where a laser or laser system classification is not available or is changed through modification of its use or engineering controls, the laser or laser system shall be classified by the Laser Safety Officer (LSO) in accordance with the criteria of ANSI Z136.1.

Class 1 lasers and laser systems require no control measures and Class 2 requires only an appropriate label. Class 3 and Class 4 lasers and laser systems require stringent engineering controls, eye protection, extensive personnel training, and strict supervision by an LSO. In the absence of federal or other binding legislation, good practice and prudence dictate that the practices, procedures, and recommendations of ANSI Z136.1 (latest revision) be adopted and considered as mandatory for local policy.

Laser or laser systems certified for a specific class and used in compliance with control measures specified for that class, normally do not require measurement of laser radiation or personnel monitoring. Evaluation of laser radiation when required should be performed by qualified health physicists or specialists with equivalent training and experience. Medical examinations for Class 3 and 4 laser operators should be conducted including patient and family eye histories and an opthalmological exam prior to exposure. Examinations at periodic intervals and at termination of employment are prudent.

Electrical hazards associated with lasers must be carefully evaluated. High-voltage power supplies required to drive high-power lasers should be properly protected from personnel exposure and in conformance with appropriate codes.

8 EDDY CURRENT INSPECTION

8.1 Process Description

In eddy current inspection (ET), eddy currents are induced in the test piece and changes resulting in these currents from discontinuities or other physical differences are measured. The basic components of this nondestructive system are (1) an oscillator, (2) a test coil, (3) a bridge circuit, (4) signal processing circuits, and (5) a readout. Figure 22.8 depicts a flow diagram for eddy current inspection.

The oscillator provides an alternating current to the test coil, which then produces an alternating magnetic field. This magnetic field induces eddy currents in any electrically conductive material in close proximity. These eddy currents are altered by physical characteristics and differences. These current changes are picked up by the test coil, transmitted to the bridge circuit where they are converted to signals and

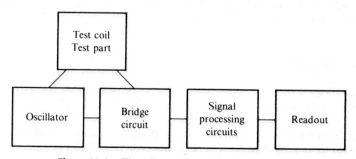

Figure 22.8. Flow diagram for eddy current system.

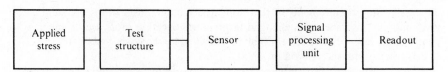

Figure 22.9. Flow diagram of acoustical emission system.

forwarded to the signal processing circuits for analysis and recording on strip charts or X-Y recorders. ET may be used to detect cracks, voids, inclusions, corrosion, and so on to measure thickness of plating or nonconductive coating, and to measure certain material composition and characteristics.

8.2 Industrial Hygiene Aspects

With the exception of precleaning, the eddy current inspection process is considered free of inherent health hazards. Electrical components should be properly protected to prevent personnel exposure and potential shock hazards.

9 ACOUSTICAL EMISSION INSPECTION

9.1 Process Description

Acoustical emission (AE) inspection is similar to ultrasonic inspection; however, the sound originates internally from dynamic changes within the test material. Figure 22.9 depicts a flow diagram for acoustical emission. The initial development or growth of cracks, slips, or plastic deformation produces sound energy that is detected by piezoelectric sensors, amplified, analyzed, and displayed for comparison with standards for the specific material. AE has great potential in real-time testing for aircraft structural fatigue flaws, pressure vessels, nuclear reactor components, welds, wire rope, cranes, hoists, ceramics, and so on.

9.2 Industrial Hygiene Aspects

With the exception of precleaning, the acoustical emission inspection process is considered free of inherent health hazards.

This chapter is a revision of one prepared by Roland E. Byrd, deceased, for Volume 2, *Industrial Hygiene Aspects of Plant Operations,* in recognition of his contribution to the profession of industrial hygiene.

GENERAL REFERENCES

American National Standards Institute, *Radiological Safety Standard for the Design of Radiographic and Fluoroscopic Industrial X-ray Equipment,* ANSI/NBS 123-1976, New York: ANSI, 1976.

American National Standards Institute, *Aerospace Materials Inspection Procedures,* Catalog of American National Standards, New York: ANSI, 1981.

American National Standards Institute, *Safe Use of Lasers,* ANSI Z136.1-1986, New York: ANSI, 1986.

Burgess, W. A., *Recognition of Health Hazards in Industry,* New York: Wiley, 1981.

Committee on Industrial Ventilation, American Conference of Governmental Industrial Hygienists, *Industrial Ventilation: A Manual of Recommended Practice,* 19th ed., Cincinnati: ACGIH, 1986.

Current Intelligence Bulletin 2, Trichloroethylene, National Institute of Occupational Safety and Health (NIOSH), Publication No. 78–127, 1975.

Current Intelligence Bulletin 20, Tetrachloroethylene (Perchloroethylene), National Institute of Occupational Safety and Health (NIOSH), Publication No. 78–112, 1978.

Current Intelligence Bulletin 39, The Glycol Ethers, National Institute of Occupational Safety and Health (NIOSH), Publication No. 83–112, 1983.

Current Intelligence Bulletin 40, Methylene Chloride, National Institute of Occupational Safety and Health (NIOSH), Publication NO. 86–114, 1986.

Hinsley, J. F., *Non-destructive Testing,* London: MacDonald & Evans, London, 1959.

Lamble, J. H., *Principles and Practice of Non-destructive Testing,* London: Heywood, 1962.

McMaster, R. C., *Non-destructive Testing Handbook,* New York: Ronald Press, 1959.

Occupational Safety and Health Standards (OSHA), Ionizing Radiation, Code of Federal Regulation, Title 29, Part 1910.96 (29 CFR 1910.96), Washington, D.C.: U.S. Government Printing Office, 1988.

U.S. Nuclear Regulatory Commission Standards for Protection Against Radiation, Rules of General Applicability to Domestic Licensing, Code of Federal Regulation, Title 10, Parts 20 and 30, (10 CFR 20 and 30), Washington, D.C.: U.S. Government Printing Office, 1988.

CHAPTER TWENTY-THREE

Painting and Coating

Jack E. Peterson
Lawrence W. Keller

1 INTRODUCTION

Paint and similar finishes (varnish, lacquer, enamel, etc.) are applied to surfaces mainly for protection and for esthetics; minor uses include color coding and the like. Non-paint-like materials such as oils and greases are usually used only for protection; waxes

457

are applied for both esthetics and protection. Surfaces can be protected and/or beautified by the application of metals (cladding, plating) or other inorganic materials such as glass and porcelain enamel. Here, only substances having an organic component are considered. The terms "coating" is used as more generic than "painting" simply because paints are different from varnishes, lacquers, inks, and other coating materials. Only industrial uses of paints and coatings are considered; to broaden the scope only a little would be to increase the length of this section enormously.

1.1 Technology

In the *Paintings/Coatings Dictionary* of the Federation of Societies for Coatings Technology, coatings are defined as liquid or liquefiable materials that can be converted to solid protective, decorative, or functional adherent films after application as a thin layer. Paints and similar materials are applied to surfaces in several ways, the most usual are by spray, brush, roller, dip, and occasionally, wipe. The coating and/or the object coated (substrate) may be heated and whether or not that is done, the coated object may be subjected to a heat treatment. The coated surface may then be "finished" by rubbing, buffing, or other means of removing minor blemishes and improving the apparent continuity of the coating.

Materials applied as powders form one class of coatings; those applied as liquids, another. Liquid coatings can be further subdivided a number of ways, one is by means of the solvent or vehicle that may be either water or an organic compound of some kind. A further subdivision is by means of how the film is fomed. "Thermoplastic" films are formed by solvent evaporation, coalescence, or heat fluxing of materials that retain their essential composition throughout the process. "Convertible" films, on the other hand, are formed by reaction of a coating component either with another component or with oxygen or water vapor in the air. The reactions are irreversible and their products are dissimilar to the starting materials.

Objects coated may range in size from screw heads and smaller to railcars and larger. Substrates may be several kinds of metals or woods or plastics in any combination, and the final result of the process may be intended for sale directly or may be a component of a larger object that, itself, may be coated. Presenting material applicable to all painting and coating is impossible; instead, this chapter concentrates on those aspects that are inherently most hazardous and/or have caused the most health problems in the coatings industry.

2 PAINT COMPOSITION

Paint composition is extremely variable and complex. Paint is a performance product that relies heavily upon physical as well as chemical interactions of its ingredients. This complexity and variability is evident in modern coating manufacturing facilities that may inventory 3000–5000 raw materials for use in manufacture of thousands of formulated products.

Paint composition is more easily considered in terms of the major ingredient categories: film formers, pigments, solvents, and additives. Each of these ingredient categories may be further subdivided, as illustrated in Fig. 23.1.

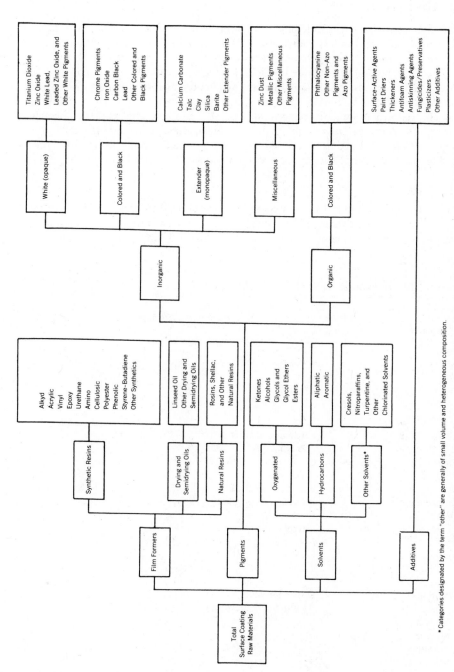

Figure 23.1. Manufacture of paint and allied coating products. (Courtesy NIOSH)

* Categories designated by the term "other" are generally of small volume and heterogeneous composition.

459

2.1 Film Formers

Film formers are the nonvolatile binders that function to uniformly coat the substrate. They are usually more or less transparent and fusable. When combined with a pigment, they function to bind the pigment particles in the film to the substrate. Film formers may be either natural or synthetic in origin. Examples of natural film formers include linseed and tung oils, rosins, and shellac. Synthetic film formers or resins include alkyds, acrylics, vinyls, epoxies, phenolics, melamines, styrenes, urethanes, and others. The majority of modern coatings, at least 94%, are based on synthetic polymer systems, also termed resins. Resins are usually nonvolatile due to their high molecular weight, which may range from several hundred to several thousand. Resins are rarely used in their pure form, but, rather are combined with a thinning solvent.

2.2 Resins

Resins are usually referred to by a common name derived from the chemistry used to produce them or the functional chemical structure such as those already mentioned. The inventory of chemical substances compiled by the U.S. Environmental Protection

Table 23.1. Common Resins and Their Composition

Resin	Origin	Composition
Natural		
Linseed	Cotton seed	Triglyceride
Safflowr	Carthamus tinctorius	Triglyceride
Soya oil	Soybean	Triglyceride
Shellac		
Synthetic		
Alkyd	Polyhydric alcohol + Polybasic acid	Complex ester
Amino	Condensation of amine + aldehyde	Methylol
Phenolic	Phenol + aldehyde	Novolac's
Polyurethane	Isocyanate + alcohol	Urethane
Epoxy	Epichlorohydrin + bisphenol	Diglycydal ether
Acrylic	Acrylic acids + esters	
Polyester	Dicarboxylic acid + Dihydroxy alcohol	Special type of alkyd
Emulsion	Dispersion of insoluble liquids	Copolymer
Vinyl acetate		
Styrene		
Acrylics		
Butadienes		
Vinyl		Polyvinyl acetate Polyvinyl chloride Polyvinyl alcohol

Agency has adopted the convention of naming resins according to their reactants, for example, polymer of ABC. Table 23.1 lists some common resins and their composition.

2.3 Pigments

Pigments have many functions is a coating system, depending on such factors as optical characteristics, particle size, shape, distribution, and dispersability. Pigments may be divided into two main classes: inorganic and organic. The inorganic pigments are the most common owing largely to their stability to light and contribution to corrosion resistance. Inorganic pigments may also serve as extenders that function as fillers and rheology control agents. The term "extender" also refers to the use of these pigments to "extend" other more costly pigments as a means of cost control.

The most common pigment utilized in paint manufacture is titanium dioxide as a replacement for white lead. Sixty percent of the global titanium dioxide production is used in paint manufacture where it may be found in one-third of paint formulations. Other inorganic pigments are listed in Table 23.2.

Organic pigments have been widely used in household coatings where metal content is a problem. Organic pigments are derived from azo, polycyclic, and other compounds. This class of pigments may also be chosen for their brighter color and transparency. They have the disadvantage of poorer heat and light stability. Specific organic pigments are best identified by their color index, which is based on their chemical type and constitution. This systematic classification is vital; the current reference fills five volumes.

2.4 Solvents

Solvents function to maintain the polymer in a solution or suspension. The solvent also functions to control viscosity for ease of production and application. The choice

Table 23.2. Common Inorganic Pigments

Pigment Type	Examples
Extender	Talc, clay, mica, magnesium carbonate, and calcium silicate
True pigments	Rutill, anatase
Titanium dioxide	Iron oxides (yellow, red)
Iron	Lead chromates
Other	Zinc chromates
	Chrome green
	Chromium oxide
	Cadmium sulfide yellow
	Chromium sulfoselenide
	Ultramarine blue
	Zinc sulfide
	Red and white lead
	Vermillon
	Carbon black
	Metals (aluminum, chromium, etc.)
	Zinc phosphate

Table 23.3. Common Pigments

Family	Type	Organic Pigments							
		Yellow	Orange	Red	Brown	Violet	Blue	Green	Black
Azo	Monoazo	x	x	x	x	x			
	Diazo	x	x	x	x		x		
	Azo condensation	x	x	x	x				
	Azo salt	x	x	x	x				
	Azo metal complex	x		x				x	
	Benzimidazolone	x	x	x	x				
Polycyclic	Phthalocyanine						x	x	
	Anthraquinone	x	x	x	x	x	x	x	
	Quinacridone	x		x		x	x		
	Dioxazine				x	x			
	Perylene			x	x				
	Perinone		x						
	Pyranthrone		x						
	Indigoid			x	x	x	x		
	Thioindigo					x			
	Isoindolinone	x	x	x					
Miscellaneous	Nitro				x				
	Nitroso							x	
	Quinoline	x							
	Azine								x
	Basic dye complex	x		x	x	x	x	x	x

Adapted Courtesy of Oil and Color Chemists Association, Australia.

of solvents is usually based on evaporation rate and solvation properties. Proper solvent selection is used by the paint formulator to control other properties or as a coupling aid for immiscible liquids. A listing of the more common solvents is found in Tables 23.3 and 23.4. Although a common differentiation is made between solvent and water-based coatings, virtually all paint contains some organic solvent. The trend in modern coating systems is to minimize the volitile organic solvent content for both health and environmental reasons.

2.5 Additives

Additives are a small but important portion of paint formulation that function as processing aids, to contribute to stability in the container, or impart some particular flow characteristic. Additives and resins are usually considered highly proprietary, making exact composition difficult to obtain.

Coating manufacture utilizes common equipment for many types of coatings, making capital requirements modest by comparison to the chemical process industries. Many small manufacturers purchase resins and formulate paint to supply regional markets. Proprietary resin manufacture does constitute an important part of this industry.

Table 23.4. Common Solvents

Solvent Type	Common Examples
Hydrocarbon	
Aliphatic	Aliphatic naphtha, hexane, cyclohexane, isobutane, heptane, isoparafin, hydrogenated oils.
Aromatic	Toluene, xylene, ethyl benzene., aromatic naphtha
Distillates	VM + P naphtha, Stoddard solvent, mineral spirits
Oxygenated Solvents	
Alcohols	Ethanol, propanol, butanol (isomers), hexanol, hydroxy toluene, amyl
Glycol	Ethylene glycol, propylene glycol, 2-ethoxy ethanol, 2-propoxy ethanol, 2-butoxy ethanol, hexylene glycol
Glycol ether acetates	Ethylene glycol monoethyl ether acetate, propylene glycol monomethyl ether acetate
Ketones	Acetone, butanone, pentanone, 2-hexanone, cyclohyexanone
Ester	Methyl amyl acetate
	Isobutyl isobutyrate
	Ethyl acetate
	Ethylene glycol diacetate
	N-butyl acetate
	Amyl acetate
Ether	Dioxane
	Tetrahydrofaran
Chlorinated	Methylene chlroide
	1,1,1-Trichloroethane

3 PARTS PREPARATION

3.1 Nature and Purpose

For a coating of any kind to adhere to a surface, that surface must have several properties, some of which require preparation. First and foremost, the surface must be clean, but "clean" can have several meanings depending on the kind of coating to be applied. Clean almost always means that the surface must be free of easily removed material such as dust and flakes of rust. Otherwise the coating would adhere to that material and not to the intended surface and thus would be easy to remove.

Clean also always means that the surface must be free of substances not wetted by the coating carrier or solvent. Again, the problem is lack of adherence to those areas unwetted by the coating and is exemplified by trying to cover a grease spot with water-base paint.

Even if a foreign substance on the surface is wetted by the coating solvent, it may interfere with proper adhesion or with film formation, in which case it must be removed prior to the coating step. Lubricating oils and greases used in manufacturing processes almost always must be completely removed prior to painting any kind of surface, for instance.

Cleanliness requirements vary considerably among coating types simply because some kinds of coatings are much more tolerant of minor amounts of contamination than are others. Completed parts to be coated with an oil/wax mixture to retard corrosion can be quite thoroughly contaminated with machining oil, for instance, without affecting the efficacy of the applied coating. The method used to apply the coating also influences the requirements; for instance, if paint is to be sprayed with an air gun, some loose dust may be tolerated because the air blast can be counted on to remove it.

3.1.1 Compatibility. The surface to be coated must be compatible with the coating to be used. That is, the solvent or carrier of the coating must not dissolve the substrate but must be able to wet it. In addition, for good adhesion, the surface usually must be abraded slightly—or at least, not be glossy.

3.1.2 Methods. Methods used for surface preparation prior to coating can be variously classed, but abrasive and solvent (using solvent to include acid and alkaline washes as well as soap and water and organic solvents) may be descriptive. Abrasion methods are used to remove relatively large amounts of contamination as well as some of the surface, itself, while solvent methods usually leave the surface almost intact while removing most or all of the contamination.

3.2 Health Hazards

Abrasion, rather than solvent cleaning, is the preferred method for wood surfaces and for precoated surfaces of any nature. Abrasive papers (sandpaper, garnet paper, and the like) and cloths are used to smooth surfaces and also to remove the gloss from previous coats to allow better adhesion of further coats. In either case, the person doing the job and others in the immediate vicinity may be exposed to dust from the abraded surface. Wood dusts of many kinds have been implicated as causes of asthma and some of nasal carcinoma, and paint dusts contain paint pigments, compounds of lead, chromium, and cadmium that may be hazardous. If metallic lead has been used to make the surface more even and smooth (auto bodies, especially), then abrasion can result in lead dust reaching the breathing zone.

3.2.1 Dusts. Most particulate matter inhaled by people during their work is physiologically inert. Some materials, on the other hand, have toxic effects. Overexposure to lead can affect the blood-forming (hematopoietic) system, the central and peripheral nervous systems, the kidneys, and the reproductive systems.

Some compounds of chromium can cause lung cancer; these and other compounds can also sensitize the skin and other organs and cause a peculiar dermatitis. Cadmium and its compounds cause kidney injury and may also be carcinogenic. These and other pigment materials are discussed more fully in Section 4.2.2.

3.2.2 Noise. Mechanical equipment used in abrasion processes is almost always noisy and a noise problem may be aggravated by ringing of the part abraded (especially if the part is metallic, but even wood can act as a sounding board).

3.2.3 Strain. Quite often, hand sanding is an arduous task made lighter with mechanical equipment that can cause problems unassociated with noise. Especially when the arm and wrist must be held in an unusual position for long periods of time, "carpal tunnel syndrome" may result. This is a set of symptoms, including pain and tingling sensations in the fingers and hand, caused by compression of a nerve in the wrist.

Hand work in preparation of larger surfaces for coating can require the abilities of a contortionist. Assumption of and work in odd positions can result in muscular strain with associated pain and perhaps cramping.

3.2.4 Solvent Vapor. Especially when bare metal is to be painted, but also for some precoated surfaces, the last tep in preparation may be a solvent wipe. Where this is so, exposure to solvent vapor may occur, as may skin and/or eye contact with the material. Substances used for this purpose include all those used as thinners (see Section 4.2) as well as chlorinated hydrocarbons such as 1,1,1-trichloroethane and trichloroethylene. In special cases, chlorofluorocarbons may be used.

3.3 Health Hazard Controls

3.3.1 Dusts. In some industries (notably the automotive), dust inhalation during surface preparation is prevented almost entirely through use of elaborate local exhaust ventilation equipment coupled with protective clothing and supplied-air respirators. In many industries, disposable respirators may be the only protection employed although external local exhaust hoods are often used. Especially where there is much dusting, high-velocity, low-volume (HVLV), local (tool-oriented) exhaust ventilation may be applied. Even though HVLV hoods and exhaust lines make tools heavier and somewhat more awkward to use, these penalties may be completely offset by discomfort of the personal protective equipment that must be used otherwise to avoid overexposure.

Especially in wood work but also more and more in metal finishing, exhaust air laden with particulate material from surface preparation is cleaned and then returned to the shop rather than being conducted outside. Most of the dust particles produced by abrasion are rather large and easily removed completely from an airstream with, for instance, a bag filter assembly. This air recirculation technique saves energy and costs in heated and air conditioned plants. For successful operation, monitoring of the return airstream for particulate material is almost always necessary, as is the automatic operation of a damper or other control to assure that contaminated air is discharged outside rather than into the work environment.

3.3.2 Noise. Engineering control of noise exposure has succeeded in the design of some abrasive blast cabinets, especially those operated in a batch mode where loading doors can be sealed, thus ensuring integrity of sound barriers. Hand-operated equipment from chipping guns to orbital sanders, however, has been much more difficult to quiet even though some progress has been made. Electrical orbital and belt sanders can be so quiet that the only noise perceived is that resulting from contact of the abrasive with the substrate. Oscillating and vibratory sanders, on the other hand, are almost always noisy. Similarly, pneumatic tools of any sort can be very noisy if their exhausts are unmuffled and their turbines (if present) are unshielded, but relatively quiet otherwise. With effective devices available from several manufacturers, there appears to be little excuse for excessive noise from pneumatic tool air discharge today. Mostly because of the varied nature of the work and of the substrate surfaces, however, noise exposure in the surface preparation business is controlled with ear protection devices where protection is necessary.

3.3.3 Solvent Vapor. Solvent vapor exposures tend to be intermittent when a final wipedown is all that is required and often no control other than reasonable general ventilation is present or needed. However, when exposure may be close to continuous, then local exhaust ventilation may be necessary. One way of achieving that with little expenditure it to take advantage of the air flow into a nearby paint spray booth, using walls or curtians to direct that air past the final wipe station. The dilution that takes place downwind of the solvent wiping station is usually great enough to assure that a paint sprayer (if present) has only a nominal solvent exposure.

When final wiping must be done under circumstances where the exposure can be excessive (inside an enclosure, for instance), then ventilation may well not be a practical means of preventing overexposure, and some kind of respirator must be used. For all the solvents normally used for this purpose, an air-purifying respirator canister(s) should be the "organic vapor" variety. Supplied-air respirators are often a better choice, however.

4 APPLICATION OF PAINTS AND COATINGS

4.1 Technology

Most industrial painting and other coating is done with a spray of some sort, whether the coating is a liquid or powder. Spraying is the most rapid way of applying a coating that results in a high-quality finish. Only very occasionally are other application methods used and then usually because the job is intermittent, can be automated easily, or a single heavy (and thus low finish quality) coat can take the place of several lighter coats that would be applied with a spray gun. The frame of an automobile body, for instance, may be dipped in rather than sprayed with a rust-protective coating. This can be done automatically; coverage is complete, and quality of the surface is unimportant.

4.1.1 Spray Application. Of the methods used for application of coating materials, only spraying results in potential or actual inhalation exposures to all materials in the coating. Other methods used result in inhalation of the solvent vapors only; other components of the coating do not become airborne in high enough concentrations and/or small enough particle sizes to cause such problems. Because spraying is by far the most often-used method and also is by far the most hazardous, it will be examined in some detail in this chapter; for the most part, other methods are mentioned only to contrast them with spraying.

4.1.2 Spraying Methods. Spray application methods are usually categorized as compressed air, airless, or electrostatic.

COMPRESSED AIR. In this method, compressed air provides the energy to atomize the coating material, producing particles in the correct size range to adhere to the substrate surface. This is done by mixing the compressed air either inside the spray nozzle (internal mix) or outside of it (external mix). Each method has its own advantages but each can produce a very high quality finish.

Coating material can be supplied to the compressed air spray gun either through the siphoning action of the compressed air or by a separate pumping system. Siphon-fed systems are more common mainly because they are less expensive, but produce more overspray than pressure-fed systems. Either method can be used to supply heated coating material to the gun. Heating is used to reduce viscosity of coating systems that can therefore be useful at high solids ratios (lower solvent concentrations), creating less overspray (less wastage, lower potential hazard), while applying a heavier coat that dries rapidly.

AIRLESS. Use of compressed air for spraying can be eliminated completely by supplying the coating to the gun at quite high pressure and forcing it through a very small orifice. This airless method requires a much simpler spray gun than does the compressed air method and produces much less wasted overspray. Airless spraying is well adapted to the use of heated coating material; reduced viscosity calls for lower hydraulic pressure but greater atomization and smaller spray particles.

Other than the usual hazards associated with inhalation of coating sprays, the airless method carries with it the additional hazard of hypodermal injection because of the high pressures employed.

ELECTROSTATIC. If the coating droplets and the substrate carry opposite electrical charges and the voltage is sufficiently high, then almost all of the droplets will be attracted to the substrate and losses by overspray will be minimized. This method requires a means of charging either the fluid stream or the droplets as they form in addition to a means of atomizing the coating material. Atomization can be done electrostatically using a simple highly charged spinning disk with the coating introduced at the center and droplets being formed at the edge mainly by electrostatic repulsion of like charges. Of course, both compressed air and airless methods can be used, also, to produce atomized sprays. In all cases, the material to be coated must be electrically conductive. This method, too, is well adapted to use of heated high solids coatings.

4.1.3 Powder Coating. The ultimate in high solids is to have no solvent at all. Powder coating technology is still in its infancy but is being pressed because of the regulatory pressure to reduce emissions of volatile organic compounds to the atmosphere. Powder coating is done by applying powdered coating material to a surface hot enough to melt the coating and cause it to flow and coalesce into a continuous film or through flame spraying where the powder is applied molten. Most powder coating is done with electrostatic or fluidized-bed methods; the usual amount of overspray is small. The chief problems are those related to cost of the coating material (almost always a convertible formulation based on epoxy, polyurethane, or polyester resins), the energy cost of application and, of course, the fact that the substrate must usually be heated above the fusing point of the coating. Uniform, durable, high-quality finishes can be obtained with this method under conditions of minimal risks to workers.

4.2 Associated Health Hazards

Paints consist of four general categories of materials; other types of coating usually have fewer components. Those associated with paints are (a) binder or film former, (b) pigment(s), (c) volatile solvent, and (d) additives. As previously mentioned, only in paint (or powder) spraying are workers exposed to every constituent of the coating; other methods of application, in general, cause exposures only to the solvent or solvent system.

4.2.1 Binders. Binders are liquids, or solids in solution, in the paint formulation that become solid films either through solvent evaporation, coalescence, or chemical reaction. When solvent evaporation is the cause of film formation, the coating is usually called a lacquer rather than a paint. Exceptions are the latex-based coatings, which are called paints even though films are formed when evaporation of the solvent (water) allows coalescence of the latex particles and subsequent fusion of resin solids.

ALKYDS. Binders are either synthetic or natural resins or so-called drying oils (oils that "dry" by reaction with oxygen in the air). The most-used binder is alkyd resin made by condensation of a polycarboxylic acid such as phthalic acid with a mono-glyceride produced by reaction of a polyalcohol such as glycerol or pentaerythritol with a fatty acid. Alkyds are often mixed with a drying oil for paint formulations and are then called "oil-modified polyester resins." The oil-modified materials cure mainly by oxidation of the drying oil; polyesters can be used as one of the components of a two-component polyurethane system. Whether oil-modified or not, alkyds have caused few problems in use; their toxicities and hazards are low by all routes of contact.

ACRYLIC AND VINYL. Acrylic and vinyl resins are also frequently used. Acrylics are based on polymers and copolymers of acrylic and methacrylic acid esters. Vinyl resins, on the other hand, are polymers and copolymers of vinyl chloride, vinylidene chloride, vinyl alcohol, vinyl acetate, and other vinyl esters. Thermoplastic (materials that melt and flow upon application of heat) acrylics are used for lacquers while thermosetting (materials that are chemically changed by the action of heat and which do not melt) acrylics are usually reacted with an epoxy or amino resin in a single-component system. Diisocyanate-based resins (polyurethanes but called urethanes in the trade) are two-component systems that may have an acrylic resin as one of the components.

Single-component acrylic, vinyl, epoxy, or polyurethane systems consist of high-molecular-weight polymers, and such materials have low toxicities because they are physiologically inert. Two-component systems, on the other hand, consist of a partially polymerized material for one of the components and a catalyst or curing agent for the other. Either or both can be toxic.

The chief problem with both epoxy and polyurethane resin systems is that of sensitization. Epoxies are notorious as skin sensitizers and polyurethanes are sensitizers of the respiratory tract (from inhalation of the reactive component, a diisocyanate of some sort, usually toluene diisocyanate). Both kinds of material, however, are capable of sensitizing either the skin or the respiratory tract or both following either single rather massive overexposure or repeated exposure to even small amounts of the active material.

Epoxy resin coatings are usually classified as two-package or amine-cured systems, epoxy baking enamels, and epoxy esters. The two-package systems cure by reacting the epoxy resin component with the curing agent, commonly a polyamino compound by thorough mixing just prior to use, perhaps assisted by heat following application of the coating, but usually not. Typical polyamines are diethylene triamine, triethylene tetramine, and tetraethylene pentamine that react with terminal epoxide groups of the resin. Polyamines may be partially reacted with low-molecular-weight epoxy resins prior to use to reduce the amount of free (and relatively volatile) amine in the package. Rather than polyamines, the curing agent can consist of amine-terminated polyamide resin formed by reacting a polyamine with, usually, a dimerized vegetable oil. Polyamides have a longer shelf- and pot-life than do the polyamines.

All of the polyamines and their adducts are capable of sensitizing the respiratory tract following either single massive overexposure or prolonged, repeated contact with much lower concentrations. In some systems, the polyamine or adduct is dissolved in a reactive diluent such as a glycidyl ether that is capable of causing skin sensitization just as is an epoxy resin.

High-molecular-weight epoxy resins can be reacted with thermosetting phenolic or amino resins to form one of the components of a baking finish. The other half of the system, the curing agent, usually consists of dicyandiamide, a polyanhydride, or an aromatic amine of some sort blended with powdered resin, pigment, and additives. After mixing, the coating is applied and then cured by baking in an oven.

The catalysts used for baking enamels are less capable of causing sensitization of either the skin or the respiratory tract than the polyamines and are much less hazardous to handle and use. They, however, are blended with a lower molecular weight epoxy resin that is capable of sensitizing the skin.

Epoxy esters are formed by reacting an epoxy resin with a drying oil; in that reaction, all free epoxide groups are consumed so that the resulting ester takes on the toxic properties of a complete polymer, being physiologically inert. The film dries by air oxidation of unsaturated components of the drying oil molecules.

Polyurethane binding films are formed by the reaction of a diisocyanate with a material having hydroxyl groups, commonly called a polyol. Most widely used of the diisocyanates is toluene diisocyanate, but many others have come into prominence including aliphatics such as hexamethylene diisocyanate. Because of its relatively low volatility (and other properties), dilphenylmethane diisocyanate (methylene bisphenyl isocyanate, MDI) is replacing toluene diisocyanate (TDI) in many applications. Diisocyanates are not used free in the formation of coatings; instead they are reacted with a polyol to produce a prepolymer that contains excess isocyanate reactive groups. The prepolymer, then, is one component of a two-component system. Otherwise, the prepolymer may be reacted with a drying oil to form an oil-modified one-package system having no free isocyanate groups and that cures by oxidation of the drying oil. One-package oil-modified urethane binders are essentially inert physiologically.

If the prepolymer is allowed to remain unreacted with a polyol, it can be used as a one-package, moisture-cured system. In this case, the free isocyanate grouops react with moisture from the air to cure and form a hard film. In much the same manner as oil-modified polyurethanes are formed, so-called blocked systems may be produced by reacting the prepolymer with phenol, caprolactam, cresol, methyl ethyl ketoxime, or another agent that will break down and/or vaporize when heated sufficiently, to free terminal isocyanate groups. Blocked urethane systems are one-package because the second component, usually a polyester, is mixed with the prepolymer and only reacts with it when heated above about 160°C. Here, again, because the isocyanate groups are blocked, the material is easy to handle safely at room temperature because of its physiological inertness.

Two-package systems are used with urethanes as with epoxies. The second component always contains a catalyst such as dibutyl tin laurate, zinc octoate, or methyl diethanolamine along with a polyol or polyamine to react with the terminal isocyanate groups. Although polyols are usually inert physiologically, the catalysts used may not be and the prepolymer may contain unreacted diisocyanate capable of causing skin and respiratory tract sensitization.

Lacquers can be made from polyurethane systems by reacting the prepolymer with a cellulose ester or thermoplastic acrylic resin and dissolving the product in a solvent. These are one-component systems that typically present only the hazards of the solvent system to the user.

OTHERS. Additional binders include cellulosic (based on esters such as cellulose acetate, nitrate, and acetate butyrate), phenolic, chlorinated rubber, styrene–butadiene, and linseed and tung oils.

The cellulose esters are typically used in lacquers, forming a film by solvent evaporation. The binders are of very low toxicity and hazard; they present no handling problems.

Drying oils such as linseed and tung "dry" by reaction of oxygen from the air with double bonds contained in their molecules. Drying is accelerated by heat and by the addition of "driers" such as cobalt naphthenate that catalyze the reaction. The chief hazard of drying oils is that of fire. A rag soaked with the oil and wadded up will allow the exothermic oxidation to proceed under circumstances where heat can, perhaps, build up. As heat accelerates the reaction, the kindling point of the cloth may be reached, to cause a fire.

Chlorinated rubber, neoprene, and the styrene–butadiene binders are used as latexes where the solvent is water and the film is formed by evaporation of the water and coalescence of latex particles. The monomers are completely polymerized prior to being incorporated into the coating and thus are essentially inert physiologically.

4.2.2 Pigments. All paints contain pigments that are finely ground water-insoluble solids. Paints without pigments are called varnishes; pigmented varnishes are called enamels. Pigments have several uses in paints, includng opacity (hiding), color, corrosion inhibition, reinforcement of the binder, and general filler and extender.

WHITE LEAD. The first pigment used on a large sale in paint was white lead, a lead hydroxycarbonate [the usual chemical formula used is $Pb(OH)_2 \cdot 2PbCO_3$] made by reacting lead acetate with carbon dioxide produced by burning coke or through fermentation. This material is relatively inexpensive even though the processes used are very time-consuming (from two weeks to a few months per batch). White lead has good hiding power and excellent color retention except in the presence of hydrogen sulfide (which turns it splotchy black). Its toxicity, however, is high, and its sweetish taste makes it attractive to children who may chew objects painted with it. For this reason, its general use as a paint pigment has been essentially banned.

TITANIUM DIOXIDE AND LITHOPONE. For most uses, titanium dioxide, TiO_2, has supplanted white lead as a paint pigment. Although more expensive, titanium dioxide has better hiding power than white lead, and its color is not affected by hydrogen sulfide or any other normal or even abnormal atmospheric constituent. Furthermore, it is essentially inert, physiologically. Titanium dioxide is one of the "inert" pigments in that it does not react with other components of the paint formulation. Lithopone (a mixture of barium sulfate and zinc sulfide) is another inert white pigment that competed with white lead at one time but that has been largely supplanted by titanium dioxide. Its components are essentially inert physiologically.

REACTIVE PIGMENTS. Reactive white pigments include white lead as well as lead sulfate and silicate along with antimony and zinc oxides. These materials react chemically with acidic and some other binders and thus are incorporated more firmly into the paint film than the inert pigments. Despite this advantage, their use continues to decline with time.

Zinc oxide as a powdered pigment may pose inhalation problems during its manufacture and incorporation into the coating, but usually does not present hazards in application of that coating.

Antimony trioxide is rarely used as a paint pigment because of its relatively high cost and relatively high toxicity, the effects including electrocardiogram alterations, dermatitis, mucous membrane irritation, and pneumoconiosis.

EXTENDERS. Materials used to help control gloss, texture, or viscosity and that have an effect upon the appearance of the paint film are the extenders, chiefly calcium carbonate, talc, clay, silica (amorphous and crystalline), along with calcium silicate and sulfate and barium sulfate. None of these materials poses any particular problem in the application of coatings if for no other reason than their concentration in the formulation is almost always low. Except for silica, these are essentially "nuisance" materials, and even with silica present, other components of the formulation are almost always more hazardous; if those hazards are properly controlled, so will be those of the extenders.

COLORED PIGMENTS. Colored pigments have included almost every conceivable insoluble colored inorganic compound and many organic dyes bound to inorganic bases. Chief of the colored inorganic pigments are iron oxide, many chromium compounds such as lead and zinc chromates, and compounds of cadmium, molybdenum, cobalt, and others. Red lead (a mixture of lead oxides, see Section 3.2.1) has long been used for corrosion inhibition.

Dusts of cadmium compounds have caused kidney injury in man and experimental animals, leading to the excretion of low-molecular-weight protein in the urine. These compounds, in general, have been water soluble; the effects, if any, of water-insoluble pigments have not been documented.

There are four categories of chromium and its compounds insofar as toxic hazards are concerned, namely, the metal, trivalent water-soluble compounds, hexavalent water-soluble compounds, and hexavalent water-insoluble compounds. Chromium-containing pigments used in paints are almost universally hexavalent water-insoluble forms. The metal is relatively inert, physiologically, as is the trivalent (chromous) oxide. The hexavalent oxide is probably capable of causing sensitization cf the skin and respiratory tract, perforation of the nasal septum, ulcerative dermatitis, and other problems (but probably not cancer), while some hexavalent water-insoluble pigments are probably carcinogenic with varying degrees of activity.

METALLIC POWDERS. Powders of aluminum (mainly) and zinc are used as pigments to give the final paint film a metallic luster and to aid in protection against corrosion. Both materials are "nuisance dusts" insofar as toxicity and hazard are concerned.

4.2.3 Solvents. Paint and other coating solvents have several functions. Foremost is to dissolve and/or make the binder more fluid. Because of their use in viscosity adjustment, solvents are often referred to as thinners. They evaporate and thus aid (or cause) film formation. Solvents are chosen for paint and other coating formulations mainly on the basis of price, evaporation rate, and compliance with federal and state regulations (not necessarily in that order) with the set of materials from which to choose being limited to those that are able to dissolve the binder satisfactorily. Other considerations include odor intensity and character as well as toxicity. Almost all paint solvents are flammable, methylene chlroide and water being the chief exceptions, and therefore, this property is usually given little attention when a choice is to be made.

The chief coating solvents have been and still are aliphatic hydrocarbons (mineral spirits, varnish maker's and painter's—VM&P—naptha, etc.). These materials are used because of their low price, good solvency, and the ability to control evaporation rate within a rather large range by choosing the boiling range of the mixture used. In general, aliphatic hydrocarbons are among the least photochemically reactive organic compounds and this property, also, has influenced their use.

Next most popular are the aromatic hydrocarbons—toulene, the xylenes, and mixtures that may include these materials as well as higher-molecular-weight substances. Benzene is rarely found in coatings because of its high toxicity. The aromatics, as a group, are slightly more photochemically reactive than the aliphatics and, therefore, their use has been limited.

Other solvents commonly used in paint and other coating formulations include several kinds of oxygenated hydrocarbons such as alcohols and glycols, ketones, and esters.

All commonly used solvents share two health hazards, those of narcosis and of skin irritation caused by defatting. The principal problem associated with overexposure to materials that cause narcosis is that of inattention. Skin irritation from the common fat/grease solvents is quite common. These materials, in general, are not severe irritants, but, instead, they cause problems by defatting the skin.

ACETONE. Acetone vapor is only slightly irritating to the eyes and respiratory tract; in higher concentrations it is slightly narcotic. As with ethanol, acetone is a normal metabolite in biological oxidation; it therefore, does not accumulate in the body. There are no effects resulting from prolonged and repeated exposure that are different from those that are experienced from short-term exposure.

The closed-cup flash point for acetone is 1.4°F(-17°C); its explosive range in air is 2.9–10.3%.

BUTYL ACETATES. *n*-Butyl acetate is simlilar to the other low aliphatic esters in its toxicity and hazards. It is irritating to both the eyes and the upper respiratory tract. This material, however, is probably narcotic in high concentrations. *n*-Butyl acetate is much better known, toxicologically, than isobutyl acetate. Effects of both materials are quite similar. Closed-cup flash points are 72°F(22°C) for *n*-butyl and 64°F(18°C) for isobutyl acetates. Explosive ranges in air for the two materials are 1.7–7.6% for *n*-butyl and 2.4–10.5% for isobutyl acetate.

BUTYL ALCOHOL. Of the three butyl alcohols, *n*-butyl (*n*-butanol) is the most commonly encountered and also more toxic than *tert*-butyl and *sec*-butyl alcohols. Furthermore, *n*-butanol can be absorbed through the intact skin in acutely toxic amounts while skin absorption does not appear to be a problem with the other isomers. It is also irritating to the eyes.

Both *sec*-butanol and *tert*-butanol are more narcotic than is *n*-butanol and are more volatile. Flammability data for these three alcohols are given in Table 23.5.

ETHYL ACETATE. Ethyl acetate is one of the most used lacquer and ink solvents and has an excellent record of very few deleterious effects on people who use it. At 400 ppm, its odor (mild, fruity, pleasant) is strong, and unacclimated people may experience some upper respiratory tract irritation in the first few minutes of exposure. Ethyl acetate is a mild narcotic; exposure to very high concentrations may cause some dizziness or sleepiness. The literature indicates that some people have become sensitized (respiratory tract) to this material, but that certainly is not a common problem or even a likely one.

Ethyl acetate can defat the skin, but is not particularly irritating otherwise. Its volatility usually prevents prolonged skin exposure. It is quite flammable with a closed-cup flash point of 24°F(-4°C) and an explosive range in air of 2.2–11.0%.

ETHYL ALCOHOL. Ethyl alcohol (ethanol) is best known as a beverage. Overexposure by ingestion causes generalized depression of the central nervous system, including narcosis.

Ethanol is the least toxic of the alcohols because of its relatively rapid and complete metabolism to carbon dioxide and water. In high concentrations on the order of 5000 ppm or greater, however, it can cause irritation of the eyes and upper respiratory tract. Its closed-cup flash point is 55°F(13°C), and its explosive range is 3.3–19.0% in air.

Table 23.5. Flammability Data

Material	Flash Point (CC,°F)	LEL[a] (%)	UEL[b] (%)
n-Butanol	84	1.4	11.2
sec-Butanol	75	1.7	9.8
tert-Butanol	52	2.4	8.0

[a]LEL = Lower Explosive Limit.
[b]UEL = Upper Explosive Limit.

METHYLENE CHLORIDE. Of all the chlorinated aliphatic hydrocarbons, methylene chloride (dichloromethane) is one of the two that are least toxic (the other is 1,1,1-trichloroethane). This material has limited ability to injure the liver and kidneys; instead, it is converted in the body to, among other things, carbon monoxide. Following exposure to methylene chloride, the carboxyhemoglobin level in blood rises just as if the exposure had been to carbon monoxide, but slower. The rise in carboxyhemoglobin is slower, but so is the fall and, in consequence, the exposure to higher than normal carboxyhemoglobin levels is more prolonged than if carbon monoxide had been inhaled. Discovery of the body's ability to convert methylene chloride to carboxyhemoglobin caused a reassessment of the various control concentrations for this material. In addition to that effect, methylene chloride is a weak narcotic (causing drowsiness); it can, of course, defat the skin.

Methylene chloride has no flash point but can be exploded in concentrations ranging from 15.5 (155,000 ppm) to 66% (660,000 ppm) in air.

METHYL ETHYL KETONE. Usually alled MEK, methyl ethyl ketone (butanone) is one of the least toxic of the commonly used solvents (its lower homolog, acetone, is still less toxic). Only a vary mild narcotic, MEK is sufficiently irritating to the eyes and respiratory tract to prevent voluntary overexposure to hazardous concentrations. At 200 ppm, there may be some mild eye irritation. Gross chronic overexposure including skin contact to an unknown extent has caused numbness of the fingers and arms in people also exposed to methyl n-butyl ketone.

MEK can defat the skin but is not particularly irritating otherwise. Its volatility usually prevents prolonged skin exposure. With a closed-cup flash point of 21°F(-6°C) and an explosive range of 1.8–10% in air, its fire hazard can be great.

METHYL ISOBUTYL KETONE. Methyl isobutyl ketone is usually called MiBK or MIBK. For many uses, it has replaced methyl n-butyl ketone (MBK) because that material has been found to induce peripheral neuropathy upon chronic overexposure.

MIBK (hexone, 4-methyl-2-pentanone) is moderately irritating to the eyes, upper respiratory tract, and skin. At 100 ppm it is not particularly irritating, but at 200 ppm, eye irritation is apparent, and the odor is objectionable to many people; at 400 ppm irritation to the eyes and upper respiratory tract is severe. The warning properties (irritation, odor) of MIBK appear to be adequate to prevent unknowing overexposure. There are few, if any, indications that chronic exposure to MIBK has effects different from those of acute exposure.

MIBK can defat the skin and, in addition, is sufficiently irritating to cause dermatitis upon prolonged, repeated contact. It has a closed-cup flash point of 73°F(23°C) and an explosive range in air of 1.4–7.5%.

MINERAL SPIRITS. Mineral spirits indicate a derivative of petroleum. Usually the materials so-designated are largely aliphatic hydrocarbons such as the hexanes, heptanes, and octanes. Mineral spirits is difficult to describe quantitatively or qualitatively because of the very complex mixtures usually involved. If the benzene and hexane concentrations are low, the inhalation hazard is probably limited to narcosis and some irritation.

Of the aliphatic hydrocarbons, hexane stands out as by far the most toxic. Chronic inhalation of hexane can cause peripheral neuritis that can progress to neuropathy (death of sensory nerves in the fingers, toes, etc.). When a mixture of aliphatic hydrocarbons is used (as in paint thinners), the analytical problem can be formidable; one of the ways of handling that problem is to express "total aliphatics" as if all were hexane whether or not hexane itself is present.

Aliphatic hydrocarbons other than hexane used in mineral sirits, VM&P (varnish maker's and painter's) naptha, and so on are capable of causing narcosis and perhaps some irritation if the exposure is excessive. For VM&P naptha, the closed-cup flash point is 28°F(-2°C), and the explosive range in air is 0.9–6.0%.

PROPYL ACETATES. Both n-propyl acetate and isopropyl acetate are much more irritating to the eyes than to the respiratory tract. At 200 ppm for n-propyl and 250 ppm for isopropyl, they will probably irritate the eyes of the unacclimated. At those concentrations, the odors are pleasant. Eye irritation, however, should act to prevent voluntary overexposure to concentrations that could prove injurious.

n-Propyl acetate has a closed-cup flash point of 58°F(14°C); that for isopropyl acetate is 64°F(18°C). Explosive ranges for these materials are 2.0–8.0% for n-propyl and 2.4–10.5% for isopropyl acetate.

PROPYL ALCOHOLS. Propyl alcohol (n-propanol) is irritating to the eyes and respiratory tract in concentrations above about 400 ppm. At higher concentrations it probably can cause narcosis.

Isopropanol enjoys widespread household use as "rubbing alcohol." At 400 ppm it is mildly irritating to the eyes and upper respiratory tract to the uninitiated. At twice that concentration, the irritation is severe enough to be uncomfortable. Although isopropanol is a narcotic in high concentration, chronic exposure to 400 ppm produces no such effects.

Isopropanol (and probably n-propanol) can be absorbed to a minimal extent through the intact skin, but the surface area contacted must be large and the exposure severe (the concentration high and over a prolonged period of time) for the effects to be pronounced. It can defat the skin but otherwise is not particularly irritating. Closed-cup flash points for these alcohols are 77°F(25°C) for n-propyl and 53°F(12°C) for isopropyl. Their explosive ranges in air are 2.1–13.5% for n-propyl and 2.0–12.0% for isopropyl alcohol.

TOLUENE. Also incorrectly called "toluol," toluene is a fairly strong narcotic and a mild respiratory tract irritant. Exposure to toluene for a few hours to a concentration of 200 ppm is likely to result in drowsiness and some deficit in manual coordination. Effects from chronic exposure to such concentrations are likely to be similar to those of a single exposure. The chronic inhalation toxicity of toluene is about the same as the acute (short-term, single exposure) inhalation toxicity.

Prolonged, repeated skin contact may result in defatting dermatitis. Its closed-cup flash point is 40°F(4°C), and its explosive range is 1.2–7.1% in air.

XYLENE(S). Commercial xylene is a mixture of the three isomers of dimethyl benzene with m-xylene predominating. The mixture or any of its components may incorrectly be called "xylol." All of the xylenes are similar in their toxic properties, both qualitatively and quantitatively. These materials are fairly strong narcotics but are much more irritating to the eyes and upper respiratory tract than toluene. That irritation is very noticeable at 200 ppm, sufficiently so that the warning properties of xylene are considered adequate to prevent voluntary overexposure to an extent tht narcosis is a problem. Prolonged overexposure may result in some gastric distress along with mild and reversible blood changes. In that manner, the chronic inhalation toxicity of xylene differs somewhat from the acute inhalation toxicity. Xylenes are more severe skin irritants than is toluene.

Prolonged contact with the liquid may cause blistering in addition to typical defatting dermatitis. Flammable properties of these isomers are not greatly influenced by structure. For all of them a closed-cup flash point of 85°F(30°C) applies as does an explosive range of 1.1–7.0% in air.

4.2.4 Additives. Materials are added to paint to enhance drying, reduce formation of "skin" in the package, and to inhibit mildew formation in the package and on the dried film. In addition, material may be added to enhance rheological properties, that is, to reduce sagging but enhance leveling. Other substances are added to latex paints to maintain pigment dispersion, reduce foaming, and promote coalescence of latex particles.

The most used drier is cobalt naphthenate. Manganese soaps of naphthenic acid or tall oil fatty acid are frequently used. Other driers are compounds of lead, barium, calcium, cerium, iron, zinc, and zirconium. All driers are used in very low concentration in the finished paint, with 1.0% being the maximum for lead. With the possible exception of lead compounds, none of the metal driers poses much of a health hazard to users. See Section 3.2.1 for a discussion of possible problems with lead.

Oximes and substituted phenols are used as antioxidant antiskinning agents in paint formulations. These materials pose no problems for paint users.

Many different fungicides have been incorporated into paint, especially that used in warm climates, to inhibit mildew formation. Most such substances, however, are confined to house paints and rarely are found or needed in usual industrial applications. When used, the fungicide most frequently found in paints is a phenyl mercury compound of some sort at a mercury concentration of up to 2.0% based on total paint weight.

Soaps, modified drying oils, and clays are used to modify the flow properties of paint. None is present in a concentration high enough to pose any special health problems.

Latex paint additives include all those mentioned as well as various soaps and detergents to help maintain pigment and binder dispersion along with one or more antifoam agents needed because of the excess surfactant present. Other additives may include materials such as alkylmercury compounds and chlorinated phenols as preservatives and vairous proteins and substances such as methyl cellulose used to thicken the paint. Most toxic of the additives are the fungicides, but they are not present in high enough concentration to materially increase the hazards of the paint.

4.3 Health Hazard Controls

The control methods most used in the industrial application of paint and other coatings are substitution of materials and/or equipment and local exhaust ventilation. To some extent, other control methods such as process substitution (brushing rather than spraying, e.g.) and isolation (especially with personal protective equipment including gloves, cap, coveralls or uniforms, and respirators) are used when the more popular methods are not applicable for one reason or another.

4.3.1 Substitution. Because substitution of one material, piece of equipment, or process for another can often completely eliminate a health hazard, possibly by exchanging it for a less significant one, perhaps at little or no expense, it is often the most effective means of control. Substitution, however, is usually the most difficult control method to apply for several reasons. First, it requires an intimate knowledge not only about materials, equipment, and processes being used and of their proposed substitutes, but also of the relative hazards associated with both the original and its proposed substitute. Second, originality is required, the inventiveness to understand the problem and to see that an answer exists by doing something different. Third, inertia must be overcome. Substitution always requires changes in the way of doing things and people resist change, sometimes strenuously. Finally, for effective substitution, usually one person must have the knowledge required, the inventiveness to use it, and the authority or persuasiveness to overcome the ever-present inertia. Such a combination is not often found.

MATERIAL. Material substitution for hazard control has been used very successfully in the coatings industry. The three best examples are removal of benzene from cleaning and binder solvents, removal of lead from coatings in consumer (and many other) products, and the growing use of water-base paints. Benzene is still used to some extent in commercial paint stripping operations, however, and lead is still used for its corrosion inhibition properties in noncommercial applications (on structures, e.g.).

Substitution of one pigmenting material for another appears always to be possible, but often to be impractical. Substitute pigments based on organic dyes, for instance, can nearly always take the place of the more colored toxic lead compounds but often at a higher cost and at the expense of reduced durability. Much the same can be said concerning efforts being made to eliminate use of colored pigments based on cadmium, chromium, antimony, and arsenic.

Reduced durability is only one problem associated with organic pigments; another is that, in contrast to some of the inorganics, they offer far less corrosion resistance. Corrosion inhibition has been the main use of zinc chromate pigment. Recently, metallic zinc powder has been substituted for zinc chromate, protecting by the galvanic action of the less toxic zinc powder rather than by the inhibiting action of the more toxic zinc chromate. This is an example of ingenuity at work.

Organic solvent substitution has taken place along two main lines. First, a real effort has been made to essentially eliminate use of benzene as a solvent in any application. Even though this material is an excellent solvent with many uses, it is far too toxic and hazardous for such uses to continue. Substitutes are always available that have far less toxicity and hazard, although often less solvency and perhaps somewhat greater cost for that reason alone. And, although never used to any great extent as a binder solvent, carbon tetrachloride has virtually been eliminated as a fat-grease solvent in cleaning applications because of its high toxicity.

The other line of attack on organic solvents has been mounted by the Environmental Protection Agency (EPA). Here, the problem has not been toxic effects on workers but rather, effects on the environment in general. Many of the hydrocarbons and their derivatives participate in the formation of photochemical smog and may interfere with formation or stability of the ozone layer in the stratosphere. EPA's first effort, then, was to reduce emissions of those materials that act most rapidly to produce smog, mainly unsaturated substances. However, because organic materials released to the atmosphere eventually become oxidized, all eventually participate in photochemical smog reactions. Therefore, the second phase of this attack has been aimed at reducing the totality of emissions of organic material to the air. To this end, pressure at reducing the amount of volatile organic material in coatings of all sorts has mounted. As a result largely of EPA action, the composition of coatings solvents has changed and the solids content of the coatings has increased. Both kinds of changes have required research. Neither kind of change, however, necessarily adds to the direct cost of the finished product. In fact, increasing the solids content of a formulation can result in substantial cost savings.

One of the prices paid for increasing the solids content of paints from the "normal" of about 20% to the "high solids" of about 60% is an increase in viscosity of the paint. To offset that increase, resins had to be developed with lower molecular weight, but, because long-chain molecules and cross linking were still required for longevity and stability of the final film, those resins had to be more reactive than formerly. This, in turn, caused a decrease in shelf life or a switch to two-component systems with their inherently greater health hazards for workers. In many cases, the reduced solvent vapor inhalation hazard as a result of lower solvent concentrations in the paint and consequently in the air was paid for by a much greater increase in the hazards associated with the binder (see Section 4.2.1).

As the solids content of the paint increases, so does the viscosity, even with lower-molecular-weight molecules in the binder system. This, in turn, calls for greater air pressures (for air guns) and, consequently, more overspray. More overspray increases losses, and, in addition, increases inhalation hazards. Those problems have caused increased use of heated paint systems and a trend toward increased use of airless spraying and electrostatic systems.

When the solids content reaches 100%, powder-coating technology comes into its own. Here, the solvent vapor inhalation problem has been eliminated entirely and the ultimate in heated paint spraying is attained with flame spraying. The largest user of this method is the appliance industry.

Rather than decreasing the concentration of organic solvents in paints, one can switch to water-based coatings. Although organic co-solvents are used to a certain extent in most water-based coatings, their concentration is kept low to reduce costs. These coatings have much lower health and safety hazards than either solvent-based or powder systems and partially for those reasons their use is on the increase.

EQUIPMENT. Substituting airless for air spray reduces the inhalation hazards associated with overspray as does the use of any electrostatic method when compared to normal systems. An air gun produces overspray that may average 50% of the amount of coating sprayed whereas airless techniques have reduced that amount dramatically. In an electrostatic system, the amount of overspray can be almost negligible. However, even though electrostatic processes can be used with success in many areas, they cannot be used in all. In particular, electrostatic forces mitigate against the coating of indentations just as they assure the coating of sharp edges.

With any system that uses solvent, less solvent is needed for thinning if the coating is heated. This in turn reduces exposures to the solvent and usually to overspray also because the viscosity reduction from heating calls for less spray pressure and consequently less bounce of spray from the substrate.

PROCESS. All methods of applying coatings other than spraying entail less exposure to the solvent and essentially no inhalation exposure to the solids, but these other methods are almost always either much slower than spraying or produce coatings that are unacceptable in appearance. Thus their use is quite limited.

4.3.2 Isolation. The ultimate in equipment substitution is reached with automatic spray where there is no exposure at all, the ultimate in isolation as a method of exposure control as well. Automatic spray systems are being installed in an ever-increasing number of factories not mainly to reduce health hazards, but to reduce costs. These systems are usually electrostatic so that a good coating is achieved even without the aim and skill of a human operator. Quite often, automatic spraying is used for the first coat(s), with the finish coat(s) being applied by an operator. This method is especially useful on an assembly line when items are painted as they pass by fixed or reciprocating spray guns. Another technique is to use robots. In either case, distance from the process is being used as a method of isolation.

Where the operator cannot be isolated by distance, other forms of isolation are usually practiced. Almost all painters wear protective clothing, for instance, to reduce skin contact with the materials they handle. Barrier creams are also used as are gloves (usually cloth) and safety glasses with or without side shields.

Isolating the respiratory tract from the environment is done with respirators of one sort or another. Although many types are available, the "painter's respirator" is most used. This is an air-purifying device, usually a half-mask having one or two organic vapor cartridges protected by filter pads that intercept the spray droplets. It is not at all unusual, however, to find respirators either chosen or used improperly, or both.

4.3.3 Ventilation. Instead of, or in addition to, other methods of controlling health hazards associated with coating, both general and local exhaust ventilation are used extensively. Ventilation as a control method is, in fact, so ubiquitous that when coating application is mentioned, spray booth probably comes to mind more often than any other control.

GENERAL VENTILATION. Figure 23.2 illustrates the correct use of general exhaust ventilation to aid in controlling exposures to coating components. The amount of air flow used is designed to assure that solvent vapor concentrations will not exceed the

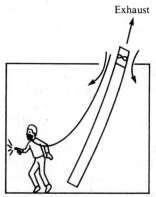

Exhaust

Note: Air supplied respirator must be worn.

Figure 23.2. Ventilation of enclosed spaces. (Courtesy NIOSH)

limits of the respirator worn by the worker. With air being exhausted at the floor level and being supplied from overhead, good mixing throughout the enclosed space is assured and vapor concentrations calculated from a knowledge of the rate of solvent usage will give an adequate estimate. Other configurations of exhaust and supply (typically using a pedestal fan in a doorway) almost guarantee much higher breathing zone concentrations than those calculated on the basis of good mixing within the enclosure.

When the spraying is done in an enclosed, isolated space such as a ship hold or storage tank, then the respirator used should be of the supplied-air variety. The minimum amount of air to exhaust for ventilation can be calculated based on the lower explosive limit of the solvent and its rate of use with the method to be found in the *ACGIH Industrial Ventilation Manual.*

LOCAL EXHAUST VENTILATION. Local exhaust ventilation hoods ("spray booths," "hoods") for coatings application are available commercially in many shapes, sizes, and method of trapping overspray or can be designed and/or constructed for specific purposes. They usually have configurations such as those illustrated in Fig. 23.3. Similar booths are used when the spraying is automatic.

Dry-type booths use filters to intercept and trap particles of overspray while water-wash booths use a flow of water over a solid surface to accomplish the same thing. Filters became clogged with time and must be replaced or the volume of air exhausted through the booth will diminish to the point where excessive amounts of overspray and/or solvent vapor reach the breathing zone of the worker and/or escape from the booth. Water-wash booths avoid that problem by substituting another, that of handling wet sludge. Either type booth can be used successfully in almost all applications.

Paint spray booths range in size from 2 or 3 ft² (0.2 m²) to many feet long and high. Objects as large as railroad cars may be sprayed in booths on a production basis. Air velocities and volumes used for overspray and vapor control vary with booth size; in general, the larger the booth, the lower the air velocity needed to achieve adequate control. As the mass of moving air increases, amounts of turbulence (caused by the act of spraying, e.g.) that are major for small masses become minor in their effect and, hence, lower velocities will achieve adequate overspray and vapor control.

When spray (or other local exhaust ventilation) booths are specified, the number used most often to describe adequacy is the "face velocity." As many spray booths do not have "faces," the term is ambiguous unless it is used to specify the air velocity in a plane with the spray gun(s) in the direction of the interceptor(s). With that definition, the same term can be used for both side draft and down draft booths.

The ACGIH Committee on Industrial Ventilation has specified face velocities for spray booths that vary with the method of spraying (air or airless) and with the size of the booth (smaller or larger than 4 ft² or "very large and deep"). OSHA, in general, uses 100–150 fpm (0.5–0.75 m/sec) for all large booths with the higher value being associated with cross-drafts of 100 fpm or more. Values for bench-type booths are 150–200 fpm, again depending on the velocity of cross drafts. The ACGIH method results in Table 23.6.

A method for calculating a recommended face velocity specific for individual booths of all kinds that embodies the same consideraitons was published many years ago and is equally valid today. That method uses three factors in an equation to calculate a face velocity designed to avoid hazardous exposures of workers. The vapor control factor, M, varies from 50 to 100 depending on the hazard of the material(s) handled;

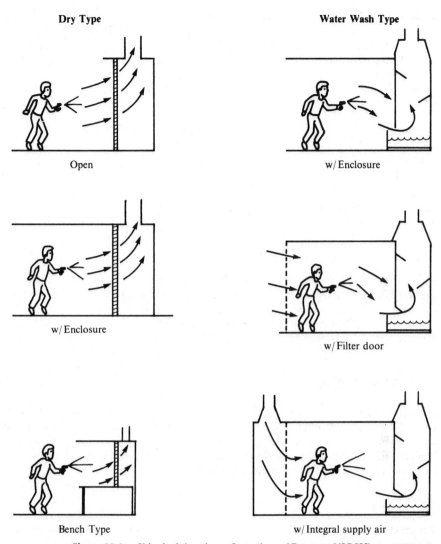

Figure 23.3. Side-draft booth configurations. (Courtesy NIOSH)

for possible carcinogenic material, the highest value is appropriate; for run-of-the-mill paints, 50 or 75 would be appropriate. An environment factor, E, is used to approximate the disturbance caused by cross drafts, foot traffic, and so forth. This is where the effect of air versus airless spraying should be taken into account; for air guns, this factor should be 0.5–1.0; for airless, 0.5 or less. A hood characteristic factor, C, is determined from a graph and is based on the size and shape of the booth face. Finally, the actual or estimated variance of face velocity data is used to account for air turbulence. For almost all paint spray booths, this figure can be estimated to be on the order of 150–360 with the lower value being used when the booth has an open

Table 23.6. Face Velocities for Spray Booths

Type	Air	Airless
Bench-type spray booth	150[a]–200[b]	100[a]–125[b]
Large spray booth		
Walk-in	100[c]	60
Operator outside	100–150	60–100
Auto spray paint booth	100	60

[a]Booth cross section less than 4 ft^2.
[b]Booth cross section more than 4 ft^2.
[c]75 for very large, deep booths.

face and is equipped with good external baffles (a rarity in spray booths). These values are then combined:

$$V_r = 0.75M(1.0 + E + C) + 2(s^2)^{0.5}$$

where V_r is the recommended face velocity in fpm, and s^2 is the variance (square of the standard deviation of individual measurements) of face velocity data. For average paint ($M = 50$), air spray with negligible external air disturbances ($E = 0.5$), use of this formula results in a V_r of 130 fpm for a 4 ft^2 booth will no baffles and a V_r of 95 fpm for a 10 by 20 ft open-face booth similarly unbaffled.

5 POSTMATERIALS TREATMENT

If the coating has been applied with a solvent-based system, then, after application that solvent must evaporate, either at room temperature or at an elevated temperature (oven drying). Convertible binder systems require further reaction, either with oxygen from the air or polymerization within the film, itself. All such reactions are accelerated by a temperature increase; baking is common.

5.1 Air Dry

Air drying has several advantages. First, it requires no use of energy. Second, heating of neither the film nor the object is required, and, therefore, the coated object can be made of temperature-sensitive materials. Third, because no oven is necessary, there is less need for extra equipment to accomplish the job; some equipment is usually still required if only to keep coated objects from being touched until the film is dry. Finally, air-dry coatings are usually less expensive than those that require heat after application.

The chief disadvantage of air drying is that this process almost always requires considerable amounts of time. That, in turn, has several consequences. If objects are being coated on an assembly line as is most common, that line must be set up to have sufficient inventory time for newly coated objects to dry completely before being taken from the line. This may call for extraordinarily long conveyor systems. Just-coated objects are sticky and the longer they stay sticky, the greater the chance that airborne substances will be trapped in the coating, perhaps to detract substantially from its appearance. Fresh coating films are quite vulnerable in other ways; a casual touch from a finger or a coat sleeve may result in the object being put through a coating-removal step (abrasive blasting is common) to then be recoated.

Finally, in most air drying systems there is absolutely no control over the solvent vapor evaporated from the film. This may be a distinct disadvantage or advantage, depending on how "fugitive" emissions are viewed by governmental agencies at the time.

5.2 Baking

Baking coated objects has many advantages. Coated objects in an oven are protected against casual contact and, if air entering the oven is filtered, even against dust and other airborne contaminants. Baking is a rapid process when compared to air drying, and thus inventory lines can be shorter and much more compact. Baking enamels may be considerably more durable than air-dry enamels or paints because of the nature of their binders. Finally, baking allows essentially complete control over solvent emissions, and such control may be either necessary because of governmental regulations or desirable from a community relations standpoint.

The chief disadvantages of oven drying or baking are (1) the capital equipment required, (2) the need to heat the film (infrared) and/or the object, (3) the generally greater toxicity and hazards associated with the reactive resins used for baked binders, and (4) that careful control of the process is required. That the oven stack becomes a "point source" of volatile organic compound emissions may be either an advantage or a disadvantage; this method does allow easy treatment of the contaminated air if necessary.

5.3 Health Hazards

The only health hazards associated with either air drying or baking relate to solvent vapor exposure. Ovens and other baking equipment can usually be vented to the out-of-doors rather easily, thus eliminating a possible in-plant problem. Air drying, on the other hand, may assure that there will be nonnegligible solvent vapor exposures within the plant.

5.3.1 Health Hazard Controls. If baking causes a problem, treatment of contaminated air exiting from the oven is probably necessary. Because that air is almost always warm, if not hot, thermal or catalytic decomposition of the organic materials present may be used with success. Sorption processes are available but are rarely used, probably because of the relatively low concentration of organic material in the effluent.

Air drying may be done in a ventilated enclosure rather than in the general plant atmosphere, especially if vapor released from coated objects is viewed as or found to be potentially hazardous. Otherwise, the usual control is simply that of general plant ventilation, which may be exhaust, supply, or both.

6 INSPECTION

Coated objects may undergo little or no inspection if the quality of the coating is relatively unimportant or may be subject to minute inspection typical of that used in automobile finishing lines. In either case, inspection is usually almost always completely visual and perhaps tactile. No health hazards have been associated with this portion of the painting/coating process.

7 GENERAL HAZARDS AND CONTROLS

7.1 Safety

Fire and explosion is an ever-present concern in most paint making operations involving volatile solvents. Grounding and bonding must be employed to prevent buildup of static electricity during all transfer procedures.

7.2 Housekeeping

Housekeeping is probably the single most important factor in hazard prevention. Batch operations such as paint and resin manufacture require continual attention to this point. Process controls in the form of effective ventilation systems are also required. It is interesting to note that the *Industrial Ventilation Manual* does not address any process equipment used in paint or resin manufacture. The industrial hygienist and ventilation engineer must collaborate closely in ventilation system design. Success has been achieved in integrating bag disposal with ventilation to prevent hazards associated with elimination of empty bags. This strategy employs a hood at the emission point leading to a duct, which is sized to convey empty bags. The bags are separated into trash compactors with dust and vapor removed from the airstream by baghouses and other emission control devices.

7.3 Waste Disposal and Segregation

Waste disposal requiring segregation of wastes is an important part of the resin and paint manufacturing process. Vigilance is required to identify the type of wastes to avoid unknown exposures.

7.4 Training and Education

Hazard identification and communication is a challenge in the paint and coatings industry where large numbers of materials are found. It is vital that materials that appear to be similar physically are identified as to their potential hazards. A simple system, the Hazardous Materials identification System, has been devised in the paint and coatings industry and is deemed to be appropriate for inplant hazard communication purposes under the Occupational Safety and Health Administration Hazard Communication Standard when coupled with material safety data sheets and training programs.

GENERAL REFERENCES

Burgess, W. A., *Recognition of Health Hazards in Industry. A Review of Materials and Processes* New York: Wiley-Interscience, 1981.

Chemical Process Industries, 4th ed, R. N. Schreve and J. A. Brink, Jr., eds., New York: McGraw Hill, 1977.

Industrial Ventilation: A Manual of Recommended Practice, 19th ed, American Conference of Governmental Industrial Hygienist, Lansing, MI: Edwards Brothers, 1986.

Keller, L. W., Schaper, K. L. and Johnson C. D., A Hazardous Materials Identification System for the Coatings and Resin Industry, *Am. Indus. Hyg. Assoc. J.,* 41:901–907, 1980.

Paint/Coatings Dictionary, Federation of Societies for Coatings Technology, 1978.

Peterson, J. E., An Approach to a Rational Method of Recommending Face Velocities for Laboratory Hoods. *Am. Indus. Hyg. Assoc J.,* 20:259–266, 1959.

Recommendations for Control of Occupational Safety and Health Hazards . . . Manufacturer of Paint and Allied Coatings Products, U.S. Department of Health and Human Services, Public Health Service Centers for Disease Control, National Institute for Occupational Safety and Health, Division of Standards Development and Technology Transfer, September, 1984.

Surface Coatings, Vol. I and II; Oil and Colour Chemists' Association, Australia, London: Chapman and Hall, 1983; NPCA Data Bank, National Paint and Coatings Association, Washington, D.C.:, 1985.

Tasca, C. and Colombi, A. M. D, private communication, 1988.

CHAPTER TWENTY-FOUR

Pulverizing and Micronizing

Paul F. Woolrich

1 INTRODUCTION

Terms often used interchangeably to describe size reduction of particulates in industry include crushing, grinding, pulverizing, cutting, disintegrating, powdering, comminuting, milling, shreding, micronizing, and micropulverizing.

Particles can be broken down by being squeezed, hammered or cut, crushed, rolled, pressed, ground, or by particle impact. In the chemical and other industries, the method of pulverizing and micronizing depends on the ultimate use and appearance of the product.

- Better blending is achieved if particles are smaller. In the pharmaceutical industry, particles are micronized to prevent streaking and spotting and to ensure that each tablet contains accurate and uniform dosage.
- Pulverizing and micronizing produces finer particles that are more soluble.
- Fine particles make better suspensions and slurries, and as such can be pumped more efficiently, reducing handling.
- There is more intimate contact (increased surface area) in chemical reactions if materials are finer.

2 TYPES OF PULVERIZERS AND MICRONIZERS

The most common type of mill for pulverizing and micronizing utilzes the hammer or cut principle. The Fitzpatrick (Fitzmill), the Microatomizer, and the Micropulverizer fall into this category.

The speed of pulverization is dependent not only on the rate of feed and the speed of the hammers but also on the size of the particles desired. Finer particles, hard or nonfragile materials, and sticky materials take longer.

The hammer mill rotor functions as an inefficient blower, drawing air into the feed inlet and discharging air at the product outlet.

Ball or rod mills (the material is rolled) are generally used for wet grinding. In wet ball grinding, particles do not agglomerate and pack together, permitting more intimate grinding. Wet milling is appropriate when it is desired to grind the material in the vehicle or solvent intended as the dispersant. Paint pigments are ground in linseed oil in a ball or roll mill. Pharmaceuticals are sometimes milled in oils in a device known as an attritor, producing a product containing product particles less than 10 microns in size.

Other wet or fluid milling can be accomplished in a device consisting of a 2- or 3-in. (7-cm) tubular "racetrack." High-pressure air causes the particles to collide against one another and the walls of the racetrack. Centrifugal force classifies the size fractions, the larger particles remaining in the track until broken up and inertially separated at the outlet. A product such as vitamins containing 2% moisture will be dried to 0.5% moisture due to the effect of the dry airstream. Such pulverizers or micronizers are useful for ultrafine grinding.

3 INDUSTRIAL HYGIENE CONTROL

The principal industrial hygiene problems associated with pulverizing and micronizing are dust control andl noise. Control of dust will be dealt with in this chapter.

Hammer mills are different than grinding and crushing mills, used for crushing in the heavy-chemical industry. Grinders tend to overheat inasmuch as chemicals block the grinder teeth. This is sometimes remedied by restricting their use to high-melting-point chemicals or minerals and even cooling the housing. The hammer mill acts as a fan and entrains air during grinding. The entrained air may keep the product cool and prevent blockage and fires. The solid particles in this entrained air may constitute a health problem, and it is often necessary to vent this air and collect the dust it contains. An alternative is to discharge the hammer mill into an airtight hopper.

Some micropulverizers are equipped with a unit bag filter mounted directly to the mill discharge. This unit collector works satisfactorily but requires head room that may not be available. Other micropulverizers are used inside HEPA-filtered laminar air flow modules.

Cage-type mills also entrain a considerable air volume and may present a problem due to pressure on the discharge of the mill. The dust control solution to this type of problem is minimizing the air entrainment by providing a positive sealed feed to the mill.

In some cases, the material is ground directly into a blender. A tight connection to the blender is required, and the dust-laden air may be vented from the mill housing through a stocking filter without auxiliary air moving equipment. These are usually batch operations common to fine chemical and pharmaceutical industries wherein only a small amount of material is ground at any one time and the equipment is used for a variety of products.

In cases where the material is ground into a storage bin or hopper, the air may be vented through a stocking filter from the mill housing, storage bin, or hopper either without auxiliary air movement equipment or incorporated in the main collection system for control of drum dumping and loading. The face velocity at the hood opening should be 125–200 fpm (38–60 mpm); the duct to fan velocity is 3500–4000 fpm (1000 mpm).

Slot-type ventiliation has been found to be effective for dust control during scooping and shoveling of solid material from drums. A common design consists of a portable hood constructed with a flanged circular slot that can be clipped to the drum top to exhaust the dust radially.

The ball mill is used in the chemical industry both for grinding and for processing. Ball mills may be loaded and discharged through a side opening like a manhole or more commonly thorugh trunnions (hollow shapes) at each end. When the ball mill is used for processing purposes, gases and vapors may be liberated, ventilating through the trunnion. Except when toxic materials are encountered, it is usually sufficient to vent the trunnion to the atmosphere by means of a pipe vent unless there is a possibility of solids blocking it, in which case the trunnion must be left open for cleaning purposes. In ventilation for dust control of the ball mill trunnion, the hood enclosure must provide access for rodding (trunnion cleaning) purposes, and an additional volume of air must be exhausted to compensate for the volume of gases or vapors evolved during the process.

Typical ventilation requirements for ball mill operations, thus include:

- Exhaust hood ventilation for dumping solid material into loading hopper. Flexible connection to ball mill manhole.
- Exhaust enclosure for trunnion. Sufficient air volume to maintain 100 fpm (30 mpm) across opening plus volume of gas or vapor evolved.
- Exhaust hood ventilation for drum filling. Flexible connection to ball mill manhole.
- The drum dumping hood and drum loading hood will be used intermittently so the system volume requirements may be determined by the maximum working opening on one hood, plus the volume requirements of the other branches. A control velocity of 125–200 fpm (38–60 mpm) should be maintained across working openings.

The Fitzpatrick (Fitz) mill is commonly used in the fine chemical and pharmaceutical industries where more than one product is processed or where a fixed charging station is impractical. Slot-type ventilation for dust control at the Fitzmill feed trough or tray while material to be processed is scooped or dumped into the feed trough is quite effective. The slot hood should be portable and easily detachable to accommodate various charging stations and cleaning of the mill between lots or batches. When utilizing ventilation on the Fitzpatrick mill during charging, valuable dust can be collected on fabric filters and salvaged.

4 INDUSTRIAL HYGIENE PRACTICE FOR MICRONIZING VERY TOXIC OR POTENT MATERIAL

The following precautions should be instituted:

- Direct contact with dust should be avoided.
- Atmospheric contamination due to dust or solvent vapors should be kept to a minimum.
- Wherever possible, the dustiest processes should be automated and enclosed.
- Dusty operations should be carried out in sealed chambers, rooms or boxes under negative pressure, or on benches fitted with local exhaust ventilation.
- Dust deposits should be collected by vacuum cleaning.
- Workers should wear protective outer garments, head protection, and hand protection.
- To prevent contamination of the work premises, the walls and equipment should be washed with appropriate solutions or detergents.
- Requirements for employees:
  - Wear a clean change of clothing each day.
  - Remove and leave protective clothing and equipment when leaving the work area at the end of the day.
  - Wash hands and face after removing the clothing and equipment and before engaging in other activities.
  - Should receive training and indoctrination.

The toxic or potent material micronizing module described here is designed to:

- Protect personnel from the pharmacological, toxicological, and allergenic effects of the dust of active substances carried by the air.
- Protect the environment, the surrounding equipment, and products from contamination and cross contamination.
- Protect the processed product from cross contamination through surrounding activities and products.

4.1 Module Description

The module is served by a separate heating, ventilating, and air-conditioning system. This system includes roughing filters, recirculating air HEPA filters, air volume monitor, and control device, supply fan with inlet vanes, heating and chilled water cooling coils, a duct distribution system, and final supply HEPA filters. The supply air is delivered into a HEPA filter supply ceiling system with sufficient air flow volume and velocity to provide a laminar flow of air from the entire ceiling, down across the work area, across dust-producing equipment and into roughing return air filters. The air system filtration then consists of throwaway-type roughing filters at the wall of the room, a bank of HEPA (99.9% DOP) prefilters and a bank of HEPA (99.97% DOP) final filters. Approximately 90% of the total air can be recirculated, 10% will come from outdoor air. An exhaust air system provides negative pressurization for the processing and support rooms.

4.2 Operation and Control Sequence

The recirculating and exhaust systems operate continuously. The desired total air quantity is controlled by an air volume monitoring device controlling fan inlet vanes. This controller will automatically modulate the inlet vanes. As filter resistance increases, the air monitor control will maintain the desired total cfm. The total air quantity per system is approximately 6000 cfm (170 m^3pm). The system is capable of being operated at 100% exhaust, 100% makeup, and no return air, if required for short periods of time.

4.3 Operator Protection

The room occupant or operator is fully suited in a disposable protective suit with air supply breathing hood. The filtered laminar flow air supplied through the ceiling drives dust generated by the operation down and into the wall filter system. Operations, maintenance, filter service, and room cleanup is done in the suited air supply hood. The operator will degown, dispose of the suit, and shower before exiting the module.

GENERAL REFERENCE

Industrial Ventilation, a Manual of Practice, 19th ed., American Conference of Governmental Industrial Hygienists, Committee on Industrial Ventilation, Lansing, MI: Edwards Bros., 1986.

Spray Drying

Paul F. Woolrich

1 INTRODUCTION

Spray drying is a means of producing a dry granular or powdered product from a solution or slurry. Spray drying is a process in which the solution or slurry is reduced to a fine spray, mixed with a flow of hot gas, and the resulting dried powder separated from the gas. The hot gas supplies the heat of evaporation and carries off the moisture. Drying is virtually instantaneous, with the elapsed time for the entire operation being 5–30 sec.

The drying functions may include moving the air, cleaning the air, heating the air, atomizing the liquid, mixing the liquid in hot air, removing the dry material from the air, additional drying of the product, cooling the product, pulverizing, and sizing the product.

2 CLASSIFICATION OF SPRAY DRYERS

Spray dryers may be classified according to:

- Method of atomizing liquid or slurry
  High-pressure nozzles

493

Centrifugal spinning disks
Two fluid systems—air, steam
- Heat sources
  Steam
  Gas
  Fuel oil
  Electricity
- Method of heating air
  Direct—gas or fuel oil
  Indirect—heat exchanger plates or coils (closed-cycle systems)
- Position of drying chamber
  Vertical
  Horizontal
- Number of drying chambers
- Direction of gas flow
  Countercurrent
  Parallel
  Right angle
- Pressure in dryer (position of fan)
  Atmospheric (slight pressure)
  Vacuum
- Method of separating product from gas
  Cyclone
  Multiclone
  Bag filter
  Electrostatic precipitators
- Treatment and movement of gas
  Recirculation
  Used and exhausted
- Removal of powder from drying chamber
  Conveyor
  Sweep conveyor
  Conveyed to inertial collector
- Method of heat transfer
  Convection
  Radiation
- Atmosphere in dryer
  Nitrogen
  Air
  Other inert gas
- Direction of air flow
  Updraft
  Downdraft
  Horizontal
  Mixed
- Shape of drying chamber
  Silo

Box
Tear-drop or conical
• Product being dried

3 ADVANTAGES AND DISADVANTAGES

The principle advantage of spray drying relates to its overall economy of operation to produce a desirable dried product. This method of drying can be applied to any liquid (even pastes) that can be pumped. Some of the characteristics of spray drying are such that what may be an advantage for one material or process may be a disadvantage for another. The variables in spray drying are so complex as to preclude establishing final design methods; even closely related chemicals do not act in a similar way. It is prudent to develop designs on most products only after performance tests are completed.

3.1 Characteristics

The following characteristics of spray drying determine whether the process is an advantage or disadvantage for a particular material:

• Single-step operation occurs from a liquid feed to dry product.
• The process is continuous or a batch operation.
• Maintenance costs are low.
• Labor costs are low (only one operator required).
• Materials spray-dry to uniform spherical particles.
• Has low bulk density because of porous or hollow nature of particles.
• Has higher rate of solubility because of porous or hollow particles.
• Contamination of products minimized.
• Low hold up of materials in equipment.
• Can be started up and shut down quickly.
• Can be adapted to a closed cycle.
• A spray dryer designed for specific purpose cannot be used for every other product. Some materials require the design only after performance testing.
• It is possible to coat, encapsulate, or prill materials.
• It is not amenable to drying volatile solvents.

4 APPLICATIONS

Spray dryers have been successfully applied for a wide range of processes and products:

• Milk and food products
• Detergents and stain removers
• Heat-sensitive materials
• Clay, silicates, ceramics, and phosphates

- Bleaches, sulfite liquor
- Alumina

5   INDUSTRIAL HYGIENE AND THE SPRAY DRYER

From the industrial hygiene point of view, it is preferable to have the main blower draw air through the system, creating a negative pressure. This prevents escape of dust from leaks in the system and permits ports to be opened for inspection. Conversely, the main disadvantages of a pressure operation is dusting and resulting loss of product through leakage, and the dryer interior cannot be conveniently viewed. Almost without exception, spray dryers should operate without product buildup on the chamber walls. The flow characteristics of a few materials are such that they do not adhere to equipment surfaces, even at a high moisture content. Others can be operated economically with high feed rates at the expense of wall buildup and resulting shutdown for cleanup. If, however, the product is toxic or noted for its skin or respiratory sensitization characteristics, exposures during cleanup can be a serious problem.

5.1   Preloading and Loading

The material to be charged by fine spray into the spray dryer usually presents no significant industrial hygiene problem since it is a liquid or slurry handled in a closed system. The application of the dryer does not include volatile solvents, so vapor exposure is minimal. Handling a liquid that has lung or skin sensitization characteristics can create obvious problems, requiring the use of engineering controls as well as personal protective equipment and clothing.

5.2   Drying Cycle

During the actual drying cycle, which consists of mixing the atomized liquid with hot air in the chamber, problems arise only from leakage in the chamber. This can be minimized by having the main blower draw air through the system to create a negative pressure in the chamber.

   All flame safety controls needed for a furnace are required by governmental or insurance safety codes. It is necessary to interlock safety and operating controls to prevent improper starting, to shut down equipment in case of furnace failure, accidental failure of feed supply, and power failure to the blower.

5.3   Cleanout

The viscosity and flow characteristics of some proteinaceous materials are such that the material adheres to the chamber walls, and thus cleanout becomes necessary.

   Entering a spray dryer chamber is ""confined-space entry." The term confined or enclosed space means any space having a limited egress that is subject to the accumulation of toxic or flammable contaminants or the development of an oxygen-deficient atmosphere. The Occupational Safety and Health Administration (OSHA) regulations apply to work in dangerous or potentially dangerous confined or enclosed spaces. These regulations address air quality in the worker's breathing zone and are summarized as follows:

- Testing of the atmosphere for presence of oxygen and absence of toxic or combustible gases.
- Establishment of emergency rescue procedures.
- Completion and posting of tank or vessel entry permit.
- Lockout or blankout of electrical circuits and piping.
- Placement of standby personnel to observe the worker(s) in the vessel.

If difficult to handle proteinaceous material is being removed from the chamber, additional precautions must be taken because of the physiologic effects of such materials. Human skin and mucous membranes can react to these proteins when in the concentrated state, and these reactions have occurred in a significant number of instances in plants under conditions permitting exposure. Employees may develop an irritation of the exposed skin and mucous membranes, while some highly susceptible individuals could develop a lung reaction not unlike hay fever allergy to pollen—an effect that may intensify as exposure continues. It is this latter physiologic effect that has prompted concentration of industrial hygiene effort toward reduction of exposures. This effort has resulted in the development of engineering control and personal hygiene guidelines as well as studies aimed at utilization of premix, prells, or slurries in lieu of powders as the final product form.

Employees who handle these materials or are involved in chamber cleanout should be equipped with the following clothing and equipment:

- Coveralls and rubber, cotton, or plastic gloves taped at the wrist or of elbow length.
- Properly fitted and approved respirator or air-supplied hood.
- Goggles.
- Boots.

For additional information, see Chapter 3, Section 5.6, and Chapter 11, Section 4.1.

5.4 Barrel, Drum, and Bag Packing

The packaging of finished products into containers is often accompanied by emissions of dust and vapor that may constitute a health problem. Methods for control of toxic materials liberated are discussed next.

The most effective method of controlling dust evolved during packaging of dry solids, the hooded enclosure, may also be utilized for weighing the product, weight adjustment, and the like. Control velocities of 125–200 fpm (38–60 mpm) should be maintained across the observation and working openings, and transport velocities of 3500–4000 fpm (1100 mpm) should be maintained in the duct work.

Many chemical products are packed in open-top bags from a weigh hopper or by chute utilizing a rotary valve or slide gate. Weight adjustment is often necessary. After weight adjustment the bags are closed by sewing or by wire fasteners. A health or nuisance problem may exist during the filling, check weighing, and the bag closing operations.

Several types of bag holders are used in filling open-top bags. These holders are not dust tight, however. Dust from these operations is best controlled by the use of a hooded enclosure. Control velocities of 125–200 fpm (38–60 mpm) should be maintained across the open face of the hoods and transport velocities of 3500–4000 fpm (1100 mpm) should be maintained in the duct work.

5.5   Contract Operations

An important consideration for spray dryer operators is the underwriting difficulties that spray drying of toxic and/or sensitizing chemicals presents for insurability.

The usual insurance coverage for specialty chemical companies is not only for Workman's Compensation but also *Comprehensive General Liability Including Products*. Thus, insurance underwriters are sensitive to their insured's legal liability relative to individual pollution and production and product processing negligence that allegedly results in contamination or injury to the general public.

Underwriters may refuse to provide catastrophe-type coverage for pollution claims that may be alleged.

For additional information, see Chapter 5.

GENERAL REFERENCES

Kent, J. A. (ed). *Riegel's Handbook of Industrial Chemistry,* 7th Ed, New York: Van Nostrand Reinnard, 1974.

Kirk–Othmer (eds.), *Encyclopedia of Chemical Technology,* New York: Wiley-Interscience 1964.

Perry, R. H., C. H. Chilton, *Chemical Engineer's Handbook,* 5th ed., New York: McGraw-Hill, 1973.

CHAPTER TWENTY-SIX

Vacuum Drying and Freeze Drying

Paul F. Woolrich

1   INTRODUCTION

Drying under vacuum is employed extensively in the chemical, pharmaceutical, food, ceramic, and even electronic and precision instrument industries. Vacuum drying implies that the removal of moisture is carried out in the drying chamber below atmospheric pressure. Since the boiling point of a liquid or solvent varies with the absolute pressure, it is possible to provide low-temperature evaporation from the product and low product temperature.

Vacuum drying is employed for materials whose essential characteristics would be damaged or altered by exposure to high temperature or atmospheric conditions. Materials that ignite, explode, or deteriorate in the presence of air require vacuum or inert gas drying conditions. Materials that contain solvents of appreciable value warrant recovery and are particularly suited for vacuum drying. Vacuum drying also reduces exposure to plant personnel from toxic liquids being evaporated.

2 TYPES OF VACUUM DRYERS

2.1 Tray or Shelf Dryer

Tray or shelf drying (Fig. 26.1) is a batch operation suited for small quantities of product. The drying is relatively slow, and the labor cost of loading and unloading the trays is high.

The vacuum shelf dryer consists of a chamber with a full-opening front door and a suitably sized vacuum port for vapor removal. The proper sizing of the vacuum line is critical since the inability to remove the moisture with a minimal pressure drop will raise the absolute pressure in the chamber to cause overheating of the product.

The dryer shelf construction must provide unrestricted flow of the heating medium to provide uniform drying, and the chamber must be vacuum tight. Air in leakage will create high velocities that blow the dried product from the trays and cause oxidation of the product. Aside from a well-insulated chamber housing, the dryer consists of fans, heating means, and a support arrangement for the material. These dryers are frequently used in the chemical industry for small batch operations as they are efficient and permit close control of drying. The most frequent application of tray or shelf dryers is for drying:

- Granular, paste, slurry, and liquid chemicals.
- Dyes and pigments.
- Pharmaceuticals.
- Ceramics.
- Other chemical solids.

2.2 Cabinet or Compartment Dryer

Cabinet or compartment dryers are similar to the tray dryer without the shelves or supporting structure for trays. They are used for batches of larger or irregular shapes. The heating, temperature, and air circulation problems are similar to those of the tray or shelf dryer. A unit with a single drying chamber is known as a cabinet dryer; with multiple chambers it is known as a compartment.

2.3 Tray-Truck Dryer

Tray-truck dryers are used for larger capacities than tray, shelf, or cabinet dryers (Fig. 26.2). Multiple tiers of trays can be loaded on a truck frame, and the truck is pushed

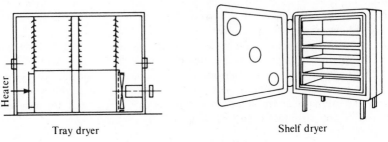

Tray dryer   Shelf dryer

Figure 26.1.   Tray and shelf dryers.

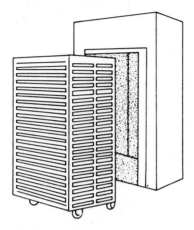

Figure 26.2. Truck dryer.

into the dryer. The trucks may be handled by a monorail system, run on tracks, or rolled into place with flat swivel wheels as directed by guides.

Products dried in tray-truck dryers include the same chemicals dried in tray or shelf dryers as well as:

- Vegetables and fruits.
- Large ceramic pieces.
- Rayon skeins or cakes.
- Painted objects.
- Lumber in racks.
- Heavy chemicals.
- Processed food products.

3 VACUUM FREEZE DRYING

Vacuum freeze drying is vacuum drying at controlled temperatures and pressures utilizing the process of sublimation, that is, conversion from solid to vapor without melting. The absolute pressure in the chamber and the heat input are controlled so that the solvent (normally water) is evaporated from the frozen state.

Freeze drying is utilized in the food and pharmaceutical industries to:

- Avoid chemical change in heat-labile components.
- Preserve food flavor.
- Prevent oxidation.
- Provide entirely reversible rehydration.

The disadvantages are the high cost of the equipment and the relatively low productivity of each machine. Freeze drying costs about 5 to 10 times conventional drying for foods and pharmaceuticals.

Figure 26.3. Pharmaceutical dryer.

Freeze drying is complimentary (not competitive) with quick or flash freezing. With many products, the freeze dehydration is just an extension of quick or flash freezing.

3.1 Freeze-Dry Process

Freezing can be done in a separate or the same chamber in which drying is done. Rapid freezing ($-20°F$ or $-29°C$ and lower) is desirable to provide formation of small ice crystals because small crystals provide least change in the properties of the product and a product that will reconstitute more easily. Once the frozen product is in the drying chamber, the chamber is evacuated; heat is introduced by electric heating platens, which can be decreased in temperature as drying progresses. The air and vapor are passed over a refrigerated coil to condense the vapor in the vacuum system.

The freeze-drying of pharmaceuticals is normally conducted in small batches on horizontally mounted shelves spaced to accept a range of vial or tray sizes (Fig. 26.3). The medical profession uses a great number of freeze-dried products including viruses, serums, bacterial sources, antibotics, bone grafts, vital body tissue, and other freeze-dried biological materials.

4 INDUSTRIAL HYGIENE CONTROL

The major health problems associated with vacuum drying occur when the dried material is dumped into drums or hoppers. To some extent, this is also the case with freeze drying, but freeze-dried granules or prills do not dust as much as powders. The principal problem associated with freeze drying relates more to good manufacturing practice that prevents cross contamination of products during the freeze-dry operations.

The loading of filtered materials containing residual solvent vapors into trays for drying may also constitute a health problem. Beyond tray loading, the solvents present a minimal problem in-plant due to the closed process. However, solvent vapors extracted during the drying operation may require removal prior to exhausting to the outside environment.

4.1 Ventilation for Dryer Tray Loading

It is desirable to perform the tray loading operation close to the filtration operation. It may be possible to drop the filtered material into a hopper beneath the filter. The loading of trays and the placing of trays onto the dryer trucks may then be performed under hooded enclosures with mechanical exhaust ventilation. Control velocities of 100 fpm (30 mpm) minimum should be maintained across the open face of the hoods, and transport velocities of 2000 fpm (610 mpm) minimum should be maintained in the system ductwork.

It is often necessary, however, to perform the tray loading operation at a location distant from filtration operations. The material is usually transported to the loading point in barrels or tote boxes. Figures 26.4 and 26.5 show typical hood enclosures for loading trays and placing trays into dryer trucks. Where a tray dryer is used, the tray loading station should be located adjacent to the dryer.

Figure 26.4 shows a design of a dryer tray loading station for the control of solvent vapors evolving from filtered material to be loaded into trays for drying. The drum or tote box of filtered material is placed in the enclosure through a hinged door. The material is shoveled or scooped into the trays within the hood, and the trays subsequently loaded directly into the dryer. A control velocity of 100 fpm (30 mpm) is maintained across all working openings.

For valuable products, the container needs to be up-ended and completely emptied to the tray. If the drum or tote box are more than 12–18 in. (30–45 cm) deep, the operator will not be able to scoop from the bottom of the container, and means must be provided to up-end the container for emptying. If the container and contents weigh more than about 20 lb, a mechanical means for up-ending is necessary. If up-ending is manual, the face of the hood needs to be designed so that the operator can perform this task without putting his or her head into the hood. Small pneumatic conveying systems are commercially available and may be quite useful for such transfers if such handling does not damage the product.

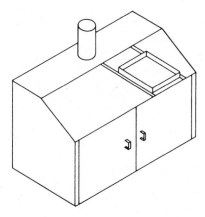

Figure 26.4. Dryer tray loading station.

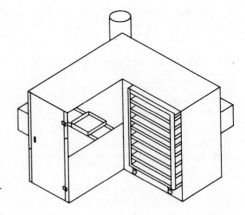

Figure 26.5. Truck dryer tray loading station.

Figure 26.5 shows a typical dryer loading station to control residual solvent vapors from materials to be loaded into trays for drying. As in Fig. 26.4, the drum or tote box of filtered material is placed in the enclosure through a hinged door, and the material is shoveled or scooped into trays that are then loaded onto the dryer truck. A control velocity of 100 fpm (30 mpm) is maintained across all working openings.

4.2 Ventilation for Dryer Pan Dumping

The dust arising from dryer tray dumping operations may require control utilizing mechanically exhausted tray dumping and dust collection systems. Figures 26.6 and 26.7 show typical tray dumping hood enclosures. Control velocities of 125–200 fpm (38–60 mpm) are maintained across the maximum working opening, and a transport velocity of 3500–4000 fpm (1200 mpm) is maintained in the exhaust duct work. A bag collector can be utilized in conjunction with the hood and exhaust system. The recommended filter rate is 3–4 cfm (0.1 m³) per square foot (0.09 m²) of filter bag area.

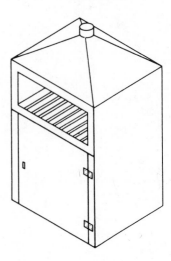

Figure 26.6. Typical pan dumping hood.

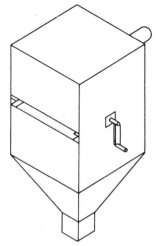

Figure 26.7. Mechanically exhausted pan dumper.

Figure 26.6 shows a typical pan dumping hood used in conjunction with a unit collector. The dryer pans are up-ended inside the hood enclosure, and the material is collected in drums located underneath the grate. This pan dumper may be adapted to feed directly into a hopper. A roller conveyor installed in the enclosure facilitates the removal of the filled container.

Figure 26.7 shows an alternate design of a mechanically exhausted pan dumper to control dust. In this design, the pan is inserted through a slot onto an angle frame, and the pan is turned mechanically using a handle. A smaller air volume is required for control as only a minimum opening is required to insert the dryer pans. Cross drafts are also minimized.

This design may be used for dumping into a mill hopper as shown, or modified for dumping into bins or drums. The control velocity of 125–200 fpm (38 mpm) across the working openings and 3500–4000 fpm (1200 mpm) duct velocity should be maintained.

GENERAL REFERENCES

Kent, J. A. (ed.), *Riegel's Handbook of Industrial Chemistry,* 7th ed., New York: Van Nostrand Reinhard, 1974.

Kirk–Othmer (eds.), *Encyclopedia of Chemical Technology,* New York: Wiley-Interscience 1964.

Perry, R. H., and C. H. Chilton, *Chemical Engineer's Handbook,* 5th ed, New York: McGraw-Hill, 1973.

CHAPTER TWENTY-SEVEN

Welding Operations

Thomas J. Slavin

1 WELDING PROCESSES

Welding is a process of joining metals by fusion in order to create a bond that is as strong as the parent materials and that has the same properties. Welding usually

Master Chart of Welding and Allied Processes

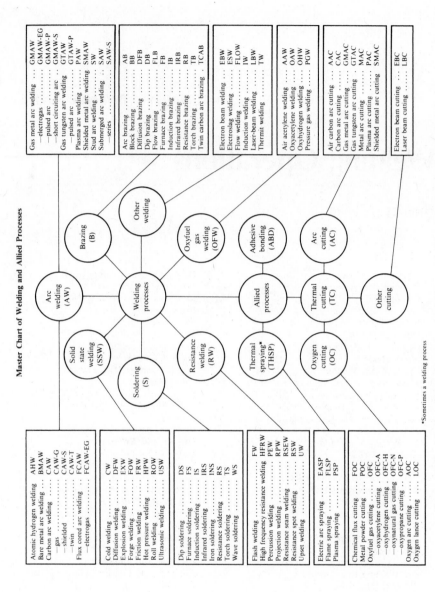

Gas metal arc welding ... GMAW
—electrogas .... GMAW-EG
—pulsed arc .... GMAW-P
—short circuiting arc .. GMAW-S
Gas tungsten arc welding . GTAW
—pulsed arc .... GTAW-P
Plasma arc welding ...... PAW
Shielded metal arc welding . SMAW
Stud arc welding .... SW
Submerged arc welding ... SAW
—series .... SAW-S

Arc brazing .... AB
Block brazing .... BB
Diffusion brazing ...... DFB
Dip brazing .... DB
Flow brazing .... FLB
Furnace brazing .... FB
Induction brazing ..... IB
Infrared brazing ...... IRB
Resistance brazing .... RB
Torch brazing .... TB
Twin carbon arc brazing . TCAB

Electron beam welding ... EBW
Electroslag welding ..... ESW
Flow welding .... FLOW
Induction welding ...... IW
Laser-beam welding ..... LBW
Thermit welding .... TW

Air acetylene welding .... AAW
Oxyacetylene welding .... OAW
Oxyhydrogen welding .... OHW
Pressure gas welding .... PGW

Air carbon arc cutting .... AAC
Carbon arc cutting ...... CAC
Gas metal arc cutting .... GMAC
Gas tungsten arc cutting .. GTAC
Metal arc cutting ...... MAC
Plasma arc cutting ..... PAC
Shielded metal arc cutting . SMAC

Electron beam cutting .... EBC
Laser beam cutting ...... LBC

Atomic hydrogen welding .. AHW
Bare metal arc welding ... BMAW
Carbon arc welding ..... CAW
—gas .... CAW-G
—shielded .... CAW-S
—twin .... CAW-T
Flux cored arc welding ... FCAW
—electrogas .... FCAW-EG

Cold welding .... CW
Diffusion welding ...... DFW
Explosion welding ..... EXW
Forge welding .... FOW
Friction welding ...... FRW
Hot pressure welding ... HPW
Roll welding .... ROW
Ultrasonic welding .... USW

Dip soldering .... DS
Furnace soldering .... FS
Induction soldering ... IS
Infrared soldering .... IRS
Iron soldering .... INS
Resistance soldering .. RS
Torch soldering ...... TS
Wave soldering ...... WS

Flash welding .... FW
High frequency resistance welding . HFRW
Percussion welding .... PEW
Projection welding .... RPW
Resistance seam welding . RSEW
Resistance spot welding . RSW
Upset welding .... UW

Electric arc spraying .... EASP
Flame spraying .... FLSP
Plasma spraying ...... PSP

Chemical flux cutting ... FOC
Metal powder cutting ... POC
Oxyfuel gas cutting .... OFC
—oxyacetylene cutting . OFC-A
—oxyhydrogen cutting . OFC-H
—oxynatural gas cutting . OFC-N
—oxypropane cutting .. OFC-P
Oxygen arc cutting .... AOC
Oxygen lance cutting ... LOC

Arc welding (AW)
Brazing (B)
Other welding
Solid state welding (SSW)
Welding processes
Oxyfuel gas welding (OFW)
Soldering (S)
Resistance welding (RW)
Adhesive bonding (ABD)
Thermal spraying* (THSP)
Allied processes
Arc cutting (AC)
Thermal cutting (TC)
Oxygen cutting (OC)
Other cutting

*Sometimes a welding process

Figure 27.1. Master chart of welding and allied processes showing standard abbreviations.
(Courtesy American Welding Society)

requires heat to bring the metals to be joined to a molten or plastic state and also to break down the oxides on the metal surfaces. Heat for welding metals comes from a gas flame, electric arc, or electrical resistance. Additional metal for filling in and reinforcing the weld usually comes from the electrode itself or from a filler metal wire or rod, but some applications require no additional filler metal.

The primary advantage of welding is the ability to completely bond metals with a joint as strong as the original or parent metals. Welds are superior to bolts and rivets because they can be made gas and water tight and can form smooth surfaces because they do not require flanges or overlapping sections for joining. Welding processes are extremely versatile; many different types, sizes, and shapes of metal can be welded, and welding can be done in locations that are hard to reach and under conditions that are demanding. Parts that would be difficult to cast or machine as a unit may be assembled by welding. For example, it may be easier and less expensive to weld a gear onto a shaft than to machine the entire part out of one piece of metal.

Figure 27.1 lists the many welding and allied processes that have been developed. Several of the welding processes have been modified to spray, cut, or remove metal rather than join it. Only the most common welding and cutting processes will be discussed here.

Each welding process has its purpose and limitations, and no single process is suited for all applications. Some of the factors dictating the use of a particular process are:

- Type of metal—nonferrous metals are more difficult and expensive to join.
- Thickness—penetration must be sufficient on thick pieces and burn through must not occur on thin stock.
- Versatility—processes may have to be portable or suited for vertical welds.
- Shielding—surface geometry or wind currents interfere with some types of shielding.
- Time—time for setup, welding, and cleanup vary for different processes and applications.
- Weld properties—corrosion resistance, surface hardness, strength, or water tightness may require special filler metals, procedures, or processes.

2 WELDING PREPARATION

2.1 Cleaning Surfaces to Be Welded

Most welding processes require clean, well-aligned surfaces for good-quality welds. However, welding on rusty, oily, or dirty surfaces is possible with appropriate fluxing and shielding; and cutting can also be done on poor surfaces. Cleaning surfaces before welding is preferred to trying to compensate in the welding process. Clean surfaces produce better weld quality and permit better fume control.

Depending on the contaminant, cleaning parts is done by abrasive blasting, sanding, solvent degreasing, or treatment with acid or alkali materials. These processes are discussed in detail under metal cleaning operations. See Chapter 17.

Special care must be taken when using chlorinated solvents, particularly trichloroethylene and perchloroethylene for degreasing prior to welding. Most chlorinated solvents will break down in the presence of heat or ultraviolet radiation from a welding arc to form highly toxic compounds such as phosgene, chlorine, and hydrogen chloride.

The threshold limit value (TLV) of the chlorinated solvent is not a good indicator of the decomposition hazard; decomposition products can produce extreme irritation and nausea if trichloroethylene is present at just a few parts per million.

Therefore, welding and chlorinated solvent degreasing operations must be separated, and cross contamination must be prevented. There should be no wet rags or open containers of chlorinated solvent in any welding area.

2.2 Asbestos Exposure

Prior to welding it is sometimes necessary to protect nearby valves, controls, or sensitive equipment against excessive heat by packing or covering with insulation. Before its carcinogenic properties were known, asbestos was widely used for this purpose. Many substitutes, such as ceramic fiber wrapping tapes, are now available and should be used.

A more frequent and serious exposure to asbestos occurs when removing asbestos insulation from pipes, ovens, boilers, structural steel, or other insulated surface prior to welding or cutting. Protecting worker health requires a number of steps:

- Inspecting the job to identify suspected asbestos-containing materials.
- Testing to verify whether the materials contain asbestos or proceeding as if they do.
- Following the provisions of the Occupational Safety and Health Administration's (OSHA's) asbestos standard including the requirements to isolate the job, train those involved, use wet methods, monitor dust levels, wear respirators and personal protective equipment, provide medical examinations, and properly dispose of waste and keep records.

2.3 Piping, Tanks, and Containers

Welding on previously used piping, tanks, or containers can be quite hazardous if traces of the former contents remain. The contents may be toxic or flammable or the intense heat and radiation may cause a hazardous chemical reaction. Therefore, the previous contents must be determined and reviewed for potential hazards. Appropriate cleaning, flushing, or inerting (filling with water, sand, or an inert gas stream) can then be done. Valves that stop the flow of material to tanks or piping should be locked out or the pipes should be disconnected and blanked to prevent inadvertent flow to the welding area.

2.4 Confined Spaces

Welding operations present special problems and call for special precautions when performed in confined spaces. There are several potential problems ordinarily associated with any work in confined spaces:

Toxic gases
Flammable gases and vapors
Oxygen deficiency
Limited egress

Tight quarters
Poor ventilation

Welding operations may add new problems:

- Heat and ignition sources
- Toxic fumes and gases from the welding process
- Leakage of gases (fuel, oxygen, or shielding gases) from welding equipment

To minimize these problems, a confined-spaces entry program should include training, procedures for testing, provisions for emergency rescue, and a permit system for entry. A permit system is a way to notify those who need to know about the confined-space entry so that proper precautions may be taken. A hot-work permit should be required for welding operations even when the confined space does not otherwise require a permit to enter. Constant contact should be maintained with welders in confined spaces.

The preentry procedures should include a review of the current and past usage or contents of the space. Welding in a confined space requires mechanical ventilation. If adequate ventilation cannot be maintained, then respiratory protection should be used and testing for combustible gases should be done if fuel gases are used and for oxygen level if inert gases are used. No hot work should be allowed in any confined space when the atmospheric level of a combustible gas exceeds 10% of the lower explosive limit (LEL) or when airborne dusts or mists present a potential explosion hazard. Because of the possibility of a gas leak in a confined space, all compressed gas tanks should be kept outside of confined spaces. Valves should be shut off outside the confined space, and preferably torches should be removed during periods of nonuse such as lunch breaks.

2.5 Coated or Plated Surfaces

Welding on coated or plated surfaces may generate highly toxic fumes. Paints may contain toxic pigments. For example, lead is commonly found in exterior structural coatings such as those used on bridges. Cadmium and zinc plating are often used for corrosion resistance. Teflon coatings may break down under heat to produce a fume that can cause polymer fume fever.

Before welding through a coated, painted, or plated surface the composition of the coating should be determined or the coating should be removed. Otherwise, the welder should take the same precautions during welding as taken for welding on Teflon-covered, lead-painted, or cadmium-plated surfaces.

3 ARC WELDING

All arc welding processes create an electric arc between the work pieces and the welding electrode. The heat of the arc melts the work pieces to be joined and also melts the electrode (or a filler metal in the case of nonconsumable electrodes). Shielding of the arc is needed to prevent the formation of metal oxides and nitrides that would weaken the weld.

3.1 Types of Arc Welding

3.1.1 Shielded Metal Arc Welding (SMAW). Shielded metal arc welding, also called manual arc or stick welding, is the most common welding process and accounts for more than half of all welding done. The electrode is a coated rod that melts into the pool of metal formed by the arc, as shown in Fig. 27.2.

Shielding is provided by ingredients in the electrode coating that decompose and generate a gaseous flow that keeps air from the arc. The coating contains fluxing and slag-forming agents and may also contain ionizing ingredients to make the electrode operate more smoothly, alloying elements for strength, and iron powder to improve the productivity of the electrode.

Coatings can be generally categorized as neutral, acid, rutile (titanium dioxide), and basic (lime–fluoride), all of which can be used for welding common carbon steel. Low-hydrogen electrodes with basic lime–fluoride coatings are commonly used for welding alloy steels containing nickel, chromium, copper, molybdenum, or higher percentages of manganese.

The manually controlled SMAW process produces a high-quality weld at a fast rate. It can be adapted for use in all positions and on a great variety of ferrous and nonferrous metals and alloys. It is rugged, simple, portable, and quite flexible, requiring the welder to take only a ground cable and electrode holder along to the welding location. This process can weld virtually any metal thickness from 18 gauge on up, but if the base metal is over $\frac{1}{4}$-in. (0.6 cm) thick, the facing edges must be beveled and multiple passes must be made to complete the weld. Slag removal is needed between each weld pass. The major limitations of this process are the high degree of skill required and the need to interrupt production to replace electrodes frequently.

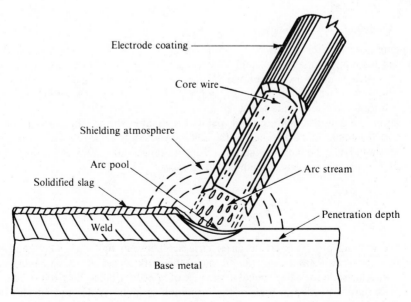

Figure 27.2. Shielded metal arc welding. (Courtesy American Welding Society)

3.1.2 Gas Metal Arc Welding (GMAW). After SMAW the most common type of welding process is gas metal arc welding, sometimes called MIG (metal inert gas) welding because of the inert gases that were used with the process at first. In the GMAW process, a wire electrode is continuously fed from a spool through the electrode holder to form the arc and provide the filler metal, as shown in Fig. 27.3. Shielding is provided by a flow of gas through the electrode holder. Besides the inert gases argon and helium, the shielding gas may be or may contain various concentrations of carbon dioxide, nitrogen, or other noninert gases that affect arc and metal transfer characteristics and joint properties. Argon, helium, or mixtures of these gases are normally used for nonferrous metals; carbon dioxide is usually used for steel. Carbon dioxide plus argon is used for stainless steel and sometimes for other steels as well.

Gas metal arc welding is faster than SMAW because the welder does not stop to replace the electrode. It is ideally suited to many production or automated welding tasks. GMAW can be used to join all types of base metals. The electrode is usually of the same composition as the base metal. For carbon–steel welding the electrode may be bare or copper-coated steel wire.

Advantages of GMAW are that it makes top-quality welds on almost all metals

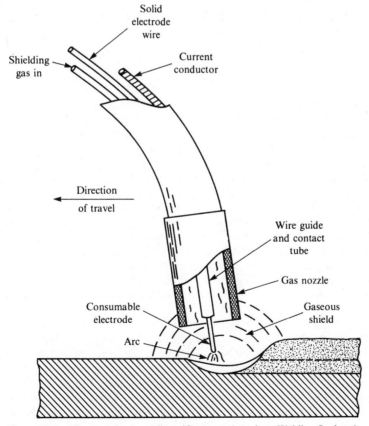

Figure 27.3.   Gas metal arc welding. (Courtesy American Welding Society.)

and alloys. Welding is possible in almost all positions. Minimum cleaning is required after welding and no slag is produced. The GMAW process also provides deep penetration and requires a small weld size compared to the SMAW process. Thus the GMAW can produce a weld of the same strength as the SMAW process but with lower current, less metal, and at a faster rate.

Drawbacks to the process include the fact that the inert gases are expensive, although carbon dioxide is very reasonable. The GMAW equipment is much more complicated and less portable than the SMAW equipment. There can also be a problem with wind or drafts interfering with the shielding gas. Surface geometry variations such as moving from a fillet weld to a butt weld may require gas flow adjustments.

3.1.3 Flux-Cored Arc Welding (FCAW). Flux-cored arc welding is similar to GMAW in that the electrode is fed from a spool through the electrode holder. The electrode is a hollow wire with a flux-filled center that helps provide the necessary shielding for the arc. The process was developed with an external shielding gas, but it is now possible to provide all of the shielding from the core.

With external gas shielding (carbon dioxide) the process provides smooth, sound welds, deep penetration, and high-quality weld metal. The gasless process provides moderate penetration but is not bothered by drafts or breezes. With either version, welding is possible in any position and can produce any type of weld joint. The types of materials that can be welded are limited to low-alloy steels and some stainless steels. FCAW can be used on a broad range of material thicknesses from $\frac{1}{16}$ in. (0.15 cm) on up.

3.1.4 Gas Tungsten Arc Welding (GTAW). In gas tungsten arc welding an arc is created between the base metal and a nonconsumable tungsten electrode. A filler metal in the form of a rod or wire is usually, but not always, fed into the arc. The process, shown in Fig. 27.4, is also called tungsten inert gas, or TIG welding, because an inert gas (argon, helium, or a mixture of these gases) is used for shielding. Small additions of oxygen, hydrogen, or other gases to the principal shielding gas are occasionally used.

GTAW makes top-quality welds in almost all metals and alloys of practical significance. There is little or no postweld cleanup and no slag to be trapped in the weld. There is no weld spatter because filler metal is fed directly to the weld pool at the end of the arc. Welding is possible in all positions and on a wide range of thicknesses, but GTAW is especially good for making high-quality welds on thinner materials that other processes may burn through. Because of the high-quality weld, it is used to make the critical root pass on carbon-steel pipe joints.

Because of the slow welding speed and high cost of the shielding gas, GTAW is used mainly where there is a technical advantage, such as for thin materials or for reactive metals.

3.1.5 Plasma Arc Welding (PAW). The plasma arc welding process is related to the GTAW process but involves higher temperatures and some other significant differences. The plasma is a mixture of electrons and gaseous ions formed when the plasma gas (usually argon or nitrogen or a mixture of hydrogen with one of these gases) is forced through the high-temperature arc. The plasma forms a shield around the arc that may be supplemented by an auxiliary source of shielding gas outside the plasma gas.

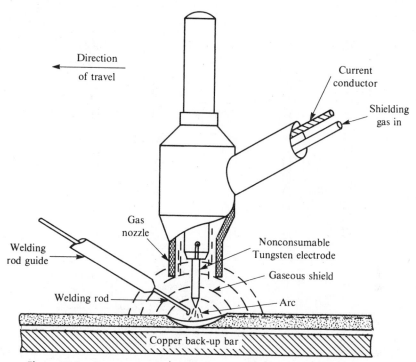

Figure 27.4. Gas tungsten arc welding. (Courtesy American Welding Society)

There are two types of plasma arc, as shown in Fig. 27.5, both of which have a constricted arc in contrast to the open or unconfined arc used in the GTAW process. In the transferred mode, the plasma arc is formed between the nonconsumable electrode and the workpiece, producing maximum heat transfer that is useful for cutting and welding. For cutting operations (Section 8.2), oxygen may be fed into the arc as it leaves the nozzle. For welding operations, filler metal may be fed into the arc near the work.

In the nontransferred mode the arc is established between the nonconsumable electrode and the side of the nozzle. This mode is useful for plasma spraying (PSP) where metal or ceramic powder is injected into the plasma stream and propelled onto the workpiece to form a well-bonded homogeneous coating.

PAW has a high initial cost, but it has several important advantages over the GTAW process for welding:

- For thin materials (less than $\frac{3}{32}$ in. or 0.2 cm), changes in arc length have less effect on heat input and weld bead shape. Thus problems of burn through and lack of penetration are minimized.
- For thick materials, the high-velocity and high-temperature plasma arc allows deep or full penetration with a narrow weld pool. This makes single-pass welding possible and thus greatly reduces weld time.
- Higher arc temperatures make it possible to weld refractory materials.

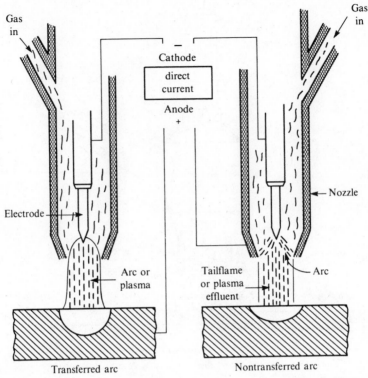

Figure 27.5. Plasma arc welding process (transferred arc is used for welding). (Courtesy American Welding Society)

For cutting operations PAW is faster than oxyfuel gas processes, leaves a narrower kerf, and can be used on any metal.

3.1.6 Submerged Arc Welding (SAW). In the submerged arc welding process the arc is established between a bare consumable electrode and the base metal and is surrounded by a pool of molten slag. A blanket of granular, fusible flux, often containing alloying elements, shields the arc and molten weld metal. The molten flux provides a current path between the electrode and base metal. The process is shown in Fig. 27.6.

A similar process used for vertical welding is the electroslag welding process wherein a pool of molten slag floats on the welded metal beneath the electrode. The arc is created between the electrode and the slag pool, which becomes the source of heat.

In both the submerged arc and the electroslag welding processes, the arc is completely covered and is therefore, not visible. This eliminates most of the radiation, smoke, sparks, and fumes associated with other arc welding processes. These processes are popular for welding relatively thick plates at high metal deposition rates. Other methods would have to use multiple welding passes to deposit the same amount of metal, each pass requiring removal of solidified slag before rewelding. Submerged arc and electroslag welding arc used mainly for plain carbon and low-alloy steel plate,

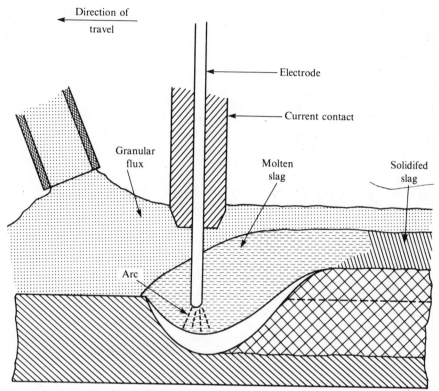

Figure 27.6. Submerged arc welding operation on heavy plate. (Courtesy American Welding Society)

but can be used for welding high-alloy ferrous metals and certain nonferrous metals as well.

3.1.7 Arc Spot Welding. A variation of the gas metal arc welding process can be used for spot welding of one plate on top of another. A weld nugget that penetrates both plates is formed by a special gun and timing device. The spot weld sequence follows:

- A preheat arc melts a pool of metal in the weld zone.
- Metal is blown out of the weld zone to create a recess extending into the lower plate.
- Sides of the recess are melted to form a new molten puddle.
- The weld nugget is completed.
- The arc is interrupted.

3.1.8 Stud Arc Welding (SW). Stud welding usually involves loading a fastener such as a stud into a spring-type chuck at the end of a welding gun. When the trigger is pulled, the stud retracts, and an arc is formed between the end of the stud and base metal to which the stud is to be attached. After a preset time interval, the current is interrupted and the stud is forced into the pool of molten metal. The arc is usually contained by a ceramic ferrule that makes the fillet appear neat. At the end of the stud is a flux pocket that serves to clean the surface before welding.

3.1.9 Carbon Arc Welding (CAW) and Atomic Hydrogen Welding (AHW). These processes are not very common anymore. In carbon arc welding, a heavy carbon electrode is used to create the arc to the workpiece. The carbon evaporates to form a shield or in some cases the electrode coating burns to help shield the arc.

Atomic hydrogen welding uses two tungsten electrodes and creates an arc between them. Hydrogen gas is fed into the arc and dissociates into hydrogen ions. The heat for metal fusion comes from the recombination of the hydrogen atoms at the workpiece surface together with convection and radiation from the arc.

Gas metal arc welding has largely replaced these two processes.

3.2 Arc Welding Fumes and Gases

The major health hazards associated with arc welding operations are fumes and gases. The extremely high temperature of the welding arc (above 6000°F, or 3300°C) is above the boiling points of most metals commonly encountered in welding operations. The metals and oxides vaporized at the welding arc rapidly condense to form particles. These welding fumes can be expected to contain compounds that represent all of the ingredients of the base metal, electrode, or filler metal and any fluxes or coatings. As much as 2% of the weight of electrode consumed can be generated as fume. Gases may be introduced for shielding or formed around the arc by reaction with the shielding gas, by ultraviolet radiation or by decomposition of fluxes or electrode coatings. Some of the welding fume and gas constituents are shown on Fig. 27.7. Reactions with oil, dirt, paint, cleaning solvents, and so on can produce additional types of fumes and gases.

3.2.1 Arc Welding Fume and Gas Generation. Many factors affect the amount and nature of welding fume constituents including the composition of base and filler metals, coatings and fluxes, current, voltage, shielding gas, electrode feed rate, arc length, and polarity. The variation of fume content under different operating conditions of the same arc welding process on the same material can be as great as between totally different processes or materials.

Compared to other processes, the shielded metal arc welding (SMAW) process produces more fume per amount of metal deposited, usually in the range of 1–2% by weight. When welding on carbon and low-alloy steel, iron oxide (present as both Fe_2O_3 and Fe_3O_4) may represent as little as 35% of the fume. Welding on stainless and chromium steels and nickel alloys produces significant levels of hexavalent chromium, nickel, and copper.

In addition to oxides from the base metal, fumes from the covered electrodes used in SMAW processes contain iron, manganese, fluoride, nickel, molybdenum, chromium, and copper. Fluoride may be the most important of these because of its toxicity and high content (10–20%) in the welding fume, particularly with basic low-hydrogen coverings. Ozone, nitrogen oxides, and carbon monoxide are not usually formed in significant quantities. The gas metal arc welding (GMAW) process produces less fume per amount of metal deposited than the SMAW process. However, because the arc

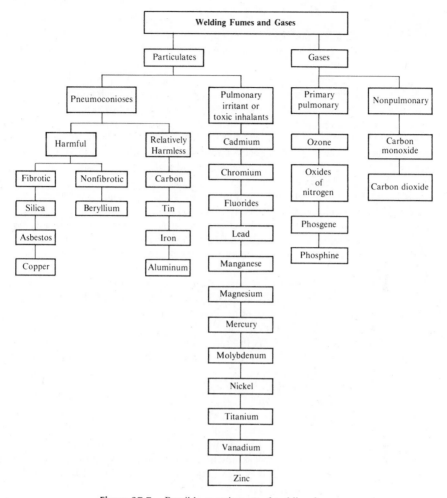

Figure 27.7. Possible constituents of welding fumes.

time (portion of a welder's time that is actually spent welding) is often much greater with GMAW, the time-weighted average exposure level may be comparable. Increasing the arc length increases both the amount of metal deposited and the amount of fume produced.

The GMAW fume contains more metal oxides than other welding processes, frequently over 80% iron oxide. Thus, metals that are present in the electrode or base metal are likely to be significant components of the fume. Chromium, nickel, and copper are important constituents of fume when welding on stainless or high-alloy steels or with copper-coated electrodes.

The GMAW process does not usually produce gases in problem concentrations. Exceptions are ozone and nitrogen dioxide levels when welding stainless steel or aluminum (due to surface reflectance) or when using argon as a shielding gas (due to its particular spectral emission properties). Carbon monoxide is produced when using carbon dioxide shielding gas. The flux-cored arc welding (FCAW) process produces more fume than other welding processes because of the high currents used. Because of the high metal deposition rates (more than double the rate of the SMAW process) the amount of fume per weight of metal deposited is no greater than for the SMAW process. However, welder exposure to fumes and gases may be much higher. Shielding gas is sometimes used with FCAW electrodes and tends to reduce fume generation.

The core of flux material in the electrode contains several compounds, most notably fluorides, which, together with oxides from the base metal, contribute to the welding fume. Significant concentration of gases, particularly carbon monoxide and nitrogen oxides, may often be found. Carbon monoxide comes from decomposition of organic matter in the flux and from CO_2 used as a shielding gas. Because the gas tungsten arc welding (GTAW) process is usually used on stainless and high-alloy steels, the fumes are likely to contain chromium and nickel. The GTAW process is also used on aluminum, which produces significant ozone and nitrogen dioxide levels. The tungsten electrode is not consumed, but filler metals are added to the process and contribute to the fume content. The total fume level is lower than for the other processes discussed.

Submerged arc welding (SAW) generates low levels of fume because the arc is covered. However the flux may produce significant levels of fluoride fume.

Besides the type of process used, some general comments apply to other factors affecting arc welding fume generation:

Current—higher currents produce more fume.

Voltage/arc length—longer arcs and corresponding higher voltages produce more fume.

Electrode feed rate—faster feed rates increase fume generation rates as well as metal deposition rates.

Polarity—when dc power is used, electrons flow in one direction only, from negative to positive poles. When the electrode is used as the positive pole (straight polarity), the electrode burns faster, the base metal is not penetrated as deeply and the fume generation rate is higher than when the base metal is used as the positive pole (reverse polarity).

Arc welding processes can be applied to a great variety of base metals and can be employed under many conditions that may create hazards not covered in the preceding discussion. Some of these hazardous conditions are listed in Table 27.1.

3.2.2 Arc Welding Fume and Gas Evaluation. Because the variable composition of welding fumes makes evaluation complex, the American Conference of Governmental Industrial Hygienists (ACGIH) has established a TLV for "welding fumes—total particulate (not otherwise classified)" as one indicator of the hazard. Appendix B2 of the 1988–1989 TLV booklet (ACGIH, Cincinnati, OH) provides a very informative explanation:

Welding fumes cannot be classified simply. The composition and quantity of both are dependent on the alloy being welded and the process and electrodes used. Reliable analysis of fumes cannot be made without considering the nature of the welding process and system being examined; reactive metals and alloys

Table 27.1 Contaminants with TLVs That May Be Exceeded Even Though Total Welding Fume Is Below 5 mg/m³

Contaminant	1988 TLV	Associated Conditions
Metals		
Beryllium[a,b,c]	0.002 mg/m³	Beryllium-copper alloys including those used for spark proof tools
Cadmium[b,d,e]	C 0.05 mg/m³	Silver brazing solder; cadmium-plated and painted surfaces
Chromium[b,f]	0.05 mg/m³	Stainless steel metal and electrodes; nickel alloys; chromium-plated or painted surfaces
Cobalt[c]	0.05 mg/m³	Cobalt tool steels; cobalt pigments
Copper	0.2 mg/m³	Brass and bronze filler and base metal; Monel alloys; plated surfaces; copper-coated wire electrodes
Lead[g]	0.15 mg/m³	Leaded steel, bronze, and brass; lead-painted surfaces; solder
Manganese	1 mg/m³	Small amounts are present in most carbon/stainless alloys and electrodes
Mercury[b]	0.1 mg/m³	Marine paints
Nickel[c]	1 mg/m³	Stainless steel or nickel alloys or nickel based electrodes; hard chrome-plated surfaces
Silver	0.1 mg/m³	Silver solder, silver plating
Vanadium[h]	0.05 mg/m³	Vanadium steel alloys and filler wire
Other particulate matter		
Asbestos[b,f]		Insulation on pipes, furnaces, and building structural members
Amosite[f,i,j]	0.5 fiber/cc	
Chrysotile[f,i,j]	2 fibers/cc	
Crocidolite[f,i,j]	0.2 fiber/cc	
Other forms[f,i,j]	2 fibers/cc	
Fluorides[k]	2.5 mg/m³	Basic coated electrodes; brazing flux
Polytetrafluoroethylene decomposition products[l]		Nonstick coatings
Quartz[b,m]	0.1 mg/m³	Electrode coatings; uncleaned castings
Rosin core solder pyrolysis products, as formaldehyde	0.1 mg/m³	
Gases		
Carbon monoxide[b]	50 ppm	Carbon dioxide shielding gas; organic materials in flux; oily stock; gasoline-operated welding equipment

Table 27.1 (*Continued*)

Contaminant	1988 TLV	Associated Conditions
Formaldehyde[a,n]	1 ppm	Oily stock; paint resins
Hydrogen chloride[d]	C 5 ppm	Chlorinated solvents
Nitrogen dioxide[b]	3 ppm	Open welding arcs, especially argon shielded welding on aluminum or stainless steel; confined spaces
Ozone[o]	0.1 ppm	Welding arcs, especially argon-shielded welding on aluminum, magnesium, titanium or stainless steel
Phosgene	0.1 ppm	Chlorinated solvents

[a]Suspected human carcinogen.
[b]OSHA and/or NIOSH has a PEL or REL lower than the TLV.
[c]Identified as a suspected or confirmed human carcinogen by other sources.
[d]C = ceiling concentration, which should not be exceeded during any part of the working exposure.
[e]ACGIH has announced an intended change to a TLV–TWA of 0.01 mg/m^3 and a designation as a suspected human carcinogen.
[f]Confirmed human carcinogen.
[g]OSHA PEL is 0.05 mg/m^3
[h]As vanadium pentoxide
[i]Fibers longer than 5 microns
[j]OSHA PEL is 0.2 fibers/cc
[k]As Fluorine
[l]No TLV established; air concentration should be minimal
[m]Respirable dust
[n]OSHA PEL is 0.5 ppm
[o]ACGIH has announced an intended change to make this TLV a ceiling level

such as aluminum and titanium are arc-welded in a protective, inert atmosphere such as argon. These arcs create relatively little fume, but an intense radiation which can produce ozone. Similar processes are used to arc-weld steels, also creating a relatively low level of fumes. Ferrous alloys also are arc-welded in oxidizing environments which generate considerable fume, and can produce carbon monoxide instead of ozone. Such fumes generally are composed of discrete particles of amorphous slags containing iron, manganese, silicon and other metallic constituents depending on the alloy system involved. Chromium and nickel compounds are found in fumes when stainless steels are arc-welded. Some coated and flux-cored electrodes are formulated with fluorides and the fumes associated with them can contain significantly more fluorides than oxides. Because of the above factors, arc-welding fumes frequently must be tested for individual constituents which are likely to be present to determine whether specific TLVs are exceeded. Conclusions based on total fume concentration are generally adequate if no toxic elements are present in welding rod, metal, or metal coating and conditions are not conducive to the formation of toxic gases.

Most welding, even with primitive ventilation, does not produce exposures inside the welding helmet above 5 mg/m^3. That which does, should be controlled.

There are many conditions under which a welding gas or a particular component of the welding fume may exceed its TLV even though the total welding fume level is below the TLV. In these cases, the total fume level is not the best indicator of the hazard. There are summarized in Table 27.1.

Some of the more common or more hazardous welding fume and gas components deserve special comment. For example, the most common welding fume constituent, iron oxide, has a TLV of 5 mg/m^3, the same as for total welding fume. It is not likely, therefore, that the iron oxide TLV will be exceeded unless the total welding fume TLV is also exceeded.

Iron oxide, which is present in the form of both Fe_2O_3 and Fe_3O_4, produces a benign pneumoconiosis called siderosis. Iron is only slowly removed from the lung, and the fume particles tend to accumulate to form a clouded picture on an X ray. Although not disabling, this condition may obscure the presentation of more serious diseases. Evidence of fibrosis in workers exposed to iron oxide has been presented, but this appears to be due to a combination of iron oxide and silica called silicasiderosis. Silicon is present in many steels as ferrosilicon, and silicon dioxide is a common ingredient of electrode coatings, but the relationship to the silicasiderosis hazard has not been well studied.

Another common component of welding fumes is zinc oxide, which, like iron oxide, has a TLV of 5 mg/m^3. High concentrations of freshly generated zinc oxide fumes can cause a temporary condition referred to as metal fume fever, zinc chill, or brass chill (from nonferrous foundry work). The condition is symptomized by a metallic taste at the back of the throat and a slight headache followed several hours later by cold, shivering, and symptoms very much like the flu. The phenomenon is usually associated with welding on galvanized stock and involves only fresh fumes that have a small particle size and are easily respirable. Zinc oxide fumes tend to quickly combine or flocculate, and after several minutes the larger particles are no longer able to penetrate to the deep lung. It is important that a proper diagnosis be made since more serious metal poisonings often display the same or similar symptoms.

Cadmium fumes can be generated when welding on cadmium-plated parts (plating produces a bright metallic surface and corrosion resistance) or through surfaces that have been painted with cadmium-pigmented paints. Cadmium is one of the most hazardous contaminants that may be encountered in arc welding operations because severe and possibly fatal lung irritation may result from fume concentrations that do not produce warning signs of irritation. The beginning symptoms may be mistaken for those of zinc fume fever, including a period of apparent recovery before the more serious effects show themselves several hours later. Chronic exposure to lower levels of cadmium fume can cause lung injury (emphysema) and kidney damage.

Ozone is formed by electric arcs and by ultraviolet photochemical reactions that take place mostly within a few inches of the arc where the UV radiation is the most intense. Ozone is a highly reactive irritating gas that readily breaks down or combines harmlessly with other fume constituents as it moves away from the arc. All of the arc welding processes can produce some ozone, but submerged arc and electroslag welding generate very little because the arc is covered. The highest relative concentrations are produced during welding on stainless steel or aluminum (due to surface reflectance) or when using argon as a shielding gas (due to its particular spectral emission properties).

Fluoride and its compounds are found when using electrodes with basic low-hydrogen coverings or fluoride fluxes. Low-hydrogen electrodes contain calcium fluoride in their covering and are used for high-strength welds on difficult-to-weld materials such as heavy armor plate. Fluoride fumes may produce respiratory tract irritation with chills, fever, shortness of breath, and cough. Excretion is slower than absorption and repeated exposure causes a buildup that can eventually lead to bone changes.

Arc welding samples of manual or semiautomatic processes for comparison to the TLV must be collected inside the welder's helmet. This usually requires a specially modified helmet designed to hold the filter close to the mouth, as shown in Fig. 27.8. The filter could also be taped inside the helmet. Most welding hoods have enough clearance for a standard two-piece filter cassette in front of the welder's face. The TLV has been established on the basis of fume levels inside the hood and the American Welding Society's standard method for sampling and analysis of welding fumes also specifies this method.

Comparisons of fume concentration outside with those inside the welding helmet do not show much agreement. Outside levels may be 5–15 times higher than inside levels. Even samples taken from the welder's collar slightly behind the helmet during welding show considerably higher levels than inside the helmet. Therefore, meaningful comparisons to the TLV and accurate measurements of breathing zone exposures require careful attention to the standardized sampling location. However, for auto-mated arc welding operations or processes where the operator does not lean over the arc when welding, the sampling location is not as critical.

Sampling for gases is more difficult than sampling for fumes because it is harder to approximate breathing zone conditions. For example, ozone sampling presents a particular problem due to the considerable ozone decay at increasing distance from the arc. Placing midget impingers inside the welding helmet is not practical. Collecting samples too close or too far from the arc may not accurately reflect the true breathing zone concentration. Similarly, a sampling strategy must also account for the direction of the plume and of thermal currents when selecting the sampling location or collection technique.

Properly evaluating an arc welding process and preparing for sampling requires a review of material safety data sheets and other available information to find out which

Figure 27.8.   Helmet modified for air sampling.

base metals, coverings, electrodes, filler metals, shielding gases, and other contaminants may be present. However, some generalizations can be made about evaluations that should be made for specific arc welding processes, materials, and conditions. Table 27.2 lists the air sampling that should be done for several common welding operations. Conditions listed in Table 27.1 would dictate additional types of sampling.

3.2.3 Arc Welding Fume and Gas Control. Satisfactory control of arc welding fumes can often be achieved by natural ventilation for both shop and field welding jobs. Many SMAW operations on carbon steel require substantial setup and cleanup time, which reduces arc time and total fume generated. Manual GMAW generally permits longer periods of uninterrupted arc time than SMAW, but welding on carbon steel can often be done without requiring local exhuast ventilation for either operation. In these cases, control of breathing zone welding fume concentrations depends largely on the volume of the workspace and on good working habits that keep the welder's head out of the rising hot plume. Of the arc welding processes, flux-cored processes generally produce the greatest amounts of fume while submerged arc may produce almost none.

Pedestal fans (man coolers) are also common in many welding shops. While they do not remove welding fumes or gases, they can move the plume away from the welder's breathing zone. Such a fan is most effective when placed at the side of the welder so that it moves the air across the welder's front. When the fan is at the welder's back, the welder's body blocks the air flow and creates eddy currents that may increase the level of fumes in the breathing zone. To be effective, pedestal fans should produce a velocity of 100 fpm past the welder at a height of about 2 ft above the work area.

Where welding activity is concentrated or welding is done under conditions likely to produce fumes containing constituents listed in Table 27.1 or where the natural ventilation is restricted, then local exhaust ventilation is needed. The most common approaches to local fume control are gun-mounted exhaust systems, freely suspended or counterbalanced hoods, and cross-draft and down-draft tables.

Air volume and energy requirements are minimized when the exhaust is as close as possible to the source of the fume. An excellent example of this principle is the electrode holder with built-in exhaust for wire-fed arc welding operations. These welding guns feed the electrode, apply a shielding gas, and remove fumes, all at the same time. Careful adjustment is required to balance the exhaust and shielding gas flows, and this creates some difficulty in adapting to different surface angles (e.g., changing from butt to fillet welding on a complex part). Other problems, such as

Table 27.2 Air Sampling for Specific Arc Welding Processes or Conditions

Process or Condition	Evaluation
SMAW, GMAW, and FCAW on ferrous alloys	Total fume
Arc welding on stainless or high-alloy steel	Hexavalent chromium, nickel
Arc welding on aluminum or titanium	Ozone, nitrogen oxides
GMAW with copper-covered electrode	Total fume, copper
Carbon dioxide shielding gas	Carbon monoxide
Argon shielding gas	Ozone
SMAW or FCAW with basic flux	Fluoride

additional weight, excessive heat at the tip, or plugging of exhaust hoses sometimes limit the use of this technique.

Several counterbalanced exhaust systems have been developed to make it easy to position a suspended exhaust hood near the weld. Some portable units can be easily used for field welding. Local exhaust systems must provide a capture velocity of 100 fpm to control the welding fume. Effective capture of welding fume should take advantage of the plume's tendency to rise. Down-draft exhaust systems can be effective but often fail because they must first overcome the thermal inertia before fumes can be pulled into the exhaust system. Cross-draft systems are most effective when welding operations can be done on smaller parts or where the part can be turned to keep the welding in front of the stationary exhaust hood.

Welding shops sometimes use dust collectors to clean and recirculate general welding room air. These are located near the ceiling and depend on thermal currents to bring fumes upward. While often effective at reducing visible general room fume levels, they have little effect on levels of gases or on fume levels in the welder's breathing zone.

3.3 Arc Welding Radiation

Arc welding processes produce visible, ultraviolet, and infrared radiation with the UV being of greatest concern. All three forms of radiation may be hazardous both to the welder and to others in the area. The intense UV radiation from the arc can burn the skin like a sunburn.

The effect on the eyes is more severe, causing an acute inflammation of the cornea that feels like sand or ground glass in the eyeball. This temporary but painful effect occurs about 6 hours after the exposure and often happens to unprotected bystanders who accidentally or unwarily view the arc rather than to the welder who is protected and knows when the arc is going to be struck.

The amount of UV energy emitted by the arc depends on the characteristics of the welding process: type of process, current, arc length, welding metals, shielding gas, and fume concentration. Submerged arc and electroslag welding do not emit UV radiation under normal operating conditions. In contrast, the plasma arc emits a much greater amount of UV radiation than do the other arc welding processes. Intense UV radiation is also emitted during gas metal arc welding on aluminum using argon shielding gas and during gas tungsten arc welding. Some of the cleaner welding processes such as GTAW allow greater radiation emission in part because they produce fewer fumes to shield or absorb the radiation.

Measurement and evaluation of the UV radiation hazard is rather difficult. The TLV for UV recognizes the varying biological effectiveness of different UV wavelengths, but no commercially available measurement instruments properly weights the overall effect of the different wavelengths. Therefore, several readings are necessary and each of these must be weighted and combined to evaluate the overall effect.

Control of UV radiation exposure requires protection of both skin and eyes. The clothing that should be worn to protect against hot sparks will provide adequate skin protection against the UV radiation. Face and eye protection should take the form of welding helmet with properly shaded lenses. Factors that determine the amount of lens shading include type of process, electrode size, arc current, thickness of stock, and visual acuity. As a rule of thumb, the welder should start with a shade that is too dark to see the weld zone and then change to a lighter shade that allows sufficient visual acuity. Table 27.3 contains some suggested shade numbers for welding operations.

Curtains or screens should be used to protect welding helpers and other bystanders from radiation produced by the arc. Some plastic curtains are opaque to UV radiation but allow some visible and near-infrared to pass through. Color alone is not an adequate basis for curtain selection since ultraviolet stabilizers or dyes can be added to many plastic resins to improve their absorption properties. Whether these curtains provide adequate protection in a particular instance depends on the curtain, the welding process, and its several parameters and on the distance and duration of arc viewing. Curtains or screens should be flame resistant and should permit circulation of air for ventilation purposes. When arc welding shop areas have painted walls, a finish with low UV reflectivity should be chosen. Pigments such as zinc oxide and titanium dioxide are recommended because they have low reflectivity while others, such as powdered or flaked metallic pigments, should be avoided because they tend to reflect UV radiation.

Table 27.3 Suggested Shade Numbers for Welding Operations (from AWS/*Safety in Welding and Cutting* Z49.1—1988)

Operation	Plate Thickness (in.)	Electrode Size (in.)	Arc Current (A)	Minimum Protective Shade No.	Shade No. (Comfort)
Shielded metal arc welding	$\frac{3}{32}$		60	7	—
	$\frac{3}{32}-\frac{5}{32}$		60–160	8	10
	$\frac{5}{32}-\frac{1}{4}$		160–250	10	12
	$\frac{1}{4}$		250–550	11	14
Gas metal arc welding					
Flux-cored arc welding			60	7	—
			60	10	11–14
Gas tungsten arc welding			150	8	10–12
			150–500	10	14
Air carbon arc cutting			500	10	12
			500–1000	11	14
Carbon arc welding				—	14
Plasma arc welding			20	6	6–8
			20–100	8	10
			100–400	10	12
			400–800	11	14
Plasma arc cutting			50	7	—
			50–300	8	9
			300–400	9	12
			400–800	10	14
Torch brazing					3 or 4
Torch soldering					2
Oxyfuel gas welding	$\frac{1}{8}$				4 or 5
	$\frac{1}{8}-\frac{1}{2}$				5 or 6
	$\frac{1}{2}$				6 or 8
Oxyfuel cutting	1				3 or 4
	1–6				4 or 5
	6				5 or 6

In addition to shielding, time and distance can be effectively used to control the exposure of helpers and bystanders to UV radiation emitted from welding arcs. Figure 27.9, from a study by the U.S. Army to determine the allowable time and distance for unprotected viewing of various types of welding arcs, shows the relationship calculated for some typical sets of welding conditions using the 1988 ACGIH TLV as

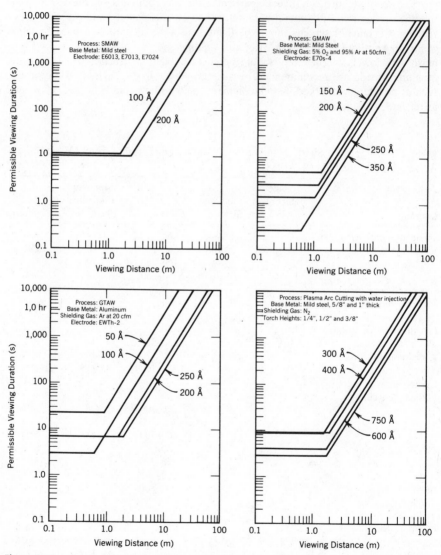

Figure 27.9. Relationship between time and distance for unprotected viewing of four typical arc welding and cutting processes. Maximum allowable UV exposure for skin and cornea at various arc currents. (From Marshall, et al.)

the exposure limit. At a distance of 10 m, for example, a bystander could look at most common arc welding or cutting operations for up to one minute without eye protection. Nevertheless, eye protection is recommended for both flying particles and for UV protection.

Infrared radiation is emitted by pools of molten weld metal and by the arc itself. Since the metal cools rapidly, the hazard exists only during welding and for a few seconds afterward. The chief hazard is to the eyes and results from chronic or repeated exposure. Infrared radiation is seldom a problem since the precautions taken to protect against UV radiation are more than adequate to protect against the infrared. An exception to this generalization must be made in the case of some plastic welding curtains, which may not always provide adequate infrared protection.

There is general agreement that welders may wear contact lenses during welding operations provided the eyes are also protected according to the guidelines in Table 27.3. There is no truth to the "welded" lens rumor that has been circulating for several years. Contact lenses cannot concentrate radiation so close to the surface of the eye and the heat of welding sources is not high enough to dry out the fluid in the eyes or alter lens materials enough to make them adhere to the eye. In fact, most contacts are concave lenses that disperse rather than concentrate light or heat waves.

3.4   Arc Welding Noise

Arc welders can be exposed to excessive noise during some arc welding operations such as plasma arc welding. Noise is due in part to the arc and to turbulence created by the expansion of metals and fluxes as they are converted from the solid to the vapor state. Gas-jet noise may also be a major factor in the noise from some forms of arc welding. Of greater concern is the noise contributed by auxiliary operations such as the impact of metal parts as they are placed into fixtures or dropped into bins and the chipping or grinding on welded parts. For the most part, the noise from the arc welding process itself is not a major problem. The helmet provides some shielding from welding arc and air noises, but this is not the case for auxiliary noise sources.

Because of the highly variable pattern of activity of most arc welders, the most practical method of measuring noise exposure is with an audiodosimeter. This is especially true where chipping, grinding, or substantial stock handling noises are involved. For automated and some semiautomated operations, it may be practical to use a sound-level meter to measure the various phases of the job and to develop an overall exposure level. As with the fume measurements, the shielding effect of the helmet must be considered. The location of the audiodosimeter microphone is very important if welder noise exposure is to be accurately measured.

4   RESISTANCE WELDING

Resistance welding is based on the principle that resistance to current flow generates heat. Since the interface of two surfaces to be joined is the point of greatest resistance in the circuit, it is also the point of greatest heat. In simple resistance welding, the metal is heated to a temperature that is high enough to cause localized fusion and formation of a weld nugget. After the current is turned off, pressure is applied to squeeze the two parts into a homogeneous mass.

4.1 Types of Resistance Welding

The principal types of resistance welding can be divided into two groups: lap welds (including spot, seam, and projection) and butt welds (including upset butt and flash butt).

4.1.1 Resistance Spot Welding (RSW). By far the most common form of resistance welding is spot welding, which joins two overlapping pieces of metal with a series of weld points. Spot welds may be made one at a time or sheet metal parts may be placed in a fixture and multiple spot welds may be made simultaneously. Robotic spot welding is common in assembly of products made of sheet metal parts such as automobiles and trucks. The time involved for spot welding is short; current flows for only a fraction of a second. For example, a spot weld can be made in two $\frac{1}{16}$-in. (0.15-cm) pieces of carbon steel in 15 cycles or 0.25 sec when using 60-cycle current. For medium- or high-carbon steels some quenching or postheating is needed following welding to strengthen the otherwise brittle weld.

4.1.2 Projection Welding (RPW). Projection welding is a variation of spot welding. Several small projections are raised on one side of a sheet or plate where it is to be welded to another. During the welding process, the heat of the welding circuit is localized at the projections. Heat and pressure cause their collapse, and the parts to be welded are brought into close contact. The process is shown in Fig. 27.10. This method is often useful for attaching small fasteners, nuts, special bolts, and studs to larger components. If water- or gas-tight joints are desired, solder must be sweated into the seam.

4.1.3 Resistance Seam Welding (RSEW). Seam welding is a continuous type of spot welding where the work is passed between copper wheels or rollers that act as electrodes. The completed weld is an overlapping series of spot welds that resemble stitches, and the process has sometimes been called stitch welding. A variation of this method may be used to form tubes from a flat coil of metal. The coil passes through a series of rollers that curl the edges up and around where they butt together to form a tube. The joint passes beneath a copper wheel that acts as an electrode to weld the edges together in a continuous seam.

4.1.4 Upset Welding (UW). In upset welding, the ends to be welded are held together under enough pressure to keep the joint from arcing. When the metal softens, it is

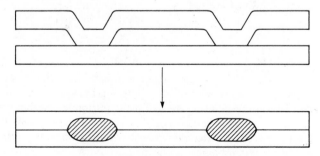

Figure 27.10. Projection welding process.

pushed together with enough force to cause the excess material to flow outward. Oxidized metal is eliminated from the joint by pressure. Excess metal must be removed by machining.

4.1.5 Flash Welding (FW). In flash welding, the ends of the stock are held in light contact. The current causes flashing and burning away of the metal. Pieces are moved together until the proper temperature is reached and then they are rapidly forced together as the current is cut off. This process is often used for end-joining of rods, tubes, bars, forgings, and the like. Excess metal must be removed by machining.

4.1.6 Percussion Welding (PEW). In percussion welding, workpieces are brought together at a rapid rate. Just before they collide, a flash of arc melts both facing surfaces. The molten surfaces are then squeezed together by the collision as some metal is forced out the sides of the joint. This type of welding is particularly good for small-diameter wires and other parts that have pinpoint projections.

4.2   Resistance Welding Hazards

Resistance welding processes are free of almost all of the hazards associated with arc welding. Fumes, gases, noise, and radiation are usually well controlled with resistance welding processes. Resistance welding may send hot sparks shooting out for distances of up to 30 ft, but these are only intermittent products because of the short weld cycle time. Even nonferrous and plated materials produce few health hazards.

Problems associated with resistance welding operations are more likely caused by auxiliary activities than by the welding process itself. For example, noise problems are more likely to be due to handling of sheet metal parts than to the welding process itself. Although not normally a problem, resistance welding does produce some fumes and gases. If welding activity is concentrated (high production, simultaneous welds, or multiple high-speed robots), if highly toxic materials are used, or if unusual conditions are encountered (oily stock, confined spaces), then exposures should be evaluated. Use of chlorinated solvents around resistance welding operations can also create problems and should not be permitted.

The flash created by loose contact points in resistance welding and by flash butt welding and percussion welding can generate UV radiation. The hazard is much less than for arc welding but still requires eye protection.

5   OXYFUEL GAS WELDING (OFW)

Oxyfuel gas welding refers to a process of heating metal for fusion purposes with an oxyacetylene flame. Oxygen and acetylene are provided from compressed gas cylinders and fed to a hand-held torch where they are mixed and burned together with oxygen from the air. In variations of the process, other fuel gases such as hydrogen, propane, or natural gas can be used in place of acetylene, but the higher temperature of the oxyacetylene flame (6300°F, or 3500°C) makes it the usual choice for OFW.

The oxyacetylene flame is more easily controlled and not as penetrating as metallic arc welding. Thus it is particularly useful for sheet metal fabrication and repair where arc welding might burn through the thin metal. However, it can also be used to weld steel over one-inch thick. The hot flame melts the two edges of the pieces to be joined together with the filler metal that usually matches or is similar to the composition of the base metals.

The gas cylinders and torch are inexpensive and portable and can be used for several operations besides welding, including brazing, braze welding, soldering, metal cutting, heat treatment, and general heat for bending, cleaning, preheating, and the like. Drawbacks to OFW include a slower rate of heating than arc welding, a wider heat-affected area that can increase distortion and weaken the metal, and more difficult fluxing and shielding. In addition, the gases are expensive and present safety problems during storage and handling (or in generating onsite as may be the case for acetylene).

5.1 Oxyfuel Gas Welding Hazards

The oxyacetylene flame is cooler than an electric arc but still generates substantial amounts of fume and radiation. The constituents of the OFW fumes reflect the base and filler metals and fluxes used. As in the case of arc welding, painted or plated surfaces may also introduce metal fumes and toxic gases depending on their composition. This is particularly important because the heat is applied to a wider surface area.

As for arc welding, the total fume level is the best indicator of the fume hazard unless those materials or conditions listed in Table 27.1 are involved. Material safety data sheets and other available information on the metals, coatings, and fluxes should help determine the types of hazards involved.

Good natural ventilation is usually adequate except when welding on or using filler metals or alloys more toxic than iron, using fluxes containing fluoride, welding on painted or plated surfaces, or in confined spaces or tight corners. Where possible, welding hoses must be removed from confined spaces when not in use to prevent a buildup of oxygen or fuel gases from leaks in the welding equipment.

When selecting eye protection for oxyfuel gas welding where the torch produces an intense yellow light, it is important to use a filter lens that absorbs the yellow or sodium line in the visible light spectrum.

6 OTHER WELDING

Two fast growing and rapidly developing welding technologies are electron beam welding and laser beam welding.

6.1 Electron Beam Welding (EBW)

Electron beam welding uses a concentrated beam of electrons to produce localized melting and joining. Welding is performed in a vacuum chamber with a special gun that emits, focuses, and accelerates the electrons. EBW is a high-energy process that provides powerful beam penetration (up to 5 in., or 12 cm) with a high depth-to-width ratio (greater than 10:1). Shrinkage and distortion are minimized and the vacuum eliminates external or atmospheric contamination.

Although the need for a vacuum chamber is inconvenient, large chambers (up to 3000 ft³, or 85 m³) allow precision welding of many large parts. Computer control allows repositioning and performance of complex welds without the need to repeatedly shutdown the process, evacuate the chamber, manually change the setup, and pump down to create the vacuum. The process is useful where high quality and consistency are necessary and special metals or alloys are used.

The process can be used with or without filler metal and may be used to join dissimilar or difficult-to-weld materials. Its unique combination of features makes EBW well suited to weld unusual part configurations. Because of its deep penetration, it can replace multipass welding processes since reduced joint completion time offsets the demanding setup requirements and vacuum pump and chamber purge times.

EBW can be used to weld at partial vacuum or even without a vacuum, but some of the advantages, such as the high penetration and depth-to-width ratio, are lost as the atmospheric pressure increases.

Ionizing radiation is produced by the electron beam itself and also when the beam strikes the target and produces X rays as a by-product. The wall thicknesses necessary for high-vacuum welding provide effective shielding for lower voltage units. Lead shielding is usually needed for accelerating potentials above 60 kV. Radiation surveys should be made to check for leakage periodically and whenever modifications are made to the equipment.

Nonvacuum systems require enclosure by lead or high-density concrete for shielding the X radiation. Safety devices are needed to make sure no one accidentally enters or is trapped in the welding area when the equipment is in operation.

Nonvacuum and partial vacuum EBW systems also require ventilation to remove gases (ozone and oxides of nitrogen) formed when the stream of electrons reacts with air.

6.2   Laser Beam Welding (LBW)

Lasers are becoming popular for welding, cutting, drilling, heat treating, and surface alloying. Laser beam welding is finding many applications in production processes because of its increased welding rate and depth, better metallurgical properties, and greater number of material combinations than other welding processes. It may be used for both seam and spot welding but requires precision control of weld parameters (such as wave wattage, weld overlap, spot focus, and pulse duration and frequency) to assure effectiveness.

Where possible, laser welding operations should be enclosed to prevent beam reflection. All access panels should have interlocks, and shutter assemblies should be failsafe. Enclosure of the LBW operations facilitates fume and gas exhaust. In fact, collection can be so efficient that partial recirculation of shielding gases is possible. An effective laser safety program should include employee training, warning signs, beam enclosure, barrier curtains, and safety glasses that are opaque to the specific laser beam output.

7   BRAZING

Brazing differs from welding in that the metals to be joined are not melted and fused. Instead a filler metal with a lower melting temperature flows by capillary action into the close-fitting joint of the metals to be joined. The surfaces must be clean so that a wetting action can take place. This involves a molecular attraction between the base metal and brazing alloy. Oxide layers inhibit wetting of the base metal and spreading of the filler metal. One of the chief functions of the flux is to remove the oxide layer and expose clean metal.

The filler metals melt at a temperature above 800°F, or 425°C (by the definition of brazing) but below the melting point of the base metals. The two most common brazing filler rods are copper–zinc (for bronze, iron, and steel) and silver alloy made of copper–zinc–cadmium (for brass, cast iron, nickel, and stainless steels). Properly designed brazing joints when made with silver brazing alloys can be much stronger than the parent material.

The oldest and most common component of brazing fluxes is borax. At the brazing temperature, most metal oxides are dissolved by borax or boric acid. Adding fluorides or fluoroborates to a borax–boric acid mixture lowers the melting point so that the flux can be used with silver solders and also extends the range of metal oxide films that can be dissolved. For aluminum, magnesium, titanium, and zirconium the most effective fluxes are mixtures of chlorides and fluorides.

Borax fluxes leave a glasslike residue after brazing that may be difficult to remove except by chipping, grinding, or quenching (thermal shock). The chloride–fluoride fluxes are water soluble and must be removed to avoid corrosion.

Brazing alloys may be introduced into the joint by face feeding or by preplacement. Face feeding is done by holding the alloy by hand and feeding it to the joint at the proper temperature. Preplacement is done by placing the filler metal into the joint before brazing in such a way that it flows into the joint at the proper temperature. Preplaced filler metals may take the form of rings, washers, shims, slugs, powder, or sprayed-on coatings.

Brazing may be done using an oxyacetylene torch to melt the filler metal, but the necessary heat can also be applied in a furnace or by induction, resistance, or dipping. Furnace brazing is applicable for high-production brazing where the parts can be assembled with the filler metal in the joint. Inert atmospheres can be used to help prevent oxidation of the metal surfaces during furnace brazing and to eliminate the need for fluxing. Dip brazing can be done using a chemical (molten salt) bath or molten metal bath. The molten metal bath is usually used for small parts such as wires.

7.1 Brazing Hazards

The principal hazards from brazing are caused by gases and fumes from the filler metals and fluxes. The higher the temperature, the more fume is produced. Cadmium, a component of silver solder alloys, and fluorides, which are found in many brazing fluxes, are the primary health concerns. The material safety data sheets or other sources of information on composition of fluxes, filler metals, and base metals should be consulted prior to the brazing job.

Materials or conditions listed in Table 27.1 will generally determine what type of sampling is needed to evaluate the exposure. Fluxes are normally the major component of brazing fume. However, metal levels reflect the composition of the filler metal and, especially where cadmium is present, may be the major health hazard.

8 OXYGEN AND ARC CUTTING

8.1 Oxygen Cutting

Oxygen cutting is the process of separating metals by means of a flame, most commonly an oxyfuel flame. When steel is heated to between 1400 (760°C) and 1600°F (870°C) in the presence of oxygen, the iron burns to form iron oxides (Fe_3O_4), which melt and run off as molten slag exposing more iron to the oxygen jet. As much as 30–40% of the metal is washed out with the slag as metallic iron. Once the flame has pierced the metal to be cut, the cutting is done mainly by the burning of the iron by the high-pressure oxygen. However, fuel gas (acetylene, hydrogen, natural gas, or propane) may still be used after that point to preheat the metal.

Although manual oxygen cutting is common, machine cutting is also quite widely used. Frequently, several torches are mounted on a fixture. As the fixture moves in response to programmed instructions or to trace a complex pattern or print, the torches move together to simultaneously cut several parts out of steel plate with great speed and accuracy.

Fuel gas preheating can be eliminated or minimized by adding a low-carbon steel wire to the cutting torch before it reaches the material. As this wire is burned a superheated iron oxide is formed that allows much faster cutting.

8.2 Plasma Arc Cutting (PAC)

An electric arc plasma is a partially ionized gas. It is formed when a strong potential gradient causes electrons to move from a cathode toward an anode through a gas and to collide with atoms in the gas, thus ionizing them or displacing their electrons. Electrons freed by this ionization are accelerated by the strong potential gradient and in turn collide with more atoms. The number of charged particles in the gas increases until a spark can be conducted between the electrodes. The spark increases current flow, electron emission, and plasma temperature that in turn allows even further increased current flow. The plasma torch can reach temperatures of 20,000–30,000°F (11,000–17,000°C) in contrast to the oxyacetylene torch flame temperature of 6000°F (3300°C). Because the plasma of ions and free electrons possesses properties not found in solids, liquids, or gases, it is sometimes referred to as a fourth state of matter.

The plasma torch may have either a transferred arc or nontransferred arc, as shown in Fig. 27.5. The transferred arc is most commonly used for metal cutting, as a greater portion of the energy is transferred directly to the workpiece. Some plasma torches use two gases. Nitrogen fed around the tungsten electrode is used as the plasma-forming gas. Oxygen is fed to the plasma below the electrode to provide a highly reactive cutting gas.

The plasma torch has several advantages over the oxyfuel torch: speed, narrower kerf, less cost, higher temperature, and applicability to any metal. This last property is due to the fact that the plasma torch uses a melting process instead of a chemical burning process.

Compact portable plasma arc cutting machines are rapidly growing in popularity. They are often used for cutting sheet metal at low power (120 V at 20 or 30 A). Many units use compressed air as the plasma gas to precisely cut a variety of metals including those with painted, coated, rusted, or dirty surfaces. In an automobile body shop, for example, a torch can cut through sheet metal without damaging paint finish beyond a sixteenth of an inch from the cut. No preheating or surface preparation is needed, and the operation can be very simple, requiring only the press of a button for control.

8.3 Other Arc Cutting Processes

Arc cutting or arc gouging can be considered a reversal of the arc welding process. An arc is established between an electrode and the base metal as in arc welding processes, but the weld pool is removed by a gas stream.

In metal arc cutting (MAC), the electrode has a heavy coating of flux. The metal center of the electrode is consumed faster than the coating, thus forming a hollow flux tube that surrounds the arc. As gases are formed by the arc inside this flux tube, they expand to create a jet that carries away molten metal. This process is most useful for small maintenance jobs.

Oxygen arc cutting (OAC) uses a hollow metal, consumable electrode to create an arc and melt the base metal. Oxygen is forced through the center of the electrode to remove the molten metal. This process is very fast and can be used on cast iron, steel, and many other metals.

Air carbon arc cutting (AAC) uses a solid carbon electrode to create the arc. Compressed air is blown along the sides of the electrode to remove the molten metal pool. This process can be used on many metals.

8.4 Cutting Hazards and Controls

In contrast to welding operations that add metal to a joint, cutting processes are designed to remove metal, much of it to the air in the form of fumes. Many of the operations are automatic and do not require any workers in the immediate area of fume generation. However, manual processes place the cutter close to a high fume-producing source.

Cutting fumes contain many of the same constituents as welding fumes, but the mixtures are generally less complex because there are fewer fluxes, electrode coatings, shielding gases, and their associated by-products. On the other hand, alloys that could not be successfully joined may be easily cut. Thus a greater variety of materials may be encountered. Iron oxide is usually the predominant contaminant when cutting ferrous metals. For high-alloy, nonferrous, painted, and coated metals the fumes reflect the materials being cut.

Unlike welding, the surface does not have to be clean to get a good quality cut, and consequently there is a greater likelihood of hazards from surface contamination. For example, cutting or gouging on castings that contain burned-in sand may also create a silica dust hazard.

Automated flame and arc cutting operations can frequently be performed over wet tables or tanks so that most of the fumes are collected when they are carried into the water beneath the stock being cut. Exhaust ventilation can also be used for fume control on stock tables for automated cutting operations by enclosing the sides of the table and pulling air from beneath the stock support surface. For plasma arc cutting, a water shower can be used around the torch to reduce fume dispersion. Manual cutting operations can employ similar fume control techniques taking advantage of the direction of the cutting swarf.

Gouging operations are much more difficult to control than cutting operations because the fume is blown out in several directions, including back toward the operator. Exhaust ventilation may help control the exposure, but personal protective equipment is usually necessary to control the respiratory hazard.

When available, material safety data sheets and other information on the materials involved will point to the most important hazards. It is important to find out as much as possible about the fumes that may be released. If still in doubt, the operation should be performed with local exhaust or with respiratory protection appropriate to the worst hazard that could be reasonably anticipated (e.g., cadmium).

Noise hazards are greater with some arc cutting operations than with corresponding welding processes. The air or gas flow is faster and the resulting noise is higher when cutting because the flow must be great enough to remove metal from the arc. In contrast, the gas is used in welding to create a calm stable environment around the arc.

Where noise is a problem (arc cutting operations), the exposure can be minimized by reducing the air or gas flow to the lowest possible level that still effectively removes metal. For large, stationary plasma arc torches, a water shower can reduce noise as well as fume. If the power source is a substantial contributor to the noise level, some rearrangement of the work area or equipment may provide distance or shielding to reduce the noise. For many manual cutting operations, there may be no noise control solutions short of personal protective equipment for the operator.

Enclosure of the operation and protective clothing and equipment are the methods needed to control radiation from arc cutting operations. For larger plasma torches with water showers, UV absorbing dyes can be added to the water to reduce the exposure. As in oxyfuel gas welding, the protective lens shade for oxyfuel gas cutting operations should account for the yellow or sodium light in the visible spectrum.

9 AUTOMATED PROCESSES

Most of the welding processes discussed can be automated. In fact, several of the processes are never truly manual, being either semiautomatic or fully automatic. Submerged arc and GMAW are examples. There are many ways of automating welding processes ranging from simple traction devices to complex multiposition robot-operated welding jobs.

Advantages of automated welding processes are:

- High weld speed and lack of process interruptions.
- Improved quality and precision.
- Consistency and uniformity.
- Reduced need for skilled operators.
- Lower breathing zone contaminant levels.
- Ability to make multiple welds simultaneously.

Disadvantages include the inability to adjust to changing conditions such as a variation in workpiece joint shape or size. Initial cost is also much higher and setup time much greater for automated operations. Thus automation is most suitable for repetitive jobs or parts with consistent dimensions.

Automated processes require movement of the welding head or the workpiece or both, and usually involve spot weld or wire-fed GMAW systems. One advantage of being able to position the joint is to increase the speed of welding. For example, arc welding in the vertical or overhead position requires special electrodes with lower deposition rates. Changing to the flat or slightly tilted position can sometimes allow the weld speed to double or triple.

The four most popular forms of automated arc welding processes are GMAW, SAW, GTAW, and PAW. Many resistance welding jobs are also automated. Robots are used for many GMAW and resistance welding operations and are especially useful because of their ability to track a moving part along an assembly line and to quickly move heavy welding units into position around a part.

9.1   Automated Welding Hazards and Controls

Although the nature of the hazards are the same, automated processes are generally less hazardous than manual processes of the same type because operators can be farther away from the fume, noise, and radiation source. On the other hand, the high percentage of actual weld time and the ability to make several welds simultaneously mean that the hazard source itself may be greater.

9.2   Measurement and Control of Automated Welding Hazards

Because automated processes are usually stationary, they frequently lend themselves to enclosure and local exhaust ventilation for effective control. In contrast to manual welding or cutting operations, exhaust hoods can be placed a few inches from the electrode or torch without interfering with visibility or operator movement. The extra weight of hoods and duct work that are attached to fixtures to follow moving torches or electrodes does not slow the operation down as might be the case with a manual process.

10   CLEANING AND FINISHING

Welding and cutting create three types of problems to be addressed when finishing a product:

- Slag and scale must be removed.
- Rough or uneven surfaces must be smoothed.
- Metal characteristics may have been altered during welding and must be restored.

Slag and scale are usually removed by chipping, either by hand (with a wire brush or chipping hammer) or with a pneumatic chipping tool. Rough surfaces left after cutting or welding can be made smooth by grinding off excess metal and/or filling in pits or depressions. Metallurgical properties that are lost or changed during welding can be restored by appropriate heat treatment methods, including some that use fuel gas torches identical to those used for cutting and welding.

After these steps have been taken, the part is ready for assembly, painting, plating, or other additional processing.

10.1   Cleaning and Finishing Hazards

The particulate matter generated during chipping is seldom a respirable dust hazard because of the large particle size. Some respirable dust will be generated, however, and this can be a problem where more toxic metals are present. Grinding operations also generate dust, but it is smaller than chipping particulate and, therefore, more respirable. Filling in depressions can be done by soldering, brazing, or using chemical compounds such as epoxy fillers. These processes may introduce metals (lead or cadmium) or chemicals that require special handling procedures that are discussed elsewhere in this text.

Ventilation systems used for welding and cutting fume control are often adequate for dust control as well. The dust particles from chipping and grinding are larger and are produced with more directional force than are the fumes from cutting or welding. Consequently they require greater capture velocity for complete control, but complete control is not usually necessary for ordinary welding materials.

Eye protection from flying particles is of particular concern during chipping and grinding operations. Welders who also chip and grind need to wear separate eye protection beneath their helmet for when they flip the helmet up to clean the weld.

Chipping and grinding are noisy operations although the exposure is usually brief or intermittant when performed in connection with welding or cutting. Some noise control can be achieved by muffling exhaust ports on pneumatic tools or running exhaust ducts away from the operator. Dampening or constraining metal vibration of large radiating surfaces can also reduce the amount of noise due to metal impact. Working in booths lined with acoustical absorbing material can reduce noise transmission to other work stations, but the combustibility of absorption material when exposed to hot flame or welding sparks must be considered. Despite the application of various noise control techniques, personal protection is often necessary for chipping and grinding operations.

Another potential chipping and grinding hazard is segmental vibration from pneumatic power tools. Because the nature of the task usually involves frequent interruptions of the exposure, the hazard is much less than for continuous chipping or grinding work.

Vibration reduction can be obtained in several ways, including process substitution (grinding for chipping), wearing padded gloves, using two handles on a tool instead of one to minimize hand gripforce, and insulating or not using the exhaust port as a handle. Cold temperatures aggravate vibration problems, and pneumatic exhaust ports get cold due to the expansion of compressed air.

11 WELDING INSPECTION

Several nondestructive test methods may be used to detect various welding defects. The most important defects to discover are improper dimensions or alignment, cracks, inadequate penetration, lack of fusion, porosity, and slag or oxide inclusion. These defects can cause performance failure or poor fit. Once a defect has been found, it can often be cut out by grinding, chipping, or gouging and then repaired by rewelding.

For welding operations the most common inspection methods are visual, magneticparticle, liquid penetrant, radiographic, and ultrasonic. Visual inspection, by far the most widely used technique, can be used at several stages of the welding process to discover most problems that are detectable from the surface of the weld (dimension, alignment, roughness, scale, cracks, lack of fusion, inadequate penetration, and porosity).

Magnetic particle testing can be used for surface and near subsurface problems, but only in ferromagnetic materials. Liquid penetration inspection is good for testing leaks and is fairly easy to interpret but can only detect surface problems. Radiographic inspection can detect both surface and internal imperfections and can be used on both magnetic and nonmagnetic materials. Ultrasonic inspection, perhaps the most sensitive and versatile method available, can be used for surface and subsurface welds on all materials.

Leak testing on tanks or vessels employs a variety of other nondestructive techniques. These include filling with oil or kerosene, submersion in a water tank, sniffing for Freon leaks, and measuring tracer gas migration with a mass spectrometer.

For a more complete description of these inspection methods and for information on their hazards and controls, see Chapter 22 on inspection methods.

GENERAL REFERENCES

Alpaugh, E. L., K. A. Phillippo, H. C. Pulsifer, Ventilation Requirements for Gas-Metal-Arc Welding versus Covered-Electrode Welding, *Am. Ind. Hyg. Assoc. J.* **29**(6), 1968.

Arc Welding and Cutting Noise, AWN, American Welding Society, Miami, 1979.

Arc Welding and Your Health, American Industrial Hygiene Association, Akron, 1984.

Characterization of Arc Welding Fumes, CAWF, American Welding Society, Miami, 1983.

Industrial Ventilation–A Manual of Recommended Practices, 19th Ed., American Conference of Governmental Industrial Hygienists, Cincinnati, 1986.

Marshall, W. J., D. H. Sliney, T. L. Lyon, N. P. Krial, and P. F. Del Valle, *Nonionizing Radiation Protection Special Study No. 42-0312-88, Evaluation of the Potential Retinal Hazards from Optical Radiation Generated by Electric Welding and Cutting Arcs, December 1975–April 1977,* U.S. Army Environmental Hygiene Agency, Aberdeen Proving Ground, MD, 1977.

Randolph, S. A., and M. R. Zavon, Guidelines for Contact Lens Use in Industry, *J. Occup. Med.* **29**(3), 1987.

Recommended Safe Practices for Electron Beam Welding and Cutting, F2.1, American Welding Society, American Welding Society, Miami, 1978.

Recommended Practices for Gas Metal Arc Welding, C5.6, American Welding Society, Miami, 1979.

Recommended Practices for Gas Tungsten Arc Welding, C5.5, American Welding Society, Miami, 1980.

Recommended Practices for Plasma Arc Cutting, C5.2, American Welding Society, Miami, 1983.

Recommended Practices for Plasma Arc Welding, C5.1, American Welding Society, Miami, 1973.

Recommended Practices for Resistance Welding, C1.1, American Welding Society, Miami, 1966.

Recommended Safe Practices for the Preparation for Welding and Cutting of Containers That Have Held Hazardous Substances, F4.1, American Welding Society, Miami, 1988.

Safety and Health in Arc Welding and Gas Welding and Cutting, NIOSH Pub. No. 78–138, 1978.

Safety in Welding and Cutting, Z49.1–88, American Welding Society, Miami, 1988.

Schwartz, M. M., *Electron Beam Welding,* Welding Research Council Bulletin No. 196, July, 1974.

Thermal Spraying: Practice, Theory, and Application, TS, American Welding Society, Miami, 1985.

The Welding Environment, American Welding Society, Miami, 1973.

Welding Fume Control, A Demonstration Project, WFDP, American Welding Society, Miami, 1982.

Welding Handbook, 8th ed., American Welding Society, Miami, 1987.

Welding Health and Safety Resource Manual, American Industrial Hygiene Association, Akron, 1984.

CHAPTER TWENTY-EIGHT

The Engineer's Responsibilities for Controlling Occupational Disease and Preventing Accidents

Dennis J. Paustenbach

1   INTRODUCTION

The early 1900s saw many engineering triumphs, ranging from the marvels of sky-high office buildings and railroads spanning continents to the more mundane but still spectacular feats of sanitary engineering that were to a large part responsible for the virtual elimination of many communicable diseases. Many problems have been solved by engineers, but as a result of their work, other problems have developed that are begging for solution. The task of designing a nuclear power plant or constructing a spacecraft to reach the planets captures our imagination; but, significant engineering challenges found next door in the common everyday workplace often escape our attention.

Fifteen years ago, safety and health concerns were not a typical agenda item in the boardrooms of America. Many engineers and managers in industry recognized that safety and health issues "should have been" considered when a new process was developed, but their "accountability" for anticipating and preventing these types of potential problems was unclear. The responsibilities of engineers have increased since that time. Now, when new projects are tackled, we think not only about what our employer expects of us but also what is expected by society. Perhaps some of our awareness is due to incidents such as the chemical plant explosion at Flixborough, England, the gas explosion at Mexico City, the space shuttle disaster, asbestos litigation, or daily newspaper reports about the hazards of toxic substances. In more recent times, the incidents at Three Mile Island, Bhopal, and Chernobyl have reminded us that not only do we have to be concerned with the workplace but also with the potential impact of an industrial site on the community.

Notwithstanding the past accomplishments of engineering, hazards to the health and well-being of workers, the community, and the environment have not always been recognized and thus unknowingly have been designed into many modern processes. In an earnest effort to protect the safety and health of workers and the community, federal and state governments have responded to societal concerns by promulgating regulation upon regulation (Table 28.1). Not all engineers, however, are aware of the applicable regulations and the agencies that enforce them. As a result, compliance is not always achieved, liabilities may be increased, and many hazards are left unchecked.

Like other professionals, engineers have certain responsibilities over and above those of the nonprofessional. Unlike some other professionals, such as policemen, lawyers, physicians, or professors, many of the responsibilities or duties of engineers have not been explicitly enumerated. However, as professionals, it is clear that engineers have responsibilities that go beyond pure engineering. Therefore, in spite of the lack of clarity in their duties, it is apparent that engineers need to be aware of those factors that might increase the likelihood of disease or accidents and to minimize those in their designs.

2 EARLY HISTORY OF ENGINEERING

The history of engineering coincides with man's material and social development. The first examples of engineering came about around 6000 B.C. when early societies gave up the nomadic life and sought to construct permanent homes for improved security. Buildings were erected, soil tilled and irrigated, foods processed, homes heated, and waterways diverted. All of these changes to the surroundings were brought about through engineering, and, unfortunately, these alterations often produced undesirable by-products. Greater congestion of people resulted in sanitation and waste disposal problems, the burning of fuel for heating dwellings polluted the air, and the tilling of soil and irrigation increased the potential for flooding (1, 2).

As early as 3000 B.C., the communities cultivating the alluvial plains of the Nile,

Table 28.1. Federal Legislation Dealing with the Manufacture, Use, Transportation, Sale, or Disposal of Hazardous Materials

Year	Legislation
1899	River and Harbor Act (RHA)
1906[a]	Federal Food, Drug and Cosmetic Act (FDCA)
1947[a]	Federal Insecticide, Fungicide, and Rodenticide Act (FIFRA)
1952	Dangerous Cargo Act (DCA)
1952[a]	Federal Water Pollution Control Act (FWPCA)
1953[a]	Flammable Fabrics Act (FFA)
1954	Atomic Energy Act (AEA)
1956[a]	Fish and Wildlife Act of 1956 (FWA)
1960[a]	Federal Hazardous Substances Labeling Act (FHSA)
1965[a]	Solid Waste Disposal Act (SWDA)
1966	Metal and Non-Metallic Mine Safety Act (MNMSA)
1969	National Environmental Policy Act (NEPA)
1969	Federal Coal Mine Health and Safety Act (CMHSA)
1970	Clean Air Act (CAA)
1970	Poison Prevention Packaging Act of 1970 (PPPA)
1970	Water Quality Improvement Act of 1970 (WQI)
1970	Federal Railroad Safety Act of 1970 (RSA)
1970	Resource Recovery Act of 1970 (RRA)
1970	Occupational Safety and Health Act (OSHA)
1972	Noise Control Act of 1972 (NCA)
1972	Federal Environmental Pesticide Control Act of 1972 (FEPCA)
1972	Hazardous Materials Transportation Act (HMTA)
1972	Consumer Product Safety Act (CPSA)
1972	Marine Protection, Research and Sanctuary Act of 1972 (MPRSA)
1972[a]	Clean Water Act (CWA)
1972[a]	Coastal Zone Management Act (CZMA)
1973	Endangered Species Act of 1973 (ESA)
1974	Safe Drinking Water Act (SDWA)
1974	Transportation Safety Act of 1974 (TSA)
1974[a]	Energy Supply and Environmental Coordination Act (ESECA)
1976	Toxic Substances Control Act (TSCA)
1976	Resource Conservation and Recovery Act of 1976 (RCRA)
1977	Federal Mine Safety and Health Act (FMS&HA)
1978	Port and Tanker Safety Act (PTSA)
1980	Comprehensive Environmental Response, Compensation and Liability Act of 1980 (Superfund) (CERCLA)
1986	Superfund emergency reauthorization art (with amendments) (SERA)

[a]Revised/Amended

the Tigrus–Euphrates, and the Indus Valleys came into being due to the technological advances that produced such an abundance of foodstuff that they could use laborers to dig canals, employ artisans to manufacture tools, and support merchants who bought other produce. The Sumerians used engineering skills (long before 3000 B.C.) to drain marshes and build canals, and they were among the first peoples to use animals to draw wheeled carts. During the same period, engineers gained sufficient public recognition for having designed and supervised the building of the temple of Deis-Al-

Bahari and several obelisks for Queen Hatshepsut (1495–1475 B.C.) that large monuments were erected in their honor. The building of Avebury and Stonehenge around 1750 B.C., as well as the construction of dams and canals by the Mycenaeans, is evidence that engineers were meeting the needs of the society (1).

Over the past 100 years, for example, engineers have contributed significantly to improving man's well-being. Engineers were involved in the eradication of hookworm in the southern United States by improving sanitation, the near elimination of typhoid fever in Chicago by reversing the flow of the Chicago River, the improvement of water quality through the introduction of chlorine into public water supplies, elimination of cholera due to improved water treatment, minimization of food spoilage and disease through refrigeration, minimization of typhoid fever in public water supplies through better design of culverts and drains, and have brought about a reduction in respiratory illness by improving ambient air quality. The important role of engineers in disease prevention was emphasized as early as 1934 in von Hovenberg's classic treatise titled "Malaria Control for Engineers" (3).

The engineering profession's sense of complete responsibility for its work, including the potential adverse effects on safety and environmental health, is of such importance that this responsibility has been cited specifically in the professional engineer's code of ethics (4, 5). In addition, more detailed responsibilities can be found in the code of ethics of many professional engineering societies such as the American Institute of Chemical Engineers (AICHE) and the American Society of Civil Engineers (ASCE).

The engineering profession continued to grow because of the demand for persons who could apply the principles of the basic sciences. Due to the obvious need for buildings, roads, and sanitation, the civil engineering profession was the first of the enginering disciplines to evolve. In about 1100 B.C., the work of Tiglathpileser I marked the birth of the second engineering discipline, military engineering. Until the early 1800s, these remained the only engineering disciplines (1).

With the onset of industrialization, civilization became much more complex, and specialties other than civil or military engineering emerged. The fundamental engineering disciplines from which the specialties were to develop are generally considered to be chemical, electrical, industrial, and mechanical engineering. In the United States, university programs in all four of these disciplines were taught in the early 1800s, and these remained the primary areas of engineering until the onset of World War I. By 1950, the second industrial revolution occurred, and there was a subsequent, almost exponential expansion of technological achievement. By 1980, following this rapid advancement in mechanization and technology, over 30 specialized fields of engineering (Fig. 28.1) had evolved from the four basic engineering disciplines of the early 1900s.

3 BROADER RESPONSIBILITIES FOR ENGINEERS

An emerging recognition of the impact on human health, adverse or favorable, that can occur when man drastically alters the environment has added a new element to the classic responsibilities of the engineer. Due to the ability to analytically measure subtle changes in the quality of food, water or air, as well as our improved ability to diagnose disease, we are now able to discover some of the less easily detected untoward effects of our engineering miracles. Having recognized these problems, engineers have a responsibility to try to prevent those diseases by engineering control (3, 6–10).

Traditionally, the ingredients for sound engineering have been sound science and

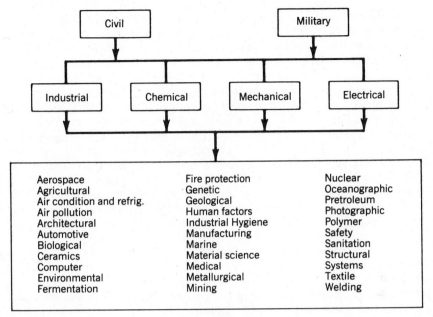

Figure 28.1. Evolution of the specialized fields of engineering resulting from the growth of scientific knowledge.

sound economics. However, since the 1950s our experience has indicated that, even when engineering contributions are well designed, environmental problems may be introduced. The exhaust gases from millions of automobiles, the stack gases of power plants, the effluent from sewage systems and waste disposal operations, strip mining, dust storms, soil erosion, nuclear waste disposal, pollution of groundwater, acid fog and rain, and the degradation of the ozone layer are but some of the hazards that have accompanied our engineering accomplishments. Having recognized that technological advances may create hazards, protection of the environment and of the worker has earned a high priority. Clearly, society has demanded that environmental and occupational health should be as much a part of the engineer's job as the economical and timely completion of the project.

Engineers, like other professionals, are expected to have a significant degree of social responsibility. It follows that engineers, as respected members of the community, should have as one of their goals the improvement of man's health and well-being. It is, therefore, not surprising that engineers have throughout history been very instrumental in helping to eliminate disease (3, 10, 11).

4 ENGINEERING, DISEASE PREVENTION, AND INDUSTRIAL HYGIENISTS

As a response to the Industrial Revolution, society recognized that the workplace can affect worker healthfulness. Along with the economic benefit of the assembly line and the movement of production away from cottage industries came the introduction of new occupational diseases. Some of these ailments and their cause included silver worker's disease, foundry worker's disease, dial painter's illness, phosphorus worker's ailment, brass worker's shakes, quarry worker's lung, and chain saw operator's syndrome, as well as a host of other acute and chronic occupationally induced illnesses. In addition, acute injuries due to safety hazards were also significant. For example, the annual death rate from industrial accidents in 1937 was 44 per 100,000 workers, which is 8 times greater than the incidence rate in 1986.

In an attempt to prevent or eliminate the cause of these disabilities and hazards, chemists, biologists, physicists, toxicologists, engineers, physicians, and others began to direct their attention to the workplace. Out of their investigations came recommendations for minimizing the risk of occupational disease. Most of the early practitioners found that by implementing so-called good hygienic practices the incidence of disease could be reduced. In the United States, these professionals were called "industrial hygienists." By the 1940s, formal training programs in industrial hygiene were available. Other industrial nations have used similar approaches to the study and prevention of occupational disease, and they have used different names to refer to these professionals (12).

The responsibilities of industrial hygienists included measuring and evaluating worker exposure to hazardous materials and physical agents; training workers in the proper use of personal protective equipment; educating them about potential workplace hazards; encouraging the use of good personal hygiene including, where indicated, the frequent washing of hands; not permitting workers to eat or smoke in the workplace; requiring employees to change contaminated clothing as needed; and

Figure 28.2. Perhaps one of the earliest engineering control devices used to prevent occupational injury or disease was the three-legged stool. If the operator went to sleep while watching for a color change in the reactor, he fell down and woke up.

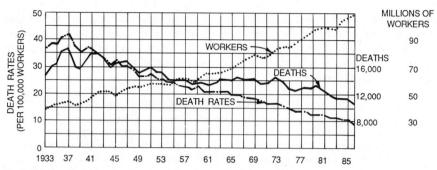

Figure 28.3. Accidental work deaths and incidence rates in the United States have fallen steadily over the past 50 years (From: *Accident Facts,* National Safety Council, Chicago, 1986.)

restricting worker exposure to overt hazards such as open vats of organic solvents. Engineers and hygienists also applied control techniques such as the use of personal protective devices, shielding, isolation, containment, and the installation and use of local exhaust ventilation. These simple procedures had a substantial effect on reducing the incidence of occupational disease. In spite of our technical sophistication, many of the procedures implemented in the early 1900s are still required today to control hazards in even the most advanced manufacturing processes (Fig. 28.2).

After industrial hygienists determined which chemical or physical agents were responsible for a malady, some engineers chose, as a course of action, to apply their special skills to the control of these workplace hazards. By the 1930s, these engineers demonstrated the importance of local exhaust ventilation and established its use as the most effective technique for minimizing worker exposure to airborne substances. They also drew attention to the importance of safe work practices, the appropriate design of machines and hand tools, use of machine enclosures for noise containment, and the reduction of ergonomic stresses (13). Through the efforts of these engineers, industrial hygienists, and safety professionals, along with the development of new personal protective devices and a stronger commitment from management, the rate of occupational disease decreased dramatically (Fig. 28.3).

5 PREVENTING DISEASE

Today, the occupational and environmental aspects of any design project may be as much a concern of the engineer as the efficiency, cost, timeliness, and overall quality of the project. Just as it is the engineer's responsibility to find ways to use scientific breakthroughs to increase profitability, engineers are likewise obligated to be aware of techniques for improving or at least protecting man's health. Growing concern over the deleterious effects of toxic substances on human health has resulted in a society demanding some protection from these agents. Since 1965, over 20 pieces of legislation have been directed at controlling the use of chemical substances in the United States (Table 28.1). Many other industrialized countries have enacted similar legislation.

Although the responsibilities of the engineer or, for that matter, any professional, are not specifically defined in the mass of environmental regulations, engineers might be held accountable if their action or lack of action causes harm to befall someone. Morton Corn, a former assistant secretary of labor and director of the Occupational Safety and Health Administration (OSHA), has offered his opinion regarding the impact of the general-duty clause of the Occupational Safety and Health Act on the engineering profession (13):

> Under Section 5(a)(1) of the Occupational Safety and Health Act, the General Duty Clause, every engineer in responsible care of a facility must treat the people and community in his responsible charge in accordance with the current state of knowledge. If anything goes wrong in the facility, he or she will be held to standard of performance, not only to the specifics of a legal standard, but to the general duty to the community and to the people who work in the facility.
>
> It has been my unfortunate privilege to work with engineers who have been brought before objective third parties to defend their performance when a tragedy occurred. Up to 9 years after the law had been enacted, they did not understand that the Occupational Safety and Health Act placed this duty on them. Under our environmental laws, the same responsibilities are placed on them as agents of the employer. The standard of care to employees and the community has lagged in the technical performance of engineering in this country for half a century.

The more than 250 technical articles and books on industrial hygiene engineering that have been published over the past 20 years probably establish a basis for the professional standards to which engineers may be accountable.

6 SOCIETAL EXPECTATIONS

Originally, the engineering profession was self-regulated. With time, machines and society became more complex, and the individuality of the engineer lessened. Although engineers were usually considered public servants, some were later seen as charlatans who were willing to sacrifice people's lives and fortunes in pursuit of self-interest (14). Following several decades of scandal, engineers in the United States initiated registration and formed technical societies. The goal of these groups was to standardize methodology, enhance public safety, and create a stable structure of technology.

It was in the 1970s that various groups brought pressure to bear on legislators to quickly generate over 30 federal laws to protect the public from the unwanted by-products of technology (Fig. 28.4). Society was apparently telling the engineering profession—among others—improve performance and do a better job of anticipating the hazards brought about by its work. Because of these higher societal expectations regarding the "standard of care," more engineers may be held accountable for what some persons will claim to be negligent behavior (15–24). If such accusations are found to be accurate, penalties are as serious as imprisonment, fines, civil liability, or, at the very least, loss of professional license.

Criminal penalties have purposefully been incorporated into recent legislation, including the Occupational Safety and Health Act and the Toxic Substances Control Act (TSCA) so as to demonstrate the accountability of persons in responsible positions. In the past, only engineers who were in management or those who were indirectly responsible for a particular task (such as consulting engineers) were held accountable,

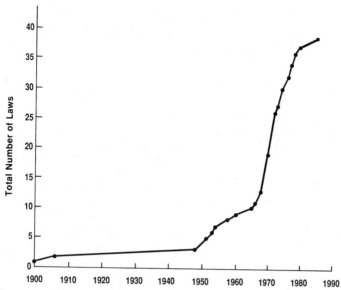

Figure 28.4. The increase in the number of federal laws regulating use, manufacture, transportation, and disposal of hazardous materials has been dramatic over the past 15 years.

but the trend has been to go further and bring suit against engineers who serve in other roles. This trend has come about due to legal considerations and the growing economic status of the profession. Frequently, successful engineers have become a "deep-pocket" target for litigation (18).

Engineering-related decisions that contain an ethical dimension are often difficult to resolve (25). Real-world decision making in the environmental arena must often be based on vague, unreliable, or imprecise data, incomplete knowledge, uncertain probability estimates, and controversial value judgments. Furthermore, decisions are influenced by many individuals of various backgrounds and training, not the least of which are politicians, lawyers, physicians, businessmen, and scientists.

Many engineers feel they constitute such a minor part of the decision-making process as to have insignificant, if any, impact on technology decisions. But, an authority on engineering ethics sees it another way (16):

> The engineer does . . . play a pivotal role. By virtue of education and experience, the engineer is better equipped than most people to foresee and appreciate problems as well as to identify and assess alternatives. This is true even when he is not professionally engaged in a technology-based enterprise; the engineer bears a very special burden. First, in the course of their work, they may be making, or influencing those who make, many decisions, large and small, that together determine critical aspects of the final product. Furthermore, it is the engineer on the job who is likely to be the first to realize that something has gone wrong. This latter point is often crucial, since the cost of rectifying serious errors generally increases steeply with time, which, in turn, reduces the chances that corrective measures will actually be taken.

Given the power of modern technology, the consequences of wrong choices or errors in execution can be massive in terms of damage to the environment and the loss of life, liberty, and property. The responsible practice of engineering is a vital element in preventing such mistakes.

7 THE ENGINEER'S SPECIAL ROLE

Although management has the primary responsibility for ensuring the safety and health of workers (and members of the community) by providing an appropriate manufacturing plant and a safe product, effective exercise of this duty depends on advice of in-house engineers and those employed by suppliers of services, equipment, and raw materials used in the business. For example, in the case of a process chemical purchased from a supplier, the supplier's duty is to notify the customer of the presence of a toxic component. The engineer responsible for the processing of this "toxic material" should be alert to the significance of this notification as well as the current techniques for minimizing the likelihood of exposure. Supporting staffs made up of many kinds of health professionals, as well as engineers, should assist in the evaluation. It is the group's role to determine what control measures are needed and to make the appropriate recommendations to management. Due to the engineer's unique training and experience, and the focal position frequently occupied by a member of this profession in corporate organizations, special responsibility is carried for being aware of common approaches to minimizing or controlling the risks identified by these other professionals.

It must be remembered that, while the role of management is to make decisions, the quality of those decisions can only be as good as the information supplied. Thus, in determining the facts and making recommendations, the engineer assumes much of the responsibility for management's ultimate decision. For example, if engineers who are responsible for designing a process fail to recognize the significance of drying a substance that is both a carcinogen and an allergen, and if they do not write detailed specifications for equipment that would prevent the escape of trace quantities of dust, or if they do not seek the advice of health professionals, how can management be accused of being at fault and irresponsible if problems later develop? (18)

The space shuttle disaster represents one of the best examples of the ethical responsibility of engineers to be sure that management is aware of the potential hazards associated with their activities, as well as the complexities of the problem. As reviewed in the press:

> The hardware problem—an O-ring seal that failed and destroyed the Challenger—remains unsolved despite efforts to redesign the booster rockets. The proposed substitute O-ring was found to leak and deteriorate in prolonged contact with rust inhibiting grease, the National Research Council recently reported. Designers have returned to the failed O-ring design—augmented by the use of heaters that, in turn, threaten to loosen other adhesive bonds. This conundrum, the council noted, exemplifies the kind of contingencies that have caught engineers unaware. (*Cleveland Plain Dealer,* Feb. 20, 1987, p. 68)

8 ACCOUNTABILITY: SOME LEGAL CONCEPTS

While actual legal responsibility is a determination that must be made based on the facts in each case, there is little doubt that engineers can, and will, be held legally responsible for their acts. The degree of accountability will clearly be affected by the particular duties of the engineer for that project as well as the predictability of the problem, but it is apparent that the engineer can be involved as a participant in expensive and difficult legal entanglements.

A brief overview of a few legal concepts might be appropriate at this point. Some excellent articles have been published in engineering journals on the topics of liability, torts, moral and legal responsibility, negligence, and product liability, which is sometimes referred to as strict liability in tort (17–24). These topics are of interest to many practicing engineers, and their very presence in these journals reflects their importance.

9 THE BASICS

Legal responsibility means that a court of law could hold an engineer responsible for harm incurred by his or her action or inaction. This responsibility can manifest itself in an award for negligence, which can result in a penalty in tort such as a judgment or award resulting in a criminal penalty, a fine, a civil liability, depending on the circumstances and the applicable law. Moral responsibility, on the other hand, has at least two general philosophical meanings. First, it is appropriate to hold a person morally responsible (accountable or answerable) for a created harmful state of affairs only if the wrong action was done under no excusing conditions. This traditional concept of moral responsibility involves a subjective test: that the engineer knows that this action is wrong and would like to restore a more just situation. A second criterion is the "future-oriented" concept of moral responsibility, and it is defined as follows: "Stating that a person is morally responsible for an action is to say that it is justifiable to blame or otherwise condemn him for having done it." The authors noted that one intent of society is that its sanctions will have some good effect on the person's or the profession's future conduct (19).

It is only partially comforting to note that ethical principles are primarily aspirational because no legal sanctions can be imposed. However, in considering the standard of care used to measure a professional's performance when sued for negligence in a malpractice suit, the questions of what the defendant knew, should have known, or should have done or not done are judged not on what this particular engineer had considered at the time but rather on what the ordinary, reasonable, and prudent engineer would have known or done. Of course, the knowledge that now exists about occupational and environmentally induced disease and its prevention makes many circumstances or hazards foreseeable that were not foreseeable years ago. Consequently, the standard of care has increased for those engineers who are responsible for specific tasks that could affect the public health. The ethical standards that a profession holds and that can be testified to at trial may very well set the mark by which one of its number may be judged. As a result, it is exceedingly difficult to clearly separate moral and legal responsibility.

The precise degree of a particular engineer's responsibility depends on the role that engineer plays in a project. For example, a design engineer working in a consulting group might have a different level of accountability than one who is involved in a production engineering function in a manufacturing facility. The magnitude of an engineer's legal liability may be greater or less than the moral liability if the responsibilities are specifically defined in his contract, such as in field engineering. On the other hand, design engineers may be held to high levels of both moral and legal responsibility because of the lack of restrictions on their job judgment (19, 26–27).

The consultant (engineer) has a legal and moral duty to recommend (design) the most appropriate action (design) in his or her professional opinion and, in case the client declines to follow that advice, to warn the client as to the possible consequences.

10 ETHICS

Engineers could ask themselves, "When might I be held responsible, and how do I know if I am acting in an ethical manner?" Lawyers see the degree of responsibility and the degree of accountability as being fairly clear: "If a person has some obligations, the neglect of which causes harm to befall someone else, it seems clear that he owes some compensation to the wronged party" (19). Criteria by which one can evaluate the ethical aspects of a decision have been offered (28), and these should be helpful to both engineers and managers.

11 TORTS

In the areas of occupational and environmental health and safety, there is some indication that torts may plague not only contract or licensed engineers but also those in-house or employed engineers who work in industry.

A tort is a private wrong, just as a crime is a public wrong, and the two sometimes overlap. Typical torts involving, for example, personal injury resulting from a defectively engineered product would be classified as (a) battery (also a crime), if the degree of culpability were determined to be intent to harm or knowledge that harm was certain, (b) negligence, if the problem was due to carelessness, or (c) strict product liability, if the defect was unreasonably dangerous even if there was no fault whatsoever. Due to the clarity of 20/20 hindsight, and to our uncertainty regarding the toxicity of many chemicals, the concept of tort liability arising from chemical toxicity presents a special problem for engineers.

As a practical matter, any flaw in design or construction that contributes to an injury or harm can lead to tort claims against the design and/or construction engineers and their employer through the vicarious liability doctrine of respondent superior, even though they performed their duties in good faith and in accordance with generally accepted engineering practices. Irrespective of the outcome of the case, the reputation of the engineer or the firm can be irreparably harmed by the mere allegation itself. It is therefore essential that the engineer carry out the duties and document the understanding of the potential hazards with great care in an attempt to "make safe" his or her every work product.

The current trend in product liability cases verifies this engineering approach to product safety by sharply focusing on the actions taken once the problem was recognized by the manufacturer. The former approach, and still the law in many states, allowed a manufacturer of a defective product to avoid liability by warning customers of the problem, as long as the warning was effective in making the product "not unreasonably dangerous." The present trend in strict product liability cases takes a hard look at whether an alternative was available to the designer that would have made the product safe. The court then tests the feasibility of the alternative by balancing the benefits (or economics) of the defective product with the inherent risk of danger (27,29).

Two cases illustrate how this feasible-alternative theory has been applied to analyze chemical hazards of products. In *Ruggeri* v. *Minnesota Mining and Manufacturing* (30) an adhesive having a flash point of $-14°F$ was too volatile and flammable for its reasonably foreseeable use in applying insulation to duct work. The manufacturer was held liable on the basis of feasible-alternative reasoning because it has developed a nonflammable adhesive in response to local codes in another jurisdiction that prohibited the use of flammable adhesives. Similarly, in *Drayton* v. *Jiffee Chemical Corp.* (31), because a liquid drain cleaner was formulated with an unnecessarily high concentration (26%) of sodium hydroxide that resulted in such immediate destruction of human tissue with no effective antidote, precautionary product labeling was insufficient to avoid liability when the feasible alternative of lower concentration of caustic was available. The message sounding from these cases is that detailed analysis of potential risks and a careful evaluation of alternatives that might make the product safer will be expected of engineers responsible for product design and manufacturers. Similar expectations may also be placed on engineers who design manufacturing facilities.

Sometimes it is extremely difficult to predict when a jury may hold a person or form liable for a particular event. The ongoing controversy regarding anyone's right to place another person at increased risk without his or her knowledge and consent (right-to-know) and the possible passage of a federal toxic tort bill make this topic a timely one for the engineering community (32,33). Recent court decisions indicate that, if liability is to be avoided, process design engineers and their employers may have to assume a responsibility beyond the scope of existing regulation or legislation. For example, only about 450 of the 10,000 commonly used chemicals are currently regulated under OSHA.

In spite of the difficulties in handling the seemingly unmanageable amount of information, management and engineers should, during the design phase of a new project, draw upon medical and industrial hygiene staffs to be sure that any hazardous materials or processes are subject to the degree of control that makes sense based on our current understanding of the potential hazards. In addition, it is important that engineers and their management be knowledgeable of those engineering control practices that are in current use in the industry so that they can feel comfortable that they have acted in as responsible a manner as possible. A clear demonstration that the employer acted in accordance with the current state of knowledge regarding a chemical's toxicity, as well as with the current state of feasible control methodology, will be helpful to any company involved in a court case similar to the one outlined in the following case study.

In 1981, a Federal District Court in Camden, New Jersey, ordered a major chemical firm to pay almost $400,000 in damages to a woman whose husband's death had been caused by a rare liver cancer, angiosarcoma, seen primarily in vinyl chloride workers. Although a financial settlement other than that awarded by the court was later reached by the two parties, this case represents a new trend toward imposing nearly strict liability when some negligence is found in cases involving toxic substances (*Grasso* v. *B. F. Goodrich*) (34).

Actually, the victim never worked for this chemical company, but he did live about two miles from the manufacturing facility. The widow's wrongful-death action was based on a negligence theory because there was no question of product liability, the husband was not a consumer of vinyl chloride, and no evidence could be produced to show an intent to harm. Therefore, the widow had to prove that the vinyl chloride caused her husband's fatal disease and that the company or its employees were remiss with respect to a duty of care owed to the man.

The causation element was provided to the satisfaction of the court by the fact that angiosarcoma has been associated with vinyl chloride exposure and that the company did manufacture the chemical at its nearby facility. This establishment of a cause–effect relationship (the most difficult hurdle in a toxic tort cause) was found despite testimony that the low concentration of vinyl chloride two miles from the plant would not be expected to pose a cancer hazard. In terms of finding the company at fault, the plaintiff was able to show that the failure to perform tests on emissions of vinyl chloride constituted a sufficient breach of a duty of care that a cause of action based on negligence was established. Given the testimony concerning the low probability of harm at the low concentrations involved, it is arguable whether tests of the emissions would have logically and reasonably led to any corrective action in any event. Nonetheless, the case illustrates the point that, when toxic substances are involved, engineers and their employers must attain the highest level of diligence or face nearly certain liability (18). Is this not an example of legalistic rationale that goes beyond any realistic engineering or technical estimate of the degree of hazard?

12 THIRD-PARTY LIABILITY

Another concern for engineers is third-party liability. Until recently, architects, engineers, and builders were not liable to third parties for negligent design or construction after the owner accepted the work. In recent years, the doctrine of privity has been abolished by most jurisdictions, so the tendency has been to hold engineers liable to third parties without limitation in time because it was held that the basis for the legal action did not come about when the work was performed but, instead, when the injury occurred. Part of the reason for such liability is that, when they construct their projects, engineers recognize that third parties will come into contact with their work, and, therefore, engineers know—or should know—that third parties can be affected by the quality of their work. The best defense that an engineer can use in such cases is that the highest standards of care, foresight, and state-of-the-art engineering controls that existed at the time of the design of the process were exercised (35).

One commentator, who has addressed the engineer's responsibility for product liability (26), noted that

> the design engineer must be aware of two sources of human error. First are those which he may commit during the design process. Second are those that the potential user may commit when he attempts to use the product.
>
> When called as a witness to defend his design decisions in a product liability suit, the design engineer must be able to show that both concepts of human fault were considered as part of the design process. Additionally, it is essential that he be familiar with methods of preventing these human errors.

It is clear that engineers are expected to be concerned with the implications of their design as well as to be familiar with the experience of other engineers who have successfully minimized human risk through their design approach. However, in spite of the apparent reasonableness of this duty on the part of the engineer, schools of engineering devote little or no time to safety or occupational disease awareness, nor to an understanding of the available engineering control techniques (36, 37).

13   POST-BHOPAL ERA

The late 1980s will probably be known as the post-Bhopal era in environmental health since it appears to have shifted public concern. Policy makers, regulators, local officials, and townspeople broadened their concern about the risks to individuals chemicals (e.g., PCBs, mercury, and dioxins) to the risks posed by an entire chemical manufacturing plant or other facilities that use chemicals. Due to these concerns, numerous companies took it upon themselves to develop strategies to identify high-risk facilities and take the necessary steps to minimize those risks.

It is likely that, in the next few years, engineers, often as managers, will be expected, either on their own or in response to regulatory requirements, to identify and often to disclose the nature of the possible risks presented by their plant sites or individual processes within them. Further, they will probably be asked to provide assurances of the adequacy of their risk management programs (38). Proposition 65, an initiative passed by the citizens of California in 1986, is among the first of such far reaching laws that require that significant and important information is shared with the community. Other initiatives will almost surely be passed in other states.

14   CAN ENGINEERS MORALLY JUSTIFY INCREASING RISK?

A basic theme underlying both legal and moral responsibility is the engineer's responsibility for anticipating hazards and judging what is acceptable risk. Some persons might accuse the engineer of being both judge and jury with respect to deciding "how much is too much risk?" For example: How many safety factors should be applied to the stress calculations of a bridge design? Should automobiles be capable of withstanding a 40-mph head-on collision? How much reinforcing should be applied to buildings in areas prone to earthquakes? There are, however, very few guidelines or standards for an engineer to follow. As in many professions, the community of engineers decides what constitutes minimum standards of performance.

Under the mandate of the Occupational Safety and Health Act of 1970, the engineer, as an agent of his employer or as a consultant, is among the professionals whose role is to aid the employer in "assuring so far as possible every working man and woman in the Nation safe and healthful working conditions." Unfortunately, "safe and healthful working conditions" cannot be defined in an absolute sense; short of prohibiting all work, some level of occupational risk must be judged both as to efficiency and equity. The competitive market outcome satisfies the efficiency criterion. Levels of health and safety, however, may be distributed unequally; in general, poor people, who will often accept a smaller monetary premium to incur risk, will be exposed to greater risks. Also, efforts to redress these perceived inequities may actually diminish the welfare of the poor by denying certain employment opportunities (39). This situation places the engineer and the employer in a predicament.

The separation of responsibilities between employer and engineer is clear. Management must decide whether the economic benefits of a project justify the expense. The employer expects that the engineer will act as a faithful agent and that "the engineers shall hold paramount the safety, health and welfare of the public in the performance of their duties" (13). Ultimately, it is the engineer who is burdened with the responsibility of deciding what risks can be reasonably accepted in the design.

The civil engineer, for example, must decide whether the costliness of building a large radius cloverleaf justifies the lives that will be saved since the curves could have simply banked at a cost of much less money. Similarly, chemical engineers and mechanical engineers must decide whether it is necessary to specify special equipment such as costly leakproof pumps, airtight centrifuges, and local exhaust ventilation whenever very toxic or carcinogenic chemicals are used in a chemical process.

15   IS IT NEGLIGENCE OR ACCEPTABLE RISK?

The courts have defined negligence as the creation of an unreasonable risk causing harm not intended. However, due to the increasingly complex nature of the engineer's tasks and responsibilities and the need to continually evaluate risk, this simple definition may not be adequate. Society, through the law, has generally decided that risks will be judged reasonable or unreasonable in the courtroom according to the reasonable-person standard. One problem with this approach is that the jury may assess, in spite of possible warnings by the judge to do otherwise, the quality of the engineer's risk decision from an "after-the-fact" viewpoint since the harm has already occurred. With respect to occupational disease prevention, this problem is particularly troublesome since the diseases of work may not manifest themselves until 10–30 years after the initial exposure.

In order for engineers to make good risk decisions, they should be sensitive to the difference between voluntary and involuntary risk. It is one thing for persons to decide to ski, race motorcycles, climb mountains, smoke, drink, or eat to excess, but it is another for someone else to impose upon a person a risk of comparable magnitude. Studies have indicated that most persons will accept voluntary risks in the range of 1 in 1000 (Table 28.2). For instance, driving a car 300,000 miles or smoking 1400 cigarettes over a lifetime produces an estimated individual fatality risk of 1 in 1000 (40). On the other hand, some people, and even some government agencies, believe it inappropriate for persons or groups to impose a situation on someone (involuntary risk) in which the individual fatality risk is greater than 1 in 1 million; other groups feel that it is appropriate for involuntary risks to be as high as 1 in 10,000 (41, 42).

Table 28.2. Activities or Exposures That Will Reduce Life Expectancy by 8 Minutes Due to the Increased Likelihood of Having Cancer[a]

Smoking 1.4 cigarettes
Living 2 months with a cigarette smoker
One X-ray (in a good hospital)
Eating 100 charcoal-broiled steaks
Eating 40 tablespoons of peanut butter
Drinking 10,000 24-ounce soft drinks from recently banned plastic bottles
Drinking 30 12-ounce cans of diet soda containing saccharin
Living 20 years near a polyvinyl chloride plant
Living 15 years within 30 miles of a nuclear power plant

[a]From R. Wilson, A Rational Approach to Reducing Cancer Risk, *New York Times*, July 7, 1978.

Engineers should be sensitive to two key issues. First, does a "significant" risk exist? Second, are there feasible alternatives for avoiding a particular risk? Since the worldwide economy weakened in the 1970s, many persons feel that they are becoming less able to quit jobs that are hazardous; consequently, society is progressively considering job-related risks nearly involuntary in nature. As a result, regulators have arbitrarily established that risks that are unknowingly or knowingly thrust upon individuals should be approximately 1 in 100,000, or about 100- to 1000-fold less hazardous than risks that an individual might assume voluntarily. This risk criterion primarily applies to persons who will receive no direct benefit from the activity (42).

How do occupational and environmental health risks fit into this scheme? In the area of disease prevention, due to the insidious and disabling nature of some ailments, the worker and society are less likely to accept significant risk. Disabling diseases due to cotton dust, coal dust, silica, and mercury usually occur without warning 10 or more years after initial exposure. All of these agents are currently regulated, and exposures have been dramatically reduced over the past 20 years. More recently, the public and the worker have turned the bulk of their attention to the prevention of cancer. Due to the likelihood of death or suffering, society has forced the scientific and engineering communities to carefully assess the need for voluntary and involuntary exposure to carcinogens.

In general, any estimate of the likelihood of an event (risk) that has not yet occurred, such as a serious nuclear accident or an earthquake, will by its nature have a high level of uncertainty. Worse yet, for the environmentally induced diseases, we usually cannot determine, with precision, how much exposure is associated with a particular level of safety. The uncertainty in the risk estimation procedure seems to make the public fearful; consequently, it is willing to accept only very low levels of risk. For example, exposing the general public to potential cancer-causing agents when it stands to gain only indirectly from the exposure (e.g., brake linings from asbestos plants) generally requires a larger safety factor in the design of the plant than that normally considered acceptable in other situations. In the case of miniscule levels of a carcinogen in drinking water, a risk of only 1 in 1 million might be all that society wishes to accept (42).

16   RISK ASSESSMENT AND THE ENGINEER

How can engineers perform their duties without fear that normal human fault will produce disastrous consequences? The answer is that society does not demand of engineers absolute perfection in their decisions or require zero risk. Engineers are expected to weigh the benefits against the risks of their actions and make reasonable judgments. Building a dam presents some risk of catastrophe, but the risks are usually justifiable in view of the enormity of the benefits. Similarly, engineers who use the best possible control techniques in an attempt to safely manufacture lifesaving pharmaceutical agents are generally considered to be acting in a responsible manner even though the manufacturing process could be responsible for producing a slight increase in the likelihood of disease in production workers. Few guidelines or standards exist for an engineer to follow when grappling with these questions, and, in fact, the community of engineers usually decides what constitutes minimum standards of performance (43).

17 WHAT ABOUT FEASIBILITY?

Existing OSHA standards require that hazardous air contaminants be controlled to levels that are equal to or below standard. They require that, before relying on personal protective equipment, "administrative or engineering controls must first be evaluated and implemented whenever feasible." New proposed standards, although they vary somewhat in wording, generally echo this preference for engineering controls as opposed to personal protective equipment (47). The reliance on control techniques such as ventilation and enclosure is a controversial one because the short-term use of personal protection equipment, usually respirators, has traditionally been considered an acceptable and appropriate method for protecting workers. Many health professionals and engineers believe that it is a waste of valuable resources to install costly controls like local exhaust ventilation if the use of a respirator for 15 min, four to six times a day, protects the worker without undue physical or emotional hardship. Respirators have allowed some firms to stay in business; if engineering controls were required, this would not have been financially possible. Consideration should be given to the development of a cost–benefit scheme wherein a limit would be placed on the amount of money required to be spent per worker on engineering controls. Above this amount, long-term use of respirators would be considered appropriate.

Neither "feasible" nor "practicable" was defined in the OSHA law. This was not an oversight by the regulatory agency, which intended these terms to be "moving targets" (sometimes called science forcing), so that industry would be encouraged to investigate new ways to minimize worker health hazards. However, as noted by Caplan (47):

> Expert witnesses retained by OSHA sometimes recite the entire textbook list of engineering control principles as being feasible engineering controls: (a) substitution of a less hazardous substance, (b) change of process, (c) isolation, enclosure, or segregation, (d) local exhaust ventilation, (e) general or dilution ventilation—without making any realistic assessment of the practicality of applying such controls to the problem at hand, or the results that would be achieved, or the cost, or the benefits.

Whenever engineers are asked to prepare a cost–benefit analysis, it is important that the engineer be sure that the analysis reflects current societal expectations for prevention of occupational disease, and that he is familiar with the currently accepted meaning of the term "feasible." The most reasonable and most thorough definition of feasible engineering controls for prevention of occupational disease has been offered by Caplan (47):

> "Feasible means that the method or equipment is available on the market and has been used before with success in the same or closely similar applications, or that the technology exists to create the equipment and implement the method with reasonable assurance of success; that the method or equipment will result in reducing the exposure to or below the time-weighted average standard or is necessary to reduce the exposure to a level where the reasonable use or administrative controls or personal protective equipment which is not unduly onerous to the employee will adequately protect the health of the employee; that the number of employees exposed, and the frequency and severity of the exposure, are included in a cost-effectiveness consideration of the implementation

of such controls; and that seriousness of the potential risk to the employee health is given due weight in all considerations.

18   THE COURTS' THOUGHTS ABOUT RISK

During the 1970s, it was apparent that society wanted regulatory agencies to ask more of industry in the area of occupational and environmental disease prevention. Dozens of books and articles described the recent results of toxicity and epidemiology studies. The mood of organized labor and federal agencies was to minimize exposures to toxicants, especially carcinogens, to levels as low as feasible or, in some cases, as low as measurable! The phrase "as low as feasible" was problematic since feasible was ill-defined and "to undetectable levels" was a moving target due to the continual improvement in analytical instrumentation. Numerous chemicals and groups of chemicals were scheduled for this degree of regulatory control until two Supreme Court rulings, one in 1980 and the other 1981, seemed to cast doubt on the wisdom of this approach. These rulings confused many practicing engineers and their employers who had thought they would always be expected to implement costly state-of-the-art safety and health control technology in their facilities.

The Supreme Court evaluated and made rulings on two substances that possessed chronic toxicity: benzene and cotton dust. Benzene had been shown to cause disorders of the blood system in humans following prolonged exposure to air concentrations of 30–50 ppm. In response to recent epidemiology studies, OSHA proposed to lower the existing permissible exposure limit (PEL) from 10 to 1.0 ppm. This move was based on the agency's position that no level of exposure to a carcinogen would be considered free of risk and because it was likely that there would be fewer cases of benzene-induced leukemia if exposures were lowered from 10 to 1.0 ppm. OSHA also claimed that engineering controls were currently available to attain these levels, although they acknowledged that these would be costly. The court supported the industry claim that the proposed limit was unnecessarily low and that 10 ppm was an appropriate limit (48).

In the Supreme Court decision involving cotton dust, OSHA proposed to lower the existing PEL from 1000 $\mu g/m^3$ to 200 $\mu g/m^3$ on the grounds that dust levels of 1000 $\mu g/m^3$ in yarn manufacturing had caused a definite increase in the incidence of bysinossis in exposed workers. OSHA claimed that the reduction to 200 $\mu g/m^3$ would markedly reduce the incidence of disease and that engineering controls were currently available for bringing about compliance. The regulated community claimed that such a standard was too stringent and that compliance was so costly as to jeopardize the survival of the cotton industry. Early estimates of costs to the textile industry for control of the cotton dust ranged from $500 million to $1 billion. Ultimately, the court decided to support OSHA's decision to reduce the exposures (49).

What is the import of these two cases? The cases were alike in that the control technology was available, albeit costly, and the regulated groups were uncertain that they could afford to comply. Engineers need to recognize that there was a basic difference in the two decisions, and the logic used by the court might affect the way engineers go about their duties.

The OSHA-proposed standard for benzene was rejected by the court not because it was costly and not because current engineering controls may not have been sufficient to attain the standard. Instead, the standard was not upheld because OSHA failed to demonstrate that lowering the air concentration from 10 to 1.0 ppm would provide additional protection to workers, that is, significantly reduce the incidence of leukemia. In fact, OSHA had failed to prove that exposures to 10 ppm could actually affect the cancer rate, as noted by Justice Stevens in the plurality opinion. In short, the Supreme Court wanted to convey the message that, although employers should be expected to protect workers' health, it is not appropriate to expend valuable resources and place financial burdens on the employer to reduce exposure unless it is clear that workers' risk of injury will also be reduced.

In the cotton dust decision, however, OSHA had demonstrated to the satisfaction of the Court that the limit for cotton dust was not sufficiently low to protect workers from disease. In addition, the Court was convinced that, if industry met the proposed standard, significantly fewer people would develop bysinossis. Testimony by industry that the proposed standard would be likely to bankrupt some manufacturers yielded the following response from Justice William Brennan, who wrote the plurality opinion (49):

> In the Occupational Safety and Health Act (OSHA) congress chose to place pre-eminent value on assuring employees a safe and healthful work environment. Congress itself defined the basic relationship between the cost and benefits by placing the benefit of worker health above all other considerations save those making attainment of the benefit achievable. Any standard based on balancing cost and benefits by the Secretary that strikes a different balance than that struck by Congress would be inconsistent with the command set forth in Section 6(b)(5). (*New York Times*, June 27, 1981)

> [Section 6(b)(5)—The Secretary, in promulgating standards dealing with toxic materials or harmful agents under this subsection, shall set the standard most adequately assures to the extent feasible, on the basis of the best available evidence, that no employee will suffer mental impairment of health or functional capacity even if such employee has regular exposure . . . for the period of his working life. In addition to the attainment of highest degree of health and safety protection for the employee, other considerations shall be the latest available scientific data in the field, the feasibility of the standards and experience gained under this and other health and safety laws. (Occupational Safety and Health Act, 84 Stat. 1954.)]

Design and production engineers should be aware of at least two concepts evident in these decisions. First, the protection of worker health must be a primary concern of the employer, and the cost–benefit relationship is a secondary concern. Second, the employer is reminded to protect the employees from workplace illnesses and injuries to the extent feasible.

19   SOCIETAL EXPECTATIONS REGARDING DISEASE PREVENTION

Most people would agree that engineers should accept a share of the responsibility for occupational safety and health; however, it may be unclear "why" so many workers and legislators have singled out occupational hazards since they probably have a small impact on the overall state of public health. In fact, from the standpoint of hazard, one should recognize that in the United States auto accidents alone accounted for 51,900 deaths in 1979, as well as 2 million disabling injuries, resulting in a total cost of $35.8 billion (50).

By contrast, in 1986, 11,500 persons were killed on the job, or about one-fourth as many persons as were killed in auto accidents, and 2.1 million sustained disabling injuries resulting in a financial cost of about $35 billion for the same year (50). In the same year, about 400,000 persons died from cancer, and another 750,000 persons died from cardiovascular diseases. Reports that claim that workplace hazards are a major cause of serious disease, especially cancer, are simply untrue. In all likelihood, minimization of workplace exposure to chemicals will not markedly affect the incidence of cancer in this country (Table 28.3). This is not entirely surprising since the age-adjusted death rates from cancer have remained nearly constant at about 140 per 100,000 population since 1935 even though during that same period (Fig. 28.5) the production rate of chemicals and allied products (and presumably the number of workers exposed) has skyrocketed from 19 to 210 billion lb per year, a 28-fold increase (51). Two respected studies in the past 4 years indicate that occupational exposures probably account for only 4–10% of the overall cancer rate, and that personal habits or lifestyle (i.e., diet, tobacco, and alcohol) are responsible for 60–70% of the problem (Table 28.3) (52).

Since it is increasingly clear that individuals are generally in control of lifestyle factors influencing their health, why has much of the industrialized world shown such interest in controlling occupational disease? Prevention through engineering controls, better equipment and process design, employee training, and personal protective equipment have been shown to reduce occupational illness and injury. Since we know how to prevent occupational diseases, unlike many other diseases, society has placed them in a special category. Society's concern, is that

One person's cost cannot be simply and evenly traded for another person's benefit, especially if the risk of harm falls on a selected group of people within an industry or laboratory. (53)

Table 28.3. Current Estimates of Proportion of Cancer Deaths Attributed to Occupation as Well as Other Factors[a]

Factor or Class of Factors	Percent of All Cancer Deaths	
	Best Estimate	Range of Acceptable Estimates
Tobacco	30	25–40
Alcohol	3	2–4
Diet	35	1–70
Food additives	<1	−5–2
Reproductive and sexual behavior	7	1–13
Occupation	4	2–8
Pollution	2	<1–5
Industrial products	<1	<1–2
Medicines and medical procedures	1	0.5–3
Geophysical factors	3	2–4
Infection	10?	1–?
Unknown	?	?

[a]From R. Doll and R. Peto, The causes of cancer: Quantitative estimates of avoidable risks of cancer in the United States today. *J. Natl. Cancer Inst.* **66**(6):1191–1308 (1981).

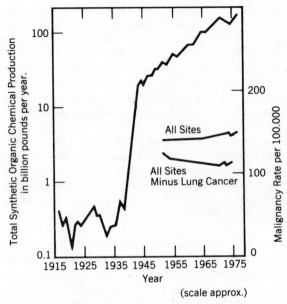

Figure 28.5. The data suggest that in spite of an increase in the presence of man-made chemicals, there has been no concurrent increase in cancer rates in the United States. (From: Cancer, Pollution, and the Workplace. A special report by the American Industrial Health Council, 1612 K Street, N.W., Washington, D.C., 1982.)

Table 28.4. Fatal Accident Frequency Rate (FAFR) for Industrial and Nonindustrial Activities in the United States and the United Kingdom[a]

FARs for Industrial Activities (UK)			FARs for Nonindustrial Activities (UK)		
Activities	Year	FAFR	Activities	Year	FAFR
Manufacturing industry	1971–74	2	Traveling by train	1963–72	3
Chemical industry	1971–74	4	Staying at home	1972	4
Construction industry	1971–74	9	Traveling by bus	1963–73	4
Mining and quarrying	1971–73	10	Traveling by car	1963–73	48
Air crew	1964–73	200	Traveling by air	1964–73	190
Professional boxers	1963–74	20,000	Canoeing	1962–72	670
			Gliding	1964–73	1,000
			Motor cycling	1963–73	1,040

FARs for Industrial Activities (US)			FARs for Nonindustrial Activities (US)		
Activities	Year	FAFR	Activities	Year	FAFR
Chemical industry	1977	3	Traveling by bus	1977	2
Manufacturing industry	1977	5	Traveling by train	1977	2
Trucking	1971–74	11	Staying at home	1977	3
Agriculture	1971–73	27	Traveling by air	1977	20
Construction	1971–74	30	Traveling by car	1977	100
Mining and quarrying	1971–73	31	Traveling by motor cycle	1977	680

[a]From Gibson, Hazard Analysis and Risk Criteria, *Chem. Eng. Prog.* **76:**46–50 (1980).

Our opinions of what is acceptable definitely depend on the degree to which we are free to opt for or decline the risk. It is one thing to choose to go skiing, drive a sports car, use a tool without safeguards, smoke cigarettes or eat the food we have sprayed ourselves; but it is quite another to breathe the air and endure the noise where we live or work (and few of us are really free to move away or change jobs). (40)

In short, society has placed the burden of providing a safe workplace on the employer even though the majority of the risks in one's life are probably left to the discretion of the individual.

However, it is important to recognize that society's value system is not fixed. The regulatory climate, which is little more than a reflection of society's expectations of the 1980s, is to some extent different than that of the 1960s. Undoubtedly, some of the support for increased regulation that existed over the past 20 years reflected a general belief that an increasingly wealthy society could "trade off" some quantity of private goods for an improvement in health and safety. However, during periods of financial duress, it is possible that we (society) may decide to lessen our demands for a low-risk existence if it is negatively impacting our personal well-being.

Even beyond the concern about personal financial impact, many persons, including many engineers, believe that regulatory bodies have gone too far in attempting to protect individuals. Such beliefs are often based on the awareness that most of life's risks are still voluntarily accepted by individuals. Examples of some of these risks are shown in Tables 28.4–28.6. With respect to occupational risks, which are generally

Table 28.5. Risk of Death by Various Causes[a]

Hazard	Total Number of Deaths		Risk of Death/Year of Continuous Exposure
All causes	1,973,003	(1973)	9.0×10^{-3}
Heart disease	757,075	(1973)	3.4×10^{-3}
Cancer	351,055	(1973)	1.6×10^{-3}
Work accidents	13,400		1.5×10^{-4}
All accidents	105,000		4.8×10^{-4}
Motor vehicles	46,200		2.1×10^{-4}
Homocides	20,465	(1973)	9.3×10^{-5}
Falls	16,300		7.4×10^{-5}
Drowning	8,100		3.7×10^{-5}
Fires, burns	6,500		3.0×10^{-5}
Poisoning by solids or liquids	3,800		1.7×10^{-5}
Suffocation, ingested objects	2,900		1.3×10^{-5}
Firearms, sporting	2,400		1.1×10^{-5}
Railroads	1,989	(1973)	9.0×10^{-6}
Civil aviation	1,757		8.0×10^{-6}
Water transport	1,725		7.8×10^{-6}
Poisoning by gases	1,700		7.7×10^{-6}
Pleasure boating	1,446		6.6×10^{-6}
Lightning	124	(1973)	5.6×10^{-7}
Hurricanes	93	(1901–1972avg.)	4.1×10^{-7}
Tornadoes	91	(1953–1971avg.)	4.1×10^{-7}
Bites and stings	48	(1970–1973avg.)	2.2×10^{-7}

[a]Based on continuous exposure of total U.S. population in 1974 except where noted; from Ref. (45).

Table 28.6. Estimated Risks for Selected Activities[a,c]

Risks	Activity
1/400	Smoking (10 cigarettes/day)
1/2,000	All accidents
1/8,000	Traffic accidents
1/30,000	Work in industry
1/50,000	Natural disasters
1/1,000,000	Driving 80.5 kilometers[b]
1/1,000,000	Being struck by lightning

[a]Risk is expressed as probability of death for an individual for a year of exposure, and orders of magnitude only are given.
[b]This risk is conveniently expressed in the form indicated rather than in terms of a year of exposure.
[c]From B. D. Flowers, Chairman, Royal Commission on the Environment; Sixth Report: Nuclear Power and the Environment, Her Majesty's Stationary Office, London, 1976.

considered to be voluntary, it has been known for many years that "work" in most industries, per se, is not as dangerous as work around the home.

20 THE NEED TO EDUCATE ENGINEERS IN OCCUPATIONAL DISEASE PREVENTION

In an earnest effort to minimize the risks associated with work, industry spends tens of millions of dollars annually on occupational safety and health. The annual cost to industry for protecting worker safety and health for the United States in 1981 has been estimated at $2 billion! About 2% of total overall capital spending is devoted to employee safety and health (54). In addition to this amount, a great deal of money is spent on local exhaust ventilation, noise control, and specialized process equipment, as well as numerous other engineering controls directed at protecting the worker. Furthermore, this expenditure represents, for the most part, retrofitting (devices or equipment added to control a hazard discovered "after" start-up of a process). In many cases, good design based on knowledge of the potential hazard would have avoided much of this cost. Unfortunately, many of these control devices are improperly designed by well meaning but inappropriately trained professionals, including engineers. Consequently, too many devices and systems are not effective in protecting the worker.

It seems clear that employers and society might be better served if engineers were made aware of their role in occupational disease prevention and were educated in the techniques for doing so. According to one safety engineering educator: "Because industry managers made no demands that engineers they employ be knowledgeable about methods of safeguarding personnel, the engineering schools did not teach the subject" (24). In addition, the author further stated that "some engineering school faculty members still ignorantly believe that safe design is nothing more than good engineering practice and avoiding failures. To them, accident prevention is telling workers to be careful and to wear hard hats and safety glasses" (24). This article also noted that in the past, "most negligence cases in product liability suits were predicted on production defects" (24).

In recent years, however, design defects have been shown to be the more significant source of accidents. Eleven years of data compiled by an association of industrial insurers indicated that most of the chemical industry accidents resulting in losses of $25,000 or more each were due to improper design. In the early 1980s, university professors in engineering and public health, as well as representatives from industry, held three symposia for the purpose of alerting engineering educators to the inadequate education of their students in occupational disease prevention (55–57).

Engineering schools should recognize that formal education in the fundamentals of occupational safety and health is really the only acceptable approach to solving the problem of inadequate training (57). The engineer's code of ethics established by the National Society of Professional Engineers states (5):

> Section 2—The engineer will have proper regard for the safety, health, and welfare of the public in the performance of his professional duties. If his engineering judgment is overruled by non-technical authority, he will clearly point out the consequences. He will notify the proper authority of any observed conditions which endanger public safety or health. (5)

It is likely that, in all cases, failure of control devices to protect workers from exposure is not an intentional, irresponsible act on the part of engineers. It appears that they are simply inadequately educated to perform the task for which employers unknowingly believe they are equipped. The result is that funds which have little hope of doing what was intended, but were appropriated by upper management who intended for them to benefit the worker, rarely perform as they were intended.

21   HOW CAN ENGINEERS MEET THESE EXPECTATIONS?

Since very few practicing engineers have had formal training in the anticipation, recognition, and control of occupational disease, what can they do to prepare for these responsibilities? First, they should increase their sensitivity to published manuscripts on occupational health that appear in their professional journals. Second, if they are in industry, they should work more closely with members of the occupational health and safety staff, who should be generally familiar with engineering control techniques that have been shown to work, as well as the best references on the topic. Third, they should contact local universities to see if courses on related topics are available. These courses may not be offered by the school of engineering so a search of the catalog may be useful.

22   WHAT AN ENGINEER SHOULD KNOW ABOUT HEALTH AND SAFETY

Educating engineers about occupational and environmental hazards has, in the past, taken place mainly on the job through association with more experienced people. This approach has merit but, at best, is haphazard and "is done at the expense of the public" (24). In recent years, there have been attempts to bring this type of training into the undergraduate engineering curriculum using different approaches. These approaches have been summarized in the literature (58–61).

The following "skills" are what the writer considers to be the overall objectives of a good engineering course on this topic and represents what an employer expects of mechanical, industrial, environmental, and especially chemical engineering graduates.

- Know how to write new equipment specifications that describe minimal performance standards so that the potential for future occupational safety and health problems will be lessened.
- Understand the limitations of local exhaust ventilation hoods and the fundamental design criteria.
- Understand the usual causes of industrial fires and explosions and the engineering techniques needed to prevent them or minimize their damaging effects.
- Understand the health problems associated with noisy equipment so that an informed decision can be reached regarding the need to buy a different kind of machine (substitution) or to enclose the machine in an acoustical booth (isolation).
- Know how to anticipate processing problems and evaluate the severity of the release of toxic substances from worn process equipment, such as valves, pumps, grinders, vessels, and dryers.
- Learn to use fault tree analysis or other approaches to anticipate situations that might produce hazardous situations due to static electricity, worn bearings, runaway reactions, chemical degradation, and the failure of automatic heat-sensing devices.
- Understand that nearly every piece of processing equipment poses a potential hazard to the worker and that the degree to which one minimizes these hazards (expense) should be directly related to the magnitude of the undesirable event.
- Understand that any material can be manufactured or handled safely despite its toxicity.
- Know about the risks of routine, nonprocessing operations such as quality control sampling and maintenance activities.
- Be able to anticipate as well as control those ergonomically (physically or emotionally) stressful manual material handling tasks known to be responsible for musculo-skeletal ailments.
- Know when respiratory protective equipment is a feasible and/or appropriate engineering control.
- Be able to recognize control panels and work stations that require excessive processing or visual data since these situations are known to increase the likelihood of human error.
- Be familiar with state-of-the-art on-line instrumentation that will enable workers to be continuously protected from exposure to airborne toxicants.

23 OVERVIEW

Companies have, by and large, done a good job in applying their expertise and experience to the prevention of accidents and occupational illness. The steady decline in serious accidents and disease since the turn of the century validates their commitment. Corporations have long recognized the high cost of building facilities that are not safe or do not meet the approval of regulatory agencies. The long list of regulations, as well as the ever increasing number of published documents describing occupational health control techniques, may now go well beyond industry's ability to incorporate these skill into on-the-job training of young engineers.

Engineers have a responsibility to use their skills to work toward a safer workplace and to prevent health hazards to the community at large. By their very nature, occupational diseases, unlike many other diseases, are man made and, in most cases, preventable. These factors have been partially responsible for the societal and regulatory pressures to minimize health hazards. Most employee exposure problems in existing processing operations could have been prevented through better application of industrial hygiene control techniques at the plant design stage. Although the task of hazard prevention seems at times to be formidable, there is an increasing body of published knowledge available to engineers that, if implemented, can improve their design. Engineers are obligated to be aware of these design concepts and use them whenever feasible.

It is not necessary for all undergraduate engineers to be experts in occupational and environmental safety and health. Rather, engineers should be able to recognize potential hazards and know how to avoid repeating mistakes made by their fellow engineers. After all, isn't this the intent of formal education: to learn from the past. Ethically, do we have any other choice (62–66)?

REFERENCES

1. Armytage, W. H. F., *A Social History of Engineering*, Boulder: Westview Press, 1976.
2. Eisenbud, M., *Environment, Technology and Health: Human Ecology in Historical Perspective*, New York: New York University Press, 1978.
3. Von Hovenberg, A., *Malaria Control For Engineers*, Amer. Soc. of Civil Engineers (ASCE), 1934.
4. Herskind, C. C., The Code of Ethics and the Professional Engineer in Industry, *Profess. Eng.* **40**, 27–29 (April 1970).
5. *Code of Ethics for Engineers*, The National Society of Professional Engineers, 1966.
6. The Social Responsibility of Engineers, *Ann. New York Acad. Sci.* **196**, 411–473 (April 10, 1973).
7. Strickland, K., The Civil Engineer/Citizen: What Role and Responsibility in Government? *Eng. Issues* **104**(Ell), 63–65 (January 1978).
8. Unger S. H., "Engineering Societies and the Responsible Engineer," in, The Social Responsibilities of Engineers (ed. Harold Fruchtbaum). *Ann. New York Acad. Sci.* **196**, 433–437 (April 10, 1973).
9. Lowrance, W., *Modern Science and Human Values*, New York: Oxford University Press, 1985.
10. McCord, C. P., *Industrial Hygiene for Engineers and Managers*, New York: Harper Bros., 1931.
11. Phelps, E. B., *Public Health Engineering: A Textbook of the Principles of Environmental Sanitation*, Vol. I, New York: Wiley, 1948.
12. McCord, C., *A Blind Hogs Acorns*, Chicago: Cloud, 1945.
13. Talty, J. T. (ed.), *Proceedings of NIOSH/University Occupational Health Engineering Control Technology Workshop*, National Institute for Occupational Safety and Health (NIOSH), National Technical Information Service, Springfield, VA, 1979.

14. Jaske, R. T., The Growing Federal Involvement in the Engineering Profession, *Chem. Eng. Progress* **74,** 22–25 (August 1978).

15. D'Anjou, E., *A Selected Annotated Bibliography of Professional Ethics and Social Responsibility in Engineering,* Center for Study of Ethics in Professions, Illinois Inst. of Tech., Chicago, 1980.

16. Unger, S. H., *Controlling Technology: Ethics and The Responsible Engineer,* New York: Holt, Rinehart and Winston, 1982.

17. Mingle, J. O., Ethical Design: The Engineer's Challenge, presented at the Annual Meeting of the American Society of Engineering Education, Lincoln, NE, 1983.

18. Paustenbach, D. J., and Dufour, J. T., The Engineer's Role in Protecting Human Health-Legal Aspects, *Mech. Eng.* **106,** 62–69 (1984).

19. Mingle, J. O., and C. E. Reagan, Legal and Moral Responsibilities of the Engineer, *Chem. Eng. Prog.* **76,** 15–23 (December, 1980).

20. Peters, L. C., G. N. Sandor, T. F. Talbot, J. F. Thorpe, R. I. Vachon, A. S. Weinstein, and R. M. Woloseqicz, Design Engineering and the Law, *Mech. Eng.* **99,** 44–49 (October, 1977).

21. Peters, G. A., New Product Safety Legal Requirements, *Profess. Safety* **24,** 39–42, 1979.

22. Reuth, N., Law and the Engineer, *Mech. Eng.* **100,** 46–47, 50–51 (February, 1978).

23. Moon, G. B., A Law Primer for the Chemical Engineer, *Chem. Eng.* **85,** 114–120 (April 10, 1978).

24. Hammer, W., Engineers Can Be Hazardous to Your Health, *Trial* **17,** 20–26 (1981).

25. Stewart, W. T. and Paustenbach, D. J., Ethical Decision Making for Engineers and Managers, *Ind. Eng.* **16,** 47–55 (April, 1984).

26. Fowler, F. D., Testimony Concerning Human Fault Concepts, *Prof. Safety,* **24,** 33–38, (March, 1979).

27. Witherell, C. E., The Products Liability Threat, *Chem. Eng.* **90,** 72–87 (January 24, 1983).

28. Nash, L. L., Ethics Without the Sermon, *Harvard Bus. Rev.* (Nov.–Dec.), 79–90 (1981).

29. *Barker* vs. *Lull Engineering Co.,* 20 Ca/3d 413, 575 P2d 443 (1978).

30. *Ruggeri* vs. *Minnesota Mining and Manufacturing,* 63 III (1978). App 3d 525, 380 NE2d 445.

31. *Drayton* vs. *Jiffee.* 395 F. (1975). Supp. 1081, later modified 591 F2d 352.

32. Victim Compensation—Paying for the Damage Caused by Toxic Wastes, *Chem. Week,* **132,** 32–38 (March 9, 1983).

33. Ember, L. R., Legal Remedies for Toxics Victims Begin Taking Shape, *Chem. Eng. News,* **61,** 11–20 (March 28, 1983).

34. Goodrich Ordered to Pay $400,000 in Liver Cancer Death, *Occ. Health Safety Newslett.* (May 1981), p. 5.

35. Moon, G. B., A Law Primer for the Chemical Engineer, *Chem. Eng.* **85,** 114–120 (April 10, 1978).

36. Flores, A., Safety in Design: An Ethical Viewpoint, *Chem. Eng. Prog.* **79,** 11–14 (November 1983).

37. Unger, S. H., The Role of Engineering Schools in Promoting Design Safety, *IEEE Tech. Soc.* **1,** 9–12 (December 1983).

38. Pollutions Beware! EPA Is Taking the Gloves Off, *Barrons Magazine,* Jan. 12, 1985, pp. 44–46.

39. Zeckhauser, R., and Nichols, A., "The Occupational Safety and Health Administration—an overview," in, *Study on Federal Regulation,* Committee on Governmental Affairs, United States Senate, Appendix to Volume VI, Framework for Regulation, 83–944, Washington, D.C.: U.S. Government Printing Office, 1978.

40. Lowrance, W. W., *Of Acceptable Risk: Science and the Determination of Safety,* Los Altos: William Kaufmann, 1976.

41. Rodricks, J. V., S. N. Brett, and G. C. Wrenn, Risk Decisions in Federal Regulatory Agencies, *Regul. Toxicol. Pharm.* **7,** 307–320 (1987).

42. Paustenbach, D. J., Risk Assessment and Engineering in the 1980's, *Mech. Eng.* **106,** 54–59 (November 1984).

43. Fernandes, J. H. Professional Responsibilities of the Standards Writer, *Mech. Eng.* **105,** 39–41 (May, 1983).

44. Gibson, S. B., Hazard Analysis and Risk Criteria, *Chem. Eng. Prog.* **76**(11), 46–50 (1980).

45. Atallah, S., Managing Industrial Risk, *Chem. Eng.* **87,** 94–100 (September 1980).

46. Lowrance, W., The Agenda for Risk Decision Making. *Environment* **25,** 4–8 (1983).

47. Caplan, K. J., "Philosophy and Management of Engineering Controls," in *Patty's Industrial Hygiene and Toxicology Vol. III,* Lewis and Lester Cralley, (eds.), New York: Wiley, 1979.

48. Supreme Courts Benzene Decision Strikes At Basic Public Health Principles, *Occ. Health Safety Letter,* p. 1, July 6, 1980.

49. The Byssinosis Case: Supreme Court Examines Cost-Benefit Issue, *Occupational Hazards,* pp. 83–87 (April, 1981).

50. National Safety Council, *Accident Facts.* Chicago, 1987.

51. Burack, W., Lies, Damned Lies, and Statistics, *Medical World News,* 1979.

52. Doll, R., and R. Peto, The Causes of which Quantitative Estimates of Available Risks of Cancer in the United States Today, *J. Natnl. Cancer Inst.* **66,** 1191–1308 (1982).

53. Ashford, N. A., Don't Ignore Legal and Ethical Concerns in Job Health Safety, *Occup. Health Safety,* **46,** 24–29 (March/April, 1977).

54. Spending on Safety and Health, *Chem. Eng.* **89,** 48 (September 3, 1982).

55. Talty, J. Integrating Health and Safety into Engineering Curricula, *Eng. Educ.* **76,** 136–139 (December 1985).

56. Talty, J. T., and J. B. Walters, Integration of Safety and Health into Business and Engineering School Curriculum. *Prof. Safety,* p. 26–32 (September 1987).

57. Peterson, J. E., "Principles of Controlling the Occupational Environment," Chapter 35, in *The Industrial Environment: Its Evaluation and Control,* Cincinnati: National Institute of Occ. Safety and Health, 1973.

58. Paustenbach, D. J., Should Engineering Schools Address Environmental and Occupational Health Issues? *J. Prof. Issues Eng.* **113,** 93–111 (1987).

59. Timmerhaus, K., Occupational Safety and Health in Engineering Schools. What Should be Done? speech presented at 1984 Conference for Engineering Educators, Purdue Univ., West Lafayette, IN, 1984.

60. Paustenbach, D. J., Discussion of Comments to Paper on Engineering Education Regarding Environmental Safety and Health, *J. Prof. Issues Eng.* (ca. January 1989).

61. Nelson, D. B., Safety and Health Awareness: What Does Industry Expect of New Chemical Engineering Graduates? *Plant Oper. Prog.* **1,** 114–116 (1982).

62. Gunn, A., and P. A. Vesilind, *Environmental Ethics for Engineers,* New York: Lewis, 1985.

63. Starr, C., Social Benefits vs. Technological Risks, *Science* **165,** 1232–1237, (1969).

64. Kardestuncer, H., *Social Consequences of Engineering,* San Francisco: Boyd and Fraser, 1980.

65. Kletz, T., Safety in Design, *Chem. Eng. Prog.* **80,** 13, (March, 1984).

66. Ruckelshaus, W. D., Better Safe than Sorry, *Environ. Health Lett.* **1** (November 15, 1983).

Index

DATE DUE

HIGHSMITH 45230